Latin American Dendroecology

Marín Pompa-García • J. Julio Camarero
Editors

Latin American Dendroecology

Combining Tree-Ring Sciences and Ecology in a Megadiverse Territory

Springer

Editors
Marín Pompa-García
Facultad de Ciencias Forestales
Universidad Juárez del Estado de Durango
Durango, Mexico

J. Julio Camarero
Instituto Pirenaico de Ecología (IPE-CSIC)
Zaragoza, Spain

ISBN 978-3-030-36929-3 ISBN 978-3-030-36930-9 (eBook)
https://doi.org/10.1007/978-3-030-36930-9

This Springer imprint is published by the registered company Springer Nature Switzerland AG
The registered company address is: Gewerbestrasse 11, 6330 Cham, Switzerland

Foreword

Trees and shrubs of Latin America from Mexico to Tierra del Fuego are critical sensors of the environment, recording key data in their growth rings about climate, disturbance, hydrology, human impacts, and chemical composition of the atmosphere and soils. Interpreting these rich data sets is a major challenge of our times as scientists, policymakers, and resource managers seek solutions for urgent problems of climate change and population pressures on the environment. This volume brings together expertise in tree-ring research applications across Latin America, highlighting the advances made by dendrochronologists of the region and their global partners. It is a welcome addition to the growing literature from this biologically and culturally valuable region.

I have had the privilege of carrying out dendroecological research in Mexico with many collaborators for several decades. In past years, tree-ring research was quite limited in Latin America. Pioneering scientists and laboratories, well represented in this book, have dramatically advanced the state of the science. The editors of this book, Marín Pompa-García and J. Julio Camarero, are two of these leading researchers who bring tree-ring expertise from within and outside the Latin American region. They have attracted contributions from well-established tree-ring laboratories as well as from new investigators.

The pace of global change is swift, and science must advance quickly to provide useful information for society. Latin American ecosystems have been woefully understudied, but new frontiers are being opened every day as new species and new ecological themes are being studied. I hope that this book proves to be useful to researchers seeking new tools for understanding ecological dynamics in the interests of conserving the high biodiversity and critical ecosystem services of Latin America. Adapting to shifting climate and sustaining the environment are essential for the growing population, already 650 million people, of the region. There are important implications outside the regional borders as well. Studies in genera such as *Nothofagus* and *Araucaria* reported here may be useful in closely related Southern Hemisphere ecosystems across the Pacific Ocean in Australia or New Zealand. Investigation of *Juniperus deppeana* and *Pinus strobiformis*, described here, is relevant to research and management of these species north of Mexico in the

United States. Finally, I hope that the frontier of studying dendrochronological potential in wet tropical tree species, another topic of this volume, serves as inspiration for other researchers in the tropics of the Americas, Africa, and Asia.

Peter Z. Fulé
Regents' Professor of Forest Ecology
Northern Arizona University
Flagstaff, AZ, USA

Preface

Dendroecology has emerged as a scientific discipline that improves our knowledge of the ecological processes faced by individual trees and forests. Recent efforts have been devoted to studying the application of tree rings to ecological problems. Specifically, the book *Dendroecology: Tree-Ring Analyses Applied to Ecological Studies*, whose scientific relevance motivated the interest of the international scientific community, was published in 2017.

In Mexico, the dendroecology network (https://dendrored.ujed.mx/) emerged as a platform linked to international researchers, supporting and encouraging the application of tree rings to ecological changes of forest ecosystems. Many scientific products have been generated under this discipline as scientific articles, congresses, conferences, books, human resource trainings, academic exchanges, etc., providing knowledge of the frontier in forest ecology. Activities within this network motivated authors to develop and write this book on dendroecology in Latin America.

Latin America constitutes a strategic region where the convergence of diverse climates, topographical conditions, soil types, and physiographies creates many types of biomes rich in biodiversity. Such abundance of tree species, many of them with dendrochronological potential, constitutes an ideal opportunity to increase their ecological knowledge in response to environmental stress, which are often threatened by anthropogenic threats including climate change. In this region, the historical interactions between human societies and the environment were also rich and diverse, albeit they are converging under the current globalization scenario, pushing some ecosystems toward their sustainability thresholds.

This book aims to fill a knowledge gap by addressing ecological problems under a dendroecological approach through applications in a wide biogeographic range including tropical and temperate forests from northern Mexico to southern Chile. It is organized according to five broad topical themes, including an Introduction; Ecological Drivers of Tree Radial Growth (Part I); Dendroecology in Neotropical Regions (Part II); Forest Dynamics, Climate, and Disturbances (Part III); and Forest Management and Conservation (Part IV). Each theme is composed of chapters (16 total) highlighting recent applications in dendroecology that often show novel methods and approaches.

Part I: Ecological Drivers of Tree Radial Growth

In Chap. 1, Rodríguez-Ramírez and Luna-Vega showed that xylem vessels can be used to detect anatomical variations in endangered species such as Mexican *Magnolia* and that temperature and precipitation are strongly associated with differences in tree-ring width between drought and non-drought years in *Magnolia vovidesii*. Their results show that anatomical adaptations need to be considered in the context of climate change, which will bring many new challenges for those depending on and managing tropical montane cloud forests worldwide.

In Chap. 2, Acosta-Hernández et al. report one of the first chronologies generated for *Pinus strobiformis* in mesic sites in northern Mexico. They found that both earlywood and latewood growth show positive relationships with temperature. Nevertheless, this tree species is overall complacent, showing a weak association of growth with climatic variables modulated by microsite conditions.

In Chap. 3, Lisi et al. deal with the dendroecology potential of *Cedrela odorata* in the Caatinga forest of northeast Brazil, a tropical dry forest area. They present the methodology used for tropical dendroecology studies and analyze the relationships between growth and environmental conditions. According to their results, true rings with marginal parenchyma bands differed from false rings by the presence or absence of vessels. They also found that the dry season duration and rainfalls from May to July drive wood formation.

In Chap. 4, González-Cásares et al. explain the responses of growth to climate and drought in two sympatric Mexican pine species. They found that *Pinus leiophylla* and *Pinus teocote* showed similar growth responses to climate and drought. However, the growth of *P. leiophylla* was more sensitive to drought than that of *P. teocote*. Furthermore, precipitation was the main factor that influenced growth in both species. Their results are relevant to forecast the impacts of global warming in similar forests prone to drought.

Part II: Dendroecology in Neotropical Regions

In Chap. 5, Giraldo et al., for the first time, characterize tree rings from the Biogeographic Chocó region (a superhumid region with annual precipitation above 7200 mm) and without water seasonality (dry or flooding periods). Unexpectedly, they found about 82% of species forming growth rings. They contrast their results with the literature around the tropics. Still, without any hydric seasonality, many tropical species may develop annual rings. The authors discuss the dendrochronological potential of tropical tree species and raise new questions on it.

In Chap. 6, Cangiano et al. address the new insights that dendrochronological studies have provided about *Prosopis* worldwide with main emphasis in Latin America and highlight the main growth driver of *Prosopis* throughout the study area. This review demonstrates the avant-garde and influential value of *Prosopis* for dendroecological research.

In Chap. 7, Camarero et al. investigate the responses to climate, drought, and ocean-atmosphere patterns of three tree species (*Acosmium cardenasii*, *Centrolobium microchaete*, *Zeyheria tuberculosa*) coexisting in a Bolivian dry tropical forest. These different growth responses to climate and drought contribute to explain the coexistence of tree species in study area. The growth patterns of the study species are valuable climate proxies for reconstructing climate and drought trends in the study area.

Part III: Forest Dynamics, Climate, and Disturbances

In Chap. 8, Srur et al. present a state of the art of those researches that apply dendroecological techniques to understand the role of climate and disturbances on forest dynamics in the Patagonian Argentina. They found some gaps of knowledge that can be further explored using dendroecological approaches.

In Chap. 9, in their analysis of the annual growth of *Juniperus deppeana* to determine its dendrochronological potential, Cerano et al. develop the first chronologies of this species in Mexico. This research allowed evaluating the influence of climatic variables on their annual growth and their potential to develop dendroecological studies. Likewise, they pointed out the potential of this species to perform hydroclimatic reconstructions in north Mexico.

In Chap. 10, Guerrido et al. describe *Nothofagus* recolonization following episodic floods caused by Perito Moreno Glacier (Argentina) by combining dendrochronology and spatial statistics. Their data indicate rapid recolonization of the bare shores following episodic floods. Species dominance and rate of colonization were found to be dependent on a combination of several variables including non-flooded forest composition, precipitation gradients, seed dispersal capacities, differences in substrate, and microclimatic conditions.

In Chap. 11, in their contribution, Pompa-García et al. demonstrate that spatial autocorrelation analysis linked with dendrochronological data represents a valuable tool for studying forest dynamics. Through a retrospective analysis of basal area increment, they tested whether the growth responses of *Picea chihuahuana*, *Abies durangensis*, and *Cupressus lusitanica* were spatially autocorrelated and used it as a proxy to evaluate how growth decreased in drought events. Their results revealed that all species presented a nonrandom growth spatial pattern with declines caused by drought, whereas only *C. lusitanica* showed no signs of growth decline.

In Chap. 12, Requena et al. present a brief review of the dendroecological studies on *Polylepis* tree species along the Central Andes, showing how their growth mainly responds to summer temperature and precipitations. They also present for the first time three new ring width chronologies for three *Polylepis* species (*P. rodolfo-vasquezii*, *P. tarapacana*, and *P. rugulosa*) from Peru and characterize their radial growth patterns.

Part IV: Forest Management and Conservation

In Chap. 13, in their review on *Araucaria araucana* forests of northwestern Patagonia, Argentina, Hadad et al. show several aspects of their dynamics along a precipitation gradient and demonstrate the importance of dendroecological studies to ascertain the influence of the regional environmental factors on forest structure and dynamics, which are relevant aspects for designing proper conservation strategies of similar endemic species.

In Chap. 14, Mundo et al. show the effect of thinning and pruning on radial growth of *Nothofagus pumilio* in Tierra del Fuego (Argentina). This is a tree species widely used in dendroclimatological studies but not yet analyzed in response to forest management. Using dendroecological methods, they demonstrate that three different combinations of silvicultural treatments produce significant growth increments after the interventions.

In Chap. 15, Vettese et al. present the first dendrochronological records of *Nothofagus antarctica* for Argentinean Patagonia and establishe the relationships between regional climatic variations and the growth of *N. antarctica*. There is a direct relationship between growth and precipitation during the current growing season, while an inverse relationship with temperature was detected. Since the middle of the twentieth century, there has been a progressive decrease in growth, consistent with a regional increase in summer temperature and a decrease in rainfall. Variations associated with documented changes in land use were also identified.

Finally, in Chap. 16, J. J. Camarero assesses the changes in radial growth and intrinsic water-use efficiency of two threatened, old-growth Chilean conifer species (*Fitzroya cupressoides*, *Pilgerodendron uviferum*) inhabiting Valdivian temperate rainforests. He demonstrates how comparing several proxies of tree performance through time improves the characterization of the long-term responses of long-living, threatened conifers facing global change.

Durango, Mexico — Marín Pompa-García
Zaragoza, Spain — J. Julio Camarero

Acknowledgments

The editors wish to acknowledge the effort of anonymous reviewers, whose recommendations improved the version of the final chapters.

Introduction

Latin America comprises a set of different traits of ideologies, traditions, beliefs, and social relationships, amalgamated within a diversity of cultural manifestations in all spheres and, at the same time, converged in one of the most biologically and culturally rich regions of the world. In Latin America, there are almost as many differences as similarities. This peculiarity confers a unique identity that is the product of its history, the sociopolitical formation process of which is still ongoing. The problems of its incorporation into the twenty-first-century world and its social and economic inequalities and ethnic diversity are paradigmatic. These features often reflect pre-columbine traits that still have not adapted to the hegemonic occidental tradition, which give Latin Americans a particular perspective on their interactions with nature. The incredibly diverse and valuable natural resources have to be sustainably managed by the descendants resulting of five centuries of mixing between American and Latin European cultures. This mixture has received many more influences from around the world but shares common languages and views of ecosystems including forests.

Given its geographic diversity, in which all climates are represented, the region possesses a megadiversity that is recognized worldwide for hosting more than half of the species of the Earth. However, grave reports have stated that the ecosystems of the region are suffering a marked degradation as a result of anthropogenic pressure and driven by climatic change, with lamentable consequences for their diversity. For example, in northern Mexico, it has been documented that extreme droughts are making the forests of temperate climates vulnerable, particularly when these droughts occur during the winter prior to the current year of growth. More recently, the losses of large areas of forest in the Amazon, caused by man-made wildfires in 2019 and aggravated by climate warming, also reflect disturbances and extreme climate conditions that cannot be ignored. Equally, the decline in the productivity of the Patagonian forests highlights the challenges that must be faced in the years to come. This combination of diverse forests and threats generates multiple responses to ecological changes, affecting the structure and functioning of Latin American forest ecosystems and cascading on global functions such as carbon capture or biodiversity maintenance.

Dendroecology is a science that has demonstrated its capacity to improve our long-term understanding of all of these ecological processes in forest and shrubland ecosystems. It incorporates attributes of greater temporal resolution (earlywood, latewood, anatomical traits, wood density, etc.) and tries to disentangle the "noise" traditionally smoothed or removed by dendroclimatologists. These traits constitute valuable dendroecological proxies since they represent reliable parameters in ecology if we aim to understand individualistic responses to global change. In 2017, the state of the art and paradigm changes in dendroecology were published, which translated into a notable improvement in our knowledge of ecophysiology, forest dynamics, disturbances, ecosystem resilience, etc. The reference for this potential is represented by the dendroecology of South America (from the Andes to Patagonia), the scientific contributions of which have refined the applications of this science and motivated spiraling scientific growth. The following chapters present scientific findings that contribute to facing the challenges that appear with the threats affecting the forests of this continent. While many questions remain to be answered, the contribution of this research significantly improves our perspective of ecological causes and effects to the benefit of these ecosystems and to the services they provide to Latin Americans and to the whole mankind.

Marín Pompa-García
Facultad de Ciencias Forestales
Universidad Juárez del Estado de Durango
Durango, Mexico

J. Julio Camarero
Instituto Pirenaico de Ecología (IPE-CSIC)
Zaragoza, Spain

Contents

Part I
Ecological Drivers of Tree Radial Growth

Chapter 1
Dendroecology as a Research Tool to Investigate Climate Change Resilience on *Magnolia vovidesii*, a Threatened Mexican Cloud Forest Tree Species of Eastern Mexico

Ernesto Chanes Rodríguez-Ramírez and Isolda Luna-Vega

Abstract *Magnolia vovidesii* has been classified as a threatened species due to its very restricted range to the tropical montane cloud forests (TMCFs) of eastern Mexico. This species, as well as many other inhabiting the Mexican cloud forest, is considered as a Miocene relict, known to be limited to three localities of Veracruz, Mexico. This species is at risk because of habitat fragmentation as a product of grazing, logging, and avocado plantations, among others. In the present research, we used dendroecological techniques to evaluate the effects of climate variability on the vessel traits and growth of *M. vovidesii* in a locality in Veracruz, Mexico. We determined the regional climatic factors influencing the resilience of this *Magnolia* species. We found that the mean radial growth of this species during the past 75 years was restrained in the study site. We detected repeated narrow tree rings during the last decades (1945–2016) which correspond to historical drought events. Climate-growth analyses carried out for the period 1941–2016 point out that wood growth depends on temperature and that drought events restrict growth. Our results suggest that the resilience of TMCF where *Magnolia vovidesii* is codominant could depend significantly on microclimatic conditions.

Keywords Vessel plasticity · Climate change · Growth rates · Anatomy · Resilience · Diffuse-porous wood · Adaptability

E. C. Rodríguez-Ramírez · I. Luna-Vega (✉)
Laboratorio de Biogeografía y Sistemática, Facultad de Ciencias, Universidad Nacional Autónoma de México, Mexico City, Mexico
e-mail: echanes@ciencias.unam.mx; isolda_luna-vega@ciencias.unam.mx

M. Pompa-García, J. J. Camarero (eds.), *Latin American Dendroecology*,
https://doi.org/10.1007/978-3-030-36930-9_1

1.1 Introduction

Tropical montane cloud forests (TMCFs) often exhibit a variety of vegetational structure and floristic composition that distinguish them from temperate and tropical forests that are less exposed to ground-level clouds with a high cover of epiphytes and tree ferns (Webster 1995; Luna-Vega et al. 2000; CONABIO 2010). Jarvis and Mulligan (2010) demonstrated that the climate of TMCFs is different from that of other tropical montane forests, especially in terms of lower mean and maximum temperature (T_{mean} and T_{max}) and precipitation ranges, mostly because they occur at higher altitude. TMCFs tend to occur at topographically exposed ridges or peaks. Cloud immersion strongly affects the ecosystem's water balance by reducing evapotranspiration and through the enhanced rainfall recycling inputs associated to wind-driven fog that is collected on the canopy's surfaces (Fahey et al. 2016; Hu and Riveros-Iregui 2016).

TMCFs benefit from high rates of rainfall during the rainy season while also taking benefit from "hidden" precipitation (e.g., mist, cloud water, and fog) during the dry season (Webster 1995; Bruijnzeel et al. 2010; Gual-Díaz and Rendón-Correa 2014; Hu and Riveros-Iregui 2016). The relationship between forest ecosystems and water flow is mediated by the configuration of the rainfall recycling (Williams-Linera and Tolome 1996; Ellison et al. 2017). Notwithstanding, increasing temperature and a higher frequency of drought events are triggering tree dieback worldwide (Breshears et al. 2009; D'Arrigo et al. 2014), evidencing that climate change could already be impacting TMCF trees (Rodríguez-Ramírez et al. 2018).

Drought events can induce tree mortality via xylem vessel hydraulic failure (Bréda et al. 2006; McDowell et al. 2008). Thus, analysis of vessel traits could indicate drought-induced adaptations from TMCF trees (Eller et al. 2017; Rodríguez-Ramírez et al. 2019b). The adjustment to climatic variability is usually controlled at xylem level by regulating vessel hydraulic plasticity (Tyree and Zimmermann 2002) given by the size, diameter, topology, density, and number of vessels and their type of interconnections (von Arx et al. 2013). As a result, vessel traits show anatomical adaptations to climate variations (Sperry 2003). Accordingly, reduced xylem vessel size in drought-treated trees has frequently been attributed to the loss of turgor pressure in expanding cambial cell derivatives (Abe and Nakai 1999; Dünisch and Bauch 2009). Thus, analysis of tree vessel traits could indicate drought-induced adaptations from TMCF trees (Eller et al. 2017; Rodríguez-Ramírez et al. 2019a).

In angiosperms, studies of drought effect on diffuse-porous wood and vessel plasticity are essential to identify the anatomical adaptations of boreal, temperate, and/or tropical species (Venegas-González et al. 2015). The vessel traits of angiosperms decrease in tree species inhabiting dry or temperate ecosystems and increase in tropical environments (Fonti et al. 2010; Islam et al. 2018a, b; Rodríguez-Ramírez et al. 2019b), because wide vessels prioritize water conductivity at the expense of high embolism risk by drought events (Fonti et al. 2010; Rahman et al. 2019; Tarelkin et al. 2019). Nevertheless, evidence of this safety-efficiency trade-off is not definite, and some species could provide a valuable, long-term perspective on plasticity in vessel safety and low hydraulic efficiency in the coming century due to deforestation and/or climate change (Bryukhanova and Fonti 2013; Granato-Souza et al. 2019).

*Magnolia*species inhabiting the Mexican TMCFs (31 species; Rivers et al. 2016) have a complex biogeographic climatic history (Cretaceous ≈ 50 to 65 Ma BP; Graham 1976). *Magnolia* species are well represented in the different fragments of Mexican TMCFs (Azuma et al. 2001). Many of these species are characterized by their high shade tolerance, by the ability to regenerate below a closed canopy, and by having a high ecological association with the canopy species (Corral-Aguirre and Sánchez-Velásquez 2006). The tree layer is sometimes dominated by *Fagus grandifolia* subsp. *mexicana* (Martínez) A.E. Murray, *Cupressus lusitanica* Mill., *Liquidambar styraciflua* L., *Pinus patula* Schltdl. & Cham., *Meliosma alba* (Schltdl.) Walp., and several oak species (e.g., *Q. delgadoana* S. Valencia, Nixon & L.M. Kelly, *Q. trinitatis* Trel.), among other tree species. Most of the Mexican *Magnolia* species are classified as endangered species under the Red List of Magnoliaceae (Rivers et al. 2016) and the Red List of Mexican cloud forest trees (Azuma et al. 2001; Rivers et al. 2016; Tombesi et al. 2018).

Magnolia's wood is mainly characterized as a diffuse-porous wood where vessels maintain the same diameter and are evenly distributed within a growth ring (Yoshizawa et al. 2000; Latte et al. 2015). Vessel hydraulic plasticity and dendroecological tools are important features that can be used to assess threatened species vulnerability to increasing drought severity, higher overall temperatures, and consequent high evaporative demand. In this chapter, we demonstrate that *Magnolia vovidesii*'s vessel plasticity is enhanced during the drought periods, manifested as the development of narrow tree rings and adaptations in vessel traits. Our goal was to identify the annual tree rings that indicate drought years in *Magnolia vovidesii* through vessel traits (e.g., size, diameter, and number). We suggest that further dendroecological studies should adopt the focus on the comparison of adaptive anatomic capacities in other *Magnolia* species as well as other TMCF trees that inhabit the Neotropics to deepen our knowledge on the effect that climate change could have in this type of vegetation.

In this case study, we show how dendroecological tools and vessel trait analysis can be used to represent the effect of climate variations on diffuse-porous wood from TMCF trees. The aim of this research was to determine the influence of past climatic fluctuations on vessel plasticity of *M. vovidesii* through dendrochronological and anatomical tools. Furthermore, evidence of plasticity changes to the vessel's anatomy during drought years could potentially be used as ecological proxies.

1.2 Climatic Effect on Tropical Montane Cloud Forest Trees

Climate change has a serious impact on the protective functions of TMCFs worldwide, as climate significantly influences the distribution, phenology, and physiology of plants. In Mexico, TMCFs occur only within microclimatic environments on montane topography (Luna-Vega et al. 2000; CONABIO 2010). Unfortunately, TMCF is extremely vulnerable to anthropogenic pressure such as logging, grazing, and land-use change for corn, avocado, and coffee plantations. This ecosystem represents only one example of the many places worldwide where anthropic effects and climate variations are quickly affecting the feedbacks between rainfall recycling and vegetation types. Climate observations show that many TMCFs worldwide have experienced a warming trend and an increase in elevation of the 0 °C isotherm during the second half of the twentieth century (Fahey et al. 2016; Hu and Riveros-Iregui 2016). Average rainfall is generally 1000–6000 mm per year, but it may fall 250–400 mm in dry season (e.g., La Mojonera locality, Zacualtipán municipality, Hidalgo State, Mexico; Rodríguez-Ramírez et al. 2018). Several studies in TMCF have highlighted the decrease in cloud cover as a result of climatic change (Gual-Díaz and Rendón-Correa 2014; Pouteau et al. 2018). This reduction is detected by the increase in the number of fog-free days or cloud formation (Pounds et al. 1999; Still et al. 1999).

Current studies have integrated climate, hydrology, and phenological features to increase the knowledge of TCMF vulnerability to climate change (Goldsmith 2013; Gotsch et al. 2014). Unfortunately, there are no specific studies evaluating the long-term response from TMCF trees to changing environmental conditions, specifically tree-ring growth and xylem vessel anatomical traits.

1.3 Effect of Climate Variation on Diffuse-Porous Wood and Vessel Traits

Vessel traits in angiosperms, especially in, have been recently used in dendroecological studies to interpret their responses to climatic variations (Albuquerque et al. 2019; Rahman et al. 2019). Efforts are being made to tease apart finer-scale effects on tree-ring growth by comparing different climate responses from weekly and daily climate records and determining the effect on the growth of individual cells in tree rings. Rodríguez-Ramírez et al. (2019a) examined the influence of temperature and precipitation in prompting changes to vessel traits as well as tree-ring width in Mexican beech (*Fagus grandifolia* subsp. *mexicana*).

They found that these present different responses to climatic variations. Islam et al. (2018a) found that differences in vessel size and frequency in *Chukrasia tabularis* A. Juss. might have caused the observed variation in hydraulic strategies between the two study species. These are phenological periods that are not usually recorded in tree-ring width and can offer a wider range of climatic data from a tree-ring series in environments with temperature seasonality (Alves and Angyalossy-Alfonso 2000; Tarelkin et al. 2019).

Diffuse-porous wood is an anatomical adaptation that grants a high performance in hydraulic conductance during drought events within TMCFs (Rodríguez-Ramírez et al. 2019b). Diffuse-porous wood from the *Magnolia* genus is mainly characterized for its evenly sized and spaced vessels within the growth ring (Latte et al. 2015). The effects of drought events on vessel plasticity for *Magnolia* species had not been analyzed until now.

1.4 Case Study: Drought Effects on the Vessel Traits of *Magnolia vovidesii* in a Tropical Montane Cloud Forest of Eastern Mexico

1.4.1 Methods

Magnolia vovidesii has been reported in three localities in Veracruz, Mexico (Huayacocotla, Coyopola, and Zongolica; Fig. 1.1). The site of study was carried out in the forest of El Batda, Huayacocotla, Veracruz (20°33'N, 98°24'W; 1829–1894 m. a.s.l), a TMCF located in the Sierra Madre Oriental mountain range in eastern Mexico. The study site is found on slightly rugged terrains.

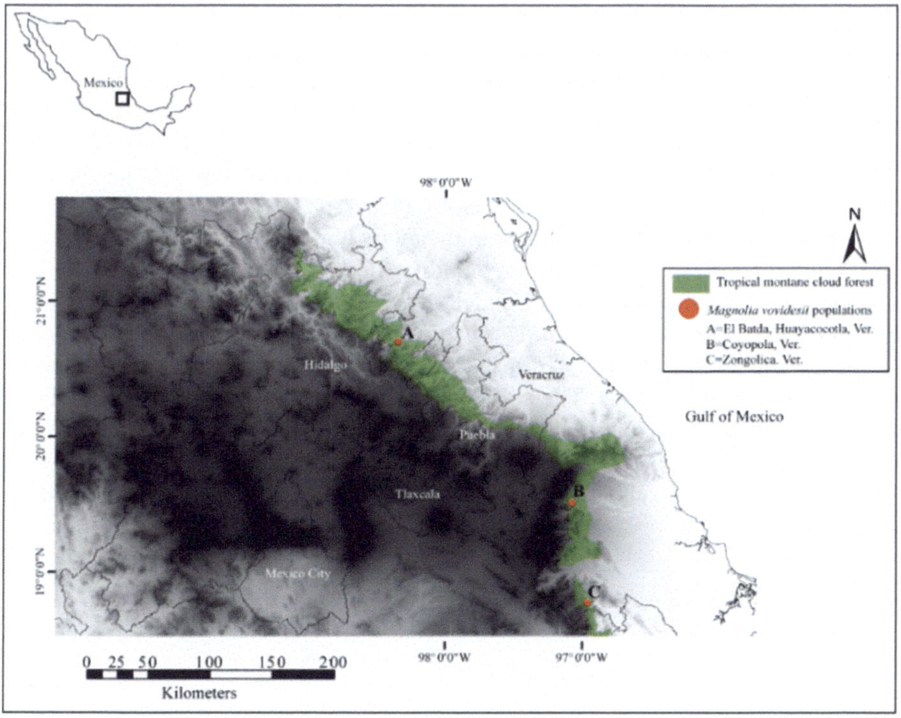

Fig. 1.1 *Magnolia vovidesii* distribution in Veracruz state, Mexico. Tropical montane cloud forest distribution based in Dinerstein et al. (2017)

Fig. 1.2 (**a**) *Magnolia vovidesii* endangered tree species. (**b**) Mature *Magnolia vovidesii* flower ("eloxochitl"). (**c**) Polyfollicle (≈20 seeds)

Within the site, *Magnolia vovidesii* trees reach up to 25 m in height (Fig. 1.2a), have long leaves (25–60) × 16–30 (35) cm, large flowers (29–39 cm in diameter) with red exocarps, a fruity fragrance, and are rich in oils. *M. vovidesii* is locally known as "eloxochitl" which means "corn flower." *Magnolia* flowers (Fig. 1.2b) are often used in traditional medicine for heart diseases. The seeds are polyfollicles (Fig. 1.2c), which when ripened expose the seeds that hang from the follicle.

In terms of species composition and vegetation structure, the study site in El Batda is predominantly dominated by *Liquidambar styraciflua* L., *Pinus patula* Schltdl. & Cham., *P. greggii* Engelm. ex Parl, and several oak species such as *Quercus meavei* Valencia-A., Sabás & Soto, *Q. delgadoana* S. Valencia, Nixon & L. M. Kelly, and *Q. trinitatis* Trel. The mid-canopy of the forest is mainly composed of *Magnolia vovidesii*, *M. schiedeana* Schltdl., *Clethra mexicana* DC., *Befaria aestuans* L. and *Podocarpus matudae* Lundell, whereas the low canopy consists mainly of several tree fern species (*Cyathea fulva* (M. Martens & Galeotti) Fée, *Cyathea bicrenata* Liebm., *Dicksonia sellowiana* var. *arachneosa* Sodiro and *Alsophila firma* (Baker) D.S. Conant). The soils of the site are humic (Th) Andosols (FAO 1998) with light sandy clay loam texture. The forest shows a temperate climate (Cwb) sensu Peel et al. (2007), characterized by mild temperatures (14.5–24.4 °C; Fig. 1.3). Humidity levels are found in the range of 60–90% (Rodríguez-Ramírez et al. 2019a).

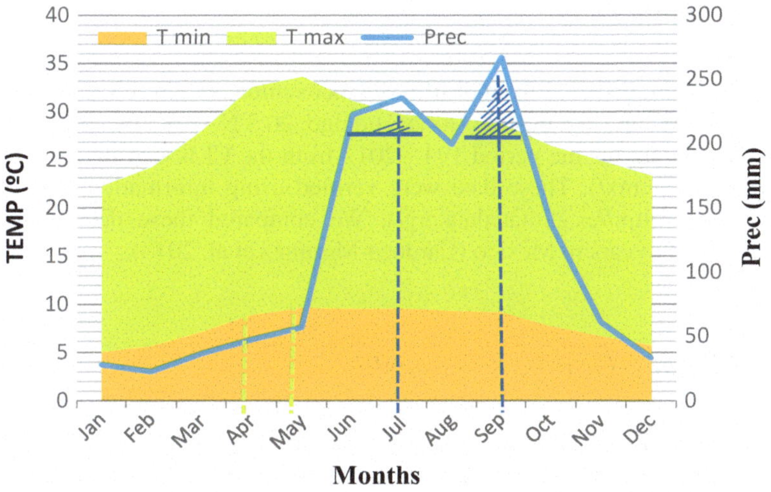

Fig. 1.3 Walter-Lieth climodiagram at the Huayacocotla meteorological station, during the period 1961–2015. Filled tips mean high levels (i.e., temperature or precipitation) and dotted lines mean specific month with high temperature (yellow) or precipitation (blue)

1.4.2 Sample Collection

We extracted two wood cores from 20 individual *Magnolia vovidesii* trees at breast height (1.3 m) with a diameter at breast height (DBH) ≥ 20 cm, with the use of a Häglof borer. The increment borings were taken in the dormant season (later November) and filled with cork plugs (treated with a mixture 80% ethanol or isopropanol and 20% tap water, or Na ethylmercurithiosalicylate "Merseptyl"), which are effective against a broad spectrum of bacteria, fungi, and viruses (Rodríguez-Ramírez et al. 2018). Wood cores were air-dried at room temperature, glued onto wooden supports, and sanded with successive coarse-grit sandpapers 100 and 360, and four finer-grit sandpapers (400, 600, 1000 and 1200) until the xylem's cellular structure was distinguishable with a 100× amplification at most (Speer 2010). Tyloses and wood dust inside vessel lumina were removed by using a hair dryer (Fonti et al. 2010; Gärtner and Nievergelt 2010; Souto-Herrero et al. 2017).

Cores were measured to the nearest 0.001 mm for each tree ring, with the TSAP-Win software and Velmex measuring machine (Velmex, Inc., Bloomfield NY, USA). Additionally, we verified the cross-dating with the software COFECHA (Grissino-Mayer 2001; Holmes 1983). A cross section from fallen *M. vovidesii* was considered a random sample of radial growth illustrative in the locality (Haghshenas et al. 2016; Rodríguez-Ramírez et al. 2019b). The series were standardized using the ARSTAN computer program (Cook and Holmes 1999). Additionally, we used mean monthly PDSI (Jun–Aug 1400–2012) to create a regional-level species index chronology using a cubic smoothing spline (10-years).

1.4.3 Climatic Data

Climate data were gathered from nearby weather stations (Huayacocotla, Veracruz; 20.5 N, −98.5 W; and Tenango de Doria, Hidalgo; 20.3 N, −98.2 W). More specifically, we used data for the period 1941–2017 from the CLICOM database (http://clicommex.cicese.mx/). These data were verified using information provided by Climate Data (http://es.climatedata.org/). We compared these data with those recorded drought years in Mexico (Cardoza-Martínez et al. 2013).

1.4.4 Climatic-Growth Relationship

Influence of climate on TRW variables was quantified by calculating Pearson's correlation coefficient between monthly mean maximum and minimum temperatures and (T_{min} and T_{max}) and monthly precipitation (P) for a 13-month window from the preceding September (−1) until the current September for the study period from 1941 to 2017. The significance level of the correlations followed the bootstrap technique (Guiot 1991); for this, we performed 10,000 iterations for each coefficient and implemented the correction proposed by Mason and Mimmack (1992) to obtain the percentile bootstrap confidence intervals. This procedure was performed with the programme R (http://www.R-202project.org/) and the Hmisc package (http://CRAN.R-project.org/package=Hmisc).

1.4.5 Historical Drought Events

We obtained historical drought years from Climate Data (http://sala.lab.asu.edu/research/drought-net/; Lemoine et al. 2016) (e.g., 1929–1930, 1940, 207 1963, 1970, 1972, 1976, 1983, 1991, 1997, 2012 and 2015–2016). We also followed this procedure for 2 consecutive years before and 2 years after "El Niño" southern oscillation events (ENSO).

1.4.6 Digitalization of Tree-Ring Width and Vessel Traits

To assess whether or not drought years have an effect on the plasticity of vessel traits [frequency, cross-sectional lumen area (henceforth, length), and diameter] of the *Magnolia vovidesii*, we selected 20 wood cores to take wood digital cores of *Magnolia vovidesii* for 2 consecutive years before, during, and after drought. For each digital image of the wood core, we selected the area occupied by each tree ring between two wood rays (an average of 7.5 mm width × 9.1 mm length). The area varied with respect to tree-ring width before, during, and after drought events [e.g., the area varied between 6.6 mm width × 8.3 mm length (ray to ray) and 2.5 mm width × 1.6 mm length (minimum area of 54.7 and 4 mm^2) (Rodríguez-Ramírez et al. 2019a)]. Digital images of wood cores were captured using a stereoscopic microscope (Leica Z16 APOA) with a 12.9–50.3 μm field of depth. Images were taken with a digital camera (Leica DFC 490) and saved as TIFF format with a 1.3 μm per pixel resolution; this technique has been used successfully with other species (Venegas-González et al. 2015). From *M. vovidesii*, a total of 108 digital images were taken. In each area selected, we quantified and manually measured all the vessel traits using the software ImageJ (Schneider et al. 2012) by means of the VesselJ plugging (García-González and Souto-Herrero 2017).

The correct recognition of all vessels was often hindered by the detection of dust spots or other undesired objects, which were manually excluded, retaining only the objects from 10,000 to 65,000 μm^2, and that had a width smaller than twice their length. Vessel outlines were improved by applying morphological operations (erode-dilate 2 × 2 one pass and calculation of the convex hull) (Fig. 1.4). Multiple comparison Tukey tests were performed to assess whether the values of vessel traits

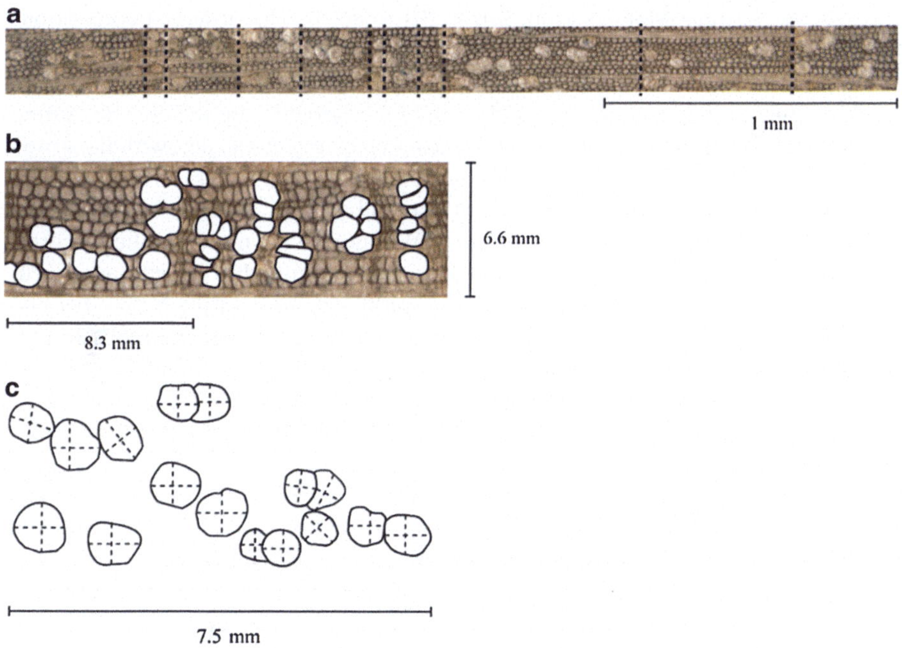

Fig. 1.4 Analysis of diffuse-porous wood of the Mexican magnolia studied. (**a**) digital image of wood core identifying tree rings (dotted lines); (**b**) micro-section of the representative wood core showing correct recognition and identification of all vessels (white circles were improved by applying morphological operations); and (**c**) measurement of vessel traits (frequency, cross-sectional lumen area, and diameter) as manually demarcated using the software ImageJ by means of the VesselJ plugging

(frequency, length, and diameter) have a significant difference between drought years and non-drought years for the *Magnolia* species studied. These analyses were performed in R (version 3.5.1) using the R-package ggplot2 and function geom_violin (Hintze and Nelson 1998).

1.5 Results and Discussion

We started by gathering information from other TMCF trees that appear to be affected by climatic variations (e.g., *Fagus* and *Castanea*). Subsequently, we focused in how the diffuse-porous wood of *Magnolia* species shows xylem vessel adaptations during drought events. Our case study supports the importance of using dendroecological tools for conservation purposes. Correlations with mean maximum temperature (T_{max}) were positive in previous Oct (−1) to Nov (−1) (e.g., during the dry cool season; Fig. 1.5) before tree-ring development. A negative correlation occurred in previous Sep (−1), Dec (−1) for radial growth chronology, and current late spring and summer (March to June). This drought-sensitive growth dynamics is supported in the literature for the genus *Magnolia* worldwide (Rivers et al. 2016; Vásquez-Morales et al. 2014, 2015, 2016). The observed correlations between previous dry-cool season from Oct (−1) to Jan (−1) monthly precipitation (*P*) and growth season were positive in all cases and mainly during the long cool summer (June to September) were linked to higher growth ring indices. Regarding T_{min} in early February to May, these were also negatively associated with the growing season. El Batda showed the highest correlation value ($r = 0.41$) from May (dry warm season) to July (long cool summer) (3 months).

In this study case, *Magnolia vovidesii* TRWs showed the most significant correlation to T_{max} and T_{min} during March and June. Maximum radial growth rates occur in September when humidity is higher but rapidly decrease during November and December. We found a positive response between radial growth and monthly precipitation (*P*). Only during the months of November, December, and January, we found that the absence of a continuous rainfall is not favorable for the growth of *M. vovidesii*. This result can be related to the end of the growth period in the TMCFs caused by a dry-cold season.

An alternative explanation to this is that from November to January, there is a lack of water availability at mid-elevation stands. This suggests that insufficient precipitation combined with higher temperatures (shown to be significant; Fig. 1.5) during the drought season induces the development of narrow rings and decreases vessel quantity, diameter, and length, due to the decline of rainfall rates (Campelo et al. 2013). Our results show that a Mexican *Magnolia* develop annual rings that can be used to confidently estimate the age of trees and rates of radial growth. We found for the first time that the annual tree rings observed in *Magnolia vovidesii* are associated with climate variability (Fig. 1.5).

The dendroecological tools allowed us to evaluate the drought effects on vessel anatomy of diffuse-porous wood up to the year of 1929. The chronologies (Fig. 1.6b)

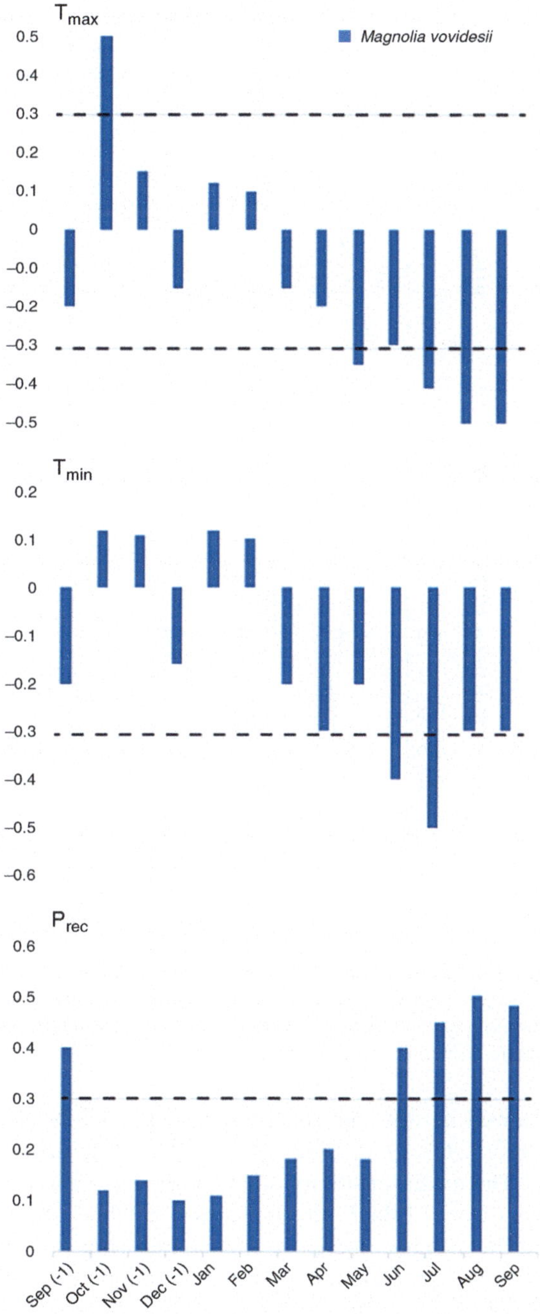

Fig. 1.5 Correlations between chronologies against the mean maximum, mean minimum temperatures (T_{max} and T_{min}) and monthly precipitation (P) for each drought year Sep (1) to Sep from the year 1941 onward. Horizontal dashed lines indicate the significance intervals ($P < 0.05$) for correlation coefficients. (−1) indicate the previous year and (+) the following years in which growth started

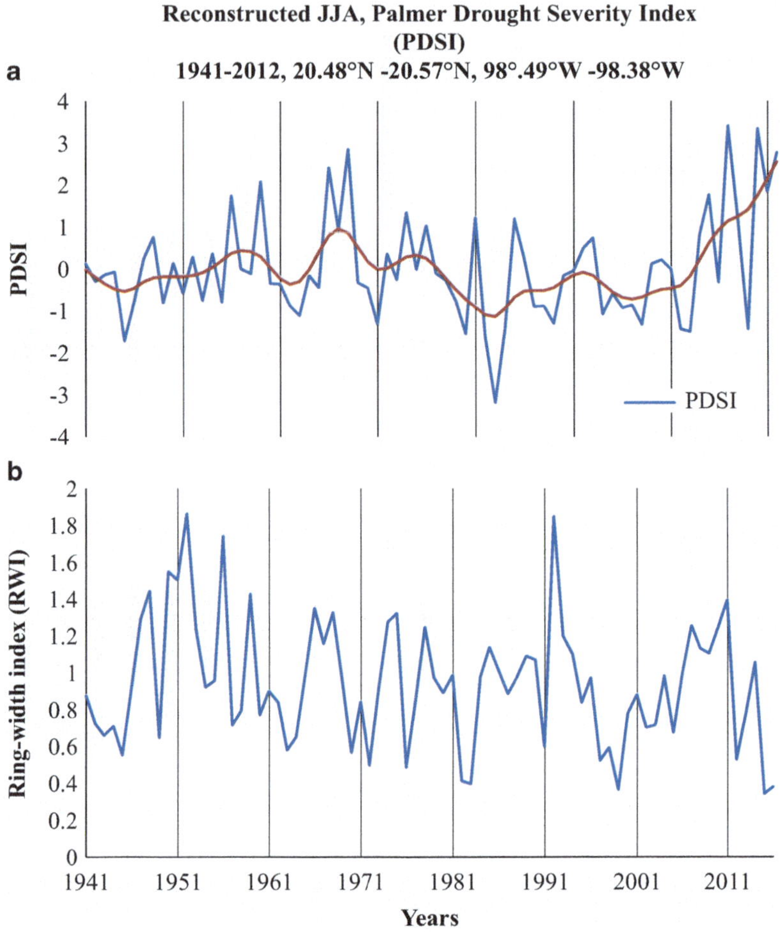

Fig. 1.6 (**a**) Response of the regional *Magnolia vovidesii* species chronologies to Reconstructed JJA, Palmer Drought Severity Index (PDSI; 1941–2012 period; Stahle et al. 2016). (**b**) Ring-width chronology of *Magnolia vovidesii* in a tropical montane cloud forest in the mountains of the Sierra Madre Oriental in eastern Mexico

showed narrow tree rings developed in strong drought years such as 1929–1930, 1940, 1963, 1970, 1972, 1976, 1983, 1991, 1997, 2012, and 2015–2016.

The mean sensitivity of *Magnolia* trees across chronologies between individual trees from El Batda site are shown in Table 1.1.

On average, for each annual tree ring, the cores contained 47.2 vessels for *M. vovidesii.* Here we show the variation of vessel traits (frequency, diameter, and length) between drought and non-drought years (Fig. 1.7). The relationship between vessel traits and drought events reflects that *Magnolia vovidesii* is capable to adapt anatomically during high hydric stress. These anatomical adjustments can allow Mexican tropical montane cloud trees of diverse functional types to continue growth

Table 1.1 Growth ring statistics for *Magnolia vovidesii* in a tropical montane cloud forest of the Sierra Madre Oriental of eastern Mexico

Statistics	*Magnolia vovidesii*
Sampled trees	20
Cross-dated series[a]	25
Master series (year)	1941–2016
Cross-dated rings[a]	2545
Series intercorrelation[b]	0.501
Mean sensitivity[a]	0.288
Autocorrelation[a]	0.501
Mean/median age (years)	75/50
Common interval	1938–2016
Signal-to-noise[c]	16.54

[a]Values obtained with COFECHA (Holmes 1983)
[b]Values statistically different using Mann-Whitney test ($P = 0.05$)
[c]Values obtained with ARSTAN (Cook and Holmes 1999)

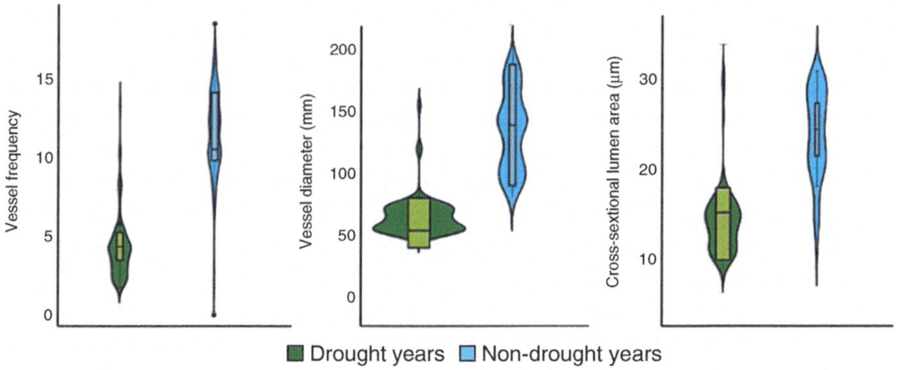

Fig. 1.7 Violin plots showing differences in vessel traits between drought and non-drought years for *Magnolia vovidesii*

during the Fall, even if they are evergreen, semi-deciduous, or deciduous trees (Aguilar-Romero et al. 2017; Gea-Izquierdo et al. 2012; Osazuwa-Peters et al. 2017; Rodríguez-Ramírez et al. 2018; Slik et al. 2018).

The strong relationship between vessel traits and drought and non-drought years (Fig. 1.7) reflects magnolia's capacity to adapt during drought periods when there is lower hydric stress, similar to other diffuse-porous deciduous species (García-González and Eckstein 2003; Fonti and García-González 2004; Tardif and Conciatori 2006; Gea-Izquierdo et al. 2012).

The dendroecological tools allowed us to evaluate the drought effects on vessel anatomy of diffuse-porous wood up to the year of 1929. We found differences in the vessel hydraulic architecture when comparing drought and non-drought years in tree-ring widths of *Magnolia vovidesii.*

Several studies suggest that recent ENSO events (from 1929 to 2016) drove the TMCFs' tree species to their anatomical and physiological limits (Gea-Izquierdo et al. 2012; Rita et al. 2015; Schöngart et al. 2017; Islam et al. 2018b). Our results demonstrate that *Magnolia vovidesii*'s vessel traits show a perceptible trend to become less in number and smaller in size during drought periods. Nonetheless, drought events could possibly not only affect *Magnolia* species' resilience as an ability to adapt to drought periods but also possibly diminish its natural regeneration.

In summary, our case study shows the application of detailed dendroecological data, combined with vessel hydraulic plasticity, is a capable tool for better understanding the drought effect on TMCF trees.

1.6 Future Directions and Challenges

Tropical montane cloud forest trees prevailed over environmental changes and survived for centuries, so we could expect that the wood anatomic features of trees are well adapted to fit different climatic variations. Tree species with contrasting vessel architecture may coexist within forest communities subject to similar hydric stress. Therefore, understanding the climate resilience of TMCF trees to drought is essential to ensure the best forest conservation of specially threatened, endemic, and relict tree species.

This case study has assessed the range of xylem plasticity to specific drought events by analyzing a whole suite of vessel traits. The adjustment of vessel traits was significantly different as a response to hydric stress from annual variability in drought periods during the last 75 years (1941–2016).

Further research is required to fully understand how climate variations (e.g., ENSO events) affect the anatomical adaptation to climate for several threatened TMCF trees. Notwithstanding, anthropic activities are having profound impacts on the health of TMCFs in two significant ways: first, habitat fragmentation through land conversion to plantation agriculture (corn, avocado, and/or coffee) or grazing (by cows, sheep, and/or goats) uses illegal logging, and second natural disturbances (e.g., tree fall-down, drought, hurricanes, frost, fungi infest, insects, and disease) influence the ecosystem. These two features affect the permanence and/or natural regeneration of *Magnolia vovidesii*.

Dendroecological analysis and anatomical traits provide essential tools that can help in the understanding of the climatic resilience and potential future impacts of climate change and global trade-offs on TMCFs worldwide.

Acknowledgments We thank Doroteo San Juan for guidance and for permitting the sampling at the study forest, Susana Guzmán Gómez (Laboratorio de Microscopía y Fotografía de la Biodiversidad II, Instituto de Biología, UNAM) for technical assistance with the digital photographs, and Carlos Andrés Solís Hay for his critical observations. The first author thanks the financial support granted by the postdoctoral fellowship CONACYT 2018–2019 and by the project PAPIIT-DGAPA IN223218.

References

Abe H, Nakai T (1999) Effect of water status within a tree on tracheid morphogenesis in *Cryptomeria japonica* D. Don. Trees 14(3):124–129. https://doi.org/10.1007/PL00009758

Aguilar-Romero R, Pineda-García F, Paz H et al (2017) Differentiation in the water-use strategies among oak species from central Mexico. Tree Physiol 37:915–925. https://doi.org/10.1093/treephys/tpx033

Albuquerque RP, Brandes AF, Lisi CS et al (2019) Tree-ring formation, radial increment and climate–growth relationship: assessing two potential tree species used in Brazilian Atlantic forest restoration projects. Trees:1–16. https://doi.org/10.1007/s00468-019-01825-6

Alves ES, Angyalossy-Alfonso V (2000) Ecological trends in the wood anatomy of some Brazilian species. 1. Growth rings and vessels. IAWA J 21(1):3–30. https://doi.org/10.1163/22941932-90000233

Azuma H, García-Franco JG, Rico-Gray V et al (2001) Molecular phylogeny of the Magnoliaceae: the biogeography of tropical and temperate disjunctions. Am J Bot 88:2275–2285. https://doi.org/10.2307/3558389

Bréda N, Huc R, Granier A et al (2006) Temperate forest trees and stands under severe drought: a review of ecophysiological responses, adaptation processes and long-term consequences. Ann For Sci 63(6):625–644. https://doi.org/10.1051/forest:2006042

Breshears DD, Myers OB, Meyer CW et al (2009) Tree die off in response to global change type drought: mortality insights from a decade of plant water potential measurements. Front Ecol Environ 7:185–189. https://doi.org/10.1890/080016

Bruijnzeel LA, Scatena FN, Hamilton LS (2010) Tropical montane cloud forests: science for conservation and management. Cambridge University Press, Cambridge, UK

Bryukhanova M, Fonti P (2013) Xylem plasticity allows rapid hydraulic adjustment to annual climatic variability. Trees 27:485–496. https://doi.org/10.1007/s00468-012-0802-8

Campelo F, Vieira J, Nabais C (2013) Tree-ring growth and intra-annual density fluctuations of *Pinus pinaster* responses to climate: does size matter? Trees 27(3):763–772. https://doi.org/10.1007/s00468-012-0831-3

Cardoza-Martínez GF, Cerano-Paredes J, Villanueva-Díaz J et al (2013) Annual precipitation reconstruction of the eastern region of Tlaxcala state. Rev Mex Cienc For 5:110–127

CONABIO (2010) El bosque mesófilo de montaña en México: amenazas y oportunidades para su conservación y manejo sostenible. Comisión Nacional para el Conocimiento y Uso de la Biodiversidad, Mexico City

Cook ER, Holmes RL (1999) Program ARSTAN-chronology development with statistical analysis (user's manual for program ARSTAN). Laboratory of Tree-Ring Research. University of Arizona, Tucson

Corral-Aguirre J, Sánchez-Velásquez LR (2006) Seed ecology and germination treatments in *Magnolia dealbata*: an endangered species. Flora 201(3):227–232. https://doi.org/10.1016/j.flora.2005.07.004

D'Arrigo R, Davi N, Jacoby G et al (2014) Dendroclimatic studies: trees growth and climate change in northern forest. American Geophysical Union, New Jersey, USA

Dünisch O, Bauch J (2009) Influence of soil substrate and drought on wood formation of Spruce (*Picea abies* [L.] Karst.) under controlled conditions. Int J Biol Chem Phys Tech Wood 48(6):447–457. https://doi.org/10.1515/hfsg.1994.48.6.447

Dinerstein E, Olson D, Joshi A et al (2017) An Ecoregion-Based Approach to Protecting Half the Terrestrial Realm. BioScience 67(6):534–545

Eller CB, Barros FV, Bitterncourt PRL et al (2017) Xylem hydraulic safety and construction cost determine tropical tree growth. Plant Cell Environ 41:548–562. https://doi.org/10.1111/pce.13106

Ellison D, Morris CE, Locatelli B et al (2017) Trees, forests and water: cool insights for a hot world. Glob Environ Chang 43:51–61. https://doi.org/10.1016/j.gloenvcha.2017.01.002

Fahey TJ, Sherman RE, Tanner EVJ (2016) Tropical montane cloud forest: environmental drivers of vegetation structure and ecosystem function. J Trop Ecol 35(5):355–367. https://doi.org/10.1017/S0266467415000176
Fonti P, García-González I (2004) Suitability of chestnut earlywood vessel chronologies for ecological studies. New Phytol 163(1):77–86. https://doi.org/10.1111/j.1469-8137.2004.01089.x
Fonti P, von Arx G, García-González I et al (2010) Studying global change through investigation of the plastic responses of xylem anatomy in tree rings. New Phytol 185:42–53. https://doi.org/10.1111/j.1469-8137.2009.03030.x
Food and Agricultural Organization of the United Nations FAO (1998) Soil map of the world: revised legend. World soil resources report 60. Available online: http://faostat.fao.org. Accessed 18 Oct 2018
García-González I, Eckstein D (2003) Climatic signal of earlywood vessels of oak on a maritime site. Tree Physiol 23(7):497–504. https://doi.org/10.1093/treephys/23.7.497
García-González I, Souto-Herrero M (2017) Earlywood vessel area of *Quercus pirenaica* Willd. Is a powerful indicator of soil water excess at growth resumption. Eur J Forest Res 136:329–344. https://doi.org/10.1007/s10342-017-1035-6
Gärtner H, Nievergelt D (2010) The core-microtome: a new tool for surface preparation on cores and time series analysis of varying cell parameters. Dendrochronologia 28:85–92. https://doi.org/10.1016/j.dendro.2009.09.002
Gea-Izquierdo G, Fonti P, Cherubini P et al (2012) Xylem hydraulic adjustment and growth response of *Quercus canariensis* Willd. to climatic variability. Tree Physiol 32:401–413. https://doi.org/10.1093/treephys/tps026
Goldsmith GR (2013) Changing directions: the atmosphere–plant–soil continuum. New Phytol 199(1):4–6. https://doi.org/10.1111/nph.12332
Gotsch SG, Asbjornsen H, Holwerda F et al (2014) Foggy days and dry nights determine crown level water balance in a seasonal tropical montane cloud forest. Plant Cell Environ 37(1):261–272. https://doi.org/10.1111/pce.12151
Graham A (1976) Studies in neotropical paleobotany II: the Miocene communities of Veracruz, Mexico. Ann Missouri Bot Gard 63(4):787–842. https://doi.org/10.2307/2395250
Granato-Souza D, Stahle DW, Barbosa AC et al (2019) Tree rings and rainfall in the equatorial Amazon. Clim Dyn 52(3–4):1857–1869. https://doi.org/10.1007/s00382-018-4227-y
Grissino-Mayer HD (2001) Evaluating crossdating accuracy: a manual and tutorial for the computer program COFECHA. Tree Ring Res 57:205–221
Gual-Díaz M, Rendón-Correa A (2014) Bosques mesófilos de montaña de México: diversidad, ecología y manejo. Comisión Nacional para el Conocimiento y Uso de la Biodiversidad, Mexico City
Guiot J (1991) The bootstrapped response function. Tree-Ring Bull 51:39–41
Haghshenas M, Mohadjer MRM, Attarod P et al (2016) Climate effect on tree-ring widths of *Fagus orientalis* in the Caspian forests, northern Iran. Forest Sci Technol 12:176–182. https://doi.org/10.1080/21580103.2016.1144542
Hintze JL, Nelson RD (1998) Violin plots: a box plot-density trace synergism. Am Stat 52:181–184. https://doi.org/10.1080/00031305.1998.10480559
Holmes RL (1983) Computer-assisted quality control in tree-ring dating and measurement. Tree Ring Bull 43:69–78
Hu J, Riveros-Iregui DA (2016) Life in the clouds: are tropical montane cloud forests responding to changes in climate? Oecologia 180:1061–1073. https://doi.org/10.1007/s00442-015-3533-x
Islam M, Rahman M, Bräuning A (2018a) Growth-ring boundary anatomy and dendrochronological potential in a moist tropical forest in northeastern Bangladesh. Tree-Ring Res 74(1):76–93. https://doi.org/10.3959/1536-1098-74.1.76
Islam M, Rahman M, Bräuning A (2018b) Long-term hydraulic adjustment of tree tropical moist forest tree species to changing climate. Front Plant Sci 4(9). https://doi.org/10.3389/fpls.2018.01761

Jarvis AJ, Mulligan M (2010) The climate of cloud forests. In: Bruijnzeel LA, Scatena FN, Hamilton LS (eds) Tropical montane cloud forests: science for conservation and management. Cambridge University Press, Cambridge, pp 39–56

Latte N, Lebourgeois F, Claessens H (2015) Increased tree-growth synchronization of beech (*Fagus sylvatica* L.) in response to climate change in northwestern Europe. Dendrochronologia 33:69–77. https://doi.org/10.1016/j.dendro.2015.01.002

Lemoine N, Sheffield J, Dukes JS et al (2016) Terrestrial precipitation analysis (TPA): a resource for characterizing long-term precipitation regimes and extremes. Methods Ecol Evol 7:1396–1401. https://doi.org/10.1111/2041-210X.12582

Luna-Vega I, Alcántara-Ayala O, Morrone JJ et al (2000) Track analysis and conservation priorities in the cloud forests of Hidalgo, Mexico. Divers Distrib 6:137–143. https://doi.org/10.1046/j.1472-4642.2000.00079.x

Mason SJ, Mimmack GM (1992) The use of bootstrap confidence intervals for the correlation coefficient in climatology. Theor Appl Climatol 45:229–233. https://doi.org/10.1007/BF00865512

McDowell N, Pockman WT, Allen C et al (2008) Mechanisms of plant survival and mortality during drought. Why do some plants survive while others succumb to drought? New Phytol 178:719–739. https://doi.org/10.1111/j.1469-8137.2008.02436.x

Osazuwa-Peters OL, Wright SJ, Zanne AE (2017) Linking wood traits to vital rates in tropical rainforest trees: insights from comparing sapling and adult wood. Am J Bot 104:1464–1473. https://doi.org/10.3732/ajb.1700242

Peel MC, Finlayson BL, McMahon TA (2007) Updated world map of the Köppen Geiger climate classification. HESS 4:439–473

Pounds JA, Fogden MP, Campbell JH (1999) Biological response to climate change on a tropical mountain. Nature 398(6728):611. https://doi.org/10.1038/19297

Pouteau R, Giambelluca TW, Ah-Peng C et al (2018) Will climate change shift the lower ecotone of tropical montane cloud forests upwards on islands? J Biogeogr 45(6):1326–1333. https://doi.org/10.1111/jbi.13228

Rahman M, Islam M, Bräuning A (2019) Species-specific growth resilience to drought in a mixed semi-deciduous tropical moist forest in South Asia. For Ecol Manag 433:487–496. https://doi.org/10.1016/j.foreco.2018.11.034

Rita A, Cherubini P, Leonardi S et al (2015) Functional adjustments of xylem anatomy to climatic variability: insights from long-term *Ilex aquifolium* tree-ring series. Tree Physiol 35:817–828. https://doi.org/10.1093/treephys/tpv055

Rivers M, Beech E, Murphy L et al (2016) The red list of Magnoliaceae revised and extended. Botanic Gardens Conservation International, Richmond

Rodríguez-Ramírez EC, Luna-Vega I, Rozas V (2018) Tree-ring research of Mexican beech (*Fagus grandifolia* subsp. *mexicana*) a relict tree endemic to eastern Mexico. Tree-Ring Res 74:94–107. https://doi.org/10.3959/1536-1098-74.1.94

Rodríguez-Ramírez EC, Terrazas T, Luna-Vega I (2019a) The influence of climate on the masting behavior of Mexican beech: growth rings and xylem anatomy. Trees 33(1):23–35. https://doi.org/10.1007/s00468-018-1755-3

Rodríguez-Ramírez EC, Vázquez-García A, Alcántara-Ayala O et al (2019b) Drought effects on vessel plasticity of two endemic *Magnolia* species in the tropical montane cloud forests of eastern Mexico. Ann For Sci. https://doi.org/10.20944/preprints201811.0282.v1

Schneider CA, Rasband WS, Eliceiri KW (2012) NIH image to ImageJ: 25 years of image analysis. Nat Methods 9:671–675. https://doi.org/10.1038/nmeth.2089

Schöngart J, Bräuning A, Barbosa ACMC et al (2017) Dendroecological studies in the neotropics: history, status and future challenges. In: Amoroso M, Daniels L, Baker P, Camarero J (eds) Dendroecology. Ecological Studies (analysis and synthesis). Springer, Cham, pp 35–73

Slik JWF, Franklin J, Arroyo-Rodríguez V et al (2018) Phylogenetic classification of the world's tropical forests. Proc Natl Acad Sci 115:1837–1842. https://doi.org/10.1073/pnas.1714977115

Souto-Herrero M, Rozas V, García-González I (2017) A 481-year chronology of oak earlywood vessels as an age-independent climatic proxy in NW Iberia. Glob Planet Change 155:20–28. https://doi.org/10.1016/j.gloplacha.2017.06.003

Speer JH (2010) Fundamentals of tree ring research. University of Arizona Press, Tucson

Sperry JS (2003) Evolution of water transport and xylem structure. Int J Plant Sci 164(S3):S115–S127. https://doi.org/10.1086/368398

Stahle DW, Cook ER, Burnette DJ et al (2016) The Mexican Drought Atlas: tree-ring reconstructions of the soil moisture balance during the late pre-Hispanic, colonial, and modern eras. Quat Sci Rev 149:34–60. https://doi.org/10.1016/j.quascirev.2016.06.018

Still CJ, Foster PN, Schneider SH (1999) Simulating the effects of climate change on tropical montane cloud forests. Nature 398(6728):608. https://doi.org/10.1038/19293

Tardif JC, Conciatori F (2006) Influence of climate on tree rings and vessel features in red oak and white oak growing near their northern distribution limit, southwestern Quebec, Canada. Can J For Res 36(9):2317–2330. https://doi.org/10.1139/x06-133

Tarelkin Y, Hufkens K, Hahn S et al (2019) Wood anatomy variability under contrasted environmental conditions of common deciduous and evergreen species from central African forests. Trees:1–17. https://doi.org/10.1007/s00468-019-01826-5

Tombesi S, Frioni T, Poni S et al (2018) Effect of water stress "memory" on plant behavior during subsequent drought stress. Environ Exp Bot 150:106–114. https://doi.org/10.1016/j.envexpbot.2018.03.009

Tyree MT, Zimmermann MH (2002) Hydraulic architecture of whole plants and plant performance. In: Tyree MT, Zimmermann MH (eds) Xylem structure and the ascent of sap. Springer, Berlin, Heidelberg, pp 175–214

von Arx G, Kueffer C, Fonti P (2013) Quantifying plasticity in vessel grouping added value from the image analysis tool ROXAS. IAWA J 34:433–445. https://doi.org/10.1163/22941932-00000035

Vásquez-Morales SG, Téllez-Valdés O, Pineda-López DRM et al (2014) Effect of climate change on the distribution of *Magnolia schiedeana*: a threatened species. Bot Sci 92:575–585. https://doi.org/10.17129/botsci.116

Vázquez-García JA, Domínguez-Yescas R, Pedraza-Ruiz R et al (2015) Magnolia rzedowskiana (Magnoliaceae), una especie nueva de la sección Macrophylla de la parte central de la Sierra Madre Oriental, México. Acta Bot Mex 112:19–36

Vázquez-García JA, Neill DA, Asanza M et al (2016) Magnolias de Ecuador: en riesgo de extinción. Universidad Estatal Amazónica, Puyo

Venegas-González A, von Arx G, Chagas MP et al (2015) Plasticity in xylem anatomical traits of two tropical species in response to intra-seasonal climate variability. Trees 29:423–435. https://doi.org/10.1007/s00468-014-1121-z

Webster GL (1995) The panorama of Neotropical cloud forest. In: Churchill SP, Balslev, Forero E, Luteyn JL (eds) Biodiversity and conservation of Neotropical montane forests. Proceedings of the Neotropical montane forest biodiversity and conservation symposium. The New York Botanical Garden, Bronx, pp 53–77

Williams-Linera G, Tolome J (1996) Litterfall, temperate and tropical dominant trees, and climate in a Mexican lower montane forest. Biotropica 28(4b):649–656. https://doi.org/10.2307/2389051

Yoshizawa N, Inami A, Miyake S et al (2000) Anatomy and lignin distribution of reaction wood in two *Magnolia* species. Wood Sci Technol 34(3):183–196. https://doi.org/10.1007/s002260000

Chapter 2
How Drought Drives Seasonal Radial Growth in *Pinus strobiformis* from Northern Mexico

Andrea Cecilia Acosta-Hernández, Marín Pompa-García, Marcos González-Cásares, José Manuel Zúñiga-Vásquez, José Ciro Hernández-Díaz, José Rodolfo Goche-Telles, José Ángel Prieto-Ruíz, and Juan Abel Nájera-Luna

Abstract Climate change and its ecological consequences are a highly topical concern, especially because of the effects of climate change on vegetation including forests. *Pinus strobiformis* is restricted by a unique set of environmental conditions including relative humidity, temperature, precipitation, elevation, and solar radiation that has no other equivalent ecological niche. In this study, the climate sensitivity of earlywood (EW) and latewood (LW) production of *P. strobiformis* was determined by analyzing the response of EW and LW widths to climate variables (temperature, precipitation, and a drought index) using correlation and regression analyses. The width of each wood type was correlated positively with the drought index and indicated that, in the case of *P. strobiformis*, the temperature in the preceding season had strong influence on seasonal growth, particularly EW production. These findings enhance our knowledge of the likely response of *P. strobiformis* growth to predicted climate change.

A. C. Acosta-Hernández (✉) · M. González-Cásares
Programa Institucional de Doctorado en Ciencias Agropecuarias y Forestales, Universidad Juárez del Estado de Durango, Durango, Mexico
e-mail: andrea.acosta@ujed.mx; magonzalez@ujed.mx

M. Pompa-García · J. M. Zúñiga-Vásquez · J. R. Goche-Telles · J. Á. Prieto-Ruíz
Facultad de Ciencias Forestales, Universidad Juárez del Estado de Durango, Durango, Mexico
e-mail: mpgarcia@ujed.mx; jgoche@ujed.mx; jprieto@ujed.mx

J. C. Hernández-Díaz
Instituto de Silvicultura e Industria de la Madera, Universidad Juárez del Estado de Durango, Durango, Mexico
e-mail: jciroh@ujed.mx

J. A. Nájera-Luna
Instituto Tecnológico de El Salto, Durango, Mexico
e-mail: jalnajera@itelsalto.edu.mx

M. Pompa-García, J. J. Camarero (eds.), *Latin American Dendroecology*,
https://doi.org/10.1007/978-3-030-36930-9_2

Keywords Dendroecology · Mexican pine · Mesic species · Tree rings

2.1 Introduction

Climate change is a subject of interest at the present time; each one of the last three decades has been successively warmer than any previous decade, and between 1983 and 2012, the warmest period of the last 1400 years was observed in the Northern Hemisphere (IPCC 2014). In several areas, changes in precipitation regimes altered hydrological systems, which, together with increases in temperature, are considered to be triggers of drought episodes (IPCC 2014; Stahle et al. 2016).

It is expected that arid conditions will intensify in the coming years, increasing the intensity and duration of droughts worldwide and strongly impacting vegetation (Dai 2013; Trenberth et al. 2014; Stahle et al. 2016). Current conditions of temperate forest composition and structure make them very susceptible to high-severity fires and drought episodes (Huffman et al. 2015), given that the majority of forest tree species are vulnerable to rapid climate changes (Hamrick 2004).

The main effects of climate change on vegetation are noticed in changes to the distribution of species as a consequence of their adaptation, changes in phenotypic plasticity or migration (González-Elizondo et al. 2005a), susceptibility to insects and diseases (Looney and Waring 2013), forest dieback (Rodríguez-Catón and Villalba 2018), and even dieback and tree mortality due to the combination of some of these factors (Allen et al. 2015; Millar and Stephenson 2015). These effects bring decreases in forest productivity and negative impacts on the carbon cycle at the regional level as a result of the reduction in photosynthesis rates (Brzostek et al. 2014; Rodríguez-Catón et al. 2016; Roman et al. 2015).

In that sense, it is essential to mention that forests play an important role in regulating global climate due to their key influence on the hydrological cycle and their importance in capturing carbon (Pan et al. 2011; Frank et al. 2015). Mexico has an important forest reserve for that is considered a hotspot of diversity of *Pinus* genus due to the number of species (40% of the world total) and endemic species that it harbors (greater than 55%) (Perry 1991; González-Elizondo et al. 2012). The species are distributed in different types of vegetation at altitudes ranging from sea level to above 4000 m (Rzedowski 1978; Gernandt and Pérez-de la Rosa 2014), and one of the most diverse areas is the Sierra Madre Occidental (González-Elizondo et al. 2012).

The Durango forests represent the most important timber resources in Mexico, given its contribution to national production (35.1% from 2007 to 2016) (SEMARNAT 2016), and they harbor a high diversity of pine species (24) (García and González 2003; González-Elizondo et al. 2012). One of these species that stands out due to its abundance, distribution, and economic and ecological importance is *Pinus strobiformis* Engelm. (García and González 2003).

P. strobiformis, commonly known as white pine, is a widely distributed conifer that grows in mesic sites, generally in humid, cool gullies with deep, well-drained soils in a subhumid temperate climate with marked seasons (Perry 1991; García and González 2003). This species develops under a unique set of environmental conditions determined by the humidity, temperature, elevation, and site protection (Park 2001; Laughlin et al. 2011) that has no other equivalent ecological niche (Aguirre-Gutiérrez et al. 2015).

It is a species that provides food and protection for wildlife (Samano and Tomback 2003; Edelman and Koprowski 2005; Contreras-Lozano et al. 2012), promotes soil retention, and influences the succession process of forests (Donnegan and Rebertus 1999; Ellison et al. 2005; Harvey et al. 2008).

P. strobiformis is currently facing a great threat from a fungal pathogen (*Cronartium ribicola* JC Fisch.) introduced in North America, which has infected approximately 40% of the populations of New Mexico (Conklin et al. 2009; Zeglen et al. 2010; Looney and Waring 2013) and also Mexican populations (Geils et al. 2010). In addition, Mexican white pines are among the taxa that, due to climate change, will face new selection pressures and may not be able to adapt (Aguirre-Gutiérrez et al. 2015). Consequently, it is important to identify the processes and response mechanisms of the species to understand the effects on its biological processes. In addition, as climate change advances, forest ecosystems in many regions are forced to experience conditions outside their natural range of variability, which increases the likelihood of widespread mortality and large-scale disturbance events (Williams et al. 2010).

Dendrochronology is the science that studies the annual growth rings of trees over time, assigning each growth ring the exact year of formation and associating the growth of trees with internal or external factors that may have influenced the tree, throughout his life (Cook and Kairiukstis 1990; Fritts 2001). One of its main fields of application is dendroecology, which is a discipline concerned with ecological questions solved using tree-ring data.

This chapter reports the effects of climatic variables on the seasonal growth of *P. strobiformis*, which is a conifer widely distributed in mesic sites in northern Mexico. The specific aim of this study was to determine the influence of climatic variables, such as the mean precipitation and maximum, mean and minimum temperature, as well as the influence that drought might have on seasonal growth (earlywood and latewood) of *P. strobiformis*. We hypothesized that this species will show a low seasonal response to drought given the species grows in mesic sites.

2.2 Materials and Methods

2.2.1 Study Area and Studied Species

The study area is located in the Ejido El Brillante, Pueblo Nuevo, Durango, Mexico, in the Sierra Madre Occidental (Fig. 2.1a), at 23.66° N and 105.44° W, at an elevation of 2829 m. According to INEGI (2014), the study area presents a cambisol-type soil, defined as a young, underdeveloped, and widely distributed soil with small

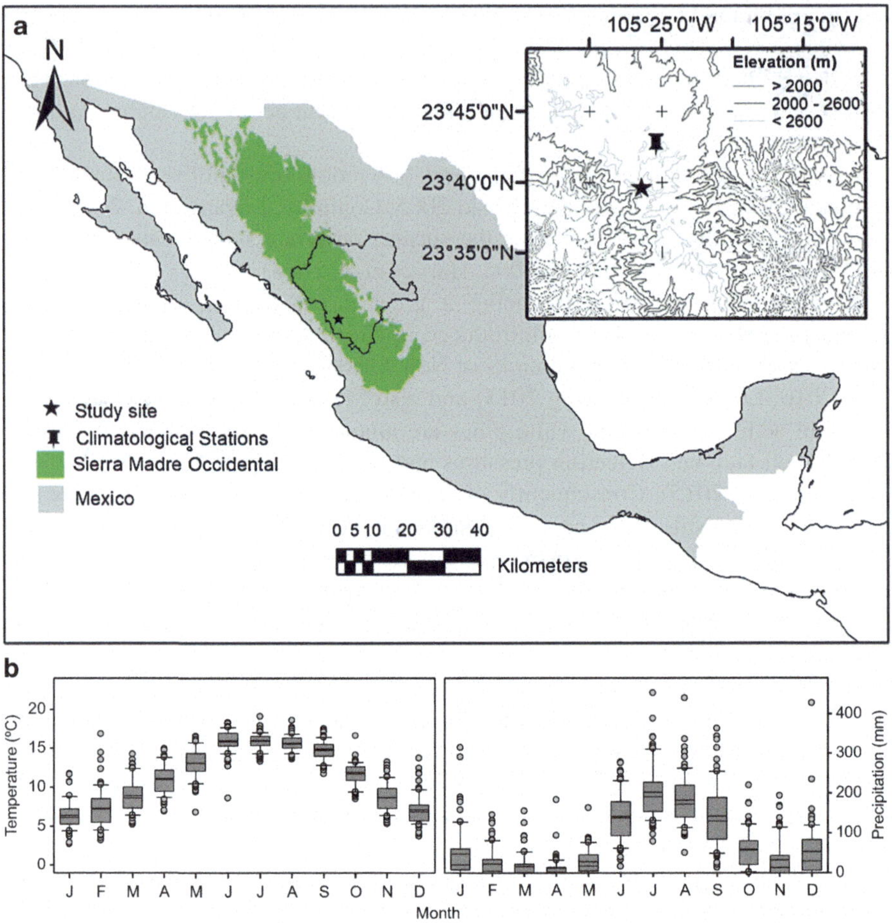

Fig. 2.1 Location of the study site (**a**, star) and monthly climate conditions obtained from the El Salto climatological station (**b**), Durango, Mexico (1946–2016). The Durango state is delimited by the black line and the green area indicates the extent of the Sierra Madre Occidental

accumulations of clay, calcium carbonate, iron, or manganese whose erosion susceptibility is considered moderate to high. It has a temperate subhumid climate, with a marked seasonality characterized by wet and cool summers derived from the influence of the North American monsoon (June–September) and winters with less intense rainfall during the cold season (November–February) (Adams and Comrie 1997).

The atmospheric patterns related to the ENSO (El Niño–Southern Oscillation) lead to hydroclimatic variability over Mexico which is modulated by sea surface temperatures in the tropical oceans of the Pacific and Atlantic Oceans (Seager et al. 2009). The average annual temperature is 11.5 °C, and the annual precipitation reaches up to 1200 mm (CNA 2016) (Fig. 2.1b).

P. strobiformis is a pine native to Mexico and the United States that lives in mesic sites at altitudes ranging from 1900 to 3500 m, in relatively deep soils, rich in humus, generally in gullies and hillsides, accompanied by species of the genera *Pseudotsuga*, *Picea*, *Abies*, *Cupressus*, and some species of *Pinus*, *Quercus*, and *Alnus* (García and González 2003; Farjon 2013). Due to its morphological characteristics, in northern Mexico, *P. strobiformis* has been often confused with *Pinus ayacahuite* (Shaw 1909; García and González 2003).

2.2.2 Sampling and Development of Chronologies

In the spring of 2017, we sampled 18 dominant individuals randomly distributed in the study area, taking two wooden çores per individual, using a Pressler borer (diameter, 5.1 mm). Additionally, the normal diameters (measured at 1.3 m) and total heights were measured (Table 2.1).

The samples were dried and mounted on wooden guides. The extracted cores were polished with coarse to fine sandpaper (80–1000 grain) to highlight their rings. The rings were counted and visually cross-dated with dendrochronological techniques (Stokes and Smiley 1968). The widths of earlywood (EW) and latewood (LW) were measured with a VELMEX system at a resolution of 0.01 mm. The dating was reviewed using the COFECHA program, which compares the series of each tree with a master chronology (Holmes 1983). To generate the chronologies of the indexed widths, standardization was carried out with a process that allows the elimination of biological and geometric growth trends not associated with climatic variables, making them comparable to each other. Detrending and standardization were done by fitting exponential negative functions so as to create width indices (RWI). The temporal autocorrelation associated with the growth of the previous year was eliminated by fitting autocorrelation models (usually of order 1), obtaining the pre-

Table 2.1 Characteristics and dendrochronological statistics of the studied variables

Variable		
Diameter at breast height (cm)	38.5 ± 1.4	
Height (m)	15.44 ± 1.1	
Age at 1.3 m (years)	56 ± 4	
Number of trees (number of radii)	18 (34)	
	EW	**LW**
Width (mm)	2.46 ± 0.03	0.71 ± 0.01
First-order autocorrelation (AC)	0.47	0.48
Mean sensitivity (MS)	0.28	0.38
Mean correlation between the trees (Rbar)	0.20	0.06
Expressed Population Signal (EPS)	0.85	0.62

EW earlywood width, *LW* latewood width
Values correspond to means ± standard errors. The dendrochronological statistics refer to the period 1945–2016

whitened or residual chronologies of EW and LW. Chronologies were developed with the dplR library (Bunn 2008, 2010) of the software R (R Core Team 2018). The mean, first-order autocorrelation (AC), mean sensitivity (MS), mean correlation between the trees (Rbar), and the signal expressed by the population (EPS) were calculated from each width chronology.

2.2.3 Influence of Climatic Variables on Seasonal Growth

The climatic variables were obtained from the El Salto climatological station (CNA 2016), which provided monthly data from the period 1946 to 2016, and it is located at 23.69° N and −105.35° W, 4.5 km from the study site. The monthly climatic variables used for the analysis were mean precipitation (PP); maximum (T_{max}), mean (T_{mean}), and minimum temperature (T_{min}); and combined variables for which the previous summer period (PS) was considered, with the variables of temperature and precipitation (PSPP, PST_{max}, PST_{mean}, and PST_{min}) including the months of June to September of the previous year; the previous winter (PW) with the four climatic variables (PWPP, PWT_{max}, PWT_{mean}, and PWT_{min}), calculated from October of the previous year to February of the current year; and the summer of the current year (CS), with the variables of temperature and precipitation (CSPP, CST_{max}, CST_{mean}, and CST_{min}) from June to September of the year of tree-ring formation.

The STATS and MASS libraries of the *R* statistical software (Venables and Ripley 2002) were used to calculate, by means of a multiple regression using the lm function, the response functions of tree growth to climate. Then, a stepwise regression was performed with a significance level of 0.10 for those climatic variables significantly correlated with the ring width indices. To reduce collinearity problems among the variables, only the combined variables were included in the stepwise regression analyses (the monthly and seasonal variables) which was verified through the VIF function of the CAR library of the R statistical software (Fox and Weisberg 2018).

2.2.4 Relationship of the Drought Index to Seasonal Growth

The influence of drought on seasonal growth was evaluated by correlating the residual series EW and LW widths with the Standardized Precipitation Evapotranspiration Index (SPEI) (Vicente-Serrano et al. 2010) using the Pearson correlation index (*r*). The SPEI considers the effects of temperature on evapotranspiration rates and quantifies the intensity of drought at different time scales. The SPEI was calculated locally from instrumental climatological data using the SPEI library of R (Beguería and Vicente-Serrano 2013) to improve the available spatial resolution (http://sac.csic.es/spei/). The correlation analysis was performed on a monthly basis. The analysis window of the correlations was from June of the prior year to September of the year of growth, based on previous dendroclimatic analyses that referred to the same study area (Pompa-García et al. 2013; Pompa-García and Camarero 2015). To determine the temporal response shown by the species to the accumulated water deficit (González-Cásares et al. 2016), the SPEI was calculated at time scales from 1 to 28 months.

Phenomena such as water stress and drought and vegetation responses to the climate dynamics can be detected and investigated using satellite-derived vegetation indices. Especially the Normalized Difference Water Index (NDWI) is widely used for monitoring the water status of vegetation and for water stress detection. It is a measure of liquid water molecules in vegetation canopies that interacted with the incoming solar radiation. To identify the water stress present in the site, the NDWI was calculated following the methodology of Gao (1996). For this purpose, Landsat 7 ETM+ data of every month of 2011 were used. The 2011 year was chosen for this analysis because it was a year with record drought (CNA 2011). Satellite data were obtained from the United States Geological Survey (USGS) website (https://glovis.usgs.gov/).

2.3 Results

2.3.1 Growth Patterns

The trees in the site had a maximum age of 72 years. The dendrochronological statistics were calculated for the best-replicated period 1945–2016. For this period, the earlywood chronology presented an EPS value of 0.85 which is recommended for dendroclimatic analysis (cf. Wigley et al. 1984), while for latewood it was lower (0.62) than that threshold (Table 2.1).

The interannual variability of EW and LW is shown in Fig. 2.2. It is easy to notice marker rings for both seasonal variables corresponding to years with favor-

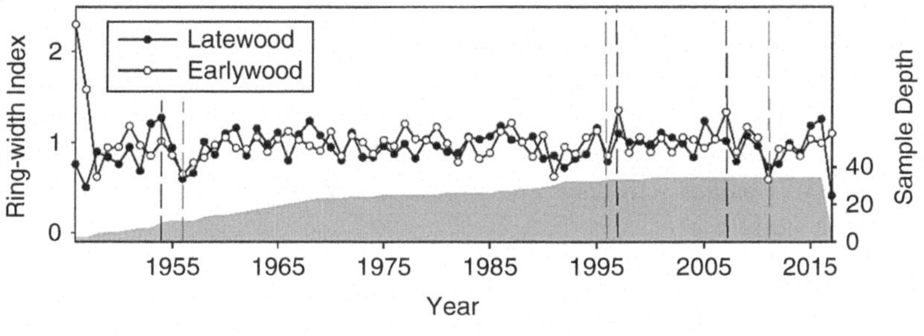

Fig. 2.2 Seasonal growth of *Pinus strobiformis*. The residual chronologies of earlywood and latewood width are shown in continuous lines. The sample depth (number of radii) is displayed by the shaded area of the graph. Dashed vertical lines show the years with extreme growth in both types of wood (the black dashed line indicates the wide rings, while the narrow rings are indicated by the gray dashed lines)

able climatic conditions, usually wet years (wide rings, black dashed vertical lines), as in 1954, 1997, and 2007, or unfavorable dry conditions (narrow rings, gray dashed vertical lines), as in 1956, 1996, and 2011 (Fig. 2.2).

2.3.2 *Climate-Growth Correlation*

Warm conditions in the previous February were negatively related to EW production (Fig. 2.3). EW presented a significant positive correlation between the residual chronology and the SPEI (March) and during the growing season (from May to June). The resulting values of the correlation coefficients (r) were, in most cases, below 0.30 (Fig. 2.3). The Pearson correlations between the residual chronology of LW showed positive correlations with precipitation in June and August of the previous year of formation ($r = 0.23$ and 0.28, respectively, $p < 0.05$), while there were negative correlations with the three temperature indicators, appearing in previous September for the maximum temperature ($r = -0.24$, $p < 0.05$) and in current June for the average and minimum temperature ($r = -0.23$ and -0.35, respectively, $p < 0.05$). Finally, LW and the SPEI showed only one significant positive association ($r = 0.25$, $p < 0.05$) in March of the growth year.

Stepwise linear regression indicated that both EW and LW production were significantly associated with some climatic variables. Table 2.2 shows the specific variables that were selected for the models of EW and LW ring widths. These models only explained 19% and 21% of the EW and LW variance, respectively. EW was only significantly associated with previous winter minimum temperature ($p < 0.05$), while LW was significantly associated with previous winter maximum and minimum temperatures as well as current summer minimum temperatures of the year of formation ($p < 0.10$).

The chronologies of EW responded to short periods of accumulated drought, from 1 to 6 months in March. This means that late winter and early spring drought conditions constrained EW formation. LW was not as sensitive to drought and presented correlation values lower than 0.20 (Fig. 2.4).

The NDWI for the month of December was the one that showed more contrasting values given the accumulated drought that it presented. The NDWI has no dimensions but varies between −1 and +1 depending not only on the water content of the leaves but also on the type of vegetation and cover. In the period of water stress, the NDWI values will decrease. Figure 2.5 shows the behavior of the water content in the study site.

2.4 Discussion

This report contains one of the first chronologies generated for *P. strobiformis* with respect to the climatic drivers of season growth, specifically of earlywood and latewood production. This species did not reach the usual threshold of 0.85 EPS for

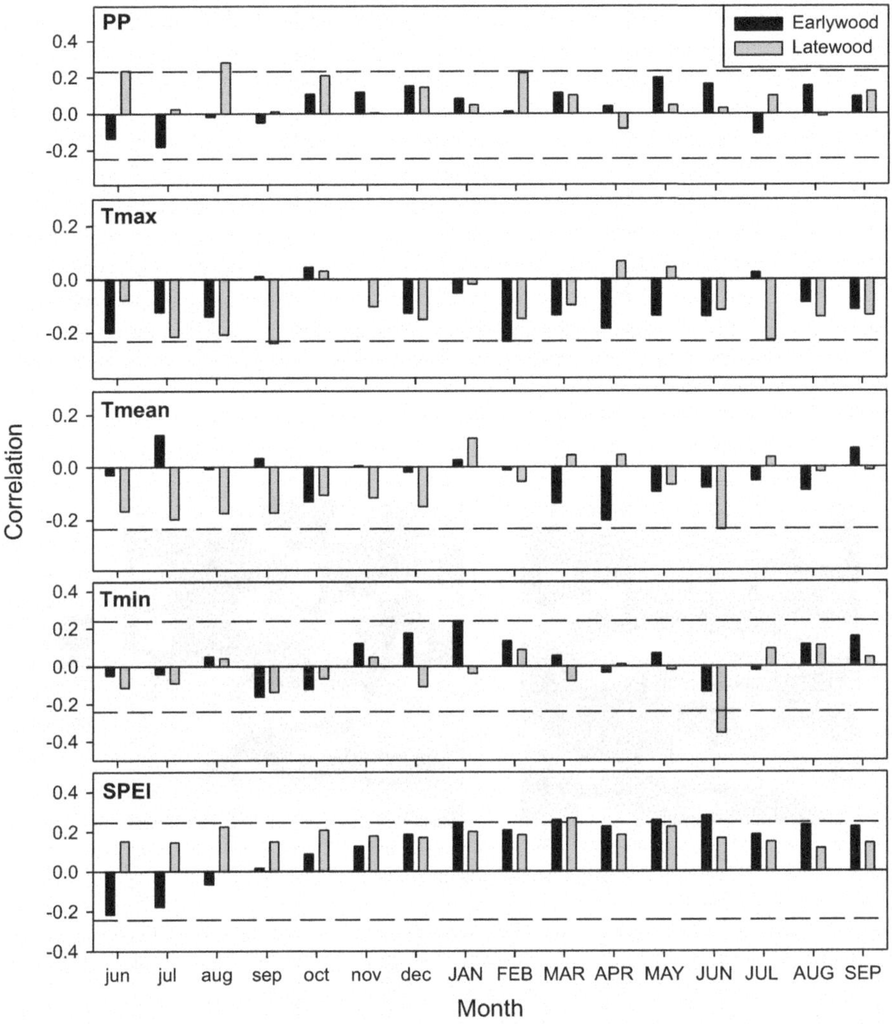

Fig. 2.3 Climate-growth associations obtained for the earlywood (EW) and latewood (LW) width chronologies. Bars show Pearson correlation coefficients calculated by relating monthly climate variables (precipitation (PP); maximum, mean, and minimum temperatures (T_{max}, T_{mean}, T_{min}); and SPEI) and residual series of width for the EW and the LW. Correlations were calculated from the previous June up to the current September (months abbreviated with lower case letters correspond to the year prior to tree-ring formation). Dashed horizontal lines, above and below, show the 0.05 significance levels

latewood (EPS = 0.62), which we attribute to mesic conditions where the species grows (Perry 1991; García and González 2003). The values of EPS obtained for LW are lower than what was reported for *Picea chihuahuana* Martinez in a site close to our study area (Cabral-Alemán et al. 2017). These results indicate that EW reflects a better climatic variability, coinciding with the results of Torbenson et al. (2016) about the important differences in the interannual variability of the EW and LW

Table 2.2 Models for the indices of earlywood (EW) and latewood (LW) widths selected by stepwise regression for climate variables at the ≤0.10 significance level

Type of wood	Variable	Parameter estimate	Standard error	Mean square error	R^2	P
EW	Intercept	1.933	0.399	0.046410	0.19	0.00002**
	PWT_{min}	0.032	0.012			0.0126**
LW	Intercept	1.263	0.417	0.038170	0.21	0.00423**
	PWT_{min}	0.021	0.012			0.07865*
	PWT_{max}	0.028	0.015			0.07771*
	CST_{min}	−0.022	0.012			0.07031*

Abbreviations: PWT_{min} minimum temperature previous winter, PST_{max} maximum temperature of the previous summer, PWT_{max} maximum temperature previous winter, CST_{min} minimum temperature of the current summer. $p < 0.10$*, $p < 0.05$**
R^2 is the determination coefficient

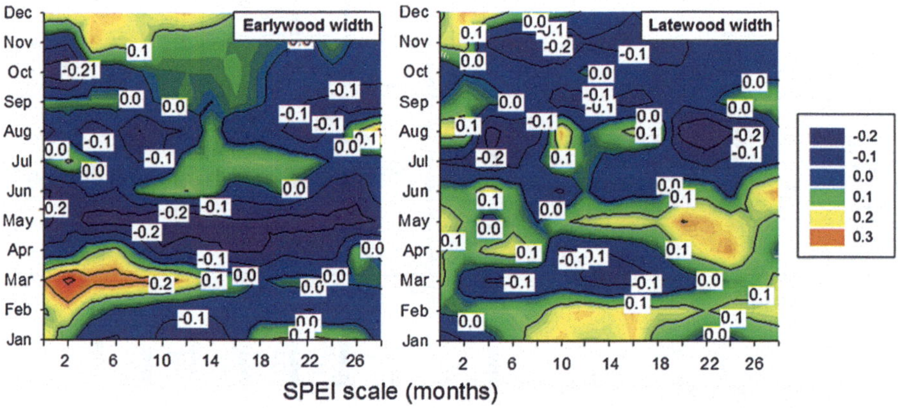

Fig. 2.4 Drought-growth associations. The plots show Pearson correlation coefficients obtained by relating the SPEI drought index at 1- to 28-month long scales (x-axis) calculated from January to December (y-axis) and the residual EW and LW width series

Fig. 2.5 The NDWI for the sampling site calculated for the month of December 2011. High values of NDWI (in green) correspond to higher water content in the vegetation and higher coverage. Low values of NDWI (in red) correspond to lower water content in the vegetation and lower coverage

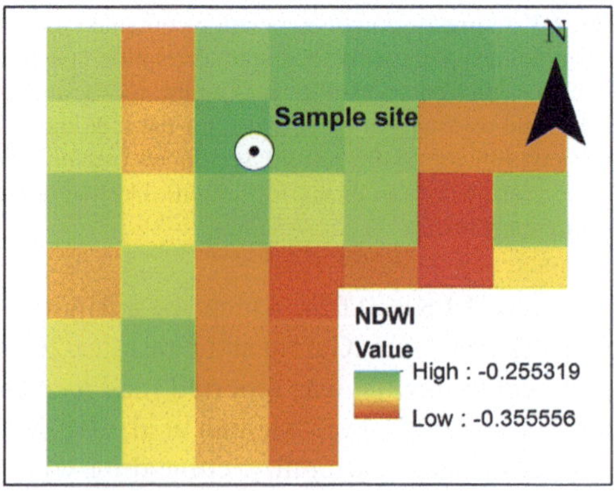

across North America. Our results from dendrochronological statistics (Table 2.1) show that the values obtained for MS were higher, while for AC and Rbar were lower than those reported in studies carried out at nearby sites, with similar characteristics for species such as *P. chihuahuana*, *Abies durangensis* Martinez, *Cupressus lusitanica* Mill., and *Pseudotsuga menziesii* (Mirb.) Franco (González-Elizondo et al. 2005b; Cabral-Alemán et al. 2017; Pompa-García et al. 2017). The lower values in Rbar (Table 2.1) may be due to different age-related growth trends, such as those observed in *Pinus sibirica* Du Tour (D'Arrigo et al. 2001).

As hypothesized, *P. strobiformis* showed a low response of seasonal growth to drought excepting in the case of EW. As mentioned in other previous studies, the different responses of EW and LW could reflect multiple climatic and ecological factors (Acosta-Hernández et al. 2017; Torbenson et al. 2016).

The Pearson correlation shows that prior spring wet conditions enhanced LW production, whereas EW growth is not associated with precipitation variable, which is different to the results obtained by Acosta-Hernández et al. (2019) and the results reported by Cabral-Alemán et al. (2017) for *P. chihuahuana* studied in nearby sites.

In regard to negative correlation in June between T_{min} and LW, it seems that it could be related to water deficit caused by rising evapotranspiration rates. However, to confirm this idea, we need a better assessment of the cambial phenology in this species including xylogenesis studies. The growth of EW responded positively to T_{min} and SPEI in January of the year of growth. These relationships indicate that wet-cool conditions increase wood production in the early growing season, agreeing with those reported in similar studies carried out in Northern Mexico (Pompa-García et al. 2017; González-Cásares et al. 2016).

The results of the linear regression highlight the importance of the previous climatic conditions of winter and summer as drivers in the formation of seasonal growth. Thus, the annual growth and latency cycle in temperate trees is integrated in such a way that environmental factors have delayed impacts on growth (Weiser 1970; Perry 1971). These results concord with Splechtna et al. (2000), who also found that summer temperatures were important for tree growth; he found that cold climate species have a greater response to temperature than precipitation in studies in *Abies lasiocarpa* forests. Wet conditions in the prior winter recharge soils and provide soil moisture in early spring when growth starts (Pompa-García et al. 2013). Warmer conditions generally increase growth due to the increase in photorespiration rates (Sage and Kubien 2007), but beyond a certain high temperature, photorespiration decreases (Way and Oren 2010). In addition, the warmth affinity of the species, its tolerance to frosts, and its discontinuous but wide distribution (Perry 1991) make *P. strobiformis* a good proxy for studying the effects of climatic changes.

In relation to drought, the growth in both seasons was positively influenced by drought conditions, with EW presenting the greatest association. The findings do not match with those obtained by Kagawa et al. (2006) since, for this species, the conditions of the summer of the year of formation of EW are those that positively influence its formation. The SPEI had high correlation values with seasonal growth at the beginning of the growing season, which indicates that this particular species has enough water to grow even in adverse conditions, and its tolerance varies with respect to the conditions in which it develops, as indicated in the literature (Sánchez-Gómez et al. 2010; Pompa-García et al. 2017). In this sense, our findings support the results related to the low impact of the drought on trees in the mesic or humid sites (Adams and Kolb 2005) as a result of the morphological and physiological adaptations that the species presents (Goodrich et al. 2016).

The response of *P. strobiformis* to short-term drought is similar to the results obtained for *P. menziesii* at a nearby site, suggesting that this species may depend on shallow water sources at the beginning of the growing season (Acosta-Hernández et al. 2019). The topography has an important role because although drought has an impact on the growth of pine species, it has a stronger effect on the trees established in sites with irregular topography since the loss of water is probably greater (Martín-Benito et al. 2008). This highlights the relevance of dry periods as drivers of forest dynamics in places where water availability is usually high, suggesting a possible tradeoff between the growth rate of seedlings and tolerance to drought (Goodrich et al. 2016).

Regarding NDWI the results found in the study site are highlighted as they reaffirm the growth response of *P. strobiformis* to short periods of accumulated drought. The NDWI values showed contrasting variability especially in the month of December of the year 2011. According to Karamihalaki et al. (2016), NDWI values mostly depend on the precipitation levels of the previous 2–4 months. The ability of NDWI to monitor drought conditions is also reaffirmed, since not only is it a sensitive index to the water content in the vegetation, but it has also been shown to provide proxy soil moisture information for those places that do not have any soil moisture measuring (Gu et al. 2008).

These results suggest that droughts in the area do not greatly affect the vegetation given the microsite conditions in which *P. strobiformis* is found, which are fundamental for underground water reserves (García and González 2003; Farjon 2013). *P. strobiformis* proved to be a complacent species with little capacity to respond to hydroclimatic variability, similar to *C. lusitanica*, which is distributed near streams, thereby ensuring that it has enough water for metabolism and carbohydrates for its growth and that it is not affected by heat (evapotranspiration demand) and dry conditions. This is consistent with the report by Thomas (2011) that found that pine species with a wide spatial distribution, such as *P. strobiformis*, can better resist some effects of climate change.

2.5 Conclusions

Results highlight the importance of dry periods as drivers of seasonal tree growth. In *P. strobiformis*, seasonal radial growth is affected mainly by extreme temperatures (maximum and minimum). Drought in the previous winter and early spring has a positive influence on earlywood production. Overall, *P. strobiformis* is a complacent species whose growth has a weak association with climatic variables.

Acknowledgments We are grateful to CONACYT for the doctoral scholarship awarded to the first author and for project CB-2013/222522 of CONACYT (National Council of Science and Technology). We also thank COCYTED (Council of Science and Technology of the State of Durango) for the resources granted to support this work. Thanks as well to the community known as "Ejido el Brillante" and to the technical head of the area (Dr. Javier Bretado Velázquez) for the support provided for data collection. We also acknowledge the support received from the Herbarium of CIIDIR-Durango (Interdisciplinary Research Center for Regional Integral Development, Durango Unit) in the measurement of the samples.

References

Acosta-Hernández AC, Pompa-García M, Camarero JJ (2017) An updated review of dendrochronological investigations in Mexico, a megadiverse country with a high potential for tree-ring sciences. Forests 8(5):160. https://doi.org/10.3390/f8050160

Acosta-Hernández AC, Camarero JJ, Pompa-García M (2019) Seasonal growth responses to climate in wet and dry conifer forests. IAWA J 40(2):311–330. https://doi.org/10.1163/22941932-40190226

Adams DK, Comrie AC (1997) The north American monsoon. Bull Am Meteorol Soc 78(10):2197–2214

Adams HD, Kolb TE (2005) Tree growth response to drought and temperature in a mountain landscape in northern Arizona, USA. J Biogeogr 32:1629–1640. https://doi.org/10.1111/j.1365-2699.2005.01292.x

Aguirre-Gutiérrez J, Serna-Chavez HM, Villalobos-Arambula AR et al (2015) Similar but not equivalent: ecological niche comparison across closely-related Mexican white pines. Divers Distrib 21(3):245–257. https://doi.org/10.1111/ddi.12268

Allen CD, Breshears DD, McDowell NG (2015) On underestimation of global vulnerability 737 to tree mortality and forest die-off from hotter drought in the Anthropocene. Ecosphere 6(8):129. https://doi.org/10.1890/ES15-00203.1

Beguería S, Vicente-Serrano SM (2013) SPEI: calculation of the standardised precipitation-evapotranspiration index. R package version 1.6. http://CRAN.R-project.org/package=SPEI

Brzostek ER, Dragoni D, Schmid HP et al (2014) Chronic water stress reduces tree growth and the carbon sink of deciduous hardwood forests. Glob Chang Biol 20(8):2531–2539. https://doi.org/10.1111/gcb.12528

Bunn AG (2008) A dendrochronology program library in R (dplR). Dendrochronologia 26(2):115–124. https://doi.org/10.1016/j.dendro.2008.01.002

Bunn AG (2010) Statistical and visual crossdating in R using the dplR library. Dendrochronologia 28(4):251–258. https://doi.org/10.1016/j.dendro.2009.12.001

Cabral-Alemán C, Pompa-García M, Acosta-Hernández AC et al (2017) Earlywood and latewood widths of *Picea chihuahuana* show contrasting sensitivity to seasonal climate. Forests 8:173. https://doi.org/10.3390/f8050173

Comisión Nacional del Agua CNA (2011) Programa Nacional Contra la Sequía. Available online: http://www.conafor.gob.mx:8080/documentos/docs/7/4637Programa%20Nacional%20Contra%20la%20Sequ%C3%ADa.pdf

Comisión Nacional del Agua CNA (2016) Datos Climáticos de Estaciones Meteorológicas de Durango: El Salto, México. Available online: http://smn1.conagua.gob.mx/index.php?option=com_content&view=article&id=180:durango&catid=14:normales-por-estacion

Conklin DA, Fairweather ML, Ryerson DE et al (2009) White pines, blister rust, and management in the Southwest. USDA Forest Service Southwestern Region R3-FH-09-01, p 16

Contreras-Lozano JA, Lazcano D, Contreras-Balderas AJ (2012) Herpetofauna of the Cerro El Potosí Natural Protected Area of Nuevo León, Mexico: status of the ecological and altitudinal distribution. Nat Areas J 32(4):377–385. https://doi.org/10.3375/043.032.0405

Cook ER, Kairiukstis L (1990) Methods of dendrochronology: applications in the environmental sciences. Kluwer Academic Publishers, Dordrecht, p 351

Dai A (2013) Increasing drought under global warming in observations and models. Nat Clim Chang 3:52–58. https://doi.org/10.1038/nclimate163

D'Arrigo R, Jacoby G, Frank D et al (2001) 1738 years of Mongolian temperature variability inferred from a tree-ring width chronology of Siberian pine. Geophys Res Lett 28(3):543–546

Donnegan J, Rebertus A (1999) Rates and mechanisms of subalpine forest succession along an environmental gradient. Ecology 80(4):1370–1384. https://doi.org/10.1890/0012-9658(1999)080[1370:RAMOSF]2.0.CO;2

Edelman AJ, Koprowski JL (2005) Diet and tree use of Abert's squirrels (*Sciurus aberti*) in a mixed-conifer forest. Southwest Nat 50(4):461–465

Ellison AM, Bank MS, Clinton BD et al (2005) Loss of foundation species: consequences for the structure and dynamics of forested ecosystems. Front Ecol Environ 3(9):479–486

Farjon A (2013) Pinus strobiformis. La Lista Roja de especies amenazadas de la UICN 2013: e.T42416A2978637. https://doi.org/10.2305/IUCN.UK.2013-1.RLTS.T42416A2978637.

Fox J, Weisberg S (2018) An R companion to applied regression, 3rd edn. Sage

Frank DA, Reichstein M, Bahn M et al (2015) Effects of climate extremes on the terrestrial carbon cycle: concepts, processes and potential future impacts. Glob Chang Biol 21:2861–2880. https://doi.org/10.1111/gcb.12916

Fritts HC (2001) Tree rings and climate. The Blackburn Press, Caldwell, New Jersey

Gao BC (1996) NDWI-A normalized difference water index for remote sensing of vegetation liquid water from space. Remote Sens Environ 58:257–266

García AA, González EMS (2003) Pináceas de Durango. Instituto de Ecología. AC Comisión Nacional Forestal, León, p 187

Geils BW, Hummer KE, Hunt RS (2010) White pines, *Ribes*, and blister rust: a review and synthesis. For Pathol 40:147–185. https://doi.org/10.1111/j.1439-0329.2010.00654.x

Gernandt DS, Pérez-de la Rosa JA (2014) Biodiversidad de Pinophyta (coníferas) en México. Rev Mex Biodivers 85:126–133. https://doi.org/10.7550/rmb.32195

González-Cásares M, Pompa-García M, Camarero-Martínez J (2016) Differences in climate-growth relationship indicate diverse drought tolerances among five pine species coexisting in Northwestern Mexico. Trees 31:531. https://doi.org/10.1007/s00468-016-1488-0

González-Elizondo MS, González-Elizondo M, López-Enriquez IL et al (2005a) Cambios y tendencias sucesionales en ecosistemas de Durango. Vid Supra 1:5–11

González-Elizondo M, Jurado E, Návar J et al (2005b) Tree-rings and climate relationships for Douglas-fir chronologies from the Sierra Madre Occidental, Mexico: a 1681–2001 rain reconstruction. Forest Ecol Manag 213:39–53. https://doi.org/10.1016/j.foreco.2005.03.012

González-Elizondo MS, González-Elizondo M, Tena-Flores JA et al (2012) Vegetación de la sierra madre occidental, México: Una síntesis. Acta Bot Mex (100):351–403

Goodrich BA, Waring KM, Kolb TE (2016) Genetic variation in *Pinus strobiformis* growth and drought tolerance from southwestern US populations. Tree Physiol 36:1219–1235. https://doi.org/10.1093/treephys/tpw052

Gu Y, Hunt E, Wardlow B et al (2008) Evaluation of MODIS NDVI and NDWI for vegetation drought monitoring using Oklahoma Mesonet soil moisture data. Geophys Res Lett 35(22). https://doi.org/10.1029/2008GL035772

Hamrick JL (2004) Response of forest trees to global environmental changes. Forest Ecol Manag 197:323–335. https://doi.org/10.1016/j.foreco.2004.05.023

Harvey AE, Byler JW, McDonald GI et al (2008) Death of an ecosystem: perspectives on Western white pine ecosystems of North America at the end of the twentieth century. Gen. Tech. Rep. RMRS-GTR-208. Fort Collins: U.S. Department of Agriculture, Forest Service, Rocky Mountain Research Station, p 10

Holmes RL (1983) Computer-assisted quality control in treering dating and measurement. Tree Ring Bull 43:69–78

Huffman DW, Zegler TJ, Fulé PZ (2015) Fire history of a mixed conifer forest on the Mogollon Rim, northern Arizona, USA. Int J Wildland Fire 24:680. https://doi.org/10.1071/WF14005
Instituto Nacional de Estadística y Geografía INEGI (2014) Conjunto de datos vectoriales Perfiles de suelos. Escala 1:1 000 000, DGG-INEGI, México. Available online: http://www.beta.inegi.org.mx/temas/mapas/edafologia/
Intergovernmental Panel on Climate Change IPCC (2014) Climate change 2014: synthesis report. Contribution of working groups I, II and III to the fifth assessment report of the Intergovernmental Panel on Climate Change [Core Writing Team, R. K. Pachauri and L. A. Meyer (eds.)]. IPCC, Geneva, p 151
Kagawa A, Sugimoto A, Maximov TC (2006) $^{13}CO_2$ pulse-labelling of photoassimilates reveals carbon allocation within and between tree rings. Plant Cell Environ 29:1571–1584. https://doi.org/10.1111/j.1365-3040.2006.01533
Karamihalaki M, Stagakis S, Sykioti O et al (2016) Monitoring drought effects on Mediterranean conifer forests using SPOT-vegetation NDVI and NDWI time series. In: Living Planet Symposium, Proceedings of the conference held, pp 9–13
Laughlin DC, Fule PZ, Huffman DW et al (2011) Climatic constraints on trait-based forest assembly. J Ecol 99(6):1489–1499
Looney CE, Waring KM (2013) *Pinus strobiformis* (southwestern white pine) stand dynamics, regeneration, and disturbance ecology: a review. Forest Ecol Manag 287:90–102. https://doi.org/10.1016/j.foreco.2012.09.020
Martín-Benito D, Cherubini P, Del Río M et al (2008) Growth response to climate and drought in *Pinus nigra* Arn. trees of different crown classes. Trees 22:363–373. https://doi.org/10.1007/s00468-007-0191-6
Millar CI, Stephenson NL (2015) Temperate forest health in an era of emerging 948 megadisturbance. Science 349(6250):823–826
Pan Y, Birdsey RA, Fang J et al (2011) A large and persistent carbon sink in the world's forests. Science 333:988–993. https://doi.org/10.1126/science.1201609
Park AD (2001) Environmental influences on post-harvest natural regeneration in Mexican pine–oak forests. For Ecol Manag 144:213–228
Perry TO (1971) Dormancy of trees in winter. Science 171(3966):29–36. https://doi.org/10.1126/science.171.3966.29
Perry JP (1991) The pines of Mexico and Central America. Timber Press, Portland, p 231
Pompa-García M, Camarero JJ (2015) Reconstructing evaporation from pine tree rings in northern Mexico. Tree-Ring Res 71(2):95–105. https://doi.org/10.3959/1536-1098-71.2.95
Pompa-García M, Cerano-Paredes J, Fulé PZ (2013) Variation in radial growth of *Pinus cooperi* response to climatic signals across an elevational gradient. Dendrochronologia 31:198–204. https://doi.org/10.1016/j.dendro.2013.05.003
Pompa-García M, González-Cásares M, Acosta-Hernández AC et al (2017) Drought influence over radial growth of Mexican conifers inhabiting mesic and xeric sites. Forests 8(5):175. https://doi.org/10.3390/f8050175
R Core Team (2018) R: a language and environment for statistical computing. R Foundation for Statistical Computing, Vienna. URL https://www.R-project.org/
Rodríguez-Catón M, Villalba R (2018) Indicadores del decaimiento en bosques de *Nothofagus pumilio* en el norte de la Patagonia, Argentina. Madera Bosques 24(2):e2421588. https://doi.org/10.21829/myb.2018.2421588
Rodríguez-Catón M, Villalba R, Morales M et al (2016) Influence of droughts on *Nothofagus pumilio* forest decline across northern Patagonia, Argentina. Ecosphere 7(7):e01390. https://doi.org/10.1002/ecs2.1390
Roman DT, Novick KA, Brzostek ER et al (2015) The 1010 role of isohydric and anisohydric species in determining ecosystem-scale response to 1011 severe drought. Oecologia 179:641. https://doi.org/10.1007/s00442-015-3380-9
Rzedowski J (1978) Vegetación de México. Limusa, México D.F

Sage RF, Kubien DS (2007) The temperature response of C_3 and C_4 photosynthesis. Plant Cell Environ 30:1086–1106. https://doi.org/10.1111/j.1365-3040.2007.01682.x
Samano S, Tomback DF (2003) Cone opening phenology, seed dispersal, and seed predation in southwestern white pine (*Pinus strobiformis*) in southern Colorado. Ecoscience 10(3):319–326
Sánchez-Gómez D, Majada J, Alía R et al (2010) Intraspecific variation in growth and allocation patterns in seedlings of *Pinus pinaster* Ait. submitted to contrasting watering regimes: can water availability explain regional variation? Ann For Sci 67:505. https://doi.org/10.1051/forest/2010007
Seager R, Ting M, Davis M et al (2009) Mexican drought: an observational modeling and tree ring study of variability and climate change. Atmósfera 22(1):1–31
Secretaría de Medio Ambiente y Recursos Naturales SEMARNAT (2016) Anuario Estadístico de la Producción Forestal 2016, 1st Edición. Secretaría de Medio Ambiente y Recursos Naturales. Available via DIALOG. https://www.gob.mx/cms/uploads/attachment/file/282951/2016.pdf. Accessed 10 Jan 2019
Shaw CH (1909) The causes of timber line on mountains; the role of snow. Plant World 12(8):169–181
Splechtna BE, Dobrys J, Klinka K (2000) Tree-ring characteristics of subalpine fir (*Abies lasiocarpa* (Hook.) Nutt.) in relation to elevation and climatic fluctuations. Ann For Sci 57(2): 89–100. https://doi.org/10.1051/forest:2000105
Stahle DW, Cook ER, Burnette DJ et al (2016) The Mexican drought atlas: tree-ring reconstructions of the soil moisture balance during the late pre-Hispanic, colonial, and modern eras. Quat Sci Rev 149:34–60. https://doi.org/10.1016/j.quascirev.2016.06.018
Stokes MA, Smiley TL (1968) An introduction to treering dating. University of Chicago Press, Chicago, p 73
Thomas CD (2011) Translocation of species, climate change, and the end of trying to recreate past ecological communities. Trends Ecol Evol 26(5):216–221. https://doi.org/10.1016/j.tree.2011.02.006
Torbenson MCA, Stahl DW, Villanueva-Díaz J et al (2016) The relationship between earlywood and latewood ring-growth across North America. Tree-Ring Res 72(2):53–66. https://doi.org/10.3959/1536-1098-72.02.53
Trenberth KE, Dai A, van der Schrier G et al (2014) Global warming and changes in drought. Nat Clim Chan 4:17–22. https://doi.org/10.1038/NCLIMATE2067
Venables WN, Ripley BD (2002) Tree-based methods. In: Modern applied statistics with S-plus. Springer, New York, pp 251–269
Vicente-Serrano SM, Beguería S, López-Moreno JI (2010) A multiscalar drought index sensitive to global warming: the standardized precipitation evapotranspiration index. J Clim 23:1696–1718. https://doi.org/10.1175/2009JCLI2909.1
Way DA, Oren R (2010) Differential responses to changes in growth temperature between trees from different functional groups and biomes: a review and synthesis of data. Tree Physiol 30(6):669–688. https://doi.org/10.1093/treephys/tpq015
Weiser CJ (1970) Cold resistance and injury in woody plants: knowledge of hardy plant adaptations to freezing stress may help us to reduce winter damage. Science 169(3952):1269–1278. https://doi.org/10.1126/science.169.3952.1269
Wigley TM, Briffa KR, Jones PD (1984) On the average value of correlated time series, with applications in dendroclimatology and hydrometeorology. J Clim Appl Meteorol 23(2):201–213
Williams AP, Allen CD, Millar CI (2010) Forest responses to increasing aridity and warmth in the southwestern United States. Proc Natl Acad Sci 107(50):21289–21294. https://doi.org/10.1073/pnas.0914211107
Zeglen S, Pronos J, Merler H (2010) Silvicultural management of white pines in western North America. For Pathol 40:347–368. https://doi.org/10.1111/j.1439-0329.2010.00662.x

Chapter 3
Dendroecological Studies with *Cedrela odorata* L., Northeastern Brazil

Claudio Sergio Lisi, Mariana Alves Pagotto, Claudio Roberto Anholetto Jr., Francisco Carvalho Nogueira Jr., Helberson Lima Santos, Clayane Matos Costa, Ítallo Romany Nunes Menezes, Fidel Alejandro Roig Juñet, and Mario Tommasiello Filho

Abstract Northeastern Brazil is home to the Caatinga Forest, characterized as a Tropical Dry Forest (TDF), and the Dense Ombrophilous Forest (the Atlantic Forest) predominates near the coast. In the Caatinga, the climate is semiarid, with <600 mm of mean annual precipitation and a mean annual air temperature of 27 °C ± 6 °C. The precipitation presents a seasonal pattern, with a rainy season concentrated in the first half of the year and at least 6 months of dry season, with volumes <50 mm/month. The climate in the Atlantic Forest has a milder air temperature and a mean rainfall of 1450 mm/year. In both regions the tree species, *Cedrela odorata* L., widely studied in dendrochronology, presents populations in farms, with several individuals per hectare. In recent decades, 10 chronologies of *C. odorata* have been produced and compared to precipitation and other environmental factors, and their competition with lianas and support of the herbivorous process has been noted. This chapter presents the potential of *C. odorata* for dendrochronology in Northeastern Brazil; the anatomy of true rings with marginal parenchyma associated with vessels differed from the false rings by the absence of vessels. Absent rings were also observed. Nine chronologies had correlations above the critical level of 0.51 (0.51–0.79) and sensitivity between 0.547 and 0.771. The correlation

C. S. Lisi (✉) · M. A. Pagotto · C. R. Anholetto Jr. · F. C. Nogueira Jr. · H. L. Santos
C. M. Costa · Í. R. N. Menezes
Laboratory of Plant Anatomy and Dendroecology, Federal University of Sergipe – UFS, São Cristóvão, Sergipe, Brazil
e-mail: anholettojr@usp.br

F. A. Roig Juñet
Laboratory of Dendrochronology and Environmental History, IANIGLA, CCT-CONICET-Mendoza, Mendoza, Argentina
e-mail: froig@mendoza-conicet.gob.ar

M. Tommasiello Filho
Forest Science Department, São Paulo University USP/ESALQ – Escola Superior de Agricultura "Luiz de Queiroz", Piracicaba, Brazil
e-mail: mtomazel@usp.br

M. Pompa-García, J. J. Camarero (eds.), *Latin American Dendroecology*,
https://doi.org/10.1007/978-3-030-36930-9_3

between all series (rbt) of all chronologies had values between 0.27 and 0.68 and expressed population signal (EPS) above 0.85 (between 0.88 and 0.98). In both biomes, several populations of *C. odorata* have resulted in climate-related chronologies that showed the seasonal rainfall from May to July induces metabolism and growth rings formation. We also presented the methodology used for tropical dendroecology studies and the relationship between plant growth and environmental conditions.

Keywords Growth rings · Conservation · Caatinga · Tropical dendrochronology

3.1 Introduction

3.1.1 Dendrochronology of Cedrela

Species of the genus *Cedrela*originate in the Neotropical region, and studies of their tree rings have been producing chronologies since the last decades of the twentieth century. Several papers were published, mainly in Argentina, Brazil, and Bolivia (Chalk 1983; Détienne 1989; Vetter and Botosso 1989; Worbes 1989, 1992, 1995; Botosso et al. 2000; Brienen and Zuidema 2005, 2006a, b; Marcati et al. 2006; Villalba et al. 2006). Studies of the wood anatomy, annual formation of tree rings, and growth rates support dendrochronology and neotropical dendroecology.

One of the first studies showing the potential of *Cedrela* for climate reconstruction was done by Villalba et al. (1985) with populations of *Cedrela angustifolia* Sessé & Moc. ex DC. from forests (Salta and Jujuy states) in Argentina. From the chronology of 39 trees collected in 1982, the authors observed high sensitivity, a strong common sign in the growth and correlation with low temperatures and high precipitation during the beginning of the growing season (September and November). Two years later, Villalba et al. (1987) presented a new analysis of two chronologies of *C. angustifolia*. The first from Cerro Chañar (Salta) was correlated with precipitation in the spring–summer period. The second chronology from the Rio Blanco region (Jujuy) is related to summer temperatures.

In 1989, in Argentina, Boninsegna et al., analyzed the formation of annual growth rings and the increment rate of 13 tree species; one of them was *Cedrela fissilis* Vell., which had 15 trees studied and the annual formation of growth rings were described. The results showed it was possible that trees ages were determinate on the increment average of this species in this region.

In the 1990s, the relationship between the climate pattern and the tree growth variation for 12 chronologies of *Juglans australis* Griseb., *C. angustifolia,* and *Cedrela lilloi* in the Argentinean mountain forests of Northwestern Argentina was analyzed. These chronologies indicated that growth ring widths are affected by late winter and early summer climatic conditions in this region. Between 60% and 80%

of the variation in annual and seasonal precipitation for the region was explained using these ring-width chronologies as predictive variables (Villalba et al. 1992).

At the beginning of the twenty-first century, Tomazello et al. (2000) indicated the potentiality of the Meliaceae family for dendrochronology in tropical and subtropical regions. A section of this paper provides a historical survey of the main research developed with *Cedrela* species. The next year, Tomazello et al. (2001) emphasized the potential of *C. angustifolia* for tropical dendrochronology, including X-ray densitometry analyses for the delimitation of growth rings.

In Brazil, Dünisch et al. (2002) studied populations of *C. odorata* from Central Amazonia to verify growth rings and the dynamics of intra-annual xylem growth (activation and inactivation changes). Dünisch et al. (2003) analyzed the growth rings in relation to precipitation from the central Amazon. The results show a significant correlation between tree rings and the precipitation in March and May of the previous growth period. Also in the Amazon, Dünisch (2005) used the ring widths to produce a growth curve of *C. fissilis* compared to El Niño-Southern Oscillation (ENSO) records showing that anomalies of this event influenced the growth variability of the studied populations, but the chronologies were not indicative of reconstruction of El-Niño events.

In Equador, the work of Bräuning et al. (2009) studied the *Cedrela montana* in the rainforest of the mountains in the south of the country. The chronology with extension up to 1840 showed statistical quality reached from 1910 when it included 13 individual series of trees. The authors did not find a good correlation between the chronology and the environmental factors indicating a relation between growth and temperature between January and April, and this factor influenced only 8% of the growth variation of these plants. This result led to the discussion of whether regional climatic data are representative for the studied population, since the species is subject to strong interference of the local climate due to mountainous topography. Another question was whether trees could also respond to other regional factors that were not reflected in the local climate data series.

Espinoza et al. (2014) studied the *C. odorata* (47 series of 27 trees) of the Peruvian Central Forest. The chronology had an extension of 215 years (1795–2009) with a close relationship between tree growth and total precipitation during the rainy season. They emphasized the influence of the precipitation of the rainy season of the previous year, the precipitation of the end of the dry season and the beginning of the rainy season of the current year.

In the same year, Becerra-Montalvo and Zevallos-Pollito (2014) used the dendrochronology of *C. odorata* to indicate the cutting cycle (minimum time required for new logging) of the species recorded in San Ignacio, Cajamarca region, Peru. The authors concluded that the region's *C. odorata* has an ideal cutting cycle of 113 years. Andreacci et al. (2014) studied two sites with *C. fissilis* from the Ombrofilous Dense (plateau) and Mixed (flat land) Forests in the Atlantic Forest of Paraná, Brazil, and found a positive climatic correlation between the chronology of the plateau trees and the precipitation and temperature, but a negative correlation with the flatland population.

In Bolivia, Paredes-Villanueva et al. (2016) analyzed the relationship between growth rates of *C. fissilis* and *C. angustifolia* with precipitation data. Three sites with variable environmental impacts were selected: dry Chiquitano-Concepción (*C. fissilis*), Amazon transitional-Guarayos (*C. fissilis*), and mountain-Postrervalle forest (*C. angustifolia*). The populations studied were sensitive to rainfall from October to April of the current year and to the period of low temperatures between May and July. Lobão et al. (2016) analyzed the periodicity of the growth of *Cedrela* sp. in different conditions of development in Acre, Brazil. As they verified a significant correlation between the different growth conditions of the trees, the researchers indicated that dendrochronology can subsidize plans and decisions in sustainable forest management programs.

Köhl et al. (2017) studied 20 *C. odorata* trees (mean 138 years) from a rainforest in Suriname. Growth rings were used as a parameter to calculate the accumulation of carbon in the individuals' life history. It was shown that in the final quarter of life of the trees there was an average accumulation of 39% of their total carbon stock, suggesting that the species maintains a high stock rate and accumulation in the adult phase of the trees. Baker (2017) tested the periodicity of the *Cedrela* sp. growth rings comparing the chronology with the radiocarbon dating. The results showed that trees from Bolivia, Ecuador, and Venezuela formed a ring per year, whereas Suriname trees formed two rings annually.

Granato-Souza et al. (2018) produced the first chronology of long-term *C. odorata* (1786–2016) that correlated sensitively with the humidity of the Equatorial Amazon. This chronology was used to reconstruct precipitation from February to November, mainly between the late eighteenth and nineteenth centuries. These dendroecological data of native species are important to understand the interannual and decadal variability of the rains, providing a valuable perspective on the hydrological history in the Amazon, as well as the forest structure and dynamics for this region.

Some studies have used chronologies and data from *Cedrela*growth rings for isotopic analyses in order to understand the efficient use of water by trees and annual precipitation reconstruction (Ballantyne et al. 2011; Brienen et al. 2012; Baker et al. 2015; Brienen et al. 2017). Theses were written on the dendrochronology and dendroecology of the *Cedrela* in Brazil, for example: reconstruction of the environmental scenario of São Paulo using dendrochronology (Ferreira 2012); management with populations of the Amazonian floodplain (Rosa 2008); dendroecology of *C. fissilis* from subtropical mountain forests of Brazil (Rauber 2010); dendrochronology, phenology, cambial activity, and wood quality in plants from Acre (Lobão 2011).

In the past decade, studies with *C. odorata* were started in the Tropical Dry Forests (TDF) of Northeastern Brazil, producing a PhD thesis with chronologies of growth rings compared to regional environmental conditions and a previous study of the $^{13}C/^{12}C$ isotopic ratio in the growth rings of TDF (Caatinga) and Atlantic Forest trees from Sergipe state (Anholletto Junior 2013). In this chapter, we aim to show some results from the studies of the *C. odorata* growth rings achieved by the team of the Tree Ring Laboratory of Federal University of Sergipe.

3.1.2 The Cedrela odorata L. *(Meliaceae)*

The *C. odorata* is known by popular names such as cedro-rosa, cedro-amargoso, cedro-cheiroso, cedro-fêmea, cedro-mogno, and cedro-vermelho. Trees reach up to 25–35 m in height, with a straight or tortuous trunk, 90–150 cm in diameter, in tropical regions where the precipitation reaches from 2500 to 4000 mm/year, and they have lower growth in drier areas, such as in the TDF (Tomazello et al. 2000). The young stems exude a characteristic odor when broken. *C. odorata* has composite leaves, paripinates, with opposing leaflets (10–16 leaflets), sessile leaves 8–15 cm long, and terminal inflorescences, pendant, 20–40 cm long, unissex flowers. The fruit capsule of 2.0–3.5 cm in length is the main differential characteristic of this species from *C. fissilis* that has larger fruit with 4.5–8.5 cm. The seeds have wings and are wind dispersed (Dávila et al. 2008).

C. odorata are deciduous, heliophyte or capable of growing under diffuse light, and are found in forests of secondary succession, usually on edges or in clearings (Amaral 2006). It flowers during the months of December–February and the fruits mature from May, when the plant is without leaves (Lorenzi 2009). In Northeastern Brazil, flowering is observed during the rainy season, the fruits mature after 4 months of flowering, and the leafless period occurs in the dry season (Fig. 3.1) (Maia 2012). The water deficit causes cambial dormancy, making the increment in tree circumference minimal or absent, forming the limits of the growth rings

Fig. 3.1 (**a**) *Cedrela odorata* from Sergipe, Brazil; (**b**) Stems; (**c**) Inflorescence; (**d**) Fruits and seeds; (**e**) Stem disk. Scales: 0.6 cm to (**d**) seeds; 2.0 cm to (**c**) inflorescence and (**d**) fruits; and 5.0 cm to (**e**) stem disk

(Worbes 1995). The wood density is 0.66 g/cm^3, it is soft and easy to cut, and used for laminates, furniture, plywood, etc.; thus, it has been exploited, reducing the natural reserves (Roweder 2011).

3.1.3 Distribution, Environments, and Study Sites of Cedrela odorata

The *C. odorata* has great ecological range, occurring from Mexico, Central America, the Caribbean Islands, and in the countries of South America, except for Chile (Pennington 1981) (Fig. 3.2a). In Brazil, it is common in the Amazon Rainforest, the TDF-Caatinga, and the Atlantic Forest (Tomazello et al. 2000) (Fig. 3.2b). In Sergipe, the populations were found from near the coast, in the Atlantic Forest until the TDF-Caatinga region, as well as in the transitional places of these forests—Agreste (data kept in the Xiloteca ASEw) (Fig. 3.2c). Its occurrence was also recorded by the National Forest Inventory—Sergipe (Serviço Florestal Brasileiro 2017).

The Semideciduous Seasonal Forest (called Atlantic Forest) in Sergipe, Brazil occupied 40% of this territory, but at the beginning of the twenty-first century only 1% of the original vegetation was registered restricted to fragmentation (Ribeiro and Siqueira 2001). In this environment, these authors recorded 469 species, belonging to 94 families, such as Myrtaceae, Fabaceae, and Rubiaceae. The Meliaceae is represented by the *C. odorata* L. (cedro), *Trichilia lepdota* Mart., and *Trichilia* sp. Patrick Browne (Serviço Florestal Brasileiro 2017).

The predominant climate from the Atlantic Forestregion, Sergipe State is "As" with dry summers (according to Köppen's climate ranking – Alvares et al. 2013). The average annual rainfall on the Atlantic coast is 1200 mm, decreasing inland up to 800 mm. This rainfall is concentrated during April and July (60% of total annual), and the dry season occurs from September to December (10% of annual rainfall). The mean annual temperature is 26 °C (18–28 °C) (Fig. 3.3). The altitude in Sergipe

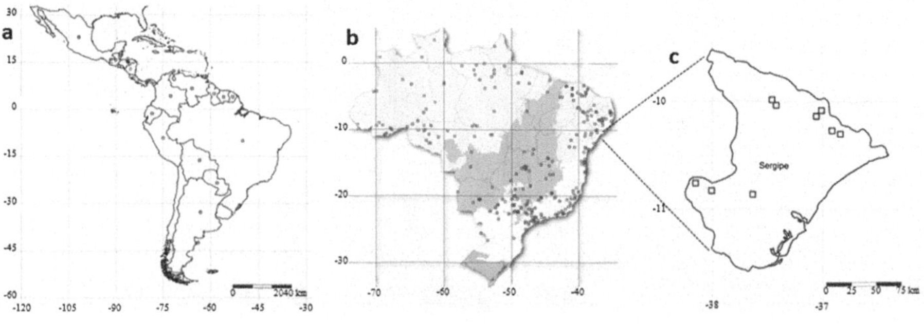

Fig. 3.2 (**a**) Neotropical occurrence of *Cedrela odorata* L. (www.tropicos.org accessed 26/02/2019); (**b**) Records in Brazil (www.cria.org.br accessed 26/02/2019); (**c**) Records in Sergipe (xiloteca ASEw)

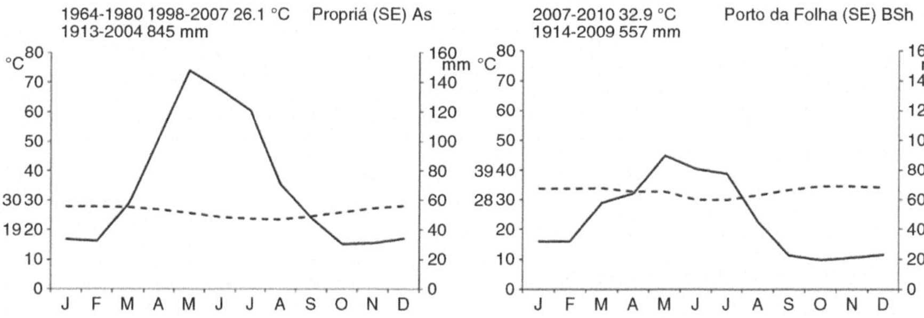

Fig. 3.3 Climatic diagrams for the periods 1913 & 2004 for the Atlantic Forest (Propriá) and 1914 & 2009 for TDF-Caatinga (Porto da Folha) study sites. The diagram consists of the monthly mean temperature (°C – dashed rows) and total precipitation (mm – solid rows). (Data was downloaded from the Instituto Nacional de Pesquisas Espaciais (www.inpe.br))

varies between 300 m a.s.l. (80% of area) and 600 m. Neosols predominate on the coast (Jacomine 2008–2009).

The TDF are represented by biomes Cerrado, Caatinga (Savanna formation), and Campos Rupestres (Janzen 1988). The Caatinga area is 844.453 km^2, present in 11% of Brazilian territory, including the northeast semi-arid region of Brazil (70% of area) and in the north of Minas Gerais State (Alves 2007). This vegetation is always subject to long dry periods and presents adaptations, such as leaf changes (Loiola et al. 2012), thorns, and small size. The Caatinga can be hyperxerophytes and hypoxerophytes. From Mariano Neto (2001), the characteristics of the hyperoxerophytic Caatinga are determined by the volume of rains below 900 mm/year, high temperatures during the dry season, aridity, and low retention of water in the soil, which is shallow and stony. In the hypoxerophytic Caatinga, the vegetation is arboreal or shrub, with species of humid forest, conferring character of transition (Alves 2008), and precipitation can reach 1200 mm/year (Cordeiro and Oliveira 2010). Common tree species in the Caatinga in Sergipe are *Anadenanthera macrocarpa*, *Schinopsis brasiliensis*, *Tabebuia caraiba*, *C. odorata*, and others.

The predominant climate in the Atlantic Forest (near the coast in Sergipe) is "As" and in the TDF-Caatinga area it is "BSh", semi-arid with dry summers (according to Köppen's climate ranking – Alvares et al. 2013), where the annual rainfall does not exceeds 900 mm. In Northeastern Brazil, the rainfall season occurs between May and July (45% of total annual), followed by a long dry season 7–8 months (Tsuchiya 1995; Sampaio 2010). In Sergipe State the rainfall starts in March or April and the total annual precipitation is <900 mm near the Atlantic Ocean coast (Propriá, SE) and <600 mm in the semi-arid region (Porto da Folha, SE) (Fig. 3.3). The hydrological deficit of the dry season tends to be high. The mean annual temperature is slightly higher than 26.5 °C (19.0–33.0 °C). Neosols are predominant in the northwest region (Jacomine 2008–2009).

The rains that occur in the Sergipe territory depend on several systems of atmospheric circulation (Ferreira and Mello 2005). The main mechanisms that act on the rains in the northeast region are: (i) El Niño Oscilação Sul events (ENOS);

(ii) The Sea Surface Temperatures (SST) of the Atlantic Ocean, associated with the trade winds, pressure, and sea level; (iii) Intertropical Convergence Zone (ICZ) on the Atlantic Ocean; (iv) Cold Fronts; and (v) High Level Cyclonic Vortices. These authors also highlight the lines of instability and the effects of sea and land breezes on precipitation.

3.2 Materials and Methods

3.2.1 Study Sites

Samples of *C. odorata* were studied from 11 sites in Sergipe (Table 3.1), with 10–45 trees/sites and between 1 and 4 rays per tree. These populations of cedro are mainly on farms, and in the conservation units of Sergipe there are few records of this species. In the study sites, it was verified that these species are close to each other, near rivers or in soils with rocks.

3.2.2 Sampling, Polishing, and Cedrela odorata *Analysis*

The samplings of *C. odorata* were made mainly by manual core increments (Suunto e Haglof) with internal diameter of 5.15 mm, but a Stihl BT45 motor drill was also used with a core internal diameter of 10–12 mm (no destructive methods), in trees with breast height diameter (BHD) 1.30 m. Wood samples of windows or doors from houses and fences from farms were also sampled.

Sandpapers (40–1500 grains/mm^2) and sanding machines with orbital and eccentric movement were used to polish the transversal surface of the samples—.

Table 3.1 Study sites of *Cedrela odorata* from Sergipe, Brazil

	Study sites	Vegetation	Latitude and Longitude
S1	Cedro de São João	AF	10°16′01.99″S 36°54′51.34″W
S2	Japoatã/Propriá	AF	10°17′48.60″S 36°50′18.32″W
S3	Poço Verde	A	10°45′35.50″S 38°09′07.95″W
S4	Porto da Folha, Area 1	TDF	10°02′12.44″S 37°24′51.53″W
S5	Porto da Folha, Area 2	TDF	09°59′02.40″S 37°27′11.44″W
S6	Porto da Folha, Area 2	TDF	09°59′35.26″S 37°26′29.15″W
S7	Tobias Barreto	TDF	10°52′11″S 38°11′06″W
S8	Canhoba	TDF	10°08′05,8″S 37°03′21,1″W
S9	Nossa Senhora de Lourdes, Area 1	TDF	10°04′52,10″S 37°00′30,13″W
S10	Nossa Senhora de Lourdes, Area 2	TDF	10°07′57,99″S 37°03′20,60″W
S11	Lagarto	A	10°51′41.9″S 37°37′51.7″W

Vegetation: *AF* Atlantic Forest, *TDF* Tropical Dry Forest (Caatinga), *A* Agreste

The core microtome WSL was used for better visualization of the anatomical elements of the wood and distinction of the narrow rings. The samples polished with the eccentric machine showed the marginal parenchyma well contrasted with the fibers, due to the powder inside the lume of cells, when examined under a 10× magnifying glass.

The growth rings were analyzed from scanned images (with HP Deskjet 4100). The images were ".tif" with a resolution of 1200 dpi. Measurements of the radial widths of the growth rings were done with ImageProPlus software (version 4.5.0.29 for Windows). The comparison of the images with the samples seen under stereomicroscope guaranteed the correct definition of the limits of the growth rings.

These measures were exported to spreadsheets (such as Excel software), where they were compared graphically and with Pearson correlation, following the cross-dating method and the synchronization procedures. This procedure proved to be efficient for the observation of true rings, verification of missing rings, and determination of false growth rings in *C. odorata*. However, the quality control of these procedures was made by COFECHA software (Holmes 1983), which indicated other possible corrections to improve the correlations between the series of growth ring measurements. For this, the measured growth rings data are converted to a new file with compact format (Tucson) using the FMT program. For Cofecha analysis, the segments of series measures are adjusted to 20 years lagged successively by 10 years because the trees were not very old. Two results found at the output of the Cofecha are the intercorrelation of the synchronized series and the sensibility of the master series.

Other statistical information has also been calculated, such as the mean inter series (trees) correlations (rbar) calculating correlations for all possible pairs of individual series of the set of series of ring measurements that present a common signal (Briffa and Jones 1990). The expressed population signal (EPS), quantifies the degree to which the chronology expresses the population chronology (Wigley et al. 1984).

The chronologies were prepared from the ARSTAN software (Cook 1985; Cook and Kairiukstis 1990). To removes the variance related to factors other than climate, the individual ring-width series were analyzed with the de-trending method using linear regression, negative curves, and/or cubic smoothing spline function with a 65% cutoff filter (according Cook and Peters 1981). Standard and residual chronologies were compared with instrumental climate records, such as annual rainfall (coincident period to chronologies was 1914–2009) and to monthly rainfall data from all study site regions (data from: Secretaria do Meio Ambiente e Recursos Hidricos – SEMARH, site: https://semarh.se.gov.br), land air temperature index (http://www.ersl.noaa.gov/psd/cgi-bin/data/timeseries), and the monthly sea surface temperature (SST) of the Atlantic Ocean, spatial coordinates between Equador-20S and 10E-30W (http://www.esrl.noaa.gov/psd/data/correlation/tsa.data), among other data (OBS: only the correlation of one chronology with the annual precipitation will be present in this chapter). Chronologies were compared to climate using response and correlation functions (Blasing et al. 1984).

For the microdensitometry X-ray analysis, the radial wood samples were cut in the transverse orientation with 1.0 mm or 2.0 mm of thickness using double circular saws equipment. These thin wood samples were stored in a conditioning chamber (12 h, 20 °C, 50% relative humidity) until reaching 12% moisture content (Tomazello et al. 2008). The analyses proceeded in a Quintek Measurement Systems (QTRX-01X EUA) (Quintek Measurement Systems, 1999). The radial scanning was carried out over the transversal surface of the wood by a collimated X-ray beam. The X-ray values that cross the sample were converted into apparent density (hereafter density) recorded every 80 μm, using QMS software (Ferreira and Tomazello 2009; Castro et al. 2014; Surdi et al. 2014).

3.3 Results and Discussion

3.3.1 Growth Rings of Cedrela odorata *in Sergipe*

The growth rings of *C. odorata* are anatomical structures of the secondary xylem which, when observed in cross-section, are noted as concentric circles around the bit (IAWA Committee 1989; Larson 1994). These structures are the result of periodic changes in the activity of the cambium, which, in turn, responds to factors such as phenology, temperature, photoperiod, precipitation, and endogenous tree rhythms (Lisi et al. 2008).

The growth rings of *C. odorata* have their limits marked by the marginal parenchyma associated with vessels (solitary, double or triple) of larger diameter in earlywood (semiporous rings) (Botosso et al. 2000). Flat fibers also occur and with thick cellular walls at the end of the latewood and the rays extend tangentially next to the marginal parenchyma (Fig. 3.4) (Anholetto 2013). In Northeastern Brazil, the

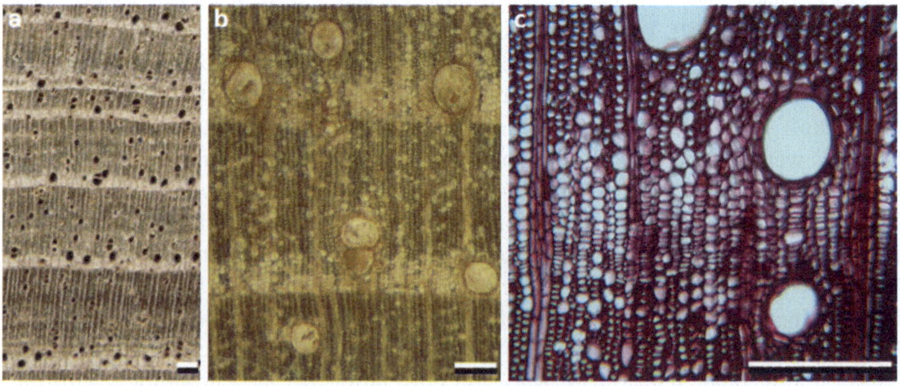

Fig. 3.4 Transversal slides of *Cedrela odorata* from Sergipe. (**a**) Semiporous tree rings and high density of fibers in latewood; (**b**) Vessels associated with marginal parenchyma and cells of parenchyma distributed between fibers; (**c**) Detail of the enlargement of the ray cells along the marginal parenchyma. Scale 0.5 mm

growth rings of *C. odorata* are well defined due to the seasonal climate of the rainy season during the summer and especially in autumn and the dry season in winter and spring, which may occur during 8–9 months of the year.

The *C. odorata* from Caatinga, Sergipe, shows beige heartwood, dark pink to light brown and beige or pinkish sapwood, and has a moderate shine, characteristic odor and taste. The heartwood color shows the marginal parenchyma and facilitates the distinction of consecutive growth rings. However, when the sample has this parenchyma darker, the color blends with the remaining tissue (Fig. 3.5). One of the reasons for these differences is the anatomy of the plant, but sometimes the way the polishing of the sample is done hinders good visualization of the limits of the growth rings. Proper polishing was achieved when cell elements (vessels, fibers, and rays) are seen clearly using a 10× magnifying glass.

Many trees of *C. odorata* in the Caatinga present eccentricity of the trunk that hinders the anatomical visualization of the growth rings, their correct measurement, and the synchronization process of the radial measurements of the series studied in the municipality of Canhoba, Sergipe (10°08′05,8″S; 37°03′21,1″W). One radius of the trunk of one tree was larger than the other, so the growth rings had different radial widths (Fig. 3.6b, c, dotted line). Between trees this is more common (Fig. 3.6a, b). This eccentric growth is common, with the marginal parenchyma coming on one side of the trunk. In the cited example, it is also possible to see two other anatomical characteristics: (i) a possible absent ring (Fig. 3.6c, indicated by

Fig. 3.5 *Cedrela odorata* samples showing different colors between heartwood and sapwood and between trees. (**a**) Marginal parenchyma with more contrast than in (**b**). Scale 5.0 mm

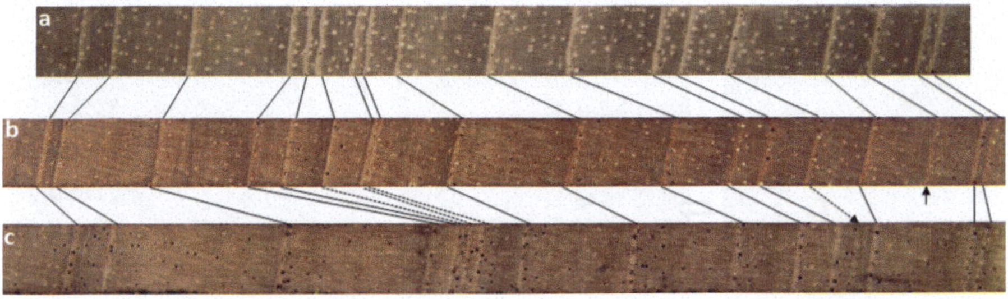

Fig. 3.6 Samples of *Cedrela odorata* from Canhoba, SE. The lines bind to the corresponding growth ring boundaries. (**a, b**) The true growth rings (*solid lines*), (**c**) Growth rings of same tree compressed in the ray (*three dotted lines*). The larger arrow indicates a ring with a parenchyma difficult to see on the ray (**c**), and the smaller arrow indicates a growth ring present in distinct tree rays (**a, b**) and absent in one of the rays of a single tree (**b, c**)

the small arrow in Fig. 3.6b) and (ii) a possible almost imperceptible growth ring (Fig. 3.6c, indicated by the long arrow) but visible on the other radius (Fig. 3.6b).

In one disk of *C. odorata,* studied true rings were seen (Fig. 3.7a) and false rings identify by confluent parenchyma lines do not continue to be associated with vessels of smaller diameter in relation to the earlywood (Fig. 3.7b). It is harder to identify these false rings in samples collected with increment core because the wood section is narrow (5–12 mm). However, the comparison of two or more samples from the same tree makes it easy to identify many of these false rings, and also during the synchronization of the radial measurements of the growth rings, it is possible to separate them through graphs or statistical analysis.

False growth rings are shown as thin layers of parenchyma circling the trunk, most often near the boundaries of the true ring, and sometimes discontinuous in some pieces (Fig. 3.7b). Thick bands of parenchyma were also observed parallel to a line of traumatic channels, with no change in the diffuse porosity of the growth ring (Fig. 3.7b, first two parenchyma bands from above).

Traumatic tissues have also been seen, which are cellular and intracellular channels that store substances as resin (Burger and Richter 1991). In the *C. odorata*

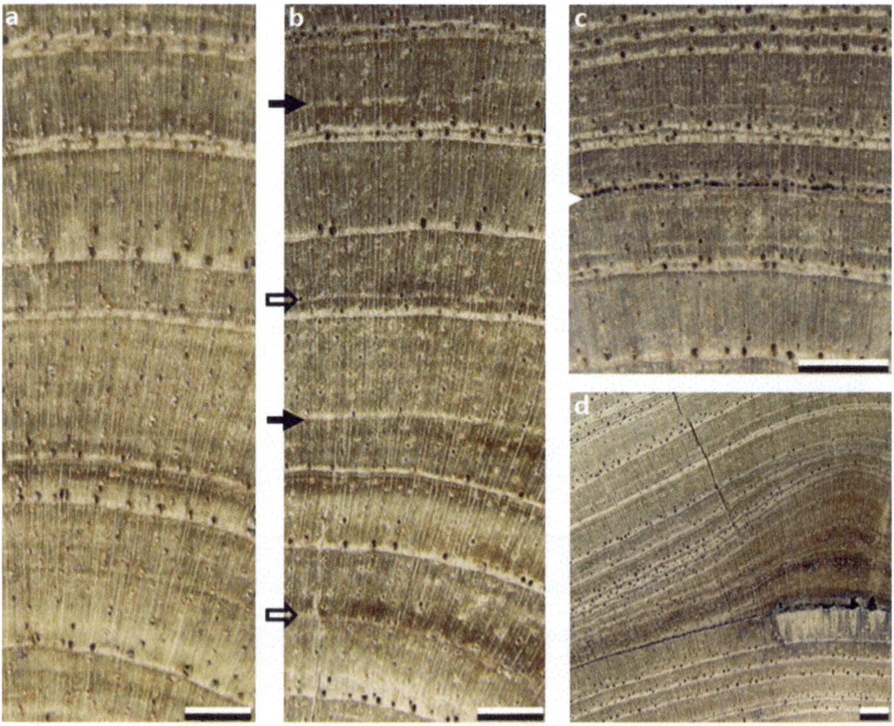

Fig. 3.7 (**a**) True growth rings; (**b**) false rings, with discontinuous confluent parenchyma (*solid arrows*) and thin parenchyma layers (*open arrows*); (**c**) traumatic tissues forming tangential lines (*white arrow*); (**d**) Cicatrization, the formation of traumatic tissues, makes the identification of growth rings difficult. Scale 3.0 mm

studied, these channels are formed near the marginal parenchyma (Fig. 3.7c) after injuries that occurred during the inactivity of the cambium. However, traumatic tissues were also seen displaced from the boundaries of growth rings (Fig. 3.7d), when the cambium was active. Traumatic tissues are an analytical problem because the radial measurements of the growth rings are altered when made close to injuries. Another error can occur when the traumatic tissues are not in the limit between two consecutive rings, because the marginal parenchyma is not well distinguished.

Scars are frequently verified on *C. odorata* trunks in Northeastern Brazil (Fig. 3.7d). They are results of xylem recovery after the tree suffers mechanical shock or insect or other animal attack or human activity. In the radial tissues where the scar occurs, it is difficult to recognize the boundaries of growth rings and/or reaction wood formed. Below we show an example with the primate *Callithrix jacchus* L. on the growth rings of *C. odorata*.

3.3.2 Chronologies, Environment, and Climate

In Sergipe, chronologies of *C. odorata* collected from 10 study sites showed intercorrelation above 0.51 (Cofecha), which was calibrated for segments of 20 measures with lag of 10 measures. The descriptive statistics of the chronologies are present in Table 3.2. The express population signal (EPS) was above the critical value of 0.85 proposed by Wigley et al. (1984) for all chronologies. The 10 chronologies resulting from the work can be seen in Fig. 3.8.

Growth ring numbers (tree ages) ranged from 36 to 110 years and most plants had less than 60 growth rings. The wood of *C. odorata* is still used by the owners of the farms, we verified that 11 studied trees of the site S5 were cut after the sampling.

Table 3.2 Study sites, sampling, and results of statistical analysis

Sites/ Vegetation	N trees T/used	N series T/used	Chronologies Years/period	Total rings	Cofecha		rbt	EPS
					Cor	Sen		
S1/AF	22/8	43/16	82/1926–2008	1507	0.586	0.569	0.52	0.95
S2/AF	17/7	35/12	42/1973–2015	360	0.601	0.547	0.45	0.91
S3/A	23/9	73/21	68/1942–2010	1116	0.532	0.695	0.27	0.88
S4/DTF	23/23	40/41	108/1902–2010	2115	0.685	0.771	0.55	0.97
S5/DTF	33/16	69/27	46/1969–2015	982	0.546	0.556	0.58	0.98
S6/DTF	20/16	40/27	56/1959–2015	949	0.565	0.666	0.37	0.93
S7/DTF	30/9	76/17	36/1977–2013	414	0.796	0.632	0.68	0.97
S8/DTF	35/18	40/33	56/1961–2017	1212	0.688	0.615	0.53	0.97
S9/DTF	20/12	38/25	60/1958–2018	1140	0.453	0.573	0.39	0.91
S10/DTF	10/8	19/16	63/1955–2018	805	0.510	0.681	0.31	0.92
Total	233/126	414/235						
% usadas	54%	56%						

Sites of Studies/Vegetation: *AF* Atlantic Forest, *DTF* Dry Tropical Forest (Caatinga), *A* Agreste, *Cor* Correlation, *Sen* Sensitivity, *rbt* Correlation between trees, *EPS* Express population signal

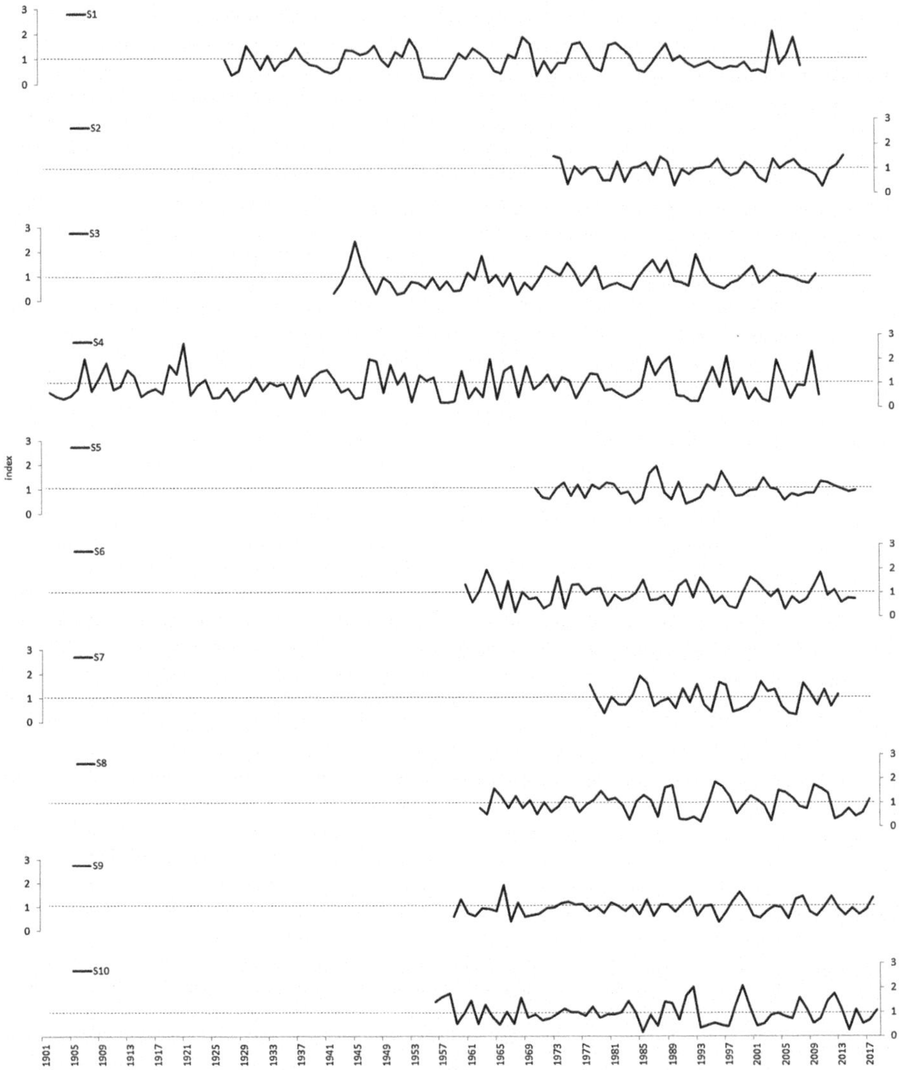

Fig. 3.8 Standard Chronologies of *Cedrela odorata* from study sites S1–S10. Dashed rows correspond to index 1

There are only a few *C. odorata*trees in the conservation units. Besides the difficulty of long chronologies, the conservation of the species is also difficult.

The intercorrelation results (between 0.453 and 0.796), sensitivity (between 0.547 and 0.771), and r bar (between 0.270 and 0.629) (Table 3.2) indicate that populations of *C. odorata* have potential for growth rings study and are sensitive to environmental conditions, especially due to the region's low annual rainfall volumes.

Table 3.3 Results of Pearson correlations between the *Cedrela odorata* chronologies by study sites (S1–S10) from Sergipe

	S1	S2	S3	S4	S5	S6	S7	S8	S9	S10
S1	1	0.292	−0.041	0.115	−0.262	−0.056	−0.209	0.091	0.245	0.332
S2		1	0.223	**0.406**	−0.096	−0.010	0.151	**0.544**	0.053	0.012
S3			1	0.087	0.087	**0.335**	0.080	0.049	−0.084	−0.224
S4				1	0.202	−0.348	0.144	**0.635**	0.155	−0.059
S5					1	0.091	0.146	0.038	−0.415	−0.445
S6						1	**0.355**	−0.191	−0.124	−0.158
S7							1	0.029	−0.245	−0.332
S8								1	−0.035	−0.198
S9									1	**0.720**
S10										1

The values in bold are higher than the confidence level of 0.361 for the common interval between the chronologies of 30 years (1978–2008)

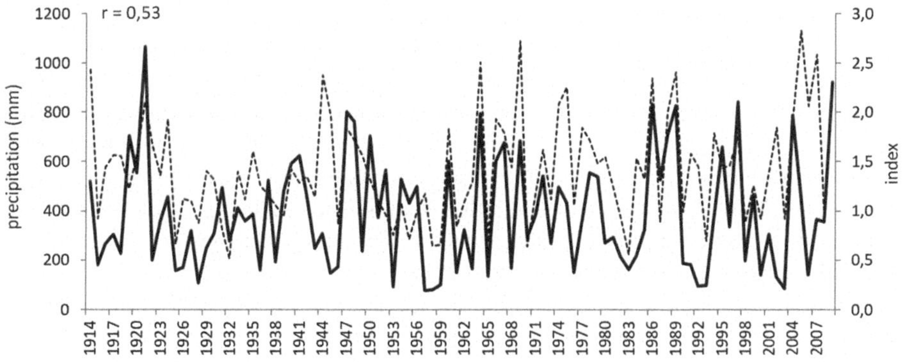

Fig. 3.9 Relation between precipitation (dashed row) and the standard chronology (solid row) of *Cedrela odorata* in site S4

Analyses of Pearson correlation (Table 3.3) indicated that only the S2 site was similar to several other study sites (S3, S4, and S8), and the values were just above 0.200. The other sites showed little similarity between them, highlighting only the sites S4 and S8 (0.597) and S9 and S10 (0.648). This may indicate two possibilities: (i) that the irregular climate of Northeastern Brazil (Ferreira and Mello 2005) causes different growth of the trees between the study sites; the rains have different volumes in each place of study; (ii) that the synchronization process needs to be improved since the false rings, for example, may not have been fully identified.

The trees of S4 are kept by the owner of the farm in an area of 50 hectares; thus, today some trees have more than 60 growth rings (reaching 110 rings) and represent a good material to study the environment of TDF-Caatinga. These ages are similar to those of *Schinopsis brasiliesis* (between 36 and 107 tree rings) (Cardoso 2014). He allowed us to use site S4 to exemplify the relation between the standard chronology of *C. odorata* and annual rainfall (Fig. 3.9). The value of 0.53 of this correlation was attributed to the irregular rains of the region, when some annual totals are much higher than those of other years (in 1983, 223 mm/year and in 1969 1087 mm/year, mean 560 mm/year), and with more dry months in the years of low precipitation. OBS: In this chapter, we chose to present only the correlations between the chronologies of site S4 with the annual precipitation. The others climatic data and correlations with others chronologies will be presented in future publications.

At site S4, woods from local trees and the region, from the time of construction of the farm were also sampled, such as wood for fences, doors, and windows of houses. Trunks of trees that died after this first step were also sawn for these uses. Sixteen pieces of wood, divided into three lots of samples, synchronized with the Master Series of live trees and correlations were: lot 01 = 0.25 (1889–1941), lot 02 = 0.45 (1902–1945), and lot 03 = 0.44 (1953–2000) (Fig. 3.10). Lot 01 is from the wood of a house on another farm, which may explain the lower correlation. Lots 02 and 03 are wood samples from the site S4, and showed higher correlations. Lot 02 has samples corresponding to the time of construction of the farm and lot 03 of a recent tree. However, lot 01 has extended the Master Series by 14 years, until 1889.

When lianas grow too much on the crown of a tree, they compete with trees for nutrients, light, and water (Gentry 1991; Brandes et al. 2010) and can decrease the number of leaves and increase the mass to be supported by the branches. The analysis of the growth rings enables us to verify decreases in the annual increments of the tree. An example of this competition can be seen in Fig. 3.11, by comparing the annual rainfall history directly with the radial measurements of the growth rings of a tree (site S2), which presented correlation 0.48, it is noted that the growth rings did not follow the annual rainfall volumes well, and that the last rings had lower values (0.65 mm/year) than the mean (2.58 mm/year).

Another interesting case is the action of the fauna on the trees. In the region of Mata Atlântica, in Sergipe, lives a primate, *Callithrix jacchus* L., with a habit of biting the bark of some species of trees such as *C. odorata* to eat gum (Castro 2003), when food resources are scarce, but plants can provide sap. The radial measures of the growth rings of a tree that showed signs of primate bites in the trunk were

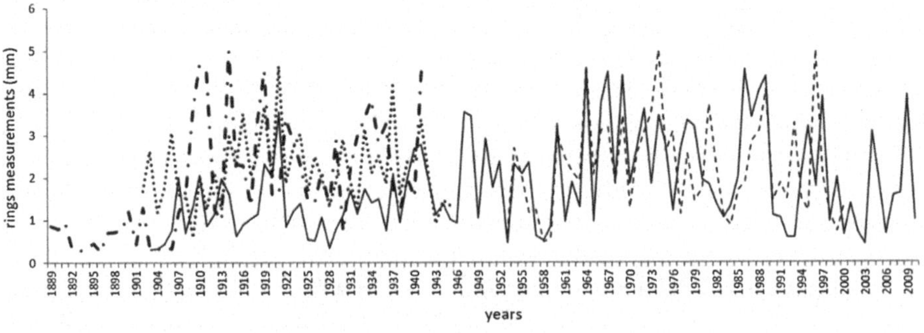

Fig. 3.10 Master Series of *Cedrela odorata* from site S4 (*solid arrow*) compared to the mean curves of the woods of rural constructions (lot 01, *dashed and dotted line*; lot 02, *dotted line*; and lot 03, *fine dashed line*)

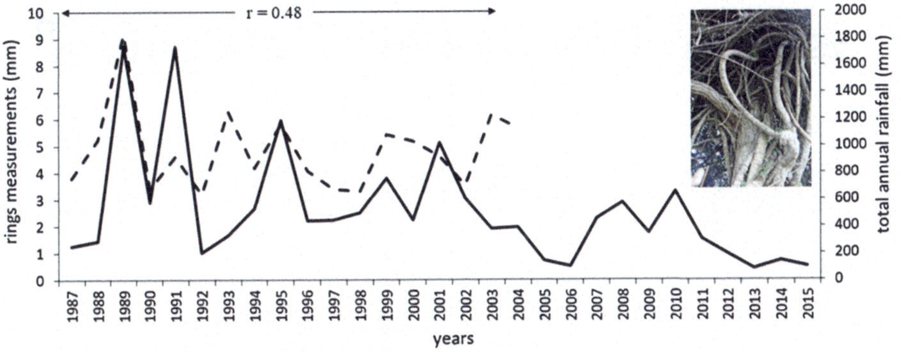

Fig. 3.11 Relation between the total annual rainfall in site S2 (*dotted line*) (source SEMARH/SE) and the growth rings of a tree in competition with lianas (*solid line*)

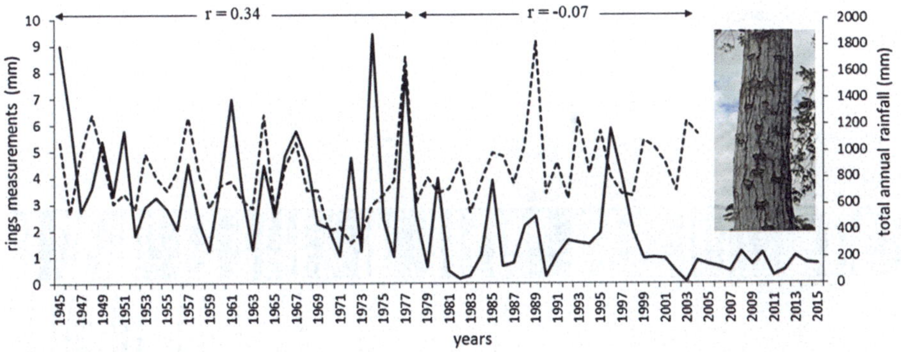

Fig. 3.12 Relation between the total annual rainfall in Propriá (*dotted line*) (source SEMARH/SE) and the growth rings measurements of a tree from site S2 bit by *Callithrix jacchus* L. (*solid line*)

compared with the historical records of the annual rains. The results showed that from the juvenile phase the trees had formed wood of reaction and traumatic tissues, probably due to healing processes; however, the correlation of 0.34 was maintained for approximately 30 years. After this period, the correlation decreased (−0.07) and the mean annual increments of 3.82 mm/year decreased until mean values of 0.74 mm/year (Fig. 3.12), when many scars were seen on the trunk and branches of the tree from the continuous bites of the primates.

3.3.3 X-Ray Analysis

This analysis was made at the Laboratory of Wood Anatomy, Identification and X-ray Densitometry, of Forest Sciences of "Escola Superior de Agricultura Luiz de Queiroz – ESALQ/USP". The study site was at Andaraí, Bahia (12°45′54.72″S

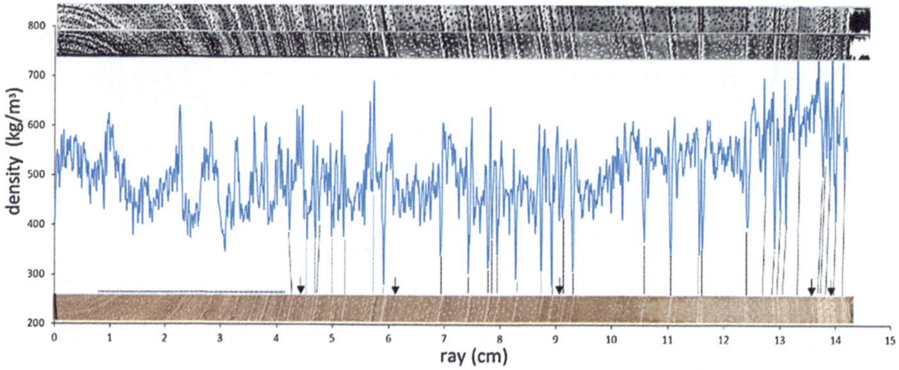

Fig. 3.13 Results of one *Cedrela odorata* X-ray microdensitometry analysis (*blue line*). Images: negative (*above*), positive (*in the middle*), and wood (*below*). The horizontal line (*dotted line*), near the wood image, indicates that the first growth rings are tangential to X-ray analysis, making it difficult to define the limits of the growth rings. Limits of the growth rings with lower density of the marginal parenchyma (*vertical dotted lines*). Confluent parenchyma lines corresponding to false rings (*arrows*)

Fig. 3.14 (**a**) Wood, (**b**) Negative, and (**c**) Positive images of a sample of *Cedrela odorata*. (**a**) Profile of polished wood with sandpapers showing the traumatic tissue (*open arrow*), a marginal parenchyma (*thick arrow*), and a confluent parenchyma (*thin arrow*). Image length = 6 cm

41°06′51.48″W) (sample example in Fig. 3.13). This analysis shows the apparent differences of wood ring density of earlywood and latewood. The ring boundaries of *C. odorata* presented an input in density in latewood because of radially flattened and concentrated fibers, and the minor density of marginal parenchyma cells in earlywood is observed. However, it is not observed in all radii analyzed.

The digital X-ray technique provides accurate negative and positive images of the sample, contributing to the definition of annual, false, and absent rings (Fig. 3.14). Marginal parenchyma is better seen in X-ray images than in wood, which makes it easier to define the limits of growth rings, as well as the distribution of vessels and fibers (darker parts).

The wood densitometry profile possibility calculates other parameters, such as mean, max, and minimum densities for every tree ring (Polge 1970; Schweingruber et al. 1978; Tomazello et al. 2008). The precise differences of wood density determined by this technique possibility define tree ring boundaries and identify the false rings. The measurement of radial dimensions of tree rings is used for the synchronization process necessary for chronologies definitions. The X-ray microdensitometry has many applications for wood technology areas, the precise values of wood apparent density are important to define the use of wood species, for example (Tomazello et al. 2008).

3.4 Conclusions

The *Cedrela odorata* from Northearstern Brazil have potential for dendrochrology. This species presents true rings and has a good relationship with climate data where the precipitation showed correlation values determined by tree rings series. Missing rings and false rings can be identified by anatomy characteristics, by synchronization of radial series of rings, and using the X-ray method. The climatic seasonality of this region, especially precipitation, with more than 6 dry months, induces the growth period in the plants and growth rings are formed. However, the irregular climate of the northeast region made it difficult to determine a good relationship between chronologies. The exploitation of timber in the past followed by protection laws allowed for the occurrence of *C. odorata* populations to form groups of plants on farms in this region.

Acknowledgements We are thankful for the support of the CNPq (Conselho Nacional de Desenvolvimento Científico e Tecnológico) and Fapitec (Fundação de Apoio à Pesquisa e a Inovação Tecnológica do Estado de Sergipe) edicts UNIVERSAL/CNPq N° 14/2011, FAPITEC/SE/FUNTEC/CNPq N° 04/2011, and UNIVERSAL- MCTI/CNPq N° 14/2014. We also thank the other team members of the UFS, ESALQ, and IANIGLA laboratories that contributed during these years to carry out these studies.

References

Alvares CA, Stape JL, Sentelhas PC et al (2013) Köppen's climate classification map for Brazil. Meteorol Z 22:711–728. https://doi.org/10.1127/0941-2948/2013/0507

Alves JJA (2007) Geoecologia da caatinga no semi-árido do nordeste brasileiro. Climatologia e Estudos de Paisagens 2(1):58–71

Alves JJA (2008) Biogeografia. João Pessoa: Editora Fotograf. 108p

Amaral VFM (2006) Multiplicação in vitro de Cedrela odorata. Dissertation, Universidade Federal de Santa Maria

Andreacci F, Botosso PC, Galvão F (2014) Sinais climáticos em anéis de crescimento de Cedrela fissilis em diferentes tipologias de florestas ombrófilas do sul do Brasil. Floresta 44(2):323–332

Anholetto Junior CR (2013) Dendroecologia e composição isotópica ($\delta^{13}C$) dos anéis de crescimento de árvores de *Cedrela odorata*, Meliaceae, na Caatinga e Mata Atlântica do Estado de Sergipe, Brasil. PHD Tesis Escola Superior de Agricultura "Luiz de Queiroz"

Anholetto CR (2013) Dendroecologia e composição isotópica (δ13C) dos anéis de crescimento de árvores de Cedrela odorata, Meliaceae, na Caatinga e Mata Atlântica do estado de Sergipe, Brasil. Dissertation, Esalq. https://doi.org/10.11606/D.11.2013.tde-30102013-105715

Baker JCA, Hunt SFP, Clerici SJ et al (2015) Oxygen isotopes in tree rings show good coherence between species and sites in Bolivia. Global Planet Change 133:298–308. https://doi.org/10.1016/j.gloplacha.2015.09.008

Baker JCA, Santos GM, Gloor M et al (2017) Does Cedrela always form annual rings? Testing ring periodicity across South America using radiocarbon dating. Trees 31(6):1999–2009. https://doi.org/10.1007/s00468-017-1604-9

Ballantyne AP, Baker PA, Chambers JQ et al (2011) Regional differences in South American monsoon precipitation inferred from the growth and isotopic composition of tropical trees. Earth Interact 15(5):1–35. https://doi.org/10.1175/2010EI277.1

Becerra-Montalvo V, Zevallos-Pollito P (2014) Determinación del turno de corta de Cedrela odorata L., Retrophyllum rospigliosii Pilger y Prumnopitys harmsiana Pilger a través del estudio dendrocronológico en San Ignacio, región Cajamarca Perú. Ceprosimad 2(2):33–47

Blasing TJ, Solomon AS, Duvick DN (1984) Response functions revisited. Tree-Ring Bull 44:1–17

Boninsegna JA, Villalba R, Amarilla L et al (1989) Studies on tree rings, growth-rates and age-size relationships of tropical tree species in Missiones, Argentina. IAWA Bull 10:161–169. https://doi.org/10.1163/22941932-90000484

Botosso PC, Vetter RE, Tomazello M (2000) Periodicidade e taxa de crescimento de árvores de cedro (*Cedrela odorata* L., Meliaceae), jacareúba (*Calophyllum angulare* A.C. Smith, Clusiaceae) e muirapiranga (*Eperua bijiga* Mart. ex Benth, Leg. Caesalpinioideae) de floresta de terra firme, em Manaus-AM. In: Roig FA (ed) Dendrocronología en América Latina. Ediunic, Mendoza, pp 357–380

Brandes AFN, Lisi CS, Barros CF (2010) Dendrochronology of lianas of the Leguminosae family from the Atlantic Forest, Brazil. Trees 25(1):133–144. https://doi.org/10.1007/s00468-010-0529-3

Bräuning A, Volland-Voigt F, Burchardt I et al (2009) Climatic control of radial growth of Cedrela montana in a humid mountain rainforest in southern Ecuador. Erdkunde 63:337–345. https://doi.org/10.3112/erdkunde.2009.04.04

Brienen RJW, Zuidema PA (2005) Relating tree growth to rainfall in Bolivian rain forests: a test for six species using tree ring analysis. Oecologia 146:1–12. https://doi.org/10.1007/s00442-005-0160-y

Brienen RJW, Zuidema PA (2006a) Lifetime growth patterns and ages of Bolivian rain forest trees obtained by tree ring analysis. J Ecol 94:481–493. https://doi.org/10.1111/j.1365-2745.2005.01080.x

Brienen RJW, Zuidema PA (2006b) The use of tree rings in tropical forest management: projecting timber yields of four Bolivian tree species. For Ecol Manage 226:256–267. https://doi.org/10.1016/j.foreco.2006.01.038

Brienen RJW, Helle G, Pons TL et al (2012) Oxygen isotopes in tree rings are a good proxy for Amazon precipitation and El Niño-Southern Oscillation variability. PNAS 109(42):16957–16962. https://doi.org/10.1073/pnas.1205977109

Brienen RJW, Gloor E, Clerici S et al (2017) Tree height strongly affects estimates of water-use efficiency responses to climate and CO2 using isotopes. Nat Commun 8(288):110. https://doi.org/10.1038/s41467-017-00225-z

Briffa KR, Jones PD (1990) Basic chronology statistics and assessment. In: Methods of dendrochronology: applications in the environmental sciences. Kluwer Academic Publishers, Dordrecht

Burger LM, Richter HG (1991) Anatomia da madeira. Editora Nobel, São Paulo

Cardoso DS (2014) Caracterização anatômica da madeira e potencial dendrocronológico de *Schinopsis brasiliensis* Engl. (Anarcadiaceae) na caatinga sergipana. Dissertation, Universidade Federal de Sergipe

Castro CSS (2003) Tamanho da área de vida e padrão de uso do espaço em grupos de saguis *Callitrix jacchus* (Linnaeus) (*Primates Callitrichidae*). Rev Bras Zool 20(1):91–96

Castro VRC, Tomazello M, Surdi PG (2014) Avaliação do perfil radial do lenho de árvores de *Pinus* (*Pinus caribaea* var. *hondurensis* Barr. et Golf.) por densitometria de raios X. Sci For 42(103):353–360

Chalk L (1983) Growth rings. In: Metcalfe CR, Chalk L (eds) Anatomy of dicotyledons. Oxford: Claredon Press, pp 250–265

Cook ER, Peters K (1981) The smoothing spline: a new approach to standardizing forest interior tree-ring with series for dendroclimatic studies. Tree-Ring Bull 41:45–53

Cook ER (1985) A time series analysis approach to tree-rings standardization. The University of Arizona, Tucson

Cook ER, Kairiukstis LA (1990) Methods of dendrochronology. Springer Netherlands, Dordrecht

Cordeiro JMP, Oliveira AG (2010) Levantamento fitogeográfico em trecho de Caatinga hipoxerófita – Sítio Canafístula, Sertãozinho-Paraíba, Brasil. OKARA: Geografia em debate 4(1–2):54–65

Dávila N, Honorio E, Baker T et al (2008) Fichas de identificación de espécies maderables de Loreto- Peru. Instituto de Investigaciones de La Amazônia Peruana, pp 12–13

Détienne P (1989) Appearence and periodicity of growth rings in some tropical woods. IAWA Bull 10:123–132. https://doi.org/10.1163/22941932-90000480

Dünisch O, Bauch J, Gasparotto L (2002) Formation of increment zones and intraannual growth dynamics in the xylem of *Swietenia macrophylla*, *Carapa guianensis*, and *Cedrela odorata* (Meliaceae). IAWA J 23(2):101–119. https://doi.org/10.1163/22941932-90000292

Dünisch O, Montoia VR, Bauch J (2003) Dendroecological investigations on *Swietenia macrophylla* King and *Cedrela odorata* L. (Meliaceae) in the central Amazon. Trees 17:244–250

Dünisch O (2005) Influence of the El-Niño southern oscillation on cambial growth of Cedrela fissilis Vell in tropical and subtropical Brazil. J Appl Bot Food Qual 79(1):5–11

Espinoza MJP, Guillen GJI, Morales MS et al (2014) Potencialidad de Cedrela odorata (Meliaceae) para estudios dendrocronológicos en la selva central del Perú. Rev Biol Trop 62(2):783–793

Ferreira AG, Mello NGS (2005) Principais sistemas atmosféricos atuantes sobre a região nordeste do Brasil e a influência dos oceanos Pacífico e Atlântico no clima da região. Revista Brasileira de Climatologia 1:15–28. https://doi.org/10.5380/abclima.v1i1.25215

Ferreira ATB, Tomazello M (2009) Caracterização dos anéis de crescimento de árvores de Pinus caribaea var. hondurensis Barr. et Golf. Por densitometria de raios X. Scientia Forestalis 37(83):287–298

Ferreira GB (2012) Análise dendroclimatológica do cedro (*Cedrela fissilis* L. Meliaaceae) para reconstrução do cenário ambiental recente da cidade de São Paulo, SP. Dissertation, Universidade de São Paulo. https://doi.org/10.11606/D.41.2012.tde-23072012-143643

Gentry AH (1991) The distribution and evolution of climbing plants. In: Putz FE, Mooney HA (eds) The biology of vines. Cambridge University Press, Cambridge

Granato-Souza D, Stahle DW, Barbosa AC et al (2018) Tree rings and rainfall in the equatorial Amazon. Climate Dynam 5:1–13. https://doi.org/10.1007/s00382-018-4227-y

Holmes RL (1983) Computer-assisted quality control in tree-ring dating and measurement. Tree-Ring Bull 43:69–78

Iawa Committee (1989) Iawa list of microscopic features for hardwood identification. IAWA Bull New Ser 10(3):219–332

Jacomine PKT (2008–2009) A nova classificação brasileira de solos. In: Anais da Academia Pernambucana de Ciência Agronômica, vol 5/6, pp 161–179

Janzen DH (1988) Tropical dry forests, the mosts endangerous major tropical ecosystem. In: Wilson EO (ed) Biodiversity. National Academy Press, Washington, DC.

Köhl M, Neupane PR, Lotfiomran N (2017) The impact of tree age on biomass growth and carbon accumulation capacity: a retrospective analysis using tree ring data of three tropical tree species grown in natural forests of Suriname. PLoS One 1:17. https://doi.org/10.1371/journal.pone.0181187

Larson PR (1994) The vascular cambium: development and structure. Springer, Berlin

Lisi CS, Tomazello-Filho M, Botosso PC et al (2008) Tree-ring formation, radial increment periodicity, and phenology of tree species from a seasonal semi-deciduous forest in southeast Brazil. IAWA J 29:189–207. https://doi.org/10.1163/22941932-90000179

Lobão MS (2011) Dendrocronologia, fenologia, atividade cambial e qualidade do lenho de árvores de *Cedrela odorata* L., *Cedrela fissilis* Vell. e *Schizolobium parahyba* var. *amazonicum* Hub. ex Ducke, no estado do Acre, Brasil. PHD Tesis Escola Superior de Agricultura "Luiz de Queiroz"
Lobão MS, Ortega GP, Amaral E et al (2016) Análise de similaridade das árvores de Cedrela sp. sob diferentes condições de crescimento no leste do estado do Acre, Brasil. Sci For 44(109):231–239. https://doi.org/10.18671/scifor.v44n109.22
Loiola MIB, Roque AA, Oliveira ACP (2012) Caatinga: vegetação do semiárido brasileiro. Ecologia 4:14–19
Lorenzi H (2009) Árvores brasileiras: manual de identificação e cultivo de plantas arbóreas nativas do Brasil, vol 2. Instituto Plantarum, Nova Odessa
Maia GN (2012) Caatinga: árvores e arbustos e suas utilidades, vol 2. Printcolor Gráfica e Editora, Fortaleza
Marcati CR, Angyalossi V, Evert RF (2006) Seasonal variation in wood formation of *Cedrela fissilis* (Meliaceae). IAWA J 27(2):199–211. https://doi.org/10.1163/22941932-90000149
Mariano Neto B (2001) Ecologia e imaginário: memória cultural, natureza e submundialização. Editora Universitária/FPB, João Pessoa
Paredes-Villanueva K, López L, Cerrillo RMN (2016) Regional chronologies of *Cedrela fissilis* and *Cedrela angustifolia* in tree forest types and their relation to climate. Trees 30:1581–1593. https://doi.org/10.1007/s00468-016-1391-8
Pennington TD (1981) Meliaceae. Flora Neotrop (New York Botanical Garden) 28:359–368
Polge H (1970) The use of X-ray densitometric methods in dendrochronology. Tree-Ring Bull 30:1–10
Quintek Measurement Systems (1999) QMS tree ring analyzer model QTRS-01X. Users Guide, Knoxville
Rauber RC (2010) Dendroecologia do cedro Cedrela fissilis Vell. (Meliaceae) em um ecótono de florestas subtropicais montanas no Brasil. Dissertation, Ufrgsl
Ribeiro FE, Siqueira ER (2001) Recursos genéticos vegetais da Mata Atlântica de Sergipe. In: Siqueira ER, Ribeiro FE (eds) Mata Atlântica de Sergipe. Embrapa Tabuleiros Costeiros, Aracaju
Rosa SA (2008) Modelos de crescimento de quatro espécies madeireiras de várzea alta da Amazônia Central por meio de métodos dendrocronológicos. Dissertation, Inpa
Roweder C (2011) Germinação de sementes e desenvolvimento de plântulas de Cedro (*Cedrella odorata* L. – MELIACEAE) e Mogno (*Swietenia macrophylla* King – MELIACEAE) em diferentes condições de luminosidade, substratos e recipientes. Dissertation, Ufac
Sampaio EVSB (2010) Caraterização do Bioma Caatinga – características e potencialidades. In: Gariglio MA, Sampaio EVSB, Cestaro LA, Kageyama PY (eds) Uso sustentável e conservação dos recursos florestais da Caatinga. Serviço Florestal Brasileiro, Brasília
Schweingruber FH, Fritts HC, Braker OU et al (1978) The x-ray technique as applied to dendroclimatology. Tree-Ring Bull 38:61–91
Serviço Florestal Brasileiro (2017) Inventário florestal nacional: Sergipe: principais resultados/ Serviço Florestal Brasileiro. MMA
Surdi PG, Bortoletto Júnior G, Castro VR et al (2014) Perfil de densidade do lenho utilizando métodos radiográficos. Sci For 42(102):229–236
Tomazello M, Botosso PC, Lisi CS (2000) Potencialidade da família Meliaceae para dendrocronologia em regiões tropicais e subtropicais. In: Roig FA (ed) Dendrocronología en América Latina. EDIUNC, Mendoza
Tomazello M, Botosso PC, Lisi CS et al (2001) *Cedrela angustifolia* Ses. et Moc. ex DC., Meliaceae: potential species for tropical endrochronology. Paleobotanist 50:47–53
Tomazello M, Brazolin S, Chagas MP et al (2008) Application of X-ray technique in nondestructive evaluation of Eucalypt wood. Maderas Cienc Tecnol 10(2):139–145
Tsuchiya A (1995) Preliminary study on the relationship between vessel growth of thorny shrubs and water balance in the semi-arid region, northeastern Brazil. Geogr Sci 50:123–131
Vetter RE, Botosso PC (1989) Remarks on age and growth rate determination of Amazonian trees. IAWA J 10(2):133–145. https://doi.org/10.1163/22941932-90000481

Villalba R, Boninsegna JA, Holmes RL (1985) *Cedrela angustifolia* and *Juglans australis*: two new tropical species useful in dendrochronology. Tree-Ring Bull 45:25–35

Villalba R, Boninsegna JA, Ripalta A (1987) Climate, site conditions and tree-growth in subtropical northwestern Argentina. Can J For Res 17(12):1527–1544. https://doi.org/10.1139/x87-235

Villalba R, Colmes RL, Boninsegna JA (1992) Spatial patterns of climate and tree-growth anomalies in subtropical northwestern Argentina. J Biogeogr 19(6):631–649. https://doi.org/10.2307/2845706

Villalba R, Delgado S, Membiela M et al (2006) Variabilidad interanual de los caracteres anatómicos en el leño de Cedrela lilloi en el noroeste de Argentina. In: Pacheco S, Brown A (eds) Interanual de los caracteres anatómicos en el leño de *Cedrela lilloi* en el NOA. LIEY-ProYungas, Tucumán

Wigley TM, Jones PD, Briffa K (1984) On the average value of correlated time series, with applications in dendroclimatology and hydrometeorology. J Clim Appl Meteorol 23:201–213. https://doi.org/10.1175/1520-0450(1984)023<0201:OTAVOC>2.0.CO;2

Worbes M (1989) Growth rings, increment and age of trees inundation forests, savanas and amount in forest in the Neotropics. IAWA Bull 10(2):109–122. https://doi.org/10.1163/22941932-90000479

Worbes M (1992) Occurrence of seasonal climate and tree-ring research in the tropics. Lundqua Rep 34:338–342

Worbes M (1995) How to measure growth dynamics in tropical trees: a review. IAWA J 16(4): 337–351. https://doi.org/10.1163/22941932-90001424

Chapter 4
Responses of Growth to Climate and Drought in Two Sympatric Mexican Pine Species

Marcos González-Cásares, Andrea Cecilia Acosta-Hernández, José Manuel Zúñiga-Vásquez, Gerardo Herrera-Soto, and J. Julio Camarero

Abstract Climate change and its ecological consequences are a highly topical concern, particularly considering the effects on forests and their capacity for growth and ability to uptake and fix carbon as wood. Exploring the influence of climate on the radial growth of sympatric tree species is important to the understanding of the impacts of climate warming on forest composition and productivity. In this study, we used dendrochronology to investigate how the growth of two sympatric pine species, *Pinus leiophylla* and *Pinus teocote*, growing in northern Mexico respond to climatic variables at inter-annual scales. In general, the response of the two species to the climatic conditions was similar. Prior humid and cold winter and current summer conditions enhanced growth, whereas warm conditions in the prior summer and late winter, related to high evapotranspiration rates, constrained growth. It is therefore necessary to focus on the response of these species to the forecasted temperature increase. Further studies could also investigate if other sympatric tree species also show similar growth responses to climate and drought.

Keywords *Pinus leiophylla* · *Pinus teocote* · Mexico · SPEI · Spatial correlation · Sensitivity climatic

---

M. González-Cásares · A. C. Acosta-Hernández (✉)
Programa Institucional de Doctorado en Ciencias Agropecuarias y Forestales, Universidad Juárez del Estado de Durango, Durango, Mexico
e-mail: magonzalez@ujed.mx; andrea.acosta@ujed.mx

J. M. Zúñiga-Vásquez · G. Herrera-Soto
Facultad de Ciencias Forestales, Universidad Juárez del Estado de Durango, Durango, Mexico

J. J. Camarero
Instituto Pirenaico de Ecología (IPE-CSIC), Zaragoza, Spain
e-mail: jjcamarero@ipe.csic.es

M. Pompa-García, J. J. Camarero (eds.), *Latin American Dendroecology*,
https://doi.org/10.1007/978-3-030-36930-9_4

4.1 Introduction

The effects of climatic change, including warmer conditions and hotter droughts, on forest ecosystems include a reduction in productivity and increased mortality rates (Allen et al. 2010, 2015). These changes, often associated with forest dieback episodes, negatively impact both carbon storage and timber production and may amplify climate-biosphere feedbacks by rising CO_2 and methane atmospheric concentrations (IPCC 2014). It is also important to understand if the distribution and abundance of species that cohabit in the sites is a function of changes in the climate (Whittaker 1953) since a change in the climate can also modify the phenology of a species in relation to other coexisting species, and this may have an impact on its ability to tolerate drought stress (Roberts et al. 2015). In addition, the responses to climatic conditions vary among tree species (González-Cásares et al. 2017), with implications for their vulnerability to drought and their potential area of distribution (Antúnez et al. 2018).

Derived from the warm conditions that have developed in the last decades which are unprecedented in the last 150 years (IPCC 2014; Dai 2013), in several North-American areas such as N. Mexico, there have been changes in precipitation regimes accompanied by temperature increases and aridification (Stahle et al. 2016). Mexico is the country with the greatest number of pine species worldwide and is considered a hotspot of diversity of *Pinus* genus (Farjon and Styles 1997). A total of 43% of the pine species of the world can be found in Mexico. Due to their high diversity in Mexico, the growth response of pine species to climatic variability has been widely studied using a dendrochronological approach. For example, there have been analyses of the response of different species like *Pinus cooperi*, *Pinus arizonica*, *Pinus durangensis*, *Pinus engelmannii*, *Pinus leiophylla*, and *Pinus lumholtzii* to climatic variables and large-scale atmospheric phenomena such as the El Niño-Southern Oscillation (Pompa-García et al. 2014; Pompa-García and Antonio-Némiga 2015; Pompa-García and Hadad 2016; González-Cásares et al. 2017; Acosta-Hernández et al. 2019), evaluations of growth responses to evaporation (Chacón-de la Cruz and Pompa-García 2015), and assessments of climate influences on wood density and carbon capture rate in *P. cooperi* (Pompa-García and Venegas-González 2016; González-Cásares et al. 2016, 2018). According to a recent review, the most studied genus using dendrochronological methods in Mexico is *Pinus* (Acosta-Hernández et al. 2017).

Increased aridification and more frequent and severe droughts are expected to induce growth decline, reduce productivity, and raise mortality in drought-prone regions of N. Mexico such as the Durango state (Allen et al. 2010; Cook and Seager 2013; Camarero et al. 2015). These changes in Mexican climate have triggered distribution shifts of tree species as a result of their different adaptation rates (González-Elizondo et al. 2005), and depending on local site conditions (Pompa-García et al. 2017). Therefore, ecological forecasts should consider the specific responsiveness to climatic variability and drought of sympatric tree species as, for instance, many Mexican pines. Sites in which several pine species coexist present an opportunity to define the seasonal response of the species to regional droughts and the increased vulnerability of some species (González-Cásares et al. 2017). Evaluation of the sensitivity of radial growth to climatic conditions and drought could reveal these differences among sympatric species (Pompa-García et al. 2017). This study took advantage of the coexistence of *Pinus leiophylla* and *Pinus teocote* in the diverse conifer forests located in northern Mexican mountains to analyze the response of their radial growth to precipitation, temperature, and drought. We hypothesized that climatic factors differentially affect even sympatric species.

4.2 Materials and Methods

4.2.1 Study Area and Tree Species

The study area is located in El Salto (Pueblo Nuevo, Durango, N. Mexico), in the Sierra Madre Occidental mountain chain (Fig. 4.1), at 23.4° N–105.2° W and an elevation of 2566–2598 m a.s.l. This site is a volcanic plateau and it is characterized by irregular topography that includes rolling highlands and deep canyons (González-Elizondo et al. 2012). The vegetation present is a mixed forest with pine and oak species which have similar ecological preferences, where the dominant element of the vegetation is several pine species (González-Elizondo et al. 2007). Among the most common pines are *Pinus teocote*, *Pinus cooperi*, *Pinus durangensis*, *Pinus leiophylla*, as well as *Pinus strobiformis* (González-Elizondo et al. 2007). The site has a temperate subhumid climate, with marked seasonality characterized by humid and cold summers derived from the influence of the North American summer Monsoon (June–September) and winters with less intense rains during the cold season (November–February) (Seager et al. 2009). It presents an annual mean temperature of 11.5 °C and mean annual precipitation of 1200 mm (CNA 2018). The site presents luvisol soil type, which is characterized by a higher clay content in the subsoil than in the ground surface, as a result of pedogenetic processes, characteristic soil of flat surfaces with little slope of temperate regions (INEGI 2001). This is an area of great ecological, economic and environmental relevance, given the diversity of species of pines present and the high timber yields of some of these forests and the ecosystem service that it provides (González-Elizondo et al. 2012). It is also the most important forest reserve in Mexico, given its contribution to timber production nationwide (SEMARNAT 2016). The site is not currently under forest use or management.

The study species are *Pinus leiophylla* Schl. & Cham. and *Pinus teocote* Schiede ex Schltdl. & Cham. These species are distributed from the southern United States to southwest Mexico, mainly in mesic sites of the Sierra Madre Occidental (Perry 1991). They are found at an elevation range of 1900–2900 m a.s.l. in deep, well-drained soils, occasionally forming monospecific stands but commonly found associated with other pines such as *P. engelmannii*, *P. arizonica*, *P. lumholtzii*, *P. montezumae*, *P. ayacahuite*, and *P. pseudostrobus* (García and González 2003).

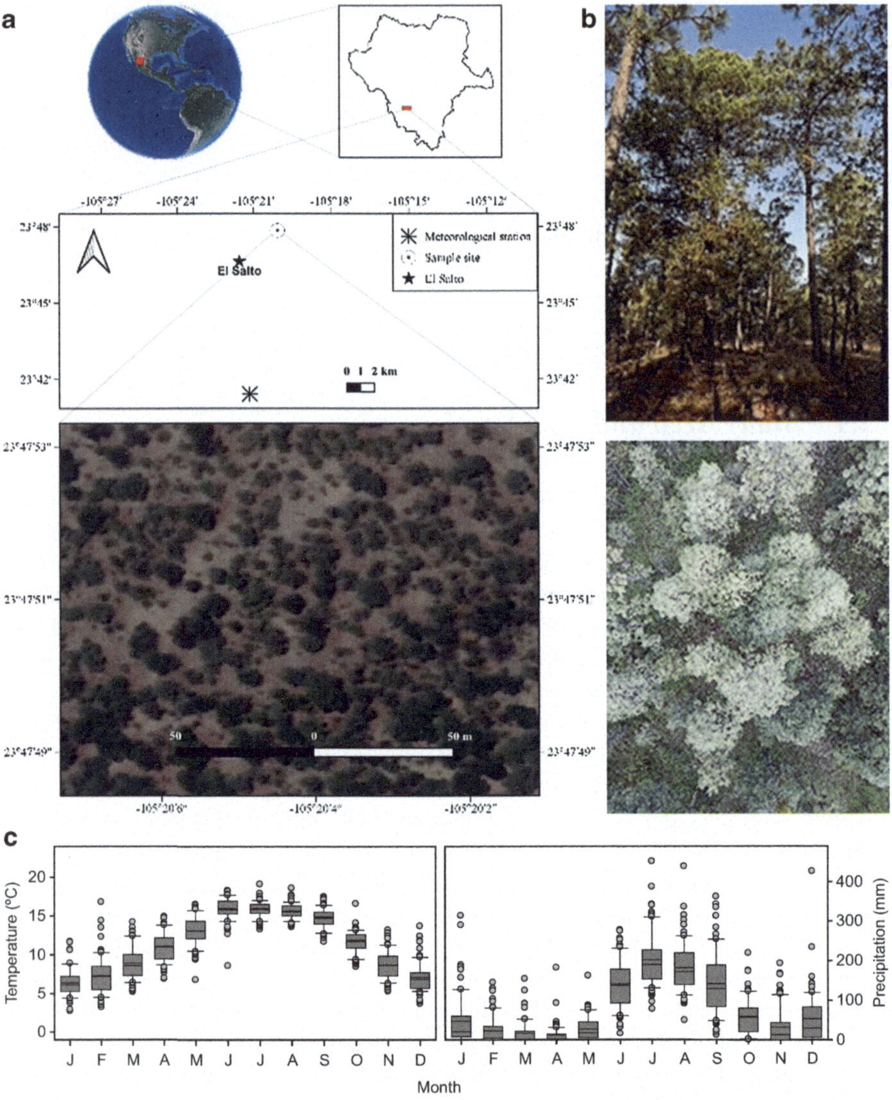

Fig. 4.1 Location of the study site and the meteorological station in northern Mexico (**a**), photograph of a sampled site and monthly climate conditions (**c**) obtained from the "El Salto" station (period 1946–2018)

4.2.2 Sampling and Chronology Development

A total of 14 and 16 dominant individuals of *P. leiophylla* and *P. teocote* were sampled, taking two wood cores per individual using a Pressler increment borer (Ø = 5.1 mm). In addition, the diameter at 1.3 m and total height were recorded for each tree.

The wood core samples were air dried and mounted on wooden groves. The cores were polished with coarse to fine sandpapers (80–1000 grit) to highlight their growth rings. The rings were visually cross-dated (Stokes and Smiley 1968) and measured with a resolution of 0.01 mm using a Velmex system. The dating was checked using the program COFECHA, which compared the series of each tree to a master chronology of each species (Holmes 1983).

To generate the chronologies of the growth ring widths of each species, standardization and detrending were performed. This process allows eliminating the biological and geometric growth trends that are not associated with climatic variables but with enlargement as the tree ages, making the data directly comparable among species (Fritts 1976). We fitted a negative exponential function to raw ring width data and obtained ring-width indices (RWI) by dividing observed width data by fitted values. Then, the temporal autocorrelation associated with the growth of the previous year was eliminated by fitting autoregressive models, and the resulting indices were averages using robust bi-weight means to obtain residual chronologies of the two species (Cook and Kairiukstis 1990). These procedures were developed with the dplR library (Bunn 2008, 2010) of the software R (R Development Core Team 2014), following Pompa-García et al. (2017).

4.2.3 Influence of Climatic Variables on Radial Growth

The influence of local climate on radial growth was evaluated through Pearson correlation coefficients (*r*) calculated between the residual chronology of each species and the following monthly climatic variables: total precipitation and mean maximum and minimum temperatures. Monthly climate data were obtained from the nearest climate station of "El Salto" over the period 1946–2018 (CNA 2018). The Expressed Population Signal was used to select the well-replicated segment of each chronology which was used in the correlation analyses. The segment with EPS > 0.85 was considered reliable following Wigley et al. (1984). The window of analyses included from the previous June up to September of the year of tree-ring formation based on previous studies (Pompa-García et al. 2017).

In order to quantify the stress by drought, we used the monthly values of the Standardized Precipitation Evapotranspiration Index (SPEI; see Vicente-Serrano et al. 2010). The SPEI considers the effect of temperature on the rate of evapotranspiration with negative and positive SPEI values reflecting dry and humid conditions, respectively. The SPEI was calculated for the same grid of 0.5° including the study site, and it was obtained from the webpage http://sac.csic.es/spei/index.html.

To analyze the influence of large-scale drought on radial growth, spatial correlations were calculated between SPEI and the RWI chronologies of both species. These correlations were calculated using the KNMI Climate Explorer webpage (https://climexp.knmi.nl/start.cgi).

4.3 Results

In total, 28 and 32 tree-ring width series were successfully cross-dated and measured in the case of *P. leiophylla* and *P. teocote*, respectively (Table 4.1). Both series covered adequately the period 1941–2018 according to their EPS values. The mean growth rates were 2.38 mm and 2.08 mm for *P. leiophylla* and *P. teocote*, respectively. The growth pattern of the species can be observed in the mean series of RWIs (Fig. 4.2). The correlation between the two pine chronologies was highly significant

Table 4.1 Dendrochronological statistics of the study species

Study species	Timespan	No. series	Mean TRW ± SD (mm)	EPS	SI
Pinus leiophylla	1941–2018	28	2.38 ± 0.87	0.96	0.70
Pinus teocote	1936–2018	32	2.08 ± 0.98	0.94	0.62

TRW tree-ring width, *EPS* expressed population signal, *SI* series intercorrelation

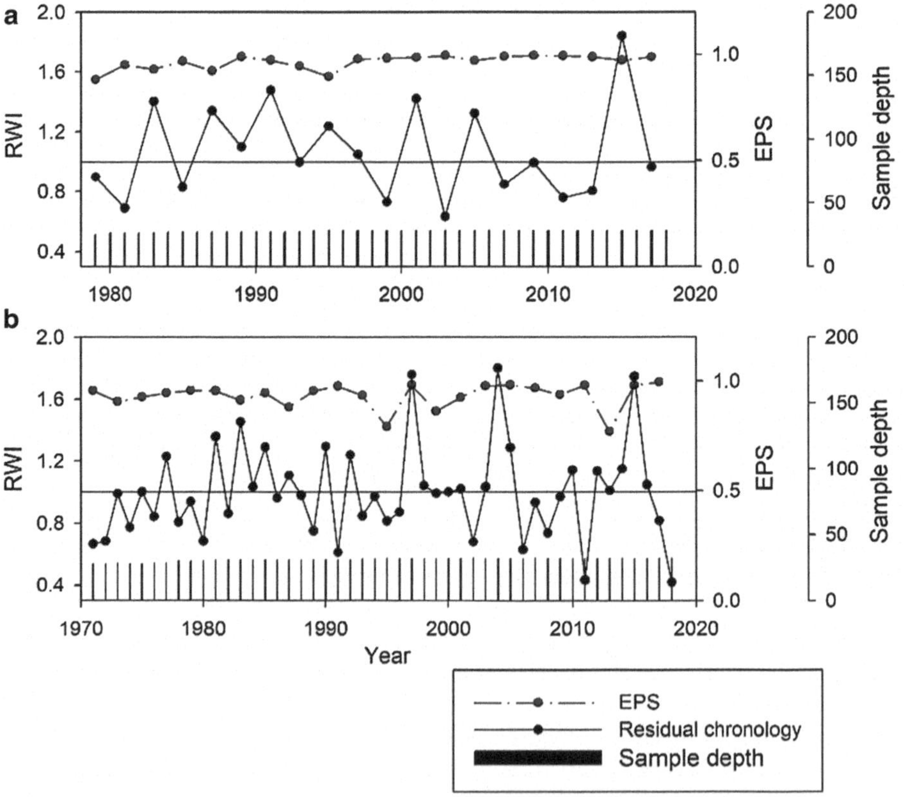

Fig. 4.2 Residual chronologies showing mean values of ring-width indices (RWI) for *Pinus leiophylla* (**a**) and *Pinus teocote* (**b**). The bars show the sample depth (right *y* axis). The dash blue lines show EPS

($r = 0.84$, $p < 0.001$). The inter-annual variation in growth was characterized by specific wide or narrow rings. For example, the year 2011 corresponded to narrow rings in both pine species, while the year 2015 corresponded to a wide ring in both species.

4.3.1 Growth-Climate and Drought Associations

Regarding *P. leiophylla*, its radial growth responded positively and significantly to the precipitation of October and November of the previous year and of January and June in the current growth year. Growth was negatively affected by the precipitation of July of the current year (Fig. 4.3a). The precipitation of the previous November and that of January to June of the current year positively and significantly influenced the growth of *P. teocote*.

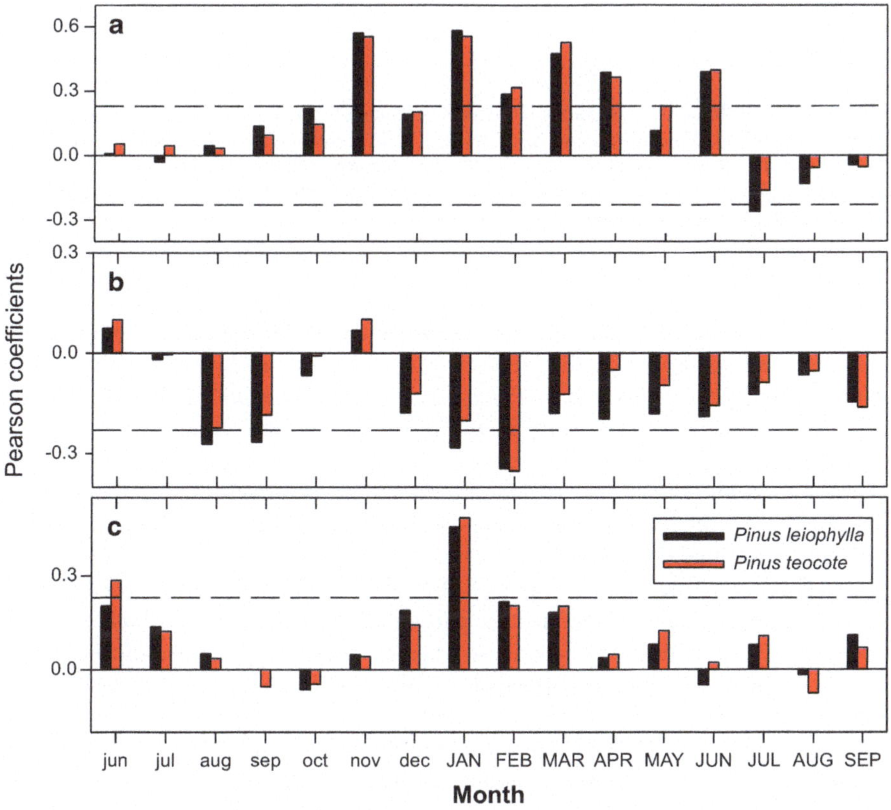

Fig. 4.3 Climate-growth associations. Bars show Pearson correlation coefficients calculated by relating and residual series of ring-width indices and monthly climate variables: (**a**) precipitation, (**b**) mean maximum temperature, and (**c**) mean minimum temperature. Correlations were calculated from June of the year prior to the growth year to September of the current growth year. The months of the previous and current years are abbreviated in lower and upper case letters, respectively. The horizontal dashed lines indicate the level of significance at $p < 0.05$

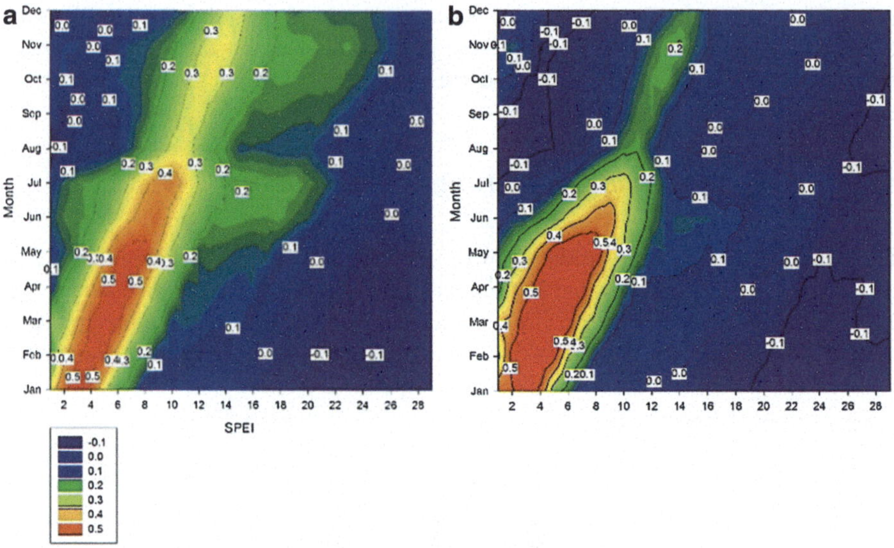

Fig. 4.4 Growth-drought associations (Pearson coefficients). The residual chronologies of ring-width indices of (**a**) *P. leiophylla* and (**b**) *P. teocote* were correlated with the SPEI drought index calculated at the scales of 1–28 months (x axis) from January to December of the current growth year (y axis). The color scale represents the Pearson coefficients of correlation with significant ($p < 0.05$) values above the threshold $r = |0.30|$

Growth responded negatively and significantly to the maximum temperature of August and September of the previous year and to that of January and February of the current year of growth in *P. leiophylla*. For *P. teocote*, those correlations were significant ($p < 0.05$) only in the prior August and the current February (Fig. 4.3b).

The minimum temperature of January influenced positively the growth of both pine species. Moreover, *P. teocote* also positively responded to this climatic variable in the previous June (Fig. 4.3c).

Dry conditions (negative values of SPEI) were associated with reduced growth rates (low RWI values) in both species, particularly in *P. leiophylla* which was very sensitive to SPEI values from January to May at scales of 2–8 months (accumulated drought), whereas *P. teocote* responded to 2–10-month scales from January to June (Fig. 4.4).

Spatial correlations based on the chronologies of the two species and the 6-month-long March SPEI (accumulated drought), a variable which presented the highest correlations with RWI (Fig. 4.4), the negative response of growth to dry conditions was concentrated mainly in the northwest and northeast parts of Mexico for both pine species (Fig. 4.5).

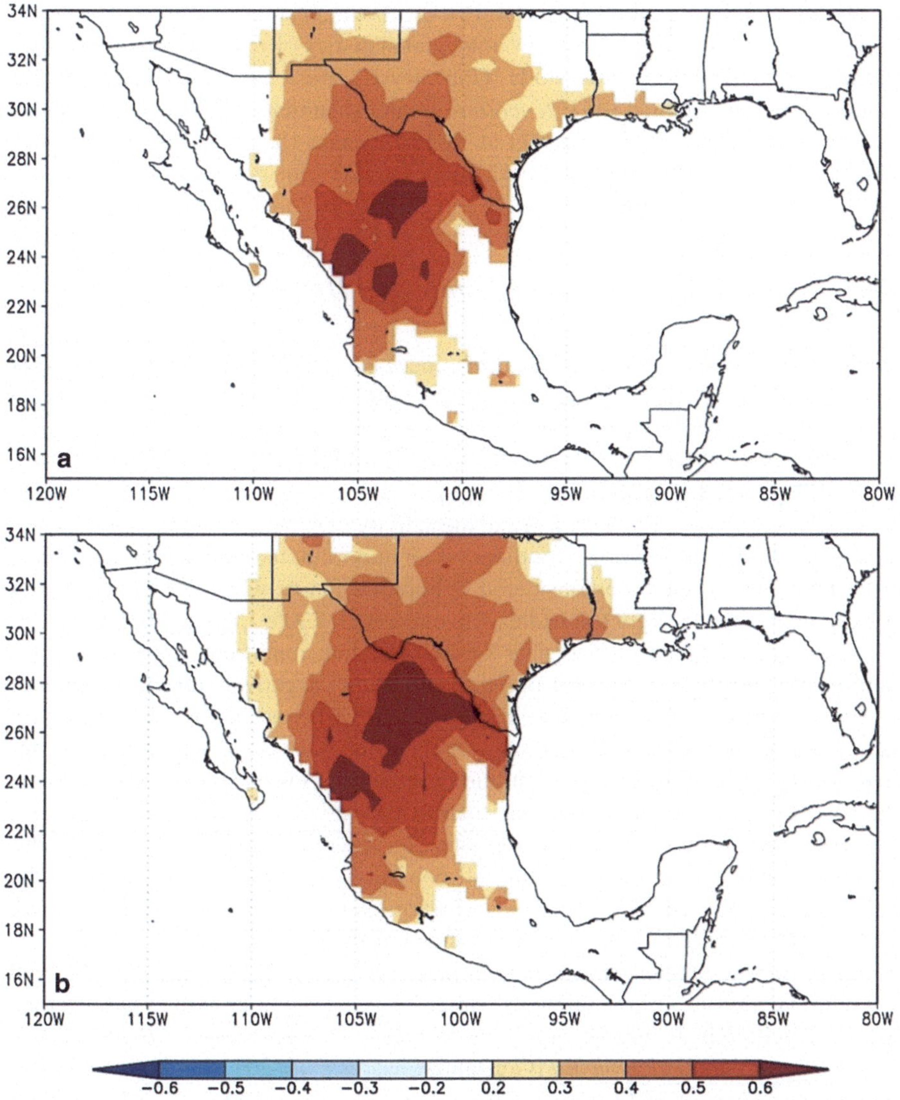

Fig. 4.5 Spatial correlations (Pearson coefficients) relating the accumulated 6-month, March SPEI with the ring-width chronologies of *Pinus leiophylla* (**a**) and *P. teocote* (**b**). In both cases, significance was below 10%. The asterisk shows the approximate location of the study area

4.4 Discussion

According to the posed hypothesis, the two studied sympatric species showed similar responses in their growth responses to climate and drought, despite some differences that could be found. Particularly, *P. leiophylla* was more responsive to precipitation and drought compared to *P. teocote* which showed a greater response to low temperatures. The dendrochronological statistics of *P. leiophylla* are similar to those reported by a previous study (González-Cásares et al. 2017). However, the mean ring width obtained for *P. teocote* is greater than that (1.73 mm) observed in the Sierra Madre Oriental (López et al. 2017) because the annual precipitation is comparatively higher in our site ($\approx$632 mm vs $\approx$ 1200 mm).

4.4.1 Growth Responses to Climate and Drought

Precipitation was the main factor that influences the radial growth in both pine species, which agrees with previous reports showing growth increases in response to higher winter rainfall (Pompa-García et al. 2013, 2017). This is due to the relevance of winter rains, which favor soil recharge and infiltration (Constante et al. 2009), while high summer rainfall in July causes soil saturation attributed to water excess, which may negatively affect growth and the activity of soil microorganisms (Bartelt-Ryser et al. 2005). In addition, cloudiness is an important factor that reduces growth given that pine species are considered heliophilic plants whose germination and growth are limited by light (González-Elizondo et al. 2005; Li et al. 2012).

In another study performed near Chihuahua, hot conditions during the prior autumn and current spring and also in the prior winter negatively affected growth of several pine species (González-Cásares et al. 2017), whereas in this study, this association was mainly observed in late winter (February). These differences are attributed mainly to the environmental conditions of the sites. On one hand, the Chihuahua site has drier and warmer conditions than in the site of the present study, with a temperate subhumid climate. The wet sites, on the other hand, have stored water reserves, thanks to the favorable rainfall conditions of the summer season (CNA 2018), which are used for dry seasons, as warm conditions in winter are increased, reserves are likely. They tend to end at the end of the season.

For studies conducted in this zone in particular, it has been found that the cold and humid conditions of the previous winter favor the radial growth of various pine species, including *P. arizonica*, *P. cembroides*, *P. engelmannii*, *P. lumholtzii*, *P. durangensis*, and *P. leiophylla* (González-Cásares et al. 2017) and other conifers as *Abies durangensis* and *Cupressus lusitanica* (Pompa-García et al. 2017). Winter precipitation produces a greater availability of water and allows recharging the soil reserves before or during the onset of cambial activity of the species, which is assumed to occur in early spring (Pompa-García and Venegas-González 2016). In most cases, maximum temperature negatively affects growth since greater temperatures lead to greater evapotranspiration rates and lower soil water availability (Williams et al. 2013). In various conifers there is a reduction in growth as a result of soil water deficit, a mechanism used to maintain a balance between water absorption capacity and transpiration demand (Barton and Teeri 1993). In *P. leiophylla* a reduction in the growth rate due to drought was found whose effects on biomass accumulation remained still 4 months after the drought occurred, suggesting that different adaptation mechanisms or lagged drought responses may exist among species of the same genus (Trinidad et al. 2002). Moreover, warmer and drier conditions affect negatively tracheid expansion, leading to smaller lumen area and thus reducing the total width of the growth ring (Pompa-García and Venegas-González 2016). The negative response of *P. leiophylla* growth to summer precipitation was also observed in another study on *Pinus cooperi* in a nearby area and could be related to increased cloudiness and decreased photosynthesis rates (Pompa-García et al. 2013).

In general, *P. leiophylla* and *P. teocote* did not present significant differences in terms of growth responses to precipitation and temperature. This agrees with studies that indicate that sympatric species present similar ecophysiological responses to climatic variability (Wieser et al. 2016). However, another study forecasted different responses of sympatric tree species to the projected climate warming which depended on their phenology, suggesting that *P. leiophylla* would be more vulnerable to hotter and drier conditions (Antúnez et al. 2017). On the other hand, a small difference in the growth-drought association was observed between the two species. Their growth presented different responses in terms of seasonality and drought duration. Specifically, *P. leiophylla* growth responded more to the SPEI at the scale of 2–8 months and from January to May, and *P. teocote* did it at the scale of 2–10 months and from January to July. The longest and strongest growth response to drought of *P. teocote* could be explained by its different phenology with respect to *P. leiophylla* which is more tolerant to drought stress (Martínez-Trinidad et al. 2002). Anyway, drought shortens the growing season in both species and reduces their growth rates and the production of xylem as has been observed in other similar species (Ziaco and Biondi 2016). Marquardt et al. (2018) also found differences in seasonal climatic response and in the duration of that response to drought in sympatric pine species from SW USA and found that growth was enhanced by wet-cool conditions in winter-spring. In a study of two concurrent pine species of the Mediterranean, *Pinus nigra* and *Pinus sylvestris*, differences were documented in the responses to climate of the two species (Martin-Benito et al. 2013). In another site in Spain subjected to Mediterranean and continental conditions, *Pinus halepensis* and *Pinus sylvestris* showed higher growth rates but a less plastic response to drought occurrence than the co-occurring *Juniperus thurifera*, which may represent a competitive advantage for the slow-growing species (Camarero et al. 2010). In another case of sympatric conifer species, *A. durangensis* and *C. lusitanica*, the former presented a high sensitivity to hydroclimatic variables, while the latter was not so responsive because it was restricted to wet sites (Pompa-García et al. 2017).

In general, *P leiophylla* and *P. teocote* presented a similar response to the assessed climatic variables (precipitation and temperature). Their sensitivity to cold humid winters is the habitual response of conifers of this drought-prone zone (see González-Cásares et al. 2017). These conditions permit low evapotranspiration and ensure the availability of water in the period prior to the growth period of the trees. The humidity facilitates the formation of a wide earlywood, while the conditions of drought act to limit growth during the late growing season (Pompa-García et al. 2017).

Despite the differences in the reproductive phenology of both study pine species that have been described (Patiño 1973; Jasso-Mata and Jiménez-Casas 1994), there is a large gap of knowledge on the phenology of their xylem development (xylogenesis) and how this affects growth responses to drought. A recent study on the wood anatomy of these species indicated that the cell wall of earlywood tracheids is twice thicker in *P. leiophylla* than in *P. teocote*, while the lumen diameter of latewood tracheids is larger in *P. teocote* (de la Paz-Pérez and Dávalos-Sotelo 2016). Future xylogenesis studies should also explain these differences in wood anatomy and how they affect growth sensitivity to drought.

In addition to climate, other factors, particularly disturbances, may affect the growth of sympatric tree species. For instance, outbreaks of insects including needle defoliators or bark beetles can affect the radial growth of *P. teocote* (López et al. 2017). Furthermore, intrinsic factors such as genetic variability can modulate the adaptive capacity of trees and their growth rates (Martínez-Berdeja et al. 2019). For instance, it has been found that these tree species show different growth and development responses across their distribution ranges as a function of local microclimate conditions (Trinidad et al. 2002; Alba-Landa et al. 2011). Future research could consider combinations of several of these extrinsic and intrinsic factors to evaluate the vulnerability of sympatric pine species to the forecasted hotter and more arid conditions as those projected for Mexico (Seager et al. 2009).

The growth response to the SPEI of sympatric species is an indicator of their sensitivity to water deficit and high temperatures at different temporal scales. It is important to further investigate this subject because the morphological and physiological responses of coexisting pine species may vary according to climatic site conditions as the forecasted climate warming. These differences remain to be analyzed at intra-annual and seasonal level through the study of the response of the earlywood and latewood or other wood anatomical variables to climate and drought.

4.5 Conclusion

The two studied sympatric pine species showed similar growth responses to climate and drought. Nevertheless, *P. leiophylla* growth was more sensitive to drought than *P. teocote*. Precipitation was the main factor that influenced the radial growth in both species. Humid and cool conditions of the prior winter and the current early spring enhanced radial growth. These results constitute a contribution on the behavior of the radial growth of these species which becomes relevant to the future impacts of global warming in similar drought-prone forest where sympatric and congeneric species exist.

Acknowledgments We are grateful to CONACYT for the doctoral scholarship awarded to the first and correspondence authors and for project CB-2013/222522 of CONACYT (National Council of Science and Technology). We also thank COCYTED (Council of Science and Technology of the State of Durango) for the resources granted to support this work. Also, we are grateful to the editors and anonymous reviewers for their useful comments and suggestions. We thank José Luis Gallardo Salazar for his assistance in edition of photographs used in this chapter.

References

Acosta-Hernández AC, Pompa-García M, Camarero JJ (2017) An updated review of dendrochronological investigations in Mexico, a megadiverse country with a high potential for tree-ring sciences. Forests 8:160. https://doi.org/10.3390/f8050160

Acosta-Hernández AC, Camarero JJ, Pompa-García M (2019) Seasonal growth responses to climate in wet and dry conifer forests. IAWA J:1–S1. https://doi.org/10.1163/22941932-40190226

Alba-Landa J, Mendizábal-Hernández LDC, Ramírez-García EO et al (2011) Tecnología para el manejo de *Pinus teocote* Schl. et Cham. Foresta Veracruzana 13(1):43–48

Allen CD, Macalady AK, Chenchouni H et al (2010) A global overview of drought and heat-induced tree mortality reveals emerging climate change risks for forests. Forest Ecol Manag 259(4):660–684. https://doi.org/10.1016/j.foreco.2009.09.001

Allen CD, Breshears DD, McDowell NG (2015) On underestimation of global vulnerability 737 to tree mortality and forest die-off from hotter drought in the Anthropocene. Ecosphere 6(8):129. https://doi.org/10.1890/ES15-00203.1

Antúnez P, Wehenkel C, López-Sánchez CA et al (2017) The role of climatic variables for estimating probability of abundance of tree species. Pol J Ecol 65(3):324–339. https://doi.org/10.3161/15052249PJE2017.65.3.002

Antúnez P, Suárez-Mota M, Valenzuela-Encinas C et al (2018) The potential distribution of tree species in three periods of time under a climate change scenario. Forests 9(10):628. https://doi.org/10.3390/f9100628

Bartelt-Ryser J, Joshi J, Schmid B et al (2005) Soil feed-backs of plant diversity on soil microbial communities and subsequent plant growth. Perspect Plant Ecol Syst 7:27–49. https://doi.org/10.1016/j.ppees.2004.11.002

Barton AM, Teeri JT (1993) The ecology of elevational positions in plants: drought resistance in five montane pine species in southeastern Arizona. Am J Bot 80:15–25. https://doi.org/10.1002/j.1537-2197.1993.tb13762.x

Bunn AG (2008) A dendrochronology program library in R (dplR). Dendrochronologia 26(2):115–124. https://doi.org/10.1016/j.dendro.2008.01.002

Bunn AG (2010) Statistical and visual crossdating in R using the dplR library. Dendrochronologia 28(4):251–258. https://doi.org/10.1016/j.dendro.2009.12.001

Camarero JJ, Olano JM, Parras A (2010) Plastic bimodal xylogenesis in conifers from continental Mediterranean climates. New Phytol 185(2):471–480. https://doi.org/10.1111/j.1469-8137.2009.03073.x

Camarero JJ, Gazol A, Sangüesa-Barreda G et al (2015) To die or not to die: early-warning signals of dieback in response to a severe drought. J Ecol 103:44–57. https://doi.org/10.1111/1365-2745.12295

Chacón-de la Cruz E, Pompa-García M (2015) Response of tree radial growth to evaporation, as indicated by early and late wood. Rev Chapingo Ser Cie 21(1):57–65. https://doi.org/10.5154/r.rchscfa.2014.10.050

Comisión Nacional del Agua CNA (2018) Datos Climáticos de Estaciones Meteorológicas de Durango: El Salto, México. Available online: http://smn1.conagua.gob.mx/index.php?option=com_content&view=article&id=180:durango&catid=14:normales-por-estacion

Constante GV, Villanueva DJ, Cerano PJ et al (2009) Dendrocronología de *Pinus cembroides* Zucc. y reconstrucción de precipitación estacional para el sureste de Coahuila. Revista Ciencia Forestal en México 34:17–38

Cook ER, Kairiukstis LA (1990) Methods of dendrochronology: applications in the environmental science. Kluwer, Alphen aan den Rijn, p 394

Cook BI, Seager R (2013) The response of the North American monsoon to increased greenhouse gas forcing. J Geophys Res Atmos 118(4):1690–1699. https://doi.org/10.1002/jgrd.50111

Dai A (2013) Increasing drought under global warming in observations and models. Nat Clim Chang 3(1):52–58. https://doi.org/10.1038/NCLIMATE1633

de la Paz-Pérez OC, Dávalos-Sotelo R (2016) Anatomía de la madera de seis especies de *Pinus* (Pinaceae) del estado de Durango, México. Madera y Bosques 22(3):113–132. https://doi.org/10.21829/myb.2016.2231460

Farjon A, Styles BT (1997) *Pinus* (Pinaceae). Flora Neotropica monograph 75. The New York Botanical Garden, New York

Fritts H (1976) Tree-rings and climate. The Blackburn Press, Caldwell, New Jersey

García AA, González EMS (2003) Pináceas de Durango. Instituto de Ecología. AC Comisión Nacional Forestal, Leon Guanajuato, p 187

González-Cásares M, Yerena-Yamaliel JI, Pompa-García M (2016) Measuring temporal wood density variation improves carbon capture estimates in Mexican forests. Acta Universitaria 26(6):25–28. https://doi.org/10.15174/au.2016.1206

González-Cásares M, Pompa-García M, Camarero JJ (2017) Differences in climate-growth relationship indicate diverse drought tolerances among five pine species coexisting in Northwestern Mexico. Trees 31:531–544. https://doi.org/10.1007/s00468-016-1488-0

González-Cásares M, Pompa-García M, Venegas-González A (2018) Climate signals from intra-annual wood density fluctuations in *Abies durangensis*. IAWA J 40(2):276–287. https://doi.org/10.1163/22941932-40190217

González-Elizondo MS, González-Elizondo M, López-Enríquez IL et al (2005) Cambios y tendencias sucesionales en ecosistemas de Durango. Vidsupra 1:5–11

González-Elizondo MS, González-Elizondo M, Márquez-Linares MA (2007) Vegetación y ecorregiones de Durango. Plaza y Valdés, México, p 219

González-Elizondo MS, González-Elizondo M, Tena-Flores JA et al (2012) Vegetación de la sierra madre occidental, México: Una síntesis. Acta Bot Mex 100:351–403

Holmes RL (1983) Computer-assisted quality control in tree ring dating and measurement. Tree Ring Bull 43:69–78

Instituto Nacional de Estadística y Geografía INEGI (2001) Conjunto de datos vectoriales edafológico, escala 1:250000 Serie II. (Continuo Nacional), escala: 1:250000. INEGI, MEX. http://www.conabio.gob.mx/informacion/gis/

Intergovernmental Panel on Climate Change IPCC (2014) Climate change 2014: synthesis report. Contribution of working groups I, II and III to the fifth assessment report of the Intergovernmental Panel on Climate Change [core writing team, Pachauri RK, Meyer LA (eds)]. IPCC, Geneva, p 151

Jasso-Mata J, Jiménez-Casas M (1994) Fenología vegetativa y reproductiva de *Pinus leiophylla* en un huerto semillero sexual. In: XI Congreso Latinoamericano de Genética y XV Congreso de Fitogenética

Li ZS, Liu GH, Fu BJ et al (2012) Anomalous temperature–growth response of *Abies faxoniana* to sustained freezing stress along elevational gradients in China's Western Sichuan Province. Trees 26:1373–1388. https://doi.org/10.1007/s00468-012-0712-9

López J, Méndez J, Zermeño A et al (2017) Impact of debarkers upon the radial increment of *Pinus teocote* Schiede. ex Schltdl. & Cham. and *Pseudotsuga menziesii* (Mirb.) Franco. Revista Mexicana de Ciencias Forestales 8(41):82–108. https://doi.org/10.29298/rmcf.v8i41.27

Marquardt PE, Miranda BR, Jennings S (2018) Variable climate response differentiates the growth of Sky Island ponderosa pines. Trees:1–16. https://doi.org/10.1007/s00468-018-1778-9

Martin-Benito D, Beeckman H, Canellas I (2013) Influence of drought on tree rings and tracheid features of *Pinus nigra* and *Pinus sylvestris* in a Mesic Mediterranean forest. Eur J For Res 132(1):33–45. https://doi.org/10.1007/s10342-012-0652-3

Martínez-Berdeja A, Hamilton JA, Bontemps A et al (2019) Evidence for population differentiation among Jeffrey and Ponderosa pines in survival, growth and phenology. For Ecol Manag 434:40–48. https://doi.org/10.1016/j.foreco.2018.12.009

Martínez-Trinidad T, Vargas-Hernández J, Muñoz-Orozco A et al (2002) Respuesta al déficit hídrico en *Pinus leiophylla*: consumo de agua y crecimiento en plántulas de diferentes poblaciones. Agrociencia 36(3):365–376

Patiño VF (1973) Flowering, fruiting, cone collection and some aspects from seeds of the Mexican pines. In: International Symposium on Seed Processing. IUFRO, Bergen

Perry J (1991) The pines of Mexico and Central America. Oregon Timber Press, Portland, p 231
Pompa-García M, Antonio-Némiga X (2015) Enso index teleconnection with seasonal precipitation in a template ecosystem of Northern Mexico. Atmósfera 28(1):43–50
Pompa-García M, Hadad MA (2016) Sensitivity of pines in Mexico to temperature varies with age. Atmósfera 29(3):209–219. https://doi.org/10.20937/ATM.2016.29.03.03
Pompa-García M, Venegas-González A (2016) Temporal variation of wood density and carbon in two elevational sites of *Pinus cooperi* in relation to climate response in Northern Mexico. PLoS One 11(6):e0156782. https://doi.org/10.1371/journal.pone.0156782.eCollection2016
Pompa-García M, Cerano-Paredes J, Fulé ZP (2013) Variation in radial growth of *Pinus cooperi* in response to climatic signals across an elevational gradient. Dendrochronologia 31:198–204. https://doi.org/10.1016/j.dendro.2013.05.003
Pompa-García M, Miranda-Aragón L, Aguirre-Salado C (2014) Tree growth response to ENSO in Durango. Mexico Int J Biometeorol 117(3-4):495–500. https://doi.org/10.1007/s00484-014-0828-2
Pompa-García M, González-Cásares M, Acosta-Hernández AC et al (2017) Drought influence over radial growth of Mexican conifers inhabiting mesic and xeric sites. Forests 8(5):175. https://doi.org/10.3390/f8050175
R Development Core Team (2014) R: a language and environment for statistical computing. R Foundation for Statistical Computing, Vienna. ISBN 3-900051-07-0. Available at: http://www.R-project.org
Roberts AM, Tansey C, Smithers RJ et al (2015) Predicting a change in the order of spring phenology in temperate forests. Glob Chang Biol 21(7):2603–2611. https://doi.org/10.1111/gcb.12896
Seager R, Ting M, Davis M et al (2009) Mexican drought: an observational modeling and tree ring study of variability and climate change. Atmósfera 22(1):1–31
Secretaría de Medio Ambiente y Recursos Naturales SEMARNAT (2016) Anuario Estadístico de la Producción Forestal 2016. In: Secretaría de Medio Ambiente y Recursos Naturales, Primera Edición edn, p 225
Stahle DW, Cook ER, Burnette DJ et al (2016) The Mexican drought atlas: tree-ring reconstructions of the soil moisture balance during the late pre-Hispanic, colonial, and modern eras. Quat Sci Rev 149:34–60. https://doi.org/10.1016/j.quascirev.2016.06.018
Stokes MA, Smiley TL (1968) An introduction to treering dating. University of Chicago Press, Chicago, p 73
Trinidad TM, Hernández JV, Upton JL et al (2002) Respuesta al déficit hídrico en *Pinus leiophylla*: Acumulación de biomasa, desarrollo de hojas secundarias y mortandad de plántulas. Terra Latinoamericana 20(3):291–301
Vicente-Serrano SM, Beguería S, López-Moreno JI (2010) A multiscalar drought index sensitive to global warming: the standardized precipitation evapotranspiration index. J Clim 23:1696–1718. https://doi.org/10.1175/2009JCLI2909.1
Whittaker RH (1953) A consideration of climax theory: the climax as a population and pattern. Ecol Monogr 23(1):41–78
Wieser G, Oberhuber W, Gruber A et al (2016) Stable water use efficiency under climate change of three sympatric conifer species at the alpine treeline. Front Plant Sci 7:799. https://doi.org/10.3389/fpls.2016.00799
Wigley TML, Briffa KR, Jones PD (1984) On the average value of correlated time series, with applications in dendroclimatology and hydrometeorology. J Clim Appl Meteorol 23(2):201–213. https://doi.org/10.1175/1520-0450(1984)023<0201:OTAVOC>2.0.CO;2
Williams AP, Allen CD, Macalady AK et al (2013) Temperature as a potent driver of regional forest drought stress and tree mortality. Nat Clim Chang 3:292–297. https://doi.org/10.1038/NCLIMATE1693
Ziaco E, Biondi F (2016) Tree growth, cambial phenology, and wood anatomy of limber pine at a Great Basin (USA) mountain observatory. Trees 30(5):1507–1521. https://doi.org/10.1007/s00468-016-1384-7

Part II
Dendroecology in Neotropical Regions

Chapter 5
Dendrochronological Potential of Trees from America's Rainiest Region

Jorge A. Giraldo, Jorge I. del Valle, Carlos A. Sierra, and Omar Melo

Abstract Hydric seasonality, dry or flooding periods, has been the explanation given by dendrochronologists for the formation of growth rings in tropical trees. However, under tropical hyper-humid conditions, there are species with growth periodicity and, therefore, with visible growth rings useful for dendrochronology. We classify, for the first time, the anatomical structures of the growth rings of the tropical trees from the Biogeographic *Chocó* Region. An annual rainfall >7200 mm without hydric seasonality characterizes the studied forest. We classify the tree rings as present (well or poorly defined) or absent. We also took into account the feasibility of tree-ring dating (dendrochronological potential). We characterized 81 species of trees belonging to 38 families. About 82% of species had growth rings, 46% well defined, and 36% poorly defined, and 18% with absent rings. Dendrochronological potential was high (14%), medium (25%), low (43%), and null (18%). We contrast our results with vast literature around tropics. Our results suggest that still without any hydric seasonality, many tropical tree species develop growth rings. Our findings drive to new questions: What is the periodicity of tree rings in these non-seasonal hyper-humid environments? What other subtle seasonality could be involved in the growth periodicity of these species, or are they genetic?

Keywords Tree rings · Non-seasonal rainfall · Biogeographic *Chocó* Region · Colombia

---

J. A. Giraldo (✉)
Universidad Nacional de Colombia, Sede Medellín, Medellín, Antioquia, Colombia
e-mail: jagiral1@unal.edu.co

J. I. del Valle
Departamento de Ciencias Forestales, Universidad Nacional de Colombia, Sede Medellín, Medellín, Antioquia, Colombia
e-mail: jidvalle@unal.edu.co

C. A. Sierra
Max Planck Institute for Biogeochemistry, Jena, Germany
e-mail: csierra@bgc-jena.mpg.de

O. Melo
Universidad del Tolima, Ibagué, Tolima, Colombia
e-mail: omelo@ut.edu.co

M. Pompa-García, J. J. Camarero (eds.), *Latin American Dendroecology*,
https://doi.org/10.1007/978-3-030-36930-9_5

5.1 Introduction

The current paradigm in dendrochronology states a more likelihood of growth-ring formation exists in these cases: (1) in seasonally dry climates – water deficit – (Détienne and Barbier 1988; Vetter and Botosso 1989; Worbes 1995, 1999; Bullock 1997; Jalil et al. 1998; Borchert 1999; Enquist and Leffler 2001; Schöngart et al. 2002, 2017; Marcati et al. 2006; Brienen et al. 2009, 2016; Morel 2013; Groenendijk et al. 2014; Silva et al. 2019; Granato-Souza et al. 2019), (2) in periodically flooded environments which produce anoxic conditions for the root system (Worbes and Junk 1989; Schöngart et al. 2002, 2005, 2017; Worbes and Fichtler 2010; Brienen et al. 2016; Parolin et al. 2016, (3) in estuaries, due to variability in salt concentration (Menezes et al. 2003; Chowdhury et al. 2016; Phan et al. 2019), (4) in deciduous or semi-deciduous species (Borchert 1999; Lisi et al. 2008; Nath et al. 2016; Shimamoto et al. 2016), and (5) under seasonal variations in irradiance, time of sunrise or sunset, or sunlight intensity (Breitsprecher and Bethel 1990; Clark and Clark 1994; O'Brien et al. 2008; Marcati et al. 2016; Shimamoto et al. 2016; Köhl et al. 2017). According to literature cited, hydric seasonality, dry periods, or periodic floods are the dominant triggering variables for the formation of growth rings in tropical trees.

Because a high cloud cover limits the availability of photosynthetically active radiation (Clark and Clark 1994), even in non-seasonal tropical forests, light appears to be the most limiting factor during the rainy season. Cloudiness induces a reduction in photosynthetic photon flux density (PPFD): the higher the cloud cover, the greater is the reduction in PPFD (Graham et al. 2003). Besides, seasonal water availability highlighted the predominant role of water availability that determined seasonal variation in tree growth (Wagner et al. 2012). The availability of water in the soil could be more determinative than rainfall in the formation of annual rings (Détienne and Barbier 1988; Falcon-Lang 1999; Silva et al. 2019). On the other hand, excess water in the soil, cloudiness, and low solar radiation in wet tropical climates could result in reduced photosynthesis, lower rates of tree growth, and low aerial biomass in the forests (Álvarez et al. 2017).

Some researchers have found that trees of boreal origin (Laurasia) tend to form growth rings more frequently than those of tropical origin (Fahn 1967; Tomlinson and Craichead 1972; Aguilar-Rodríguez and Barajas-Morales 2005). This type of growth rings, like a fingerprint of its phytogeographic origin, implied that their formation is from the genetic background. The variables that trigger the formation of growth rings in tropical trees can affect species in the same forest differently. For instance, four species with annual growth rings from the Atlantic rainforests in Brazil responded to four different triggering variables: the period of leaves abscission, flooding, photoperiod, and endogenous periodicity (Callado et al. 2001).

Although information on growth periodicity of tree species in tropical hyperhumid, non-seasonal climates is scarce, there are species where periodic growth was observed through annual measurements of diameter (Clark and Clark 1994), phenological observations (Frankie et al. 1974), cambial activity (Breitsprecher and

Bethel 1990; O'Brien et al. 2008; Bräuning et al. 2009), and annual rings (Fichtler et al. 2003; Moreno and del Valle 2014). All of these measurements suggested the existence of a possible triggering of growth rates associated with rainfall, the duration of the day, the angle of insolation, or a combined response between rain and light levels. The evaluation of species with growth rings is the first step in selecting those potentially useful in dendrochronological studies (Tarhule and Hughes 2002; Tanaka 2005; Groenendijk et al. 2014; Silva et al. 2017; Nath et al. 2016; Islam et al. 2018). The presence of growth rings suggests the existence of growth periodicity, which is probably regulated by some environmental factor. In this study, for the first time, we perform an anatomical classification of the tree-ring structure in the rainiest region of the Americas, the Colombian Pacific, in a lowland, wet, non-seasonal tropical forest with mean annual rainfall >7200 mm. The discovery of species with dendrochronological potential in a tropical hyper-humid and non-seasonal region would be a milestone in dendrochronology.

5.2 Materials and Methods

5.2.1 Study Area

The Biogeographic *Chocó* Region extends across the Pacific of South America, from northern Ecuador to the Colombian-Panamanian Darien Gap. Most of this region, the rainiest in the New World (Poveda and Mesa 2000), is located along the Colombian Pacific coast, on the western flank of the Western Cordillera of the Andes. In some sites, mean annual precipitation exceeds 12,000 mm, with extremes of 18,449 mm in 1953 and 19,839 mm in 1936 (Mesa et al. 1997). The study was conducted in a wet tropical forest in the Holdridge system (Holdridge 1967) with 4000–8000 mm mean annual precipitation. Specifically, the study area belongs to the Low *Calima* River Basin, where the Tropical Forest Centre Pedro Antonio Pineda of the University of Tolima (3°55'N, 77°07'W) is located there (Fig. 5.1), which includes an area of influence of 66,700 ha of forest that belongs to the Community Council of the Low *Calima* River Basin.

5.2.2 Dry Month Analysis in Study Area

The climatic diagram (Fig. 5.1) used widely by dendrochronologists incorporates the xerothermic index for intertropical climates of Gaussen (Bagnouls and Gaussen 1957). According to this criterion, a month is dry if rainfall is less than twice the mean monthly temperature expressed in millimeters of rain. For our case, that would be approximately 25.9 (2) = 51.8 mm. Even in February, the least rainy month, rainfall exceeded the dry month criterion by sevenfold: 350 mm compared with

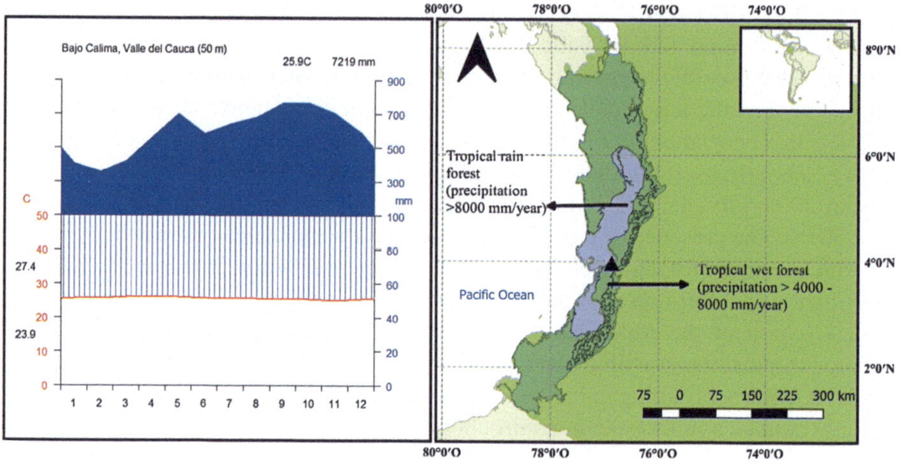

Fig. 5.1 Study area in the Biogeographic Chocó Region, located both in the wet tropical forest and tropical rain forest life zones (Holdridge 1967). Walter and Lieth's ombrothermic climate diagram (Walter et al. 1975) of the Low *Calima* Climatic Station (triangle), which belongs to the Institute of Hydrology, Meteorology, and Environmental Studies (IDEAM). The dark area indicates an excess of water year around (all months are hyper-humid)

51.8 mm. The North American climatologist Thornthwaite (1948) incorporated the concept of potential evapotranspiration (PET) calculated from air temperature and day length. Under this concept, a month was dry if the precipitation was lower than the PET. Holdridge (1967) developed an approximate formula based only on the mean monthly temperature to calculate the monthly PET. This formula that is near the thermic tropic where our area of study is located, calculates Thornthwaite's PET very well. In the study area, the mean annual temperature and the mean temperature for each month are almost the same (Fig. 5.1), them, if we assume that every month has 30.4 days, the Holdridge (1967) formula allow us to estimate the monthly PET = (58.93) (25.9)/12 = 127 mm. That is to say, the xerothermic index of Gaussen underestimated PET by about 2.4 times. However, even using the PET criterion, which is much more demanding than Gaussen's xerothermic index, rainfall in February still exceeded PET by about 2.7 times.

5.2.3 Sampling

In the study area, a local Afro-Colombian ethnic group that inhabits these forests belonging to the Community Council of the Low *Calima* River Basin, among the subsistence, economic, and cultural activities they carry out, there is a sporadically felling of trees in small areas of the forest. As virtually all the species of trees existing there has some use in the wood market, from these trees, we collected complete cross-sectional discs. Then, this sampling was reasonably at random as recommended (Speer 2010) to represent a landscape. We also collected 15 species existing in the region, but that did not appear in the sampling. They were found within the experimental area and arboretum of the Pedro Antonio Pineda Tropical Forest Center of the University of Tolima, most of them with increment borers because of restrictions of the University. For this purpose, we extracted 2–3 cores 5 mm in diameter from the trees at 1.3 m above the ground aproximately. The holes produced by the increment borer were filled with wax, and antibacterial and healing products, to avoid penetration by insects and pathogens. From each tree, we also collected leaves, flowers, and fruits, whenever present, to aid in taxonomic identification.

5.2.4 Processing of Wood Samples

Wood samples were taken to the Tropical Dendroecology Laboratory of the National University of Colombia, Medellín, and dried to a constant weight at 28 °C. Then, to highlight the growth rings, the samples were polished successively with abrasive sandpapers, from 60 to 600 grains inch^{-2}. Samples were scanned at a resolution of 1800–2400 dpi in an Epson Expression 10,000XL Scanner. The high-resolution images allowed macroscopic characteristics to be observable up to 10–15 μm. All images were processed with ImageJ 11.52i Software (Schindelin et al. 2012).

5.2.5 Observable Anatomical Characteristics

We identified the growth rings according to their type: (A) density changes due to an increase in the wall thickness of the fibers, (B) presence of marginal bands of axial parenchyma, or (C) differences in the diameter, distribution, and grouping of the pores (ring-porous, semi-ring-porous, and diffuse-porous). One or several types of these anatomical structures can be present in the same sample. We classified the presence of growth rings according to Silva et al. (2017) as well defined (+), poorly defined (±), or absent (−).

The analysis considers the difficulty or impossibility of dating the growth rings when there are many anomalies such as false rings, wedged rings, tangential discontinuity, or lost rings. The analysis included the visual dating possibility of growth rings that is an essential procedure in dendrochronology. We made this classification by visual inspection using different degrees of magnification, granting these qualitative classifications: (1) null, there were no growth rings; (2) low, several poorly defined growth rings, difficult to mark, lost rings, too many together, or frequent tangential discontinuity; (3) medium, false or lost rings were evident and not frequent; or (4) high, growth rings were clearly defined with tangential continuity and few false rings.

5.2.6 *Growth-Ring Periodicity*

We determined the growth periodicity in some sampled species by using the bomb-peak dating method (Worbes and Junk 1989; Fichtler et al. 2003). Samples were prepared according to Steinhof et al. (2017) and were analyzed in an accelerator mass spectrometry (AMS) at the radiocarbon laboratory of Max Planck Institute for Biogeochemistry in Jena, Germany. To obtain chronological dates from the radiocarbon values, we used the calibration software Cali-Bomb (Reimer et al. 2004).

5.2.7 *Pantropical Growth-Ring Potential*

An exhaustive pantropical literature review was made on the dendrochronological potential of trees covering from Northern subtropic to Southern subtropic and from sea level to high mountains. All types of known seasonality affecting the growth rings are included: temperature, droughts, and floods. We compare our results with the pantropical revision in terms of the number of studied species and percent of species with growth rings. We also determined which species in our study were reported with annual rings in studies carried out in seasonal climates from other Neotropical regions.

5.3 Results and Discussion

According to Gaussen xerothermic index (Bagnouls and Gaussen 1957) and the monthly potential evapotranspiration (Holdridge 1967), in the study area (Fig. 5.1), on average, there are no ecologically dry months. However, because it is soil water-holding capacity and not precipitation that is the crucial factor for tree growth, water limitation in some species cannot be ruled out. Trees from wet environments are more water-demanding than those from dry climates which is a species-specific trait (Silva et al. 2019).

Eighty-one species were sampled from 38 plant families after a month of sampling. From the sampling, basically at random, it was found that 58% of the species were only represented by a single tree, which shows the high richness and diversity of trees existing in the study area. The families with the most number of species included Sapotaceae (7), Fabaceae (7), Malvaceae (5), Chrysobalanaceae (4), Moraceae (4), Myristicaceae (4), and Vochysiaceae (4). Of the total, 17 endemic species (21%) were recorded from Colombia: *Jacaranda hesperia*, *Dacryodes frangens*, *Marila macrophylla*, *Clusia leptanthera*, *Vismia rufa*, *Quararibea foenigraeca*, *Sterculia aerisperma*, *Qualea lineata*, and *Vochysia artantha*, which included eight species endemic to the Low *Calima* River Basin Region (*Licania caldasiana*, *Tapura costata*, *Tachigali colombiana*, *Magnolia calimaensis*,

Pterandra ultramontana, *Henriettea maguirei*, *Coussapoa valaria*, and *Vochysia pacifica*). In addition to these, we sampled 14 endemic species of the Biogeographic *Chocó* Region (18%): *Campnosperma panamensis*, *Licania sp*, *Dipteryx oleifera*, *Humiriastrum procerum*, *Gustavia occidentalis*, *Huberodendron patinoi*, *Phragmotheca siderosa*, *Otoba latialata*, *Virola dixonii*, *Otoba lehmannii*, *Isertia pittieri*, *Pouteria buenaventurensis*, *Pouteria collina*, and *Cecropia virgusa*. The other species (not mentioned) have a much wider distribution (Table 5.1).

The macroscopic structures observed in the analyzed species (Table 5.1, Figs. 5.2, 5.3, and 5.4) can be seen with the naked eye or with only a few zooms as changes in the form of alternate bands of light and dark areas. This was interpreted as increases in wood density that were attributed to the thickening of the fiber cell walls (A), which may be gradual or abrupt (Fig. 5.2). This ring pattern was dominant in the sampled species (51%) (Fig. 5.4b). It sometimes occurred together with bands of marginal parenchyma (B) (Fig. 5.3), which was present in 36% of the species (Fig. 5.4b).

In 82% of the studied species, we found growth rings; 46% were well defined (distinct rings), and 36% were poorly defined with visible although faint rings. Only 18% of the species had no growth rings or discontinuous bands, where it is difficult to establish if growth rings were present (Fig. 5.4a and Table 5.1).

The semi-ring-porous pattern (C), that is a characteristic generally reported in species that grow in dry environments, or at least, with a dry season (Worbes and Junk 1999), was exhibited by 13% of the species in combination with the other anatomical characteristics: The combination A–C occurred in 9% of the species, the combination A-B-C occurred in 3%, and the B–C pattern only occurred in 1% (Figs. 5.4b and 5.5). Although the growth rings were well defined, the possibility of dating them was not necessarily high, as occurred in *Symphonia globulifera*, *T. costata*, *Cedrela odorata*, and other species, where the possibility of dating them was only medium (Table 5.1). On the other hand, in species with poorly defined rings, the possibility of dating them was null as in *Hebepetalum humiriifolium*, low as in *Cespedezia spathulata*, or medium as in *Pourouma bicolor*, but never high.

The possibility of dating the rings depends both on their presence and on their continuity, which is sometimes limited by competition (Fig. 5.6a), traumatic rings produced by injuries in the cambium (Fig. 5.6b), or other difficulties such as the anomalous architecture of the stems (Fig. 5.6c, d). In these cases, the rings can be identified using complete cross-sectional discs, which are recommended in tropical dendrochronology (Détienne 1995; Worbes 2002; Brienen and Zuidema 2005). Undoubtedly, the minimal observation area when corers from increment borers are used makes dating difficult. In general, in 39% of the species, the dating of growth rings was feasible. These included both medium (25%) and high (14%) dating percentages (Figs. 5.4d and 5.6). Species such as *Goupia glabra* with visible growth rings (Table 5.1) may present areas with tangential discontinuity that can only be detected with complete cross-sectional discs (Fig. 5.7a).

The results of this study show that, in contrast to the pessimistic opinion of many dendrochronologists and ecologists about the existence of growth rings in the tropics (Whitmore 1975; Schweingruber 1988; Evans and Schrag 2004; Poussart et al.

Table 5.1 Tree species sampled in the Lower *Calima* River Basin, Biogeographic *Chocó* Region, characterized according to their growth rings

Family	Species	Tree-ring visibility	Wood anatomy	Vessel type	Potential in dendrochronology	Number of trees	Type of samples	Distribution
Anacardiaceae	*Campnosperma panamensis* Standl.	–	A	Diffuse-porous	None	2	Core	Biogeographic *Chocó*
Apocynaceae	*Aspidosperma excelsum* Benth.	+	A	Diffuse-porous	Low	1	Cross-section	Wide distribution
Apocynaceae	*Aspidosperma desmanthum* Benth. ex Müll. Arg.	+	A	Diffuse-porous	Low	2	Cross-section	Wide distribution
Bignoniaceae	*Jacaranda hesperia* Dugand.	+	A	Diffuse-porous	High	2	Core	Endemic
Boraginaceae	*Bourreria costaricensis* (Standl.) A.H Gentry	±	A, C	Semi-ring-porous	Low	1	Core	Wide distribution
Burseraceae	*Dacryodes frangens* Daly & M. C. Martínez	–	A	Diffuse-porous	None	1	Cross-section	Endemic
Burseraceae	*Protium* sp.	+	A	Diffuse-porous	Medium	1	Cross-section	Wide distribution
Calophyllaceae	*Calophyllum longifolium* Willd.	±	A, B	Diffuse-porous	Low	14	Cross-section (5) – Core (9)	Wide distribution
Calophyllaceae	*Marila macrophylla* Benth.	±	A	Diffuse-porous	Low	1	Cross-section	Endemic
Chrysobalanaceae	*Hirtella americana* L.	±	A	Diffuse-porous	Low	1	Cross-section	Wide distribution
Chrysobalanaceae	*Hirtella* sp.	±	A, B	Diffuse-porous	Low	1	Cross-section	Wide distribution
Chrysobalanaceae	*Licania caldasiana* Cuatrec.	±	A, B	Diffuse-porous	Low	1	Cross-section	Endemic

Chrysobalanaceae	*Licania* sp.	–	A	Diffuse-porous	None	1	Cross-section	Biogeographic *Chocó*
Clusiaceae	*Clusia leptanthera* Cuatrec	±	A, C	Semi-ring-porous	Low	1	Cross-section	Endemic
Clusiaceae	*Symphonia globulifera* L. f.	+	A, B	Diffuse-porous	Medium	1	Cross-section	Wide distribution
Dichapetalaceae	*Tapura costata* Cuatrec.	+	B, C	Semi-ring-porous	Low	1	Cross-section	Endemic
Dichapetalaceae	*Sloanea* sp.	±	A	Diffuse-porous	Low	1	Cross-section	Wide distribution
Erythroxylaceae	NN	+	A	Diffuse-porous	Low	1	Cross-section	Wide distribution
Euphorbiaceae	*Mabea sp.* Aubl.	+	A	Diffuse-porous	High	2	Cross-section (1) – Core (1)	Wide distribution
Euphorbiaceae	*Hevea brasiliensis* (Willd. ex A. Juss.) Müll. Arg.	+	A, B	Diffuse-porous	Medium	1	Cross-section	Wide distribution
Fabaceae	*Dipteryx oleifera* Benth.	±	A, B	Diffuse-porous	Medium	1	Cross-section	Biogeographic *Chocó*
Fabaceae	*Inga acreana* Harms	+	A, B	Diffuse-porous	Medium	1	Cross-section	Wide distribution
Fabaceae	*Inga rubiginosa* (Rich.) DC.	+	A, B	Diffuse-porous	Medium	1	Cross-section	Wide distribution
Fabaceae	*Schizolobium parahyba* (Vell.) S.F. Blake	+	A, B	Diffuse-porous	Medium	1	Core	Wide distribution

(continued)

Table 5.1 (continued)

Family	Species	Tree-ring visibility	Wood anatomy	Vessel type	Potential in dendrochronology	Number of trees	Type of samples	Distribution
Fabaceae	*Pentaclethra macroloba* (Willd.) Kuntze	+	A, B	Diffuse-porous	Medium	2	Core	Wide distribution
Fabaceae	*Tachigali colombiana* Dwyer	+	A	Diffuse-porous	High	3	Cross-section	Endemic
Fabaceae	*Abarema* sp.	±	A, B	Diffuse-porous	Low	1	Core	Wide distribution
Goupiaceae	*Goupia glabra* Aubl.	+	A	Diffuse-porous	High	21	Cross-section	Wide distribution
Humiriaceae	*Humiriastrum procerum* (Little) Cuatrec.	+	A	Diffuse-porous	High	8	Cross-section	Biogeographic *Chocó*
Humiriaceae	*Sacoglottis* sp.	±	A	Diffuse-porous	Low	3	Cross-section	Wide distribution
Hypericaceae	*Vismia rufa* Cuatrec.	–	A	Diffuse-porous	None	1	Cross-section	Endemic
Lauraceae	*Nectandra* sp.	+	A	Diffuse-porous	Low	1	Cross-section	Wide distribution
Lauraceae	*Ocotea* sp.	+	A	Diffuse-porous	Low	1	Cross-section	Wide distribution
Lecythidaceae	*Couratari guianensis* Aubl.	+	A	Diffuse-porous	Low	1	Cross-section	Wide distribution
Lecythidaceae	*Cariniana pyriformis* Miers.	+	A	Diffuse-porous	High	14	Cross-section	Wide distribution
Lecythidaceae	*Gustavia occidentalis* Cuatrec.	+	A, B	Diffuse-porous	Low	1	Cross-section	Biogeographic *Chocó*
Linaceae	*Hebepetalum humiriifolium* (Planch.) Benth.	±	A	Diffuse-porous	Low	7	Cross-section	Wide distribution

Magnoliaceae	*Magnolia calimaensis* (Lozano) Govaerts	+	A, B	Diffuse-porous	Medium	1	Core	Endemic
Malpighiaceae	*Pterandra ultramontana* H.P. Riley ex Cuatrec.	+	A, C	Semi-ring-porous	Low	2	Cross-section	Endemic
Malvaceae	*Huberodendron patinoi* Cuatrec.	+	A, B	Diffuse-porous	High	1	Core	Biogeographic *Chocó*
Malvaceae	*Apeiba macropetala* Ducke	+	A, C	Semi-ring-porous	High	6	Cross-section	Wide distribution
Malvaceae	*Quararibea foenigraeca* Cuatrec.	±	A, B	Diffuse-porous	Low	1	Cross-section	Endemic
Malvaceae	*Phragmotheca siderosa* Cuatrec.	+	A	Diffuse-porous	Medium	2	Cross-section	Biogeographic *Chocó*
Malvaceae	*Sterculia aerisperma* Cuatrec.	±	A, B, C	Semi-ring-porous	Low	3	Cross-section	Endemic
Melastomataceae	*Henriettea maguirei* (Wurdack) Penneys, Michelang., Judd & Almeda	–	A, B	Diffuse-porous	None	1	Cross-section	Endemic
Melastomataceae	*Miconia* sp.	–	A, B	Diffuse-porous	None	1	Cross-section	Wide distribution
Meliaceae	*Cedrela odorata* L.[1]	+	A, B, C	Semi-ring-porous	High	1	Cross-section	Wide distribution
Metteniusaceae	*Dendrobangia boliviana* Rusby	–	A	Diffuse-porous	Low	2	Cross-section	Wide distribution

(continued)

Table 5.1 (continued)

Family	Species	Tree-ring visibility	Wood anatomy	Vessel type	Potential in dendrochronology	Number of trees	Type of samples	Distribution
Moraceae	*Castilla elastica* var. *costaricana* (Liebm.) C.C. Berg	+	A	Diffuse-porous	Medium	1	Cross-section	Wide distribution
Moraceae	*Clarisia racemosa* Ruiz & Pav	+	A, B	Diffuse-porous	Medium	1	Cross-section	Wide distribution
Moraceae	*Sorocea* sp.	±	A	Diffuse-porous	Low	1	Cross-section	Wide distribution
Moraceae	*Ficus* sp.	–	A, B	Diffuse-porous	None	1	Core	Wide distribution
Myristicaceae	*Osteophloeum platyspermum* (Spruce ex A. DC.) Warb.	±	A	Diffuse-porous	Medium	2	Cross-section	Wide distribution
Myristicaceae	*Otoba latialata* (Pittier) A.H. Gentry	+	A	Diffuse-porous	High	11	Cross-section	Biogeographic *Chocó*
Myristicaceae	*Virola dixonii* Little	+	A	Diffuse-porous	High	4	Cross-section	Biogeographic *Chocó*
Myristicaceae	*Otoba lehmannii* (A.C.Sm.) A.H.Gentry	±	A, B	Diffuse-porous	Medium	2	Core	Biogeographic *Chocó*
Myrtaceae	*Myrcia* sp.	–	A	Diffuse-porous	None	1	Cross-section	Wide distribution
Myrtaceae	*Syzygium malaccense* (L.) Merr. & L.M. Perry	–	A	Diffuse-porous	None	1	Core	Wide distribution
Ochnaceae	*Cespedesia spathulata* (Ruiz & Pav.) Planch.	±	A, C	Semi-ring-porous	Low	1	Cross-section	Wide distribution
Olacaceae	*Minquartia guianensis* Aubl.	–	A	Diffuse-porous	Low	1	Cross-section	Wide distribution

Phyllantaceae	*Hieronyma alchorneoides* ALLEMAO	+	A, C	Semi-ring-porous	Medium	3	Core	Wide distribution
Rhizophoraceae	*Cassipourea* sp.	+	A, C	Semi-ring-porous	Low	1	Cross-section	Wide distribution
Rubiaceae	*Isertia pittieri* (Standl.) Standl.	–	A	Diffuse-porous	None	1	Cross-section	Biogeographic *Chocó*
Rubiaceae	*Posoqueria latifolia* (Rudge) Schult.	+	A	Diffuse-porous	Medium	1	Cross-section	Wide distribution
Rutaceae	*Zanthoxylum riedelianum* subsp. *hygrophilum* (Cuatrec.) Reynel	±	A, B	Diffuse-porous	Low	7	Cross-section (1) – Core (6)	Wide distribution
Salicaceae	*Casearia arborea* (Rich.) Urb.	±	A	Diffuse-porous	Low	1	Cross-section	Wide distribution
Sapotaceae	*Manilkara* sp.	±	A	Diffuse-porous	Low	1	Cross-section	Wide distribution
Sapotaceae	*Pouteria buenaventurensis* (Aubrév.) Pilz	±	A	Diffuse-porous	Low	2	Cross-section	Biogeographic *Chocó*
Sapotaceae	*Pouteria engleri* Eyma	+	A	Diffuse-porous	Medium	1	Cross-section	Wide distribution
Sapotaceae	*Pouteria collina* (Little) T.D. Penn.	+	A, B	Diffuse-porous	Medium	1	Cross-section	Biogeographic *Chocó*
Sapotaceae	*Pouteria eugeniifolia* (Pierre) Baehni	–	A	Diffuse-porous	None	3	Cross-section	Wide distribution
Sapotaceae	*Pouteria* sp.	–	A, B	Diffuse-porous	None	1	Cross-section	Wide distribution

(continued)

Table 5.1 (continued)

Family	Species	Tree-ring visibility	Wood anatomy	Vessel type	Potential in dendrochronology	Number of trees	Type of samples	Distribution
Sapotaceae	*Pouteria* sp.	±	A, B	Diffuse-porous	Low	2	Cross-section	Wide distribution
Simaroubaceae	*Simarouba amara* Aubl.	±	A, B	Diffuse-porous	Low	13	Cross-section	Wide distribution
Urticaceae	*Cecropia virgusa* Cuatrec.	±	A	Diffuse-porous	Medium	15	Cross-section	Biogeographic *Chocó*
Urticaceae	*Coussapoa valaria* C.C. Berg	±	A, B	Diffuse-porous	Low	1	Cross-section	Endemic
Urticaceae	*Pourouma bicolor* subsp. *bicolor* Mart.	±	A, B	Diffuse-porous	Low	5	Cross-section	Wide distribution
Vochysiaceae	*Qualea lineata* Stafleu	+	A	Diffuse-porous	Medium	7	Cross-section	Endemic
Vochysiaceae	*Vochysia artantha* Stafleu	–	A	Diffuse-porous	None	5	Cross-section	Endemic
Vochysiaceae	*Vochysia pacifica* Cuatrec.	±	A, B	Diffuse-porous	None	2	Cross-section	Endemic
Vochysiaceae	*Vochysia* sp.	±	A, B	Diffuse-porous	None	4	Cross-section	Wide distribution

Visibility: well defined (+), poorly defined (±), absent (−). Anatomical structure: density changes due to increased fiber wall thickness (A), the presence of marginal bands of axial parenchyma (B), differences in diameter, distribution, and grouping of pores that delimited the ring (C). Porosity: diffuse-porous, semi-ring-porous or ring-porous. The possibility of dating: high, medium, low. [1]Planted tree with known age

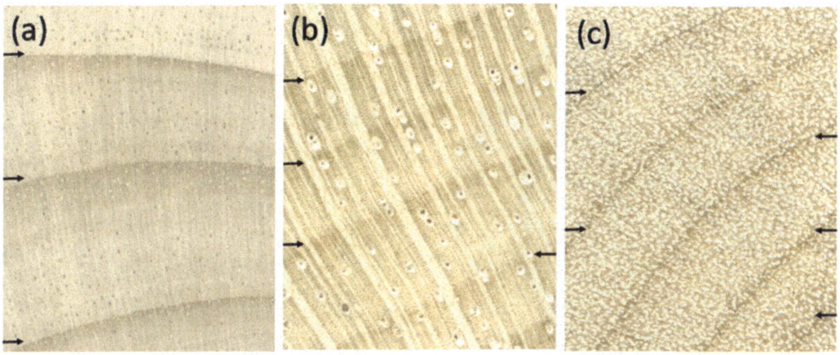

Fig. 5.2 Species of trees sampled in the Lower *Calima* River Basin, Biogeographic *Chocó* Region, with growth rings that were defined by increases in fiber wall thickness (A): (**a**) *Apeiba macropetala*, (**b**) *Phragmotheca siderosa*, (**c**) *Posoqueria latifolia*. The arrows indicate the growth rings

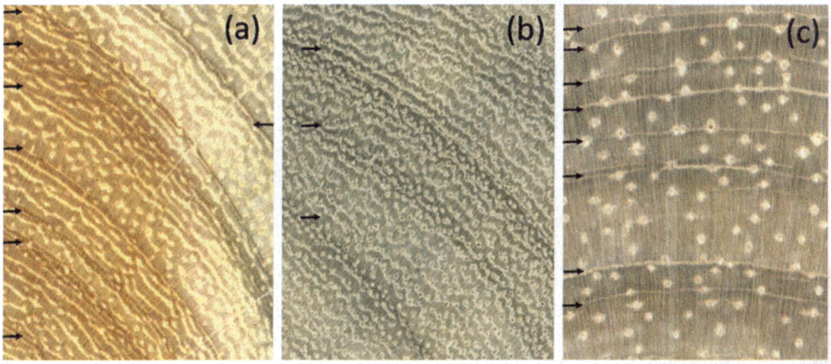

Fig. 5.3 Species of trees sampled in the Lower *Calima* River Basin, Biogeographic *Chocó* Region, with (A) growth rings defined by increases in fiber wall thickness and (B) parenchyma bands: (**a**) *Clarisia racemosa*, (**b**) *Zanthoxylum riedelianum*, (**c**) *Coussapoa valaria*. Arrows indicate growth rings

2004; Vaganov et al. 2006; Anchukaitis et al. 2008; Speer 2010; Hughes et al. 2011; Thomas 2014; Hu and Riveros 2016; Hirons and Thomas 2018; Nakai et al. 2018), in this non-seasonal hyper-humid tropical environment, many tree species have growth rings. Surprisingly, 66 species out of 81 sampled species (82%) showed growth rings that were visible with the naked eye (Table 5.1). Also, these species occurred in a rainier, non-seasonal environment than that in which any other researchers had previously attempted to study tropical growth rings (Table 5.2).

Among the 60 pantropical study areas cited in Table 5.2, which included trees that grew from the lowlands to high tropical mountains, and with all types of seasonality including droughts, floods, variation in temperature, and various combinations of those variables, these study areas included a broad latitudinal gradient going from the subtropics of the Northern Hemisphere to the subtropics of the Southern Hemisphere. In all of them, the percentage of trees with growth rings ranged from 3% to 100%. Our study ranked 17th in percentage of tree species with growth rings,

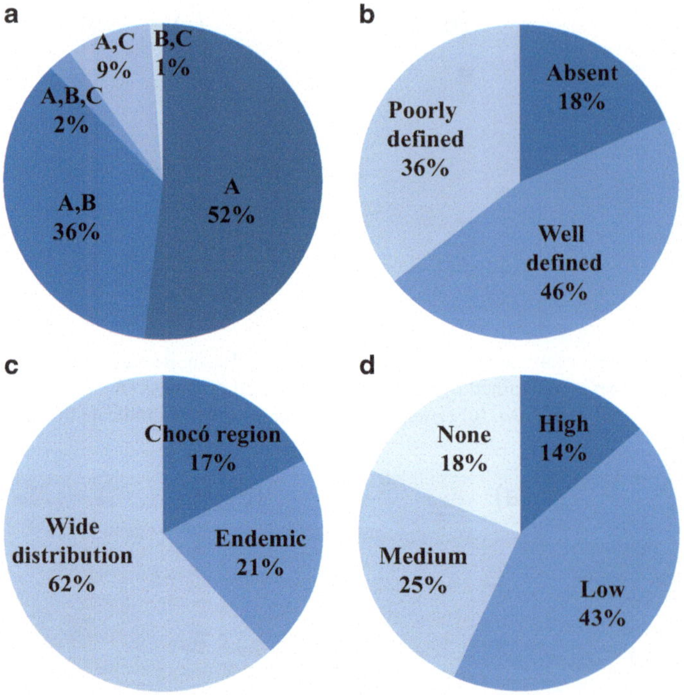

Fig. 5.4 (**a**) Percentage of tree species sampled in the Lower *Calima* River Basin, Biogeographic *Chocó* Region, with growth rings absent or present. (**b**) Percentage of growth rings according to their type: changes in density due to an increase in fiber wall thickness (A); the presence of marginal bands of axial parenchyma (B); differences in diameter, distribution, and grouping of pores that delimited the ring (C); and combinations of the above. (**c**) Percentage of species with growth rings classified according to their distribution range. (**d**) Classification of species with growth rings that could be dated

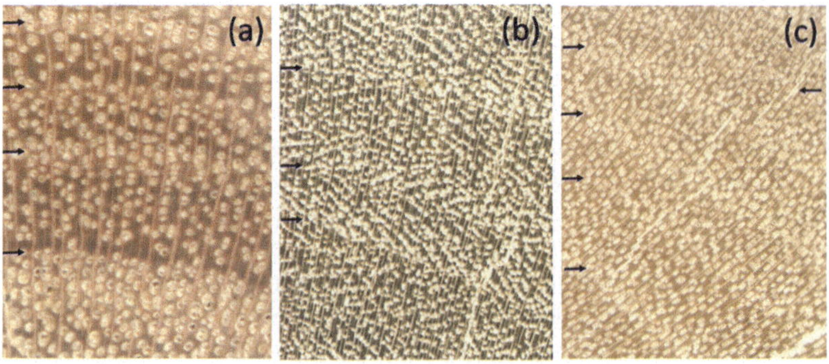

Fig. 5.5 Semi-ring-porous trees species sampled in the Lower *Calima* River Basin, Biogeographic *Chocó* Region, (Type C) or combined with another anatomical trait. (**a**) *Clusia leptanthera*, (**b**) *Tapura costata*, (**c**) *Pterandra ultramonanta*. Arrows indicate growth rings

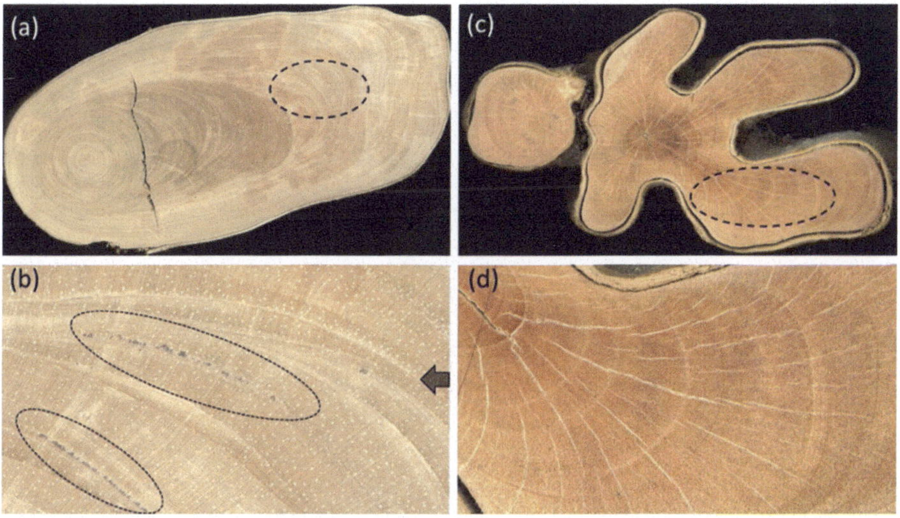

Fig. 5.6 Irregularities in growth rings due to competition and architecture that was present in some tree species sampled in the Lower *Calima* River Basin, Biogeographic *Chocó* Region: (**a**) irregular very eccentric discs with the pith very displaced in *Couratari guianenesis*; (**b**) tangential discontinuity of the growth rings, and wedging rings (fused), followed by traumatic tissue in *Couratari guianenesis*; (**c**) irregular wooden disc typical of *Aspidosperma excelsum*; (**d**) faint growth rings in *Aspidosperma excelsum*

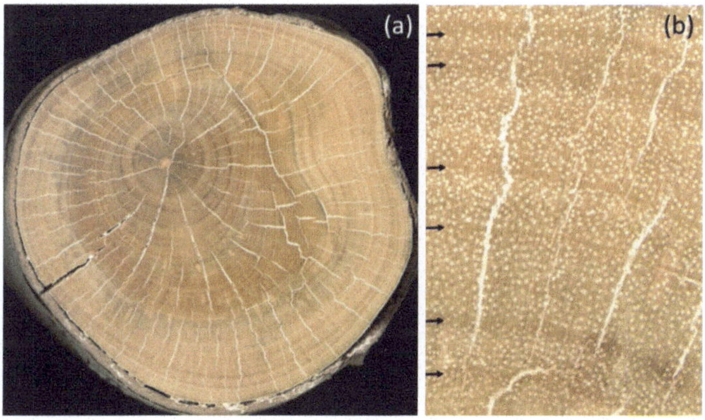

Fig. 5.7 Growth rings of *Goupia glabra*. (**a**) Complete cross-sectional disc with growth rings as seen with the naked eye. (**b**) Growth rings in a rectangular cross-section of 1.5 cm × 2.3 cm

surpassing most of the studies cited in Table 5.2. Nevertheless, the results of this chapter are one of only two without any seasonality. Although 16 studies reviewed in Table 5.2 exceeded the percentage of trees with growth rings reported here (Table 5.2), the real potential of trees with growth rings in the Low *Calima* River Basin is even more considerable as discussed below.

Table 5.2 Percentage of growth rings in tropical trees that grow in diverse environments with seasonality

Code	Species with growth rings (%)[a]	Total species studied (n)	Site	Mean rainfall (mm)	Mean temperature (°C)	Type of seasonality	Reference
1	100	58	Pantanal, Brazilian Mato Grosso, South America	1150	23–26	Rainfall, temperature. flooding	Schöngart et al. (2011)
2	100	29	Amazonian *Terra Firme* forests, South America	2000–3000	26–27	Rainfall	Vetter and Botosso (1989)
3	100	28	Pantanal, Brazilian Mato Grosso, South America	1182	20–31[b]	Rainfall, temperature, flooding	Póvoa de Mattos (1999)
4	100	17	*Terra Firme,* flooded and swampy Amazonian forests, Brazil, South America	2100–2810	27	Rainfall	Cintra et al. (2013)
5	94	48	Brazilian *Cerrado*, São Paulo, South America	1450	20–26[b]	Rainfall, temperature, flooding	Marcati et al. (2006)
6	92	24	Seasonaly semi-deciduous forests Saoutheast Brazil, South America	1370	15–23[b]	Rainfall, temperature	Lisi et al. (2008)
7	94	67	Pantanal, Brazilian Mato Grosso, South America	1150	21–31[b]	Rainfall, temperature, flooding	de Arruda et al. (2017)
8	93	27	Rema-Kalenga Wildlife Sanctuary, Bangladesh, South Asia	2370	25	Rainfall, temperature	Islam et al. (2018)
9	90	57	Diverse ecosystems, Colombia, South America	1500–4000	<24	Rainfall	Vásquez and Ramírez (2005)
10	90	27	Arid and dry savanna trees, Africa	500–1000	28	Rainfall	Fahn et al. (1981)
11	89	19	Tana River, Kenya, Africa	370	28	Flooding	Maingi (2006)
12	88	78	Floodplain forests, Amazon, Brazil, South America	1500–3000	26	Flooding	Worbes and Fichtler (2010)

13	86	183	Seasonal deciduous forests of the north and center of Peru	730–2,680[c]	16–26[c]	Rainfall	Marcelo (2017)
14	87	52	Yucatan, Mexico, North America	1036	25	Rainfall	Roig et al. (2005)
15	84	64	Amazonian *Terra firme* forests, Pará, South America	2000–3000	26	Rainfall	Trevizor (2011)
16	85	177	Central and South Brazil, South America	1500–2000	24–27	Rainfall, temperature	Alvim (1964)
17	**82**	**81**	**Biogeographic *Chocó* Region, Colombia, South América**	**7219**	**25.9**	**None**	**This chapter**
18	82	22	South East Cameroon, Africa	4082	27	Rainfall	Groenendijk et al. (2014)
19	82	38	Kodagu, Karnacata, India, South Asia	2725	14–29[b]	Temperature	Nath et al. (2016)
20	81	26	Eastern Amazon Forests, Brazil, South America	1973	23	Rainfall	Campos et al. (2008)
21	79	226	Central Amazonian Forests, Brazil, South America	3400	26	Rainfall	Tanaka (2005)
22	79	24	North Minas Gerais, Brazil, South America	814	22–26[b]	Rainfall, temperature	Barbosa et al. (2018)
23	79	111	Moist forests (Atlantic Forest), Brazil, South America	1278	15–25[b]	Rainfall, temperature	Silva et al. (2017)
24	79	179	Central Amazon, Brazil, South America	3439	26	Rainfall	Tanaka (2005)
25	78	239	All Philippines, Southeast Asia	1500–3500	26–28[b]	Rainfall, temperature	Baguinon et al. (2008)
26	78	37	Caparo Forest Reserve, Venezuela, South America	1700	25	Rainfall	Worbes (1999)
27	77	90	Amazonian forests, Brazil, South America	1500–3000	26	Rainfall	Santini (2013)
28	77	65	All Sri Lanka, South Asia	900–5.000	26–28[b]	Rainfall	Baguinon et al. (2008)

(continued)

Table 5.2 (continued)

Code	Species with growth rings (%)[a]	Total species studied (n)	Site	Mean rainfall (mm)	Mean temperature (°C)	Type of seasonality	Reference
29	77	13	Misiones, North Argentina, Subtropics, South America	1700	15–25[b]	Temperature	Boninsegna et al. (1989)
30	76	21	Lowland dry forests, Oaxaca, South of Mexico, North America	930	26	Rainfall	Brienen et al. (2009)
31	75	79	Moist forests, Los Tuxtlas Biosphere Reserve, Mexico, North America	1300–4200	24–27[b]	Rainfall, temperature	Gutiérrez (2014)
32	70	56	All Ecuador, South America	1000–3000	>12	Rainfall	MAE-FAO (2014)
33	70	27	Central Kalimantan, (Indonesia) South Borneo, Southeast Asia	2572	27	Rainfall, flooding	Worbes et al. (2017)
34	71	24	Moist forests from Africa	2000–3000	27	Rainfall	Fahn et al. (1981)
35	66	61	South Brazil, South America	1000–1500	14–23[b]	Rainfall, temperature	Alves and Angyalossy-Alfonso (2000)
65	64	135	Bolivia, South America	1000–4000	<20	–	Gutiérrez and Silva (2002)
37	61	26	Malaysia, Southwest Asia	1800–3000	26–27[b]	Rainfall	Abdul-Azim and Okada (2014)
38	59	97	Florida, USA, Subtropical North America	1500	10–26[b]	Temperature	Tomlinson and Craichead (1972)
39	58	38	Suriname, South America	2209	26	Rainfall	Lotfiomran and Köhl (2017)
40	57	60	Amazon rainforests from Brazil, South America	2000–3000	27	Rainfall	Alvim (1964)
41	56	103	Democratic Republic of Congo, Africa	1200–1700	24–26[b]	Rainfall	Tarelkin et al. (2016)

42	54	91	All Thailand, Southeast Asia	1000–2200	26–31[b]	Rainfall, temperature	Baguinon et al. (2008)
43	53	176	Southeastern Brazil, South America	1500–2000	14–23[b]	Raifall, temperature	Alves and Angyalossy-Alfonso (2000)
44	53	108	Venezuela, South America	500–4000	>12	Rainfall	León (2014)
45	52	217	North Brazil (Amazonia), South America	2000–3000	26	Flooding	Alves and Angyalossy-Alfonso (2000)
46	52	85	Lowland moist forest, Java Island, Indonesia, tropical Asia	2000–4000	28	Rainfall	Fahn et al. (1981)
47	52	50	Moist forests of Chiapas, Mexico, North America	2800	23	Rainfall	Brienen et al. (2009)
48	52	21	Lowland deciduous forest, Mexico, North America	700–1000	22–26[b]	Rainfall, temperature	Ramírez-Martínez et al. (2017)
49	50	80	Peru, Moist Central Forest, South America	2325 2250	24–27[b] 15–19[b]	Rainfall, temperature Rainfall, temperature	Beltran and Valencia (2013)
50	47	38	Northeast Brazil, South America	500–1500	26	Rainfall, flooding	Alves and Angyalossy-Alfonso (2000)
51	43	43	Premontane and lowland forests, Colombia, South America	1500–4000	>18	–	Arévalo and Londoño (2005)
52	40	25	Central-western Brazil, *Cerrado,* South America	1000–2000	20–26[b]	Raifall, temperature	Alves and Angyalossy-Alfonso (2000)
53	38	541	Pakistan, India, Sri Lanka, and Burma	—	—	—	Wheeler et al. (2007)
54	38	29	Dicot trees of a mountain moist forest, Mexico, North America	1200–1700	17	Rainfall	Aguilar-Rodríguez and Barajas-Morales (2005)
55	28	708	Dicot trees tropical Africa	—	—		Wheeler et al. (2007)

(continued)

Table 5.2 (continued)

Code	Species with growth rings (%)[a]	Total species studied (n)	Site	Mean rainfall (mm)	Mean temperature (°C)	Type of seasonality	Reference
56	24	1273	Dicot trees Southeast Asia and the Pacific	—	—	—	Wheeler et al. (2007)
57	22	110	Wet non-seasonal forests, Costa Rica, Central America	6000	25	No one	Araya (2012)
58	17	1695	Dicont trees Neotropics	—	—	—	Wheeler et al. (2007)
59	10	70	Sahel Savannah, Sudan, Africa	250–1100	28	Rainfall	Tarhule and Hughes (2002)
60	3	107	French Guiana (Guyana), South America	3200	29	Rainfall	Miller and Détienne (2001)

In bold are the results of this chapter
[a]When two or more studies have the same percentage, the one with the largest sample (n) is presented first
[b]The mean monthly temperatures of the coldest and warmest months, respectively
[c]Data taken in seven seasonal deciduous forests between 230 and 2200 m asl with average annual temperatures between 26 °C and 16 °C, respectively. The mean annual rainfalls vary from 730 to 2680 mm

5.3.1 Biodiversity of the Chocó Region

The study area of this chapter is known for a very high richness and species diversity of tree species. Faber-Langendoen and Gentry (1991) found in 1 ha of this same area 241 dicot tree species with a diameter at breast height (dbh) ≥10 cm: According to the mixing ratio, per each one tree species sampled there are only 2.4 individuals. In the sample from our study area, we found 2.9 individuals per species. If the Faber-Langendoen and Gentry (1991) sampling was random and representative for our study area, we estimate that in just 1 ha of forest in the Lower *Calima* River Basin, there would be about 113 tree species with well-defined rings (0.47 × 241). If only a small fraction of these tree species have annual growth rings, or with any other constant periodicity, we would have discovered a gold mine of potential dendrochronological species in the non-seasonal tropical wet forests of the Biogeographic *Chocó* Region. Colombia, which contains 5746 tree species, has more species of trees than any country in the world after Brazil. However, on a per square kilometer basis, Colombia has many more species than Brazil (Beech et al. 2017).

The Biogeographic *Chocó* Region is one of the most crucial tree diversity and endemism hot spots in the world (Prance 1982). Among the species with great dendrochronological potential, the following stand out due to the clarity of their rings: *Huberodendron patinoi*, *Cariniana pyriformis*, *Mabea* sp. *Inga acreana*, *Tachigali colombiana*, *Goupia glabra*, *Apeiba macropetala*, *Humiriastrum procerum*, *Nectandra* sp., *Ocotea* sp., *Clarisia racemosa*, *Otoba latialata*, *Virola dixonii*, *Posoqueria latifolia*, *Pouteria englerii*, *Pouteria collina*, and *Qualea lineata*. Nevertheless, the high diversity of tree species found in our study area has an Achilles heel: necessarily, the populations are very small difficulting to obtain enough replicas for most species.

5.3.2 About Tree-Ring Formation in Tropical Hyper-humid Regions

Based on the data in Table 5.2, and contrary to the prevailing hypothesis that periodic droughts or floods condition growth-ring formation in tropical trees, the percentage of trees with growth rings does not seem to depend only on these environmental variables. For instance, Tarelkin et al. (2016) in Democratic Republic of Congo, found no differences in the percentages of trees with distinct, indistinct, or absent growth rings between two forests with contrasts both in precipitation amount and seasonal precipitation patterns. See more cases in Table 5.2.

In our study, the hypothesis of the existence of growth rings in deciduous or semi-deciduous tree species (Borchert 1999; Lisi et al. 2008; Nath et al. 2016; Shimamoto et al. 2016) was applied only to *C. odorata*, *S. parahyba*, *D. panamensis*, *Goupia glabra*, *Couratari guianensis*, *Tachigali colombiana*, and *Humiriastrum procerum*, which are deciduous, and for *C. pyriformis*, which is a brevi-deciduous species. Together they represent less than 12% of the studied species that exhibited growth rings in this study (Table 5.1). On the contrary, our results were more in agreement with other authors (Worbes 1999; Callado et al. 2001; Carlquist 2001; Fichtler et al. 2003; Aguilar-Rodríguez and Barajas-Morales 2005; Evert 2006; Tarelkin et al. 2016), who did not find an association between deciduousness and the formation of growth rings in tropical trees. In fact, according to these authors, neither do all deciduous trees have growth rings nor do all evergreens lack growth rings.

Based on the hypothesis related to phytogeographic origin, in which growth rings are more common in species from a boreal origin (Laurasia) (Fahn 1967; Aguilar-Rodríguez and Barajas-Morales 2005; Ramírez-Martínez et al. 2017), we found that all species of plant families with an assigned origin, except for two, were from Gondwana and centered on the Amazonian region. The two families from Laurasia were Lineaceae and Salicaceae (Gentry 1982).

5.3.3 Dendrochronological Potential Studies Around Tropics: Unclear Pattern

We show a comparison from our study results with other studies from pantropical regions in Tables 5.2 and 5.3. Some examples illustrate the inexplicable differences found by the authors cited in Table 5.2. The only other case in wet tropical forests without any seasonality was that of Araya (2012), who reported 22% with growth rings, almost four times less than in our study.

Some results emerge from Table 5.2: (1) There are growth rings in trees in all tropical forests around the world; (2) there were no clear trends for ring formation in terms of dry, moist, or wet climates, seasonality of rainfall, flooding, temperature, or any combination of these variables; and (3) there was no trend with the latitudinal distance from the Equatorial line to the North and South subtropics.

It is very striking that Miller and Détienne (2001) did not find growth rings in many tree species in French Guiana; by contrast, other researchers working in the same region, and other areas of the Neotropics, have reported annual growth rings in these same species. To cite a few examples, growth rings were found by several authors for *Hymenaea courbaril*, *Carapa guianensis*, *Couratari guianensis*, *Dipteryx odorata*, *Goupia glabra*, *Jacaranda copaia*, *Simarouba amara*, *Symphonia globulifera*, and many other species as cited by Brienen et al. (2016) and by Schöngart et al. (2017).

Curiously, when we analyzed the disciplinary background of investigators, a pattern emerged. Most studies in Table 5.2 that reported >70% of trees with growth rings were tropical dendrochronologists who did not follow the IAWA Committee (1989) classification of growth rings. In sharp contrast, all studies that reported <50% were performed by wood anatomists who followed this classification.

Table 5.3 Species of this chapter reported with annual rings in other regions and in this chapter (in bold)

Species	Visibility		References	Annual rings and visibility	References
	This chapter	Other studies			
Apeiba macropetala[a]	+	±	Vásquez and Ramírez (2005)	+	Cintra et al. (2013)
		–	Araya (2012)		
Cariniana pyriformis	+	+	Arévalo and Londoño (2005)	+	Moreno and del Valle (2014)
		–	Sudworth and Mell (1911)	+	**This chapter**
		+	Vásquez and Ramírez (2005)		
Castilla elastica	+	–	Araya (2012)	+	Brienen et al. (2009)
Cedrela odorata	±	+	Alves and Angyalossy-Alfonso (2000)	+	Baker et al. (2017)
		+	Miller and Détienne (2001)	+	Köhl et al. (2017)
		+	Arévalo and Londoño (2005)	+	Lotfiomran and Köhl (2017)
		+	Vásquez and Ramírez (2005)	+	Rozendaal (2010)
				±	**This chapter**
				+	Worbes (2002)
Clarisia racemosa	+	–	Alves and Angyalossy-Alfonso (2000)	+	Rozendaal (2010)
		+	Beltran and Valencia (2013)	+	Soliz-Gamboa et al. (2011)
		±	Gutiérrez (2014)	+	Tanaka (2005)
		+	Nahuz et al. (2013)		
		+	Santini (2013)		
Couratari guianensis	+	–	Alves and Angyalossy-Alfonso (2000)	+	Morel (2013)
		+	León (2008, 2014)		
		+	Miller and Détienne (2001)		
		–	Nahuz et al. (2013)		
		+	Santini (2013)		
		±	Trevizor (2011)		
		±	Vásquez and Ramírez (2005)		

(continued)

Table 5.3 (continued)

Species	Visibility		References	Annual rings and visibility	References
	This chapter	Other studies			
Dipteryx panamensis	+	–	Miller and Détienne (2001)	+	Fichtler et al. (2003)
		±	Rojas (2000)		
		+	Vásquez and Ramírez (2005)		
		+	Wiedenhoeft (2011)		
Goupia glabra	+	+	Vetter and Botosso (1989)	±	Détienne and Barbier (1988)
		–	Miller and Détienne (2001)	+	Vetter and Botosso (1989)
					This chapter
		+	Vásquez and Ramírez (2005)	+	Köhl et al. (2017)
		±	Trevizor (2011)	+	Lotfiomran and Köhl (2017)
		–	Pons and Helle (2011)	+	Lopes de Oliveira (2010)
		±	Nahuz et al. (2013)	+	Cintra et al. (2013)
		–	Santini (2013)	+	Oliveira et al. (2014)
				+	**This chapter**
Hevea brasiliensis	+	+	Beltran and Valencia (2013)	+	Jalil et al. (1998)
		±	Gutiérrez (2014)	+	Ohashi et al. (2001)
Inga acreana	+	+	Wiedenhoeft (2011)	+	Callado et al. (2001)
					This chapter
Inga rubiginosa	+	+	Alves and Angyalossy-Alfonso (2000)	+	Tanaka (2005)
Osteophloeum platyspermum	+	+	Marcelo (2017)	+	Lopes de Oliveira (2010)
		–	MAE-FAO (2014)		
Pentaclethra macroloba	+	+	León (2008)	+	Hazlett (1987)[b]
		+	Araya (2012)		Fichtler et al. (2003)
Pourouma bicolor	±	±	Tanaka (2005)	±	Tanaka (2005)
				±	**This chapter**

(continued)

Table 5.3 (continued)

Species	Visibility		References	Annual rings and visibility	References
	This chapter	Other studies			
Schizolobium parahyba	+	+	Gutiérrez (2014)	+	Brienen et al. (2009)
		+	Vásquez and Ramírez (2005)	+	Lisi et al. (2008)
				+	Marcati et al. (2008)
Simarouba amara	±	–	Miller and Détienne (2001)	±	Détienne and Barbier (1988)
		+	Gutiérrez and Silva (2002)	±	Fichtler et al. (2003)
		–	Araya (2012)	+	López-Ayala et al. (2006)
		–	MAE-FAO (2014)		
		±	Santini (2013)		
		±	Trevizor (2011)		
		+	Vásquez and Ramírez (2005)		
Symphonia globulifera	+	+	Arévalo and Londoño (2005)	±	Détienne and Barbier (1988)
		–	Miller and Détienne (2001)	±	Détienne (1995)
		–	Araya (2012)	+	Callado et al. (2001)
		+	Santini (2013)	±	Brienen et al. (2009)
		±	Trevizor (2011)		

[a]In Tropicos (http://www.tropicos.org/), *Apeiba echinata* var. *macropetala* is a synonym for *A. macropetala*

[b]Although Hazlett (1987) did not study the growth rings of this species, he demonstrated by means of dendrometer bands that annually it considerably reduced its diameter growth during the rainiest months in a wet non-seasonal climate of Costa Rica

Silva et al. (2019) compiled many proposals for the classification of the growth rings. But, the one most used by wood anatomists appeared to be that of the IAWA Committee, although they usually made modifications to it.

The IAWA Committee (1989) classifies the growth rings from dicot wood into two types: (1) distinct growth ring boundaries and (2) indistinct or absent growth-ring boundaries. The limits of the rings are distinct if "there is an abrupt structural change in the limits between them, usually including a change in the thickness of the cell wall of the fibers and/or in the radial diameter of the fibers." The separation between these two groups is highly subjective, but the creation of a single group for indistinct and absent and the emphasis on density changes can lead to classifying species with growth rings, even annual rings, within the second group. But, also, the emphasis given to increased fiber density tends to classify tropical species that have subtle, though perceptible rings, boundaries into the second group, as is the case in the very narrow parenchyma bands typical in the wood of tropical legumes (Worbes 1995). This family of plants has the highest number of species with annual rings (Brienen et al. 2016; Schöngart et al. 2017).

Most of the wood anatomists cited in Tables 5.2 and 5.3 followed the IAWA Committee's standards, but in reality, few followed them literally, and they modified them to create three groups, (1) distinct, (2) indistinct, or (3) absent, although they may have used other terms that are roughly equivalent (Vásquez and Ramírez 2005; Marcati et al. 2006; Baguinon et al. 2008; Santini 2013; León 2014; Silva et al. 2017; Tarelkin et al. 2016; Islam et al. 2018). Most tropical dendrochronologists follow the classification of Worbes (1995), which covers four basic types of rings: (1) reduction in diameter and increase in fiber density, (2) one or more marginal parenchyma bands, (3) parenchyma bands and fibers of different diameters, or (4) varying frequency and width of vessels within the rings. Combinations among these types often occur.

Dendrochronologists have found annual rings in virtually all dicot trees from temperate and cold regions (Schweingruber 1988, 2007; Speer 2010; D'Arrigo et al. 2014; Schweingruber and Börner 2018), except in the Mediterranean where many irregularities occur (Silva et al. 2019). However, Wheeler et al. (2007) who used the IAWA Committee's standards report for the regions of Europe (excluding the Mediterranean), Temperate Asia, and North America, 85%, 84%, and 81% of trees, respectively, with distinct rings. The remaining species were in that sort of black box: indistinct or absent. Note that similar and even higher percentages are in Table 5.2 in the tropics. It seems, therefore, that even in temperate and cold regions, the IAWA Committee's criteria tend to underestimate the presence of growth rings in dicot species.

Some researchers who have studied the growth rates of tropical trees and their climatic relationships using stable isotopes of ^{13}C and ^{18}O justified their use for the lack of growth rings in these species (Evans and Schrag 2004; Poussart et al. 2004; Hietz et al. 2005; Pons and Helle 2011; Hu and Riveros 2016; Nakai et al. 2018). However, most species studied by these authors were repeatedly reported by several tropical dendrochronologists to have annual growth rings. Such is the case for *Cordia alliodora*, *Carapa guianensis*, *S. amara*, *C. odorata*, and *Swietenia macrophylla* (Brienen et al. 2016; Schöngart et al. 2017), or at least they had growth rings as in *Samanea saman* (Skolmen 1974; de Arruda et al. 2017) and *Hieronyma alchorneoides* (León 2014, and this chapter in Table 5.1).

5.3.4 Species in This Chapter with Annual Rings Reported in Other Regions

Table 5.3 presents the species of this chapter with annual rings and their visibility reported by different authors. Thirty percent of the species from this chapter have been reported previously with annual growth rings in other regions. It suggests that they probably also form annual growth rings in the area of our study in Colombia. At the time of writing, we had only received radiocarbon analyses of the growth rings of six species, all with annual rings; these species were *Cariniana pyriformis* (171 y), *G. glabra* (70 y), *Pourouma bicolor* (36 y), *Inga sp* (62 y), *Mabea sp* (80 y), and *Inga acreana* (100 y). Growth rings of the last three species had not been reported before. Additionally, *C. odorata*, one of the trees sampled, had a known planting date where the number of rings coincided with the years since planting (9 y). That is, we found numerous species with dendrochronological potential in the rainiest region of the Americas where there is no hydric seasonality, which marks a milestone that contradicts the long-held prejudice of many dendrochronologists and wood anatomists that there are no growth rings in trees from these environments. This finding opens the frontiers of dendrochronology to these tropical environments that closed until now because of hypothetical and unjustified causalities, which had not been substantiated by empirical observations in situ, as prescribed by science. In this chapter, prejudices have been replaced by rigorous straightforward observations.

Using dendrometer bands in a non-seasonal wet forest at La Selva, Costa Rica, with an annual precipitation of about 4000 mm, O'Brien et al. (2008) found an annual diameter growth periodicity in both a deciduous (*D. panamensis*) and an evergreen species (*S. amara*), although not in the evergreen species *P. macroloba* (Tables 5.1 and 5.3). These three species had been reported to have annual rings in the same forest (Fichtler et al. 2003). Perhaps the growth periodicity of *P. macroloba* was so subtle that it is undetected by the instrument used by O'Brien et al. (2008).

The evergreen *H. alchorneoides* did not show annual periodicity in the diameter growth by O'Brien et al. (2008), but in this chapter and also in León (2014), this species exhibited growth rings (Table 5.1). Something similar could have happened with *P. macroloba*, but it seems that, until now, the periodicity of its growth rings had never studied. O'Brien et al. (2008) could not identify a single triggering factor for the seasonal periodicity, phenology, and stem increments observed at La Selva. Hypothetically, they thought that deciduousness and periodic growth of species such as *D. panamensis* might have resulted from phylogenetic inertia in species that have their center of diversity in the Central and Northern Amazon, where the climate is more seasonal.

In sharp contrast to O'Brien et al. (2008), Hazlett (1987), and Breitsprecher and Bethel (1990) found evidence of periodic growth in *P. macroloba* in the same study area. Hazlett (1987) used dendrometer bands, and Breitsprecher and Bethel (1990) used radial dendrometers with data loggers that stored hourly measurements. Hazlett (1987) reported that *P. macroloba* grew at faster rates during the less rainy season of the year. But Breitsprecher and Bethel (1990) found that during these months, *P. macroloba* went into dormancy both in well-drained and in swamp forests. Perhaps because they used more precise instruments with a much higher temporal resolution, Breitsprecher and Bethel (1990) found a dormancy period in this species. *D. panamensis* was dormant during the less rainy season in 2 successive years, but only one tree was studied (Breitsprecher and Bethel 1990). This dormancy was undetected by O'Brien et al. (2008). Breitsprecher and Bethel (1990) suggested that variation in the daily amount of incoming solar radiation may be an environmental trigger of the annual growth periodicity at La Selva. Hazlett (1987), although not ruling out the annual endogenous growth control in *P. macroloba*, emphasized that during the less rainy period, there was more illumination and higher temperatures, increasing growth rates that depended mainly on current photosynthesis.

In French Guiana, under a marked annual drought and 2700 mm mean annual rainfall, Détienne and Barbier (1988) studied the diameter growth for 5 consecutive years using dendrometer bands and by marking the cambium with Mariaux windows to determine the periodicity of the growth rings in *S. amara*, *G. glabra*, and *S. globulifera*. They found a high correspondence between the monthly precipitation through the years and the monthly growth rates in *S. amara* and *G. glabra*. The relationship was less evident in *S. globulifera*. Growth rings were annual in all three species, but they were difficult to observe, especially in *G. glabra*. The authors attributed the formation of annual rings to annual droughts. By contrast, in flood-prone forests of the Mata Atlantica in Brazil, Callado et al. (2001) attributed the formation of annual rings in *S. globulifera* to endogenous periodicity.

Considering that the four species discussed in the last three paragraphs (*D. panamensis*, *S. amara*, *P. macroloba*, and *S. globulifera*) have been described previously as with annual growth rings and for which several environmental factors responsible for their formation are suggested in Table 5.3, how valid can these proposed variables for the study sites in Costa Rica at 10°26'N and in French Guiana at 5°04'N be for our study area? Our area of study is much closer to the Equatorial line (3°55'N) and, therefore, the length of the day, temperature, and sunshine irradiance are more constant throughout the year. Our study area receives almost twice the precipitation of the site in Costa Rica and nearly triples that of French Guiana. This last site also has a marked dry period, non-existent in our study area. However, all of those species have growth rings in our study area (Tables 5.1 and 5.3). Both these species, and the others in Table 5.1, respond to very subtle environmental cues, or the hypothesis for the origin of their growth rings must be modified totally. If it has accepted that the primary growth of some architectural models of tropical trees has genetically determined periodicity, endogenous secondary growth periodicity could be an alternative hypothesis for some species included in this chapter. Silva et al. (2019) proposed this classification for the growth rings: (1) obligatorily with growth rings, developing rings independently of environmental conditions; (2) facultative, growing rings depending on the variation of one or more environmental factors; and (3) obligatorily without growth rings, never forming growth rings, independently of environmental conditions. If the trees species with growth rings in Table 5.1 are not sensitive to subtle environmental cues, we must conclude that these species are obligatory with growth rings. Precise measurements with dendrometer bands suggest that there are ever-growing species that form annual rings, and so, the cambium does not enter into dormancy. It is enough for tree-ring formation in these species that the cambium reduces its activity substantially, but without a complete halt (Silva et al. 2019). This seems to be the case in *C. insignis*, *C. obtusifolia*, *Minquartia guianensis*, and *Simarouba amara* (O'Brien et al. 2008); *Schefflera morototoni*,

G. glabra, *S. globulifera*, and *S. amara* (Détienne and Barbier 1988; Morel 2013); *Shorea patoiensis*, *S. pinanga*, and *S. dasyphylla* (Fujii et al. 1999); *S. leprosula*, *S. acuminata*, *S. parvifolia*, and *Dryobalanops sumatrensis* (Sass et al. 1995; Wang and Hamzah 2018); and young trees of *Terminalia superba* (de Faÿ 1992). Perhaps this is one of the reasons why some of these species have faint rings. For this reason, a third group of species with periodic growth, but without cambial dormancy, should exist.

In 11 of the 17 species found with annual rings by dendrochronologists (Table 5.3), and with growth rings and annual growth rings in this chapter, some wood anatomists did not find growth rings at all, or they grouped them as indistinct or absent. Two US Forest Service wood anatomists, who studied the anatomy of *C. pyriformis* and *Swietenia macrophylla*, did not observe growth rings in these species (Sudworth and Mell 1911). Both species have been described repeatedly by other anatomists as having growth rings (Table 5.3) and by dendrochronologists with annual growth rings marked by increased fiber density (Brienen et al. 2016; Schöngart et al. 2017). Détienne and Barbier (1988) described as annual the growth rings of *G. glabra*, *S. amara*, and *S. globolifera* in the French Guiana forests, and Détienne (1995) also confirmed the annual rings of *S. globulifera*. But Miller and Détienne (2001), with one author in common and in the same forest, did not find growth rings in any of the three species. How do we explain such contradictory results?

A way to avoid subjectivity in the identification of growth rings is to develop computer vision methods on high-resolution scanned images for the identification and measurement of several growth-ring variables. In this way, the types of rings could be determined automatically, and their variables measured quickly and safely. Not without challenges, the development of this technique would be the entrance to the fourth revolution of knowledge for dendrochronology and wood anatomy.

Dendrochronology would learn a great deal thoroughly investigating the trees of these non-seasonal, hyper-humid tropical climates. In highly complex tropical ecosystems, there may be emerging properties and nonlinearities that seem not contemplated in dendrochronological studies of the temperate region.

Dendrochronology in the temperate region operates linearly according to Liebig's law. This law states that growth is dictated not by total resources available, but by the scarcest resource (limiting factor). However, ecology teaches us that both the deficit of a growth factor and its excess produce tension and limit growth. Therefore, a law of the minimum and a law of the maximum should exist as limiting factors for growth. These two laws are involved in Shelford's tolerance law (Niinemets and Valladares 2008): If other factors are not limiting, as the magnitude of a limiting factor increases, organisms respond by increasing their growth, up to a limit at which the response saturates. Subsequent increases in this factor produce less and less growth because the organism is growing under incremental stresses. Consequently, Shelford's law is nonlinear and depicts a unimodal curve (Fig. 5.8). Under a non-seasonal, hyper-humid tropical climate, we cannot exclude the possibility that soil moisture exerts tension on tree growth of some species. Then, the correlation between tree growth rates and annual precipitation would be negative

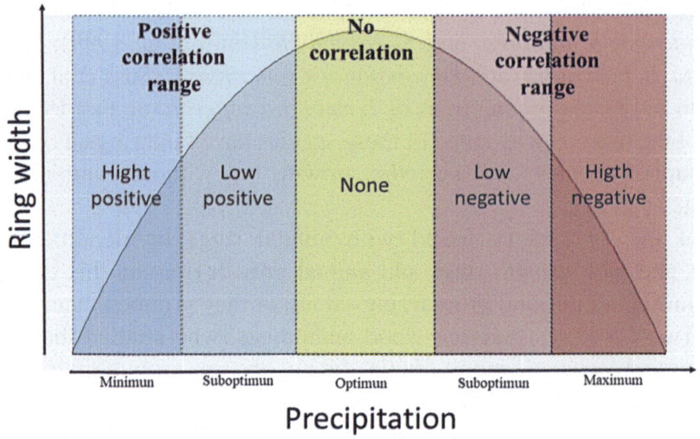

Fig. 5.8 It shows a representation of Shelford's tolerance law for dendrochronological applications. There is a limiting factor: by water deficit in the soil (left) and by an excess of water in the soil that increases the anoxic conditions for roots (right). Toward the center of the curve, the rings should be complacent. Consequently, the response of ring widths to soil moisture must be nonlinear

and not positive as the law of the minimum stipulates. In fact, this has already been demonstrated for temperature in tropical lowland trees where the correlation between ring width and mean annual temperature is often negative (Herrera and del Valle 2011; Ramírez and del Valle 2011, 2012; Moreno and del Valle 2014), in line with what we might call the law of the maximum.

However, the range of tolerance to soil moisture is different for each species, and their populations respond to the resources available depending on the assemblage of species in which they are inserted (Niinemets and Valladares 2008). Thus, for a given soil moisture content, different species may respond in different ways and even the same species: subsidy or stress. In temperate regions, the cambium of virtually all tree species synchronously enters into dormancy during autumn and is activated during spring. Therefore, there is minimal uncertainty about when trees grow or stop growing. Something similar is assumed to occur when there is hydric stress in the tropics. The cambium of trees in temperate regions responds to temperature and day length cues (Evert 2006). In seasonal tropical climates, trees respond to droughts or floods (Evert 2006; Brienen et al. 2016; Schöngart et al. 2017), although day length or irradiance may also be involved (Callado et al. 2001; Evert 2006; Silva et al. 2019). But, if in non-seasonal tropical wet climates trees form growth rings, as proved in this chapter, will there be synchrony between the different species? And to what cue or cues will these trees respond when they are only at 3°55′ from the Equatorial line?

We should not rule out that genetically controlled periodicities that regulate their frequencies may exist, at least in some species (Fahn 1967; Tomlinson and Craichead 1972; Tomlinson and Longman 1981; Jacoby 1989; Mushove et al. 1995; Callado et al. 2001; Silva et al. 2019). In perennials, the molecular and genetic factors that influence these clocks are gradually becoming disentangled with *Populus* as a prototype (Johansson 2010). Today it is known that these periodicities are controlled genetically (Lüttge and Hertel 2009; Johansson 2010). The input to the clock in *Populus* is mediated by the phytochrome-A photoreceptor to detect the circa-annual clock (Johansson 2010). A large number of genes have identified that expressed during the radial woody growth of plant stems (Liu et al. 2014). There are two types of periodicity: (1) synchronization of behavioral and physiological processes to cyclic environmental factors, or extrinsic adaptive value, and (2) coordination of internal metabolic processes, or intrinsic adaptive value (Sharma 2003). These periodicities are also called exogenously and endogenously controlled, respectively, by scientists who have developed the architectural models for trees (Hallé et al. 1978; Barthélémy et al. 1995; De Reffye et al. 2008; Nicolini et al. 2012; Palubicki 2013). In our opinion, tropical dendrochronology should devote to this task in the coming years.

Growth rings have both genetic and environmental origins, but little scientist knows about the contribution of each of these factors in the growth ring formation (Schweingruber 2007). In temperate regions, the sharp contrasts between temperatures and day lengths throughout the year make it challenging to discriminate environment influence from genetic factors. In non-seasonal, tropical hyper-humid climates that are very close to the Equatorial line, such as in our study area, mean monthly temperatures and day lengths are nearly constant the year round. Hypothetically, this circumstance suggests that if any other subtle environmental factor as cloudiness is not involved in these growth ring periodicities, genetic factors are the principal causes of the growth-ring formation in this environment.

5.4 Conclusions

We report here the existence of growth rings for trees growing under an extreme humid tropical environment without marked seasonality in precipitation, temperature, or solar irradiance. From a total of 81 trees studied, ~45% showed well-defined growth rings. The percentage of trees with growth rings was even higher than in other tropical regions with more seasonal climates. A survey of the literature suggests no clear relationship between climate and ring formation, but curiously, an association with ring-detection criteria was developed by different research communities. The large proportion of growth ring in the trees in this tropical non-seasonal region suggests that other environmental factors, or even from the genetic origin, may determine growth-ring development in trees.

Some questions to be answered by dendrochronology for non-seasonal hyper-humid tropical environments would be (1) what are the periodicities of the growth rings? (2) what is the relationship between high rainfall and growth rings? and (3) what other subtle seasonality or cues might be involved in these periodicities?

We believe that future researches in non-seasonal tropical regions such as those with extreme precipitation would help to understand better the factors controlling growth ring formation in tropical trees.

Acknowledgments This chapter was supported by project 4083 from the National University of Colombia, COLCIENCIAS with Project 1118-714-51372 and Max Planck Institute for Biogeochemistry. We acknowledge "Tropical Dendroecology Laboratory" of the Department of Forest Sciences of the National University of Colombia and MEDEL Herbarium. We thank "Pedro Antonio Pineda Tropical Forestry Center" of the University of Tolima and those people who help us during sampling: Amalia Forero, Faber Hernández, Sixto Cáseres, Andrés Caro, Diego Andrés David, and Jorge Mario Velez. Finally, we would like to thank Thomas A. Gavin, Professor Emeritus, Cornell University, for the help with editing this paper.

References

Abdul-Azim AA, Okada N (2014) Occurrence and anatomical features of growth rings in tropical rainforest trees in Peninsular Malaysia: a preliminary study. Tropics 23:15–31. https://doi.org/10.3759/tropics.23.15

Aguilar-Rodríguez S, Barajas-Morales J (2005) Anatomía de la madera de especies arbóteas de un bosque mesófilo de montaña: un enfoque ecológico-evolutivo. Bot Sci 77:51–58. https://doi.org/10.17129/botsci.1712

Álvarez E, Cayuela L, González-Caro S et al (2017) Forest biomass density across large climate gradients in northern South America is related to water availability but not with temperature. PLoS One 12:1–16. https://doi.org/10.1371/journal.pone.0171072

Alves ES, Angyalossy-Alfonso V (2000) Ecological trends in the wood anatomy of some Brazilian species. 1. Growth rings and vessels. IAWA J 21:3–30. https://doi.org/10.1078/0367-2530-0058

Alvim P (1964) Tree growth periodicity in tropical climates. In: The formation of wood in forest trees. Academic Press, New York, pp 479–495

Anchukaitis KJ, Evans MN, Wheelwright NT et al (2008) Stable isotope chronology and climate signal calibration in neotropical montane cloud forest trees. J Geophys Res Biogeosci 113:1–17. https://doi.org/10.1029/2007JG000613

Araya MÁ (2012) Manual para la identificación de maderas a nivel macroscópico de 110 especies maderables del caribe norte de Costa Rica. Instituto Tecnológico de Costa Rica, Cartago

Arévalo R, Londoño A (2005) Manual para la identificación de maderas que se comercializan en el departamento del Tolima. Corporación Autónoma Regional del Tolima (Cortolima), Ibagué, Tolima

Bagnouls B, Gaussen H (1957) Les climats biologiques et leur classification. Ann Georgr 355:19–220

Baguinon NT, Borgaonkar H, Gunatilleke N et al (2008) Collaborative studies in tropical Asian dendrochronology: Addressing challenges in climatology and forest ecology. Asia-Pacific Network for Global Change Research-APN-Final Report submitted to APN. Project: ARCP 2008-03CMY-Baguinon

Baker JCA, Santos GM, Gloor M et al (2017) Does *Cedrela* always form annual rings? Testing ring periodicity across South America using radiocarbon dating. Trees 31:1999–2009. https://doi.org/10.1007/s00468-017-1604-9

Barbosa A, Pereira G, Granato-Souza G et al (2018) Tree rings and growth trajectories of tree species from seasonally dry tropical forest. Aust J Bot 66:414–427. https://doi.org/10.1071/BT17212

Barthélémy D, Blaise F, Fourcaud T et al (1995) Modelisation et simulation de l'architecture des arbres bilan et perspectives. Rev For Française 47:71–96. https://doi.org/10.4267/2042/26721

Beech E, Rivers M, Oldfield S, Smith PP (2017) GlobalTreeSearch: the first complete global database of tree species and country distributions. J Sustain For 36:454–489. https://doi.org/10.1080/10549811.2017.1310049

Beltran L, Valencia G (2013) Anatomía de anillos de crecimiento de 80 especies arbóreas potenciales para estudios dendrocronológicos en la Selva Central, Perú. Rev Biol Trop 61:1025–1037
Boninsegna JA, Villalba R, Amarilla L et al (1989) Studies on tree rings, growth rates and age-size relationships of tropical tree species in Misiones, Argentina. IAWA J 10:161–169. https://doi.org/10.1163/22941932-90000484
Borchert R (1999) Climatic periodicity, phenology, and cambium activity in tropical dry forest trees. IAWA J 20:239–247
Bräuning A, Volland-Voigt F, Burchardt I et al (2009) Climatic control of radial growth of *Cedrela montana* in a humid mountain rainforest in southern Ecuador. Erdkunde 63:337–345. https://doi.org/10.3112/erdkunde.2009.04.04
Breitsprecher A, Bethel J (1990) Stem-growth periodicity of trees in a tropical wet forest of Costa Rica. Ecology 71:1156–1164
Brienen R, Lebrija-trejos E, Van BM et al (2009) The potential of tree rings for the study of forest succession in Southern Mexico. Biotropica 41:186–195. https://doi.org/10.1111/j.1744-7429.2008.00462.x
Brienen R, Schöngart J, Zuidema P (2016) Tree rings in the tropics: insights into the ecology and climate sensitivity of tropical trees. In: Goldstein G, Santiago SL (eds) Tropical tree physiology. Springer, Switzerland, pp 441–461
Brienen R, Zuidema PA (2005) Relating tree growth to rainfall in Bolivian rain forests: a test for six species using tree ring analysis. Oecologia 146:1–12. https://doi.org/10.1007/s00442-005-0160-y
Bullock SH (1997) Effects of seasonal rainfall on radial growth in two tropical tree species. Int J Biometeorol 41:13–16. https://doi.org/10.1007/s004840050047
Callado CH, Silva NSJS, Scarano FR et al (2001) Periodicity of growth rings in some flood-prone trees of the Atlantic Rain Forest in Rio de Janeiro, Brazil. Trees - Struct Funct 15:492–497. https://doi.org/10.1007/s00468-001-0128-4
Campos LE, Lobão MS, Rosero-Alvarado J et al (2008) Potencialidad de especies forestales para dendrocronología a traves de la caracterización anatómica de los anillos de crecimiento en la Amazonía Peruana – Brasilera. In: VII Congreso Nacional de Estudiantes Forestales, Madre de Dios, Perú
Carlquist S (2001) Comparative wood anatomy: systematic, ecological, and evolutionary aspects of dicotyledon wood. Springer, Berlin Heidelberg
Chowdhury MQ, Kitin P, De Ridder M et al (2016) Cambial dormancy induced growth rings in *Heritiera fomes* Buch.- Ham.: a proxy for exploring the dynamics of Sundarbans, Bangladesh. Trees - Struct Funct 30:227–239. https://doi.org/10.1007/s00468-015-1292-2
Cintra BBL, Schietti J, Emillio T et al (2013) Productivity of aboveground coarse wood biomass and stand age related to soil hydrology of Amazonian forests in the Purus-Madeira interfluvial area. Biogeosci Discuss 10:6417–6459. https://doi.org/10.5194/bgd-10-6417-2013
Clark DA, Clark DB (1994) Climate-induced annual variation in canopy tree growth in a Costa Rican Tropical Rain Forest. J Ecol 82:865–872
D'Arrigo R, Davi N, Jacoby G et al (2014) Dendroclimatic studies: Tree growth and climate change in Northern forests. American Geophysical Union - Wiley, Hoboken
de Arruda EC, Paixão E, Nunes da Cunha C et al (2017) Methodological baseline for dendrochronological studies in the Pantanal. In: Soares M, Gonçalves M (eds) Natural resources in wetlands: from Pantanal to Amazonia. Museu Paraense Emílio Goeldi, Belem, pp 90–114
de Faÿ E (1992) Vegetative development, primary and secondary growth of the shoot system of young *Terminalia superba* tropical trees, in a natural environment. I. Spatial variation in structure and size of axes. Ann des Sci For 49:389–402
De Reffye P, Heuvelink E, Barthélémy D, Paul-Henry C (2008) Plant growth models. In: Jorgensen SE, Fath B (eds) Encyclopedia of ecology. Elsevier, Amsterdam, pp 2824–2837
Détienne P (1995) Nature et périodicité des cernes dans quelques bois Guyanais. Bois Forêts des Trop 243:65–75
Détienne P, Barbier C (1988) Rythmes de croissance de quelques essences de Guyane française. Bois forêts des Trop 217:63–76

Enquist BJ, Leffler J (2001) Long-term tree ring chronologies from sympatric tropical dry-forest trees: individualistic responses to climatic variation. J Trop Ecol 17:4160. https://doi.org/10.1017/S0266467401001031
Evans MN, Schrag DP (2004) A stable isotope-based approach to tropical dendroclimatology. Geochim Cosmochim Acta 68:3295–3305. https://doi.org/10.1016/j.gca.2004.01.006
Evert RF (2006) Esau's plant anatomy: meristems, cells, and tissues of the plant body: their structure, function, and development. Wiley, Hoboken
Faber-Langendoen D, Gentry AH (1991) The structure and diversity of rain forests at Bajo Calima, Choco Region, Western Colombia. Biotropica 23:2–11
Fahn A (1967) Plant anatomy. Pergamon Press, Oxford
Fahn A, Burley J, Longman K et al (1981) Wood anatomy: possible contribution of wood anatomy to the determination of the age of tropical trees. In: Borman H, Berlyn G (eds) Age and growth rate of tropical trees: new directions for research. Yale University, New Haven, pp 31–55
Falcon-Lang H (1999) The early carboniferous (Courceyan –Arundian) monsoonal climate of the British Isles : evidence from growth rings in fossil woods. Geol Mag 136:1999. https://doi.org/10.1017/S0016756899002307
Fichtler E, Clark D, Worbes M (2003) Age and long-term growth of trees in an old-growth tropical rain forest, based on analyses of tree rings and ^{14}C. Biotropica 35:306–317. https://doi.org/10.1646/03027
Frankie G, Baker H, Oppler P (1974) Comparative phenological studies of trees in tropical wet and dry forests in the lowlands of Costa Rica. J Ecol 62:881–919
Fujii T, Salang A, Fujiwara T (1999) Growth periodicity in relation to the xylem development in three *Shorea* sp. (Dipterocarpaceae) growing in Sarawak. In: Wimmer R, Vetter R (eds) Tree ring analysis: biological, methodological, and environmental aspects. CABI Publishing, Oxon, pp 169–183
Gentry AH (1982) Patterns of neotropical plant species diversity. In: Evolutionary biology. Plenum Press, New York, pp 1–84
Graham EA, Mulkey SS, Kitajima K et al (2003) Cloud cover limits net CO_2 uptake and growth of a rainforest tree during tropical rainy seasons. Proc Natl Acad Sci USA 100:572–576. https://doi.org/10.1073/pnas.0133045100
Granato-Souza D, Carolina A, Campos M et al (2019) Drivers of growth variability of *Hymenaea stigonocarpa*, a widely distributed tree species in the Brazilian Cerrado. Dendrochronologia 53:73–81. https://doi.org/10.1016/j.dendro.2018.12.001
Groenendijk P, Sass-Klaassen U, Bongers F et al (2014) Potential of tree-ring analysis in a wet tropical forest: a case study on 22 commercial tree species in Central Africa. For Ecol Manag 323:65–68. https://doi.org/10.1016/j.foreco.2014.03.037
Gutiérrez G (2014) La relación entre crecimiento y clima en los árboles tropicales: Un estudio dendrocronológico en la Selva de Los Tuxtlas (Veracruz, México). Universidad Atónoma de México, Posgrado en Ciencias Biológicas
Gutiérrez V, Silva J (2002) Información técnica para el procesamiento industrial de 134 especies maderables de Bolivia
Hallé F, Oldeman R, Tomlinson P (1978) Tropical trees and forests: an architectural analysis. Springer-Verlag, Berlin
Hazlett D (1987) Seasonal cambial activity for *Pentaclethra*, *Goelthasia* and *Carapa* trees in a Costa Rican lowland forest. Biotropica 19:357–360
Herrera DA, del Valle JI (2011) Reconstruccion de los niveles del río Atrato con anillos de crecimiento de *Prioria copaifera*. Dyna 169:121–130
Hietz P, Wanek W, Dünisch O (2005) Long-term trends in cellulose ^{13}C and water-use efficiency of tropical *Cedrela* and *Swietenia* from Brazil. Tree Physiol 25:745–752
Hirons A, Thomas P (2018) Applied tree biology. Wiley, Hoboken
Holdridge LR (1967) Life zone ecology. Tropical Science Center, San José
Hu J, Riveros DA (2016) Life in the clouds : are tropical montane cloud forests responding to changes in climate ? Oecologia. https://doi.org/10.1007/s00442-015-3533-x

Hughes M, Swetnam T, Diaz H (2011) Dendroclimatology, progress and prospects. Springer, Dordrecht

IAWA Committee (1989) Iawa list of microscopic features for hardwood identification. AIWA Bull 10:219–332

Islam M, Rahman M, Bräuning A (2018) Growth-ring boundary anatomy and dendrochronological potential in a moist tropical forest in northeastern Bangladesh. Tree-Ring Res 74:1–18. https://doi.org/10.3959/1536-1098-74.1.76

Jacoby G (1989) Overview of tree-ring analysis in tropical regions. IAWA J 10:99–108

Jalil NRA, Takao I, Sahri MH et al (1998) Periodicity of xylem growth of rubberwood (*Hevea brasiliensis*) grown in Malaysia. Holzforsch - Int J Biol Chem Phys Technol Wood 52:567–572

Johansson M (2010) The circadian clock in annuals and perennials: environmental rhythms coordination of growth with environmental rhythms. Umeå Plant Science Centre Department of Plant Physiology, Umeå University, Umeå

Köhl M, Neupane PR, Lotfiomran N (2017) The impact of tree age on biomass growth and carbon accumulation capacity: a retrospective analysis using tree ring data of three tropical tree species grown in natural forests of Suriname. PLoS One 12:e0181187. https://doi.org/10.1371/journal.pone.0181187

León W (2008) Anatomía de madera en 31 especies de la subfamilia Mimosoideae (Leguminosae) en Venezuela. Colomb For 11:113–135

León W (2014) Anatomía de la madera de 108 especies de Venezuela, vol PE1. Pittieria, Mérida, Venezuela pp 1–263

Lisi C, Fihlo M, Botosso PC et al (2008) Tree-ring formation, radial increment periodicity, and phenology of tree species from a seasonal semi-deciduous forest in Southeast Brazil. IAWA J 29:189–207. https://doi.org/10.1163/22941932-90000179

Liu L, Filkov V, Groover A (2014) Modeling transcriptional networks regulating secondary growth and wood formation in forest trees. Physiol Plant 151:156–163. https://doi.org/10.1111/ppl.12113

Lopes de Oliveira C (2010) Estimativa da dinâmica de carbono na biomassa lenhosa de terra firme na Reserva de Desemvolvimento Sustentável de Amanã por métodos dendrocronológicos. Dissertação (Mestrado), Instituto Nacional de Pesquisas da Amaônia, Programa de Posgraduação, Manaus

López-Ayala JL, Valdez-Hernández JI, Terrazas T et al (2006) Anillos de crecimiento y su periodicidad en tres especies tropicales del Estado de Colima, México. Agrociencia 40:533–544

Lotfiomran N, Köhl M (2017) Retrospective analysis of growth a contribution to sustainable forest management in the tropics. IAWA J 38:297–312. https://doi.org/10.1163/22941932-20170173

Lüttge U, Hertel B (2009) Diurnal and annual rhythms in trees. Trees - Struct Funct 23:683–700. https://doi.org/10.1007/s00468-009-0324-1

MAE-FAO (Ministerio del Ambiente del Ecuador - Organización de las Naciones Unidas para la Alimentación y la Agricultura) (2014) Propiedades anatómicas, físicas y mecánicas de 93 especies forestales. FAO, Quito

Maingi JK (2006) Growth rings in tree species from the Tana River floodplain, Kenya. J East African Nat Hist 95:181–211. https://doi.org/10.2982/0012-8317(2006)95[181:GRITSF]2.0.CO;2

Marcati CR, Oliveira JS, Machado SR (2006) Growth rings in cerrado woody species: occurrence and anatomical markers. Biota Neotrop 6. https://doi.org/10.1590/s1676-06032006000300001

Marcati CRC, Dias C, Rodriguez S et al (2008) Seasonal development of secondary xylem and phloem in *Schizolobium parahyba* (Vell.) Blake (Leguminosae: Caesalpinioideae). Trees - Struct Funct 22:3–12. https://doi.org/10.1007/s00468-007-0173-8

Marcati CR, Machado SR, Podadera DS et al (2016) Cambial activity in dry and rainy season on branches from woody species growing in Brazilian Cerrado. Flora Morphol Distrib Funct Ecol Plants 223:1–10. https://doi.org/10.1016/j.flora.2016.04.008

Marcelo J (2017) Diversidade florística, dendrologia e dendroecologia de florestas estacionais decíduas do Centro e Norte do Peru. Dissertação (Doutorado), Escola Superior de Agricultura "Luiz de Queiroz-USP

Menezes M, Berger U, Worbes M (2003) Annual growth rings and long-term growth patterns of mangrove trees from the Braganca Peninsula, North Brazil. Wetl Ecol Manag 11:233–242(10). https://doi.org/10.1023/A:1025059315146

Mesa O, Poveda G, Carvajal L (1997) Introducción al clima de Colombia. Universidad Nacional de Colombia Sede Medellín, Facultad de Ciencias, Medellín

Miller R, Détienne P (2001) Major timber trees of Guyana wood anatomy. Tropenbos International, Wageningen

Morel H (2013) Dynamique de croissance radiale saisonnière et annuelle des arbres en forêt tropicale humide guyanaise. Thèse (Sciences de la Vie), Université des Antilles et de la Guyane, Cayenne, Guyane Française

Moreno MM, del Valle JI (2014) Influence of local climate and ENSO on the growth of Abarco (*Cariniana pyriformis*) in Chocó, Colombia. Trees 29:97–107. https://doi.org/10.1007/s00468-014-1094-y

Mushove PT, Prior JAB, Gumbie C et al (1995) The effects of different environments on diameter growth increments of *Colophospermum mopane* and *Combretum apiculatum*. For Ecol Manag 72:287–292. https://doi.org/10.1016/0378-1127(94)03468-C

Nahuz AR, Miranda M, Ielo P et al (2013) Catalogo de Madeiras Brasileiras para a Construcao Civil. Instituto de Pesquisas Tecnológicas do Estado de São Paulo (ITP), Sao Paulo

Nakai W, Okada N, Sano M et al (2018) Sample preparation of ring-less tropical trees for $\delta^{18}O$ measurement in isotope dendrochronology. Tropics 27:49–58. https://doi.org/10.3759/tropics.ms17-09

Nath CD, Munoz F, Pélissier R et al (2016) Growth rings in tropical trees: role of functional traits, environment, and phylogeny. Trees. https://doi.org/10.1007/s00468-016-1442-1

Nicolini E, Beauchêne J, De La Vallée BL et al (2012) Dating branch growth units in a tropical tree using morphological and anatomical markers: the case of *Parkia velutina* Benoist (Mimosoïdeae). Ann For Sci 69:543–555. https://doi.org/10.1007/s13595-011-0172-1

Niinemets O, Valladares F (2008) Environmental tolerance. In: Jorgensen P (ed) Encyclopedia of ecology. Elsevier, Oxford, pp 1370–1376

O'Brien JJ, Oberbauer SF, Clark DB et al (2008) Phenology and stem diameter increment seasonality in a Costa Rican wet tropical forest. Biotropica 40:151–159. https://doi.org/10.1111/j.1744-7429.2007.00354.x

Ohashi Y, Sahri MH, Yoshizawa N et al (2001) Annual rhythm of xylem growth in rubberwood (*Hevea brasiliensis*) trees grown in Malaysia. Holzforschung 55:151–154. https://doi.org/10.1515/HF.2001.024

Oliveira M, Mattos P, Muñoz-Braz E et al (2014) Growth pattern of *Qualea albiflora* and *Goupia glabra* in Amazon forest, Mato Grosso state, Brazil. In: The International Forestry Review. IUFRO

Palubicki W (2013) A computational study of tree architecture. University of Calgary, Calgary

Parolin P, Ferreira F, Piedade MTF et al (2016) Flood tolerant trees in seasonally inundated lowland tropical floodplains. In: Goldstein G, Santiago L (eds) Tropical tree physiology adaptations and responses in a changing environment. Springer, Switzerland, pp 127–148

Phan SM, Nguyen HTT, Nguyen TK et al (2019) Modelling above ground biomass accumulation of mangrove plantations in Vietnam. For Ecol Manag 432:376–386. https://doi.org/10.1016/j.foreco.2018.09.028

Pons TL, Helle G (2011) Identification of anatomically non-distinct annual rings in tropical trees using stable isotopes. Trees 25:83–93. https://doi.org/10.1007/s00468-010-0527-5

Poussart PF, Evans MN, Schrag DP (2004) Resolving seasonality in tropical trees: multi-decade, high-resolution oxygen and carbon isotope records from Indonesia and Thailand. Earth Planet Sci Lett 218:301–316. https://doi.org/10.1016/S0012-821X(03)00638-1

Poveda GG, Mesa OJ (2000) On the existence of Lloró (the rainiest locality on earth): enhanced ocean-land-atmosphere interaction by a low-level jet. Geophys Res Lett 27:1675–1678. https://doi.org/10.1029/1999GL006091

Póvoa de Mattos P (1999) Identificação de anéis anuais de crescimento e estimativa de idade e incremento anual em diâmetro de espécies nativas do Pantanal da Nhecolândia. Dissertação, Universidade Federal do Paraná

Prance GT (1982) A review of the phytogeographic evidences for Pleistocene climate changes in the neotropics. Ann - Missouri Bot Gard 69:594–624. https://doi.org/10.2307/2399085

Ramírez JA, del Valle JI (2011) Paleoclima de La Guajira, Colombia; según los anillos de crecimiento de *Capparis odoratissima* (Capparidaceae). Rev Biol Trop 59:1389–1405

Ramírez JA, del Valle JI (2012) Local and global climate signals form tree rings of *Parkinsonia praecox* in La Guajira, Colombia. Int J Climatol 32:1–30. https://doi.org/10.1017/CBO9781107415324.004

Ramírez-Martínez M, Terrazas T, Aguilar-Rodríguez S et al (2017) Anatomía de la madera de especies de la selva baja caducifolia de Tamaulipas, México. Madera Bosques 23:21–41. https://doi.org/10.21829/myb.2017.2321126

Reimer PJ, Brown TA, Reimer RW (2004) Discussion: reporting and calibration of post-bomb ^{14}C data. Radiocarbon 46:129–1304

Roig FA, Osornio JJJ, Diaz JV et al (2005) Anatomy of growth rings at the Yucatán Peninsula. Dendrochronologia 22:187–193. https://doi.org/10.1016/j.dendro.2005.05.007

Rojas V (2000) Identificación y clasificación de maderas. Instituto Nacional de Aprendizaje, Cartago

Rozendaal DMA (2010) Looking backwards: long-term growth patterns of Bolivian forest trees. PROMAB Scientific series 12, Riberalta, Bolivia

Santini L (2013) Descrição macroscópica e microscópica da madeira aplicada na identificação das principais espécies comercializadas no Estado de São Paulo - Programas "São Paulo Amigo da Amazônia" e "Cadmadeira." Universidade de São Paulo

Sass U, Killmann W, Eckstein D (1995) Wood formation in two species of Dipterocarpaceae in peninsular Malaysia. IAWA J 16:371–384

Schindelin J, Arganda-Carreras I, Frise E et al (2012) Fiji: an open-source platform for biological-image analysis. Nat Methods 9:676–682. https://doi.org/10.1038/nmeth.2019

Schöngart J, Piedade MTF, Ludwigshausen S et al (2002) Phenology and stem-growth periodicity of tree species in Amazonian floodplain forests. J Trop Ecol 18:581–597. https://doi.org/10.1017/S0266467402002389

Schöngart J, Piedade MTF, Wittmann F et al (2005) Wood growth patterns of *Macrolobium acaciifolium* (Benth.) Benth. (Fabaceae) in Amazonian black-water and white-water floodplain forests. Oecologia 145:454–461. https://doi.org/10.1007/s00442-005-0147-8

Schöngart J, Arieira J, Felfili Fortes C et al (2011) Age-related and stand-wise estimates of carbon stocks and sequestration in the aboveground coarse wood biomass of wetland forests in the northern Pantanal, Brazil. Biogeosciences 8:3407–3421. https://doi.org/10.5194/bg-8-3407-2011

Schöngart J, Bräuning A, Barbosa A et al (2017) Dendroecological studies in the Neotropics: history, status and future challenges. In: Amoroso M, Daniels L, Baker P, Camarero J (eds) Dendroecology: tree-ring analysis applied to ecological studies, Springer, Cham, pp 35–69

Schweingruber FH (1988) Tree Rings. Kluwer Academic Publishers, Dordrecht

Schweingruber FH (2007) Wood structure and environment. Springer, Berlin

Schweingruber FH, Börner A (2018) The plant stem. Springer, Cham

Sharma VK (2003) Adaptive significance of circadian clocks. Chronobiol Int 20:901–919. https://doi.org/10.1081/CBI-120026099

Shimamoto CY, Botosso PC, Amano E et al (2016) Stem growth rhythms in trees of a tropical rainforest in Southern Brazil. Trees 30:99–111. https://doi.org/10.1007/s00468-015-1279-z

Silva M, de Assis F, Callado CH et al (2017) Growth rings in woody species of *Ombrophilous* Dense Forest: occurrence, anatomical features and ecological considerations. Brazilian J Bot 40:281–290. https://doi.org/10.1007/s40415-016-0313-8

Silva M, Funch LS, da Silva LB (2019) The growth ring concept: seeking a broader and unambiguous approach covering tropical species. Biol Rev 94:1161–1178. https://doi.org/10.1111/brv.12495

Skolmen RG (1974) Some woods of Hawaii properties and uses of 16 commercial species. Forest Service, Berkeley

Soliz-Gamboa C, Rozendaal DMA, Ceccantini G et al (2011) Evaluating the annual nature of juvenile rings in Bolivian tropical rainforest trees. Trees 25:17–27. https://doi.org/10.1007/s00468-010-0468-z

Speer J (2010) Fundamentals of tree ring research. University of Arizona Press, Tucson

Steinhof A, Altenburg M, Machts H (2017) Sample preparation at the Jena ^{14}C laboratory. Radiocarbon 59:815–830. https://doi.org/10.1017/RDC.2017.50

Sudworth G, Mell C (1911) "Colombian mahogany" (*Cariniana pyriformis*), its characteristics and its use as a substitute for true mahogany (*Swietenia mahagoni*). Department of Agriculture, Forest Service, Whasington

Tanaka A (2005) Avaliação de anéis de crescimento de espécies florestais de terra-firme no município de Novo Aripuanã-AM. Dissertação (Doutorado), Universidade Federal do Amazonas – UFAM Instituto Nacional de Pesquisas da Amazônia – INPA

Tarelkin Y, Delvaux C, De Ridder M et al (2016) Growth-ring distinctness and boundary anatomy variability in tropical trees. IAWA J 37:275–294. https://doi.org/10.1163/22941932-20160134

Tarhule A, Hughes M (2002) Tree-ring research in semi-arid West Africa: need and potential item. Tree-Ring Res 58:31–46

Thomas P (2014) Trees: their natural history. Cambridge University Press, Cambridge

Thornthwaite CW (1948) An approach toward a rational classification of climate. Geogr Rev 38:55–94. https://doi.org/10.2307/210739

Tomlinson T, Craichead F (1972) Growth-ring studies on the native trees of subtropical Florida. In: Ghouse K, Yunus M (eds) Research trends in plant anatomy. McGraw Hill, New Delhi, pp 39–51

Tomlinson P, Longman K (1981) Growth phenology of tropical trees in relation to cambial activity. In: Bormann F, Berlyn G (eds) Age and growth rate of tropical trees: new dimensions for research. Yale University, New Haven, pp 7–19

Trevizor TT (2011) Anatomia comparada do lenho de 64 espécies arbóreas de ocorrência natural na floresta tropical Amazônica no estado do Pará Tássio. Universidade de São Paulo, Piracicaba

Vaganov E, Hughes M, Shashkin A (2006) Growth dynamics of conifer tree ring. Springer, Berlin

Vásquez A, Ramírez A (2005) Maderas comerciales en el valle de Aburrá. Área Metropolitana del Valle de Aburrá, Medellín

Vetter RE, Botosso PC (1989) Remarks on age and growth rate determination of amazonian trees. IAWA J 10:133–145. https://doi.org/10.1163/22941932-90000481

Wagner F, Rossi V, Stahl C et al (2012) Water availability is the main climate driver of neotropical tree growth. PLoS One 7. https://doi.org/10.1371/journal.pone.0034074

Walter H, Harnickell E, Mueller-Dombois D (1975) Climate-diagrams maps. Springer, Berlin

Wang KH, Hamzah MZ (2018) Different cambial activities in response to climatic factors of three Malaysian rainforest *Shorea* species with different stem diameters. Trees - Struct Funct 32:1519–1530. https://doi.org/10.1007/s00468-018-1730-z

Wheeler EA, Baas P, Rodgers S (2007) Variations in dicot wood anatomy: a global analysis based on the insidewood database. IAWA J 28:229–258. https://doi.org/10.1163/22941932-90001638

Whitmore T (1975) Tropical rain forest of the Far East. Clarendon Press, Oxford

Wiedenhoeft A (2011) Identificación de las especies maderables de centroamérica. USDA, Madison

Worbes M (1995) How to measure growth dynamics in tropical trees. IAWA J 16:337–351. https://doi.org/10.1163/22941932-90001424

Worbes M (1999) Annual growth rings, rainfall-dependent growth and long-term growth patterns of tropical trees from the Caparo Forest Reserve in Venezuela. J Ecol 87:391–403. https://doi.org/10.1046/j.1365-2745.1999.00361.x

Worbes M (2002) One hundred years of tree-ring research in the tropics – a brief history and an outlook to future challenges. Dendrochronologia 20:217–231. https://doi.org/10.1078/1125-7865-00018

Worbes M, Fichtler E (2010) Wood anatomy and tree-ring structure and their importance for tropical dendrochronology. In: Junk WJ, Piedade MTF, Wittmann F et al (eds) Amazonian floodplain forest. Springer, Dordrecht, pp 329–346

Worbes M, Junk WJ (1989) Dating tropical trees by means of 14C from bomb tests. Ecology 70:503–507

Worbes M, Junk WJ (1999) How old are tropical trees? The persistence of a myth. IAWA J 20:255–260

Worbes M, Herawati H, Martius C (2017) Tree growth rings in tropical peat swamp forests of Kalimantan, Indonesia. Forests 8:1–15. https://doi.org/10.3390/f8090336

Chapter 6
Dendroecology of *Prosopis* Species in the World: Secular Traces of Natural and Anthropic Events and Their Effects on *Prosopis* Growth

Maria Laura Cangiano, Maria Alicia Cendoya, Lucía Verónica Risio-Allione, and Stella Marys Bogino

Abstract *Prosopis* genus comprises 44 species which grow in arid and semiarid environments of America, Asia and Africa and in Oceania as introduced species. Of those 44 species, 30 grow in America. Argentina is supposed to be the origin center of the genus *Prosopis*, and that could explain the fact that this country has the highest species variability. *Prosopis* species are of great ecological and social value providing goods such as firewood, food, feed for livestock and medicines as well as services acting as watershed stabilizers and nitrogen and carbon sequesters. In this chapter, we address the new insights that dendrochronological studies have provided about *Prosopis* genus worldwide with main emphasis in Latin America. Many studies provide information about growth dynamics which are later on applied to develop management plans, to quantify the increment in dead and alive biomass throughout time or to estimate growth changes linked to social and political events. Rainfall is the main growth driver of *Prosopi*s throughout the Andes from Peru to central Argentina, whereas in the flat Pampas temperature is the main tree-ring width driver. Climate variables are hard to separate from other factors affecting growth such as geomorphology, anthropogenic impact and groundwater depth. Outside of Latin America, dendrochronology of *Prosopis* has been used for the analysis of anthropogenic contamination. Fire, the main disturbance factor in arid and semiarid environments, showed a regional dynamic as a result of human activities.

M. L. Cangiano
Department of Crop and Soil Sciences, North Carolina State University, Raleigh, NC, USA

M. A. Cendoya
Agricultural Experimental Station, The National Agricultural Technology Institute of San Luis, Villa Mercedes, San Luis, Argentina

L. V. Risio-Allione · S. M. Bogino (✉)
Department of Agricultural Sciences, National University of San Luis, Villa Mercedes, San Luis, Argentina

M. Pompa-García, J. J. Camarero (eds.), *Latin American Dendroecology*,
https://doi.org/10.1007/978-3-030-36930-9_6

Most dendroecological studies on *Prosopis* species, which allowed determining these species dynamics, concluded that *Prosopis* do not present an encroaching behavior. This review demonstrates the avant-garde and influential value of *Prosopis* genus for dendroecological research as it allows reconstructing past disturbances as fire, anthropogenic impact and changes in groundwater depth for the last 50–100 years and up to 356 years in the case of *Prosopis caldenia* Burkart in the Pampean Region.

Keywords Radial growth · Climate · Fire · Semiarid · Arid

6.1 Why the Genus *Prosopis*?

Prosopis species are trees or shrubs mainly xerophilous that grow in arid and semiarid areas in Southwest Asia, Africa, Australia (as introduced species) and predominantly in America. More than 30 species of the 44 that exist worldwide occur in South America mainly in Argentina and Paraguay with a few occurring from Chile to Ecuador and Colombia (Burkart 1976). Argentina, the center of its polymorphism, is thought to be the origin of the species (29 of the total 44 species grow in this country). The fact that the *Prosopis* species from South America contain almost all the morphological characters of the *Prosopis* growing in North America and the Old world confirms South America as the center of world origin for this genus (Bovey 2016). *Prosopis* genus is supposed to be around 25–36 million years old (late Paleocene) (Smeins 1983). It grows under harsh environmental conditions as soils with salinity levels similar to those of the seawater in India with a pH of 9 or 45 °C temperatures as in the Death Valley in California (Bovey 2016).

Prosopis species provide a vast amount of services and goods in unfavorable environments (Tello et al. 2018). *Prosopis* trees provide soil protection as they favor water infiltration avoiding soil sealing after heavy rainfall (Virginia and Jarrell 1983). At an ecosystem level *Prosopis* dominated communities create "islands of fertility" enhancing environmental heterogeneity by providing an increase in species richness (Golubov et al. 2001). As an example, 20 bee and 70 bird species can be found in the *Prosopis* woodlands of Mexico. As other *Fabaceae* species do, *Prosopis* fix atmospheric nitrogen by symbiosis with *Rhyzobium* bacteria increasing soil fertility. They also play an important role as a carbon sink under climate change conditions (Viglizzo and Frank 2006). *Prosopis'* wood is used for structural purposes and its high calorific value of approximately 5000 kcal kg^{-1} makes it an excellent fuel as well. As an example, 70% of the firewood used by rural populations in dry regions in India belongs to *Prosopis* species.

The fruits produced by *Prosopis* species are legume pods that contain 13% protein and 30% sucrose (Fig. 6.1a). Pods have been a historic source of food for human populations; they used to be a major staple for native people in the United States of America (Bovey 2016), becoming less important as human food and more impor-

Fig. 6.1 Goods and services provided by *Prosopis* forests. (**a**) Pods of *Prosopis caldenia* used for food and livestock feed. (**b**) Cattle production system in the Prosopis woodlands. (**c**) Encroachment of grassland by *Prosopis* genus in the Argentinean Pampas and (**d**) hydrological changes as a result of severe *Prosopis* woodlands deforestation in central Argentina

tant as livestock feed during the last few centuries. The leaves of some *Prosopis* species can be used as feed for animals. A minority of species are known to have leaves which are palatable to livestock and which are highly valued as a source of forage. Medicinal uses have also been recorded for *Prosopis.* Pod, leaf and bark extracts are useful for mouth and throat infections, internal diseases including general pains, parasites and urinary disorders and skin disorders. Honey is another good that can be produced in *Prosopis* woodlands. The flowers that are available between October and March in the Pampean region allow a production of 20–25 kg of honey per beehive (Cozzarín 2008). Lastly, tannins for leather industry are another *Prosopis* good.

However, from the European occupation the perception of the role of the *Prosopis* genus has changed worldwide from an essential human item to an encroaching undesirable weed (Pasiecznik et al. 2001). This encroachment started in North America in the nineteenth century with *Prosopis glandulosa* var. *glandulosa* Torr. and *Prosopis velutina.* This resulted in a great effort to eliminate these species known as the "great mesquite war." Presently, the effort to control these species has decreased due to the fact that chemical treatments are not effective and are expensive; the genus is resilient and fire is not useful after the seedling becomes 2.5 cm in diameter (Bovey 2016).

From the 1800 onwards other *Prosopis* species became encroachers. This is the case of *Prosopis juliflora* (Sw.) DC. in arid and semiarid areas of tropical Africa, Asia, the Antilles, Australia, the pacific Islands and Brazil and *Prosopis ruscifolia* Griseb. in the Chaco of Paraguay and Argentina (Bovey 2016). Nowadays, the point of view on *Prosopis* genus in all these countries has changed from undesirable to an essential component of the ecosystem. As an example, in Argentina *Prosopis* grows in plain areas as the Pampean region and Chaco which are being affected by massive deforestation (more than 1% $year^{-1}$) due to agriculture linked to new technologies (Genetically modified organisms and agrochemicals). As a result of this severe deforestation hydro ecological changes have occurred in the southern part of the *Prosopis* distribution area resulting in the appearance of new rivers and the increase of groundwater levels which have created lagoons (Contreras et al. 2012) (Fig. 6.1d).

Summing up, *Prosopis* genus is core in arid and semiarid ecosystems of the world. The perception of the genus is ambiguous going from an irreplaceable component of the environment (Fig. 6.1b) to an encroaching undesirable weed (Fig. 6.1c). Despite the sometimes pessimistic view of the genus it cannot be denied that *Prosopis* species provide innumerable and valuable goods and services and that they have an extraordinary ability to adapt to a constantly changing environment. Dendrochronology, the science that studies tree rings of woody species, can provide a better insight into the *Prosopis* genus and the factors that affect its dynamics.

6.2 Addressing the *Prosopis* Dilemma Using Dendroecological Research

The objective of this chapter is to review the contributions made by dendroecological studies in order to assess the knowledge of the *Prosopis* genus dynamics worldwide with emphasis in Latin America. Scientific publications in English and Spanish were compelled using the web browsers Google Scholar, Scielo, Scopus, Dianet and WorldWideScience.org. The key words used were dendroecology, *Prosopis*, tree rings, dendrochronology, *anillos de crecimiento*, *Dendrocronología* and *Dendroecología*. Fifty-eight scientific papers were collected from which 52 used dendrochronological techniques to address different ecological questions in relation to the *Prosopis* genus around the world (Table 6.1).

Of the total 44 *Prosopis* species that exist worldwide 15 were analyzed in the scientific publications presented here and from those 11 belonged to Latin America (Fig. 6.2). Sixty-eight percent of the studies belonged to Argentina, and a majority were on *Prosopis caldenia* Burkart and *Prosopis flexuosa* DC, which is consistent with the fact that this country is the *Prosopis* origin center. The majority of the chronologies are around 100 years old with the exception of *P. caldenia* with several chronologies longer than 200 years.

In order to detect the main topics that the 56 scientific publications try to address, a tag cloud emphasizing the most used words was built. The titles' words and key

Table 6.1 Dendroecological studies on the genus *Prosopis* in the world

Species	Authors	Time span (years) (period)	Number of trees
Prosopis africana[a]	Tarhule and Hughes 2002	Anatomy	12
Prosopis alba	Bravo et al. 2008	102 (1894–1996)	4
Prosopis alba	Juárez de Galíndez et al. 2005	113 (1884–1997)	10
Prosopis alba	Juárez de Galíndez et al. 2008	113 (1884–1997)	10
Prosopis alba	Rivera et al. 2010	76 (1932–2007)	23
Prosopis caldenia	Bogino and Jobbágy 2011	81 (1929–2009)	98
Prosopis caldenia	Bogino and Villalba 2008	100 (1901–2001)	97
Prosopis caldenia	Bogino et al. 2015	220 (1794–2013)	351
Prosopis caldenia	Bogino et al. 2019	353 (1659–2011)	11
Prosopis caldenia	Dussart et al. 1998	65 (1928–1993)	112
Prosopis caldenia	Dussart et al. 2011	266 (1743–2008)	25
Prosopis caldenia	Krebs and Fischer 1931	160 (1769–1928)	1
Prosopis caldenia	Medina 2007	208 (1788–1995)	31
Prosopis caldenia	Medina 2008	22 (1971–1992)	10
Prosopis caldenia	Medina et al. 2000	207 (1787–1993)	22
Prosopis caldenia	Risio et al. 2014	41 (1971–2012)	30
Prosopis caldenia	Risio et al. 2018	93 (1911–2004)	65
Prosopis caldenia	Tapia and Dussart 2013	262 (1742–2003)	38
Prosopis caldenia	Velasco-Sastre et al. 2018	274 (1738–2011)	86
Prosopis chilensis	Giménez et al. 2010	221 (1779–2000)	4
Prosopis chilensis	Holmgren et al. 2006	125 (1877–2001)	190
Prosopis chilensis	López et al. 2006	85 (1918–2002)	14
Prosopis cineraria[a]	Korori et al. 2013		257
Prosopis ferox	Morales and Villalba 2012	107 (1897–2003)	57
Prosopis ferox	Morales et al. 2001	117 (1883–1999)	20
Prosopis ferox	Morales et al. 2004	117 (1883–1999)	23
Prosopis ferox	Morales et al. 2005	69 (1932–2000)	120
Prosopis flexuosa	Álvarez et al. 2011a	100 (1907–2007)	39
Prosopis flexuosa	Álvarez et al. 2011b	70	30
Prosopis flexuosa[b]	Ferrero et al. 2013		15
Prosopis flexuosa	Giantomasi et al. 2009	65 (1940–2004)	21
Prosopis flexuosa	Giantomasi et al. 2013	88 (1919–2006)	56
Prosopis flexuosa	Giordano et al. 2011	70 (1938–2009)	13
Prosopis flexuosa	Perpiñal et al. 1995	67	31
Prosopis flexuosa	Piraino and Roig 2016	51 (1960–2010)	38

(continued)

Table 6.1 (continued)

Species	Authors	Time span (years) (period)	Number of trees
Prosopis flexuosa	Piraino and Roig 2017	80 (1931–2010)	21
Prosopis flexuosa	Piraino et al. 2015	96 (1916–2011)	121
Prosopis flexuosa	Piraino et al. 2017	94 (1918–2011)	80
Prosopis flexuosa	Villagra et al. 2005a	60	40
Prosopis flexuosa	Villagra ct al. 2005b	130	76
Prosopis flexuosa	Villalba and Boninsegna 1989	84	71
Prosopis glandulosa var. *glandulosa*	Flinn et al. 1994	30	75
Prosopis juliflora	Beramendi-Orosco et al. 2013	105 (1903–2004)	4
Prosopis juliflora	Carvalho et al. 2019	41	40
Prosopis juliflora[c]	Leavitt and Long 1991		
Prosopis kuntzei	Giménez et al. 1997	46	12
Prosopis kuntzei	Ríos et al. 2001	58	30
Prosopis nigra	Bravo et al. 2008	52	3
Prosopis nigra	Giménez et al. 2003	100	10
Prosopis pallida	Holmgren et al. 2006	37 (1965–2001)	228
Prosopis pallida	López et al. 2005	18	54
Prosopis pallida	López et al. 2006	95 (1908–2002)	110
Prosopis pallida	Rodríguez et al. 2005	95	
Prosopis pallida	Salazar et al. 2018	51 (1963–2014)	31
Prosopis ruscifolia	Juárez de Galíndez et al. 2007	25	67
Prosopis ruscifolia	Giménez et al. 2009	50	15
Prosopis sp.	Ghezzi and Rodríguez 2015	95	130
Prosopis sp[d]	Villalba et al. 2000		
Prosopis spicigera[a]	Rajput et al. 2008		
Prosopis tamarugo	Decuyper et al. 2016	26 (1987–2012)	8
Prosopis tamarugo	Rivera et al. 2010	76 (1932–2007)	23
Prosopis velutina	Judd et al. 1971	137	3

[a]Only anatomical studies to state the feasibility for dendrochronological studies
[b]Branch analysis
[c]Samples that could not be dated
[d]Review of dendrochronological potential that include all *Prosopis* species
The time span and number of samples are also indicated

words in English and Spanish of each publication were used. R software was used to build the cloud (R Core Team R et al. 2016) (Fig. 6.3).

The tag cloud shows that growth dynamics is a common concern in almost all references. Fires and climate appear as variables of high interest but with an emphasis

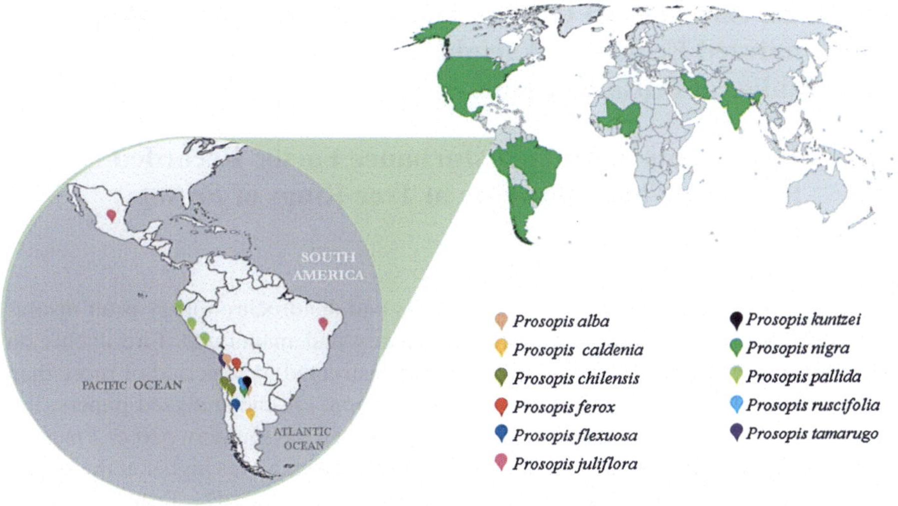

Fig. 6.2 Dendroecologial studies on *Prosopis* species in the world and Latin America. Countries with dendroecological studies on *Prosopis* genus (green color) and species detail in Latin America

Fig. 6.3 Tag cloud built using the word titles' words and keywords in English and Spanish languages of 58 scientific publications of dendroecology of *Prosopis* genus

on the regional approach. The cloud also shows specific concerns such as groundwater variability, heavy metal pollution, anthropogenic impact, *Prosopis* use in archaeology, etc. This cloud does not highlight encroachment as a problem according to the dendroecological insights.

6.3 Fire, Climate and Growth Dynamics: The Three Most Analyzed Ecological Variables in Dendroecological Studies of *Prosopis* Genus

6.3.1 Fire Events: a Main Disturbance Factor Recorded for More Than 200 years in Tree Rings of *Prosopis* Species

Fire is a widespread component of ecosystems and dendrochronology is an invaluable tool for the reconstruction of fire frequency and magnitude. Fire events on woodlands of *Prosopis caldenia* have been reconstructed for a period of more than 200 years, from indigenous to European occupation. On the analyzed papers, fire occurrences changed according to the sampling site. In the southern part of *Prosopis caldenia* woodlands, natural distribution area fire frequency was higher at the beginning of the nineteenth century and between 1880 and 1930. The former possibly coincides with the tremendous drought mentioned by Darwin (1839) when he was travelling in South America (1827–1832) and the later with the end of the indigenous occupation and the beginning of European migration settlements (Medina 2007). The lowest fire frequency was 7.13 years from the 1930s onwards in La Pampa province, similar to the mean frequency of 7.25 years from 1911 onwards in San Luis province (Medina 2007). It is interesting to note that the mean frequency is not coincident between sites, the highest mean frequency for San Luis was in the period between 1911 and 1993 (Medina et al. 2000; Bogino et al. 2015), whereas for La Pampa it was between 1795 and 1930 (Medina 2007; Velasco-Sastre et al. 2018). Human impact on fire frequency was evident in the fire chronology made on *P. caldenia* for the 1970–1992 period with a mean interval of 4.2 years as a result of the use of fire as a management tool (Medina 2008). Furthermore, this high frequency is similar to the 4.4 years mean for the period of indigenous occupation where fire was used as a strategic tool as well (Medina 2007). Although they belonged to different cultural backgrounds fire as a management tool was a commonality between indigenous and European people.

A fire chronology of *Prosopis alba* Griseb in the Arid Chaco (1913–1996) (Bravo et al. 2008) showed that fire events were more frequent than in the *P. caldenia* area (3.32 years, not available in the document, calculated by these authors). From the four species *Schinopsis lorentzii* (Griseb.) Engl., *Aspidosperma quebracho-blanco* Schlecht, *Prosopis alba* Griseb. and *Prosopis nigra* Griseb. analyzed by Bravo et al. (2008), *P. alba*'s longevity of 102 years stood out demonstrating once again the potential of the *Prosopis* genus for dendrochronological studies (Villalba and Boninsegna 1989). Chronological series of big longevity are of great importance for long term dendrochronological reconstructions. The studies reported chronological series of less than 100 years with the exception of *P. caldenia* in the province of La Pampa which chronology covers more than 200 years. To conclude, all these studies showed that fire regime is a combination of temporal and spatial components and that regardless of the cultural groups that occupied the geographical spaces, fire has always been part of these biological systems intimately linked to anthropic impact.

6.3.2 The Challenges of Separating Climate from Other Variables as Driving Factors of Growth Dynamics of *Prosopis* Species

Climate seems to be a common interest in all the studies analyzed for this chapter. All of them evaluated the association between the growth of *Prosopis* genus and some global or regional climatic variable as El Niño Southern Oscillation (ENSO), drought, aridity, rainfall, precipitation or temperature. However, each *Prosopis* species seems to have a different response to climate. For the purpose of understanding the relation between ENSO and tree rings, López et al. (2005, 2006) conducted a study of the species *Prosopis pallida* HBK and *Prosopis chilensis* Stuntz in an area strongly vulnerable to ENSO impact, the coasts of Peru and Chile in South America. Their results showed that rainfall, closely linked with ENSO, controls radial growth in both species. On the other hand, no effect of temperature could be detected in spite of the long distance between the sampling sites in tropical and subtropical semiarid environments (López et al. 2005, 2006). Another study on *P. pallida* in the north of Peru also stated the positive impact of summer rainfall on growth (Rodríguez et al. 2005). Holmgren et al. (2006) also found that ENSO episodes were the main factor affecting recruitment and growth of *P. pallida* in Peru. Nevertheless, *P. chilensis* survival was controlled by herbivory no matter if climatic conditions were the most convenient for tree establishment.

A study by Salazar et al. (2018) showed that *P. pallida* trees had a different climate-growth association depending on the sampling sites' distance from the sea (from the coast of the Pacific Ocean, 13 km from the sea, to the Andes, 109 km from the sea). No significant association was determined between growth and climate variables in the coastal sampling site. Because of their phreatic root systems the *Prosopis* species can reach the 5 m deep groundwater in the coastal site and consequently hide the climate signal. On the other hand, the populations on the site in proximity to the Andes Mountains showed sensitivity to climatic variations and ENSO events. Since groundwater is 50 m deep in this site, trees cannot access it and, consequently, they show a higher climatic sensitivity (Salazar et al. 2018).

In the Pampean Region (36–40°S) the accessibility of *Prosopis caldenia* to groundwater also determined a lack of association between growth and rainfall and a clear impact of temperature during the growing season (Bogino and Jobbágy 2011). The use of annual aridity indexes (AAIs) determined that the age and woodland management affect high frequency growth variability at the north edge of the distribution area of *P. caldenia* being AAIs no significant when trees became older (Risio et al. 2018). *P. caldenia* was the first species used in an attempt to link rainfall and tree-ring growth in the year 1931. This may be considered as the first dendrochronological study in South America (Krebs and Fischer 1931). Nowadays, the longest *Prosopis* chronology (1656–2011 period) is of *P. caldenia* (Bogino et al. 2019).

Wood anatomy (number and size of vessels) is another tree-ring characteristic that together with tree rings is useful to determine growth and ecophysiology of *Prosopis flexuosa*. High temperatures reduce vessel size and tree-ring width without affecting the number of vessels. All variables are affected positively when rainfall increases (Giantomasi et al. 2009). A study of *P. flexuosa* growing in semiarid to arid conditions (rainfall from 500 to 80 mm year^{-1}) in central Argentina covering 65–68°W of longitude determined that the climate signal was linked to the accessibility to groundwater, the rainfall signal on tree rings was significant when trees did not have access to groundwater. Groundwater accessibility hid growth-climate association (Giantomasi et al. 2013). The study of *P. flexuosa* provided new insights about tree-ring growth and climate association in relation to the tree structure (multi-stemmed trees or one-stemmed trees) suggesting a higher sensitivity to climate extremes in the case of multi-stemmed trees (Piraino and Roig 2017). A different case is that of *P pallida* and *P. chilensis* growing in Peru and Chile which show no impact of temperature in growth (López et al. 2005, 2006).

The impact of geomorphology and climate on the radial growth of *P. flexuosa* was analyzed in the Monte desert in Central Argentina (around 32°S, 67°W). Rainfall determined growth dynamics of *P. flexuosa* but the landform and soil variability set a changing climate-growth association according to the sampling site. At regional level rainfall affected growth positively at the beginning and end of the growing season. At local scale, nevertheless, this association is a complex link between soil properties and landform units such as river, paleo-river and inter-dune valley which differ by their surface water balance. The edaphic characteristics set up different soil-water dynamics which determine variable tree-ring width (Piraino et al. 2015).

P. flexuosa tree rings of individuals with different water use accessibility were used to study the association between geomorphological characteristics and growth to better understand hydrological changes as stream-flow variability in a riparian environment. The distance between the trees and the riverbank determined different growth dynamics. The best models in Piraino and Roig (2016) imply that in trees located near the river (15 m) tree growth was strongly affected by changes in the stream flow during spring and summer months. Trees located away from the river (400 m), on the other hand, were affected by August (winter in the Southern hemisphere) stream-flow and March (summer in the Southern hemisphere) rainfall only (Piraino and Roig 2016).

Dendrochronological studies of *Prosopis ferox* Griseb. in the Quebrada de Humahuaca, Argentina, (23°S 65°W) showed an association between growth and climatic variables. The mean annual growth was affected positively by rainfall and inversely by temperatures during the growing season in the last century. This was the first study to develop a chronology of *P. ferox* and showed the potential of this species for dendroclimatological studies in the Argentinean Prepuna (Morales et al. 2001). A later study was conducted on *P. ferox* at the tree-line in subtropical areas in Argentina in order to analyze the link between growth and climate under these extreme conditions. Trees at around 3500 m a.s.l showed that growth was affected inversely by temperature and positively by rainfall by the previous summer (January to March) climatic conditions for the 1934–1980 period. Summers in Northwestern Argentina are characterized by high evapotranspiration and low water availability what results in lower growth rates during this time of the year. These results indicate the complex association between climatic variables and tree-ring growth in subtropical environments challenging the statement of an upper tree-line growth limited only by maximum temperature (Morales et al. 2004).

The relation between *P. ferox*'s recruitment, climate and social change was analyzed in the Quebrada de Humahuaca, Argentina. Recruitment increased in the 1970s and 1980s which is consistent with an increase in precipitation together with the reduction of domestic grazers being these two the dominant factors driving the recovery of *P. ferox* woodlands (Morales et al. 2005). Multiannual oscillations in rainfall (at least 4 years) determined successful episodic recruitment of *P. ferox* growing in the Prepuna Argentina between 2800 and 3400 m a.s.l. Meanwhile, interannual variations in precipitation of summer months influenced year-to-year patterns of radial growth, $r = 0.56$ ($n = 96$, $p < 0.001$). The year-to-year growth variability implied a fast reaction of growth to more humid environmental conditions. The fact that the chronology covered a 97 year period (1904–2001) allowed the authors to disentangle the impact of climate versus anthropogenic activities on *P. ferox* dynamics which changed substantially in the last century. The intensive grazing, mainly by sheep and goat, that occurred in the Prepuna during the first half of the twentieth century masked the influence of precipitation on the establishment and survival of *P. ferox*, this suggests that high livestock density limited tree establishment (Morales and Villalba 2012).

A plantation of *Prosopis tamarugo* Phil. in the Atacama Desert in Chile was analyzed to determine the groundwater impact on different physiological variables including radial growth. With the increase in groundwater depth from 3 m to more than 10 m, there was an increase in drought stress which leads to a reduction in growth (Decuyper et al. 2016).

After analyzing the references about the impact of climate on the growth of *P. pallida*, *P. chilensis*, *P. ferox*, *P. caldenia* and *P. flexuosa*, rainfall showed to be a determinant factor in the Andean area with disregard of the latitudinal position of the sampling site (from 5°10'S to 36°44'S), from tropical to temperate climate. Temperature drives the growth of *P. ferox* at the tree line as well as the growth of *P. caldenia* in the pampas when trees have access to groundwater. At the same time, livestock browsing in the case of *P. ferox* and groundwater accessibility in the case of *P. caldenia* hide the positive impact of rainfall on growth. Livestock density also counteracts the positive effect that rainfall has on recruitment. Summarizing, it is almost impossible to separate the impact that climate and human activities have in the growth of these species, the latter being the reason why in several cases the climate signal is hidden. New insights suggested the need of multivariable models that include landscape physiognomy and soil properties in order to better understand the growth-climate association.

6.3.2.1 Recruitment of *Prosopis* Species: The Result of Climate Drivers or Anthropogenic Impact?

Even though *Prosopis* species are considered as undesirable invasive species by some, the tag cloud did not raise this as an important topic within dendroecological studies. *Prosopis* species are indigenous in North America and they only started behaving as weeds after the European occupation. Overgrazing and fencing, the spread of seeds by livestock, the suppression of fires, a reduction in grass competition and drought (Buffington and Herbel 1965) are among the disturbances that the occupation brought and which eventually changed these species dynamics. All these disturbances together with CO_2 enrichment of the atmosphere (Archer 1995) resulted in the encroaching behavior of the *Prosopis* genus. Chronicles from the middle of the eighteenth century in North America suggested the existence of open grazing areas in places where today there are *Prosopis* woodlands (Rogers 2000). *Prosopis* woodlands have recently expanded into grasslands from their characteristic mesic drainages and upland slopes where they had been confined during the Holocene prior to livestock introduction (Polley et al. 1994). Some references point out the *Prosopis* genus as the worst woody invader in the world (Shackleton et al. 2014).

Dendroecological studies provide new insight about the dichotomy between encroacher and essential species of *Prosopis* genus. Recruitment of some *Prosopis* species is favored by rainfall regardless of where the *Prosopis* species grows as is the case of *Prosopis ferox* in Argentina, *Prosopis pallida* in Peru and *Prosopis chilensis* in Chile. Recruitment of *P. pallida* and *P. chilensis* is favored by warmer periods during ENSO episodes for this reason reforestation plans are encouraged during these periods (López et al. 2006; Holmgren et al. 2006). In spite of the findings of these studies of a significant impact of climate on recruitment, these results cannot be generalized since *P. flexuosa* recruitment and climatic variables showed no association in Monte desert in Argentina where recruitment was the result of anthropogenic and ecological interactions (Villagra et al. 2005a).

The establishment rates of *Prosopis* genus can be severely modified by anthropogenic disturbances as stocking rate and fire events. *P. ferox* establishment was favored by the reduction of goats and sheep in the Quebrada de Humahuaca, Argentina, (Morales et al. 2005) whereas *P. caldenia* establishment was favored by cattle introduction in the Pampean Region (Dussart et al. 1998). Recruitment of *P. caldenia* has changed in relation to livestock management. In the middle of the twentieth century sheep was replaced by cattle and as a result *P. caldenia* establishment increased. An increment from 0.99 to 10 tree year^{-1} was recorded in one sampling site and from 12.7 to 48 tree year^{-1} in a second sampling site, no matter the climate conditions (Dussart et al. 1998). The association between climate and recruitment could be hidden by other variables as fire events or grass competition reduction (Bogino et al. 2015).

Prosopis species are also considered serious invaders in Africa, Asia and Australia being a foreign species in the latter. The dendroecological potential of *Prosopis* species growing in Africa under tropical climate has stimulated the development of a new area of study in order to address environmental concerns (Tarhule and Hughes 2002). Most dendroecological studies on *Prosopis* species, which allowed determining these species dynamics, concluded that *Prosopis* do not present an encroaching behavior. This can be due to the fact that the majority of this research has been done in South America where even though some references reported an increase in *Prosopis* population, the authors did not state that this increase should be regarded as the behavior of an invasive species (Dussart et al. 1998; Giménez et al. 2009). A recent publication in Brazil, nevertheless, classified *Prosopis juliflora* as an invasive tree (Carvalho et al. 2019).

6.3.3 Tree-Ring Growth Dynamics: The Most Recorded Dendroecological Variable That Quantifies Tree's Relationship with the Environment Over Time

The study of annual tree-ring growth of *Prosopis* genus provides valuable information not only for the scientific community but also for silvicultural practices. Among other things, this variable can be used for the calculation of annual biomass and consequently CO_2 capture quantifications, it also reveals the impact of climate on growth and dead of *Prosopis* trees. This information enables a better understanding of *Prosopis* trees in particular and the ecosystem in general what leads to develop accurate and sustainable woodland management practices. Because of tree-ring growth being the most analyzed dendrochronological variable worldwide it provides information about *Prosopis* genus dynamics throughout the world with an emphasis in Latin America. In the next paragraphs we present the growth dynamics data that as today has been measured for this genus.

Studies in the Central Arid Chaco (subtropical climatic conditions) have described multiple *Prosopis* species. *Prosopis alba,* one of the *Mimoseae* of maximum economic value in Argentina has a mean radial growth of 3.8–4 mm year^{-1} and a growth culmination age between 60 and 74 years (Juárez de Galíndez et al. 2008). *Prosopis nigra* (Griseb.) Hieron has a mean radial growth of 3.29 mm year^{-1} and a culmination age at 44–46 years (Giménez et al. 2003). The only chronology (1779–2000) of *Prosopis chilensis* species in Argentina is that of a historic tree at Facundo Quiroga's house which had a mean radial growth of 3.75 mm year^{-1} (SD ± 1.3 mm year^{-1}). A study in *Prosopis kuntzei* Harms reported that this species wood has a density of 1.2 kg dm^{-3} and has a mean annual radial growth of 3.22 mm year^{-1} (variability between 1.8 and 10 mm year^{-1}) (Giménez et al. 1997) whereas another study on the same species reported a mean annual radial growth of 3 mm year^{-1} and a culmination age of 60 years (Ríos et al. 2001). The growth dynamics of the *Prosopis* species described in this paragraph were used for the development of growth models which used dendrochronology as a year-to-year information provider and which contributed to the improvement of management practices.

P. ruscifolia, which is one of the most important woody invaders in Argentina, is another native species from the Semiarid Chaco. This species had significant intra-specific mean annual radial growth variability from 4.9 to 5.3 mm $year^{-1}$ and a growth culmination age between 11 and 23 years (Juárez de Galíndez et al. 2007). This growth rate is the highest for all the native woody species growing in this area. Something that makes this species particularly interesting is its ability to grow under salinity conditions and in soils with anaerobiosis. Floods favor *P. ruscifolia* establishment which modify soil properties allowing the recruitment of economically valuable species (Giménez et al. 2009).

Prosopis flexuosa is, together with *Prosopis caldenia,* one of the most studied species within the *Prosopis* genus. We found that the reported growth rates, culmination age and wood production can vary greatly according to location, topography, tree structure, anthropogenic activity and climate conditions. In the next three paragraphs we summarize the findings on this species.

In central Monte desert in Mendoza province (32°20' 68°00'W) *Prosopis flexuosa* growth varies according to the topography from 4.6 ± 0.05 mm $year^{-1}$ in the valleys to 2.9 ± 0.03 mm $year^{-1}$ in the dunes (Giordano et al. 2011). Piraino et al. (2015) and Piraino and Roig (2016) found that landform and soil characteristics in the north of Mendoza province near the Desaguadero River affected the mean annual growth rate which oscillated between 1.22 and 2.03 mm $year^{-1}$ and 1.52 and 1.56 mm $year^{-1}$. They reported as well that tree-ring width releases followed disturbances events. This information suggests that changes in growth dynamics are the result of anthropogenic activities together with landform characteristics. A study in the Telteca and Ñacuñan reserves in the Monte desert in Argentina found that mean annual radial growth of individuals between 7 and 50 years of age was 1.5–2 mm $year^{-1}$ and 1 mm $year^{-1}$, respectively, for each site (Villagra et al. 2005a). Lastly, *P. flexuosa* studies in the Chaco Arid in Córdoba province (31°22' 65°29'W) found that this species had a mean radial growth of 2.3 mm $year^{-1}$ when trees are 35 years old and they get to their growth culmination age (Perpiñal et al. 1995) and that mean radial growth of branches was 0.7–1.16 mm $year^{-1}$ (Ferrero et al. 2013).

P. flexuosa individuals can be one-stemmed or multi-stemmed. A study by Piraino and Roig (2017) comparing these two types found that the mean radial growth of one-stemmed and multi-stemmed trees was 1.74 and 1.36 mm $year^{-1}$ respectively. In addition, a study by Álvarez et al. (2011a) reported a mean radial growth of 1.23 mm $year^{-1}$ for one-stemmed trees and of 1.6 mm $year^{-1}$ for multi-stemmed trees and a culmination age of 104 and 84 years, respectively.

Villagra et al. (2005b) found that wood production of *P. flexuosa* in the Monte desert varied according to the sampling site between 0.2 and 1.3 m^3 ha^{-1} $year^{-1}$ which indicates that the economic potential of these woodlands varies greatly in relation to the ecological conditions. On the other hand, Álvarez et al. (2011a) analyzed wood production of *P. flexuosa* in four types of forests in Monte desert and they found that wood production of *P. flexuosa* was similar among all of them (121.6–173.7 kg ha^{-1} $year^{-1}$) and that there was no significant difference between one-stemmed and multi-stemmed individuals.

Prosopis caldenia has been the focus of several dendrochronological studies in the Pampean Region. Bogino and Villalba (2008) determined that the mean annual radial growth at four sampling sites (between 33°44′09 and 35°25′20 S and 65°31′47 and 65°07′18 W) ranged from 2.7 to 4.0 mm year^{-1} and that the biological rotation age was around 66 years. High groundwater levels, which cause anaerobiosis, were found to negatively affect growth and cause death of *P. caldenia* individuals. A mean radial growth of 2.91 mm year^{-1} (SD ± 0.63) versus 3.41 mm year^{-1} (SD ± 0.94) was determined for trees with and without the negative effect of anaerobiosis, respectively (Bogino and Jobbágy 2011). Anthropogenic activities have also had an effect on this species growth dynamics. A clear example is the massive deforestation brought by the European occupation which caused a reduction in tree competition and consequently a release in the mean radial growth of *P. caldenia* from 1.36–1.81 to 2.5–2.8 mm year^{-1} when analyzing the same tree (Dussart et al. 2011; Tapia and Dussart 2013). Furthermore, Velasco-Sastre et al. (2018) found that in the last 250 years the mean radial growth varied between 1.88 and 2.49 mm year^{-1} according to the sampling site in the Pampean Region. Biomass is one of the most valuable variables when analyzing forests as carbon sinks under climate change conditions. Using tree-ring growth of *P. caldenia* it was possible to estimate that the carbon stock for trees younger than 41 years old was of 20.2 Mg ha^{-1} (Risio et al. 2014). Dendrochronological studies are of great importance since tree-ring growth quantification allows the estimation of biomass throughout time.

Judd et al. (1971) studied the species *Prosopis velutina* in Arizona, USA. They found that groundwater depletion and insects or diseases were the factors affecting tree decay. Mean annual growth was determined to be around 1.5–2 mm year^{-1} (estimated by the authors of this chapter from the figures in the paper). This study is particularly interesting since it is the only one on *Prosopis* growth dynamics outside of Latin America.

To finalize this section we present some results on the effects of ENSO on the *Prosopis* genus growth dynamics. *Prosopis pallida* in Peru showed a significant variability in growth which went from 7.0 mm year^{-1} (SD ± 2.9) in ENSO years (heavy rainy year) to 2.5 mm year^{-1} (SD ± 1) in No-ENSO episodes. *Prosopis chilensis* in Chile, on the other hand, showed no response to this phenomenon with a reported mean annual growth of 1.9 mm year^{-1} (SD ± 1.1) and 1.8 mm year^{-1} (SD ± 0.9) in ENSO and No-ENSO years respectively (Holmgren et al. 2006). Three populations of *P. pallida* were selected in the Peruvian Pacific coast from West to East in order to analyze this species behavior in relation to ENSO events. The individuals showed a strong interannual growth variability determined by drought and wet episodes, showing increases in the growth rate of up to three times in wet years. A mean radial growth between 3.13 and 3.95 mm year^{-1} in trees of 44 years of age was determined. (Salazar et al. 2018).

To recapitulate, all dendrochronological studies of *Prosopis* genus were possible due to the feasibility of tree-ring detection (Villalba et al. 2000). Mean annual radial growth of *P. flexuosa, P. caldenia, P. chilensis, P. pallida, P. ruscifolia* and *P. kuntzei* was determined using dendrochronological techniques. According to their growth rate species can be grouped into those with values around 2.8 mm year^{-1} as *P. caldenia, P. pallida, P. ruscifolia* and *P. kuntzei* and those with values around 1.5 mm year^{-1} as *P. flexuosa, P. velutina* and *P. chilensis*. For the case of *P. pallida* and *P. caldenia*, higher growth rates are linked to higher mean annual rainfall no matter if the species belong to Subtropical (*P. pallida*) or Temperate (*P. caldenia*) climate. Growth rate and recruitment of *Prosopis* sp. can be favored by warmer weather but in many cases other factors as anthropogenic disturbances can significantly affect both variables as well.

6.4 Tree Rings of *Prosopis* Genus as Records of Anthropogenic Pollution Activities Worldwide

Studies conducted on *Prosopis* in the last decade show the value of tree rings as a record of pollution. Concentration of heavy metals as Zn, Cu and Pb have been measured in tree rings of *Prosopis juliflora* in Mexico (Beramendi-Orosco et al. 2013). In India industrial contaminants were proven to affect the wood anatomy of tree rings of *Prosopis spicigera* Linn. (Rajput et al. 2008). Furthermore, in Iran the pollutants produced during the Gulf War were recorded in tree rings of *Prosopis cineraria* (L.) Druce (Korori et al. 2013).

6.5 Final Words

Dendroecological studies of *Prosopis* genus started in 1931 in an effort to address the association between growth and rainfall. From that year until the present many studies have been conducted in an attempt to answer general or regional inquiries. Dendroecological research has evolved from anatomical studies in order to elucidate the feasibility of tree-ring demarcation and consequently the potential of *Prosopis* genus for dendroecological studies, to research where dendrochronology is the main focus or acts as an accessory tool. There is no doubt about the invaluable information that dendrochronological studies of *Prosopis* can provide in relation to growth dynamics, biomass accumulation, fire frequency and climate-growth association facing the increasing global concern about climate change and its consequences. This review not only demonstrates this already well-known value but also presents some evidence of the use of tree rings as a record of pollution which could be an invaluable tool for the determination of contaminants throughout time. While the increasing demand for new ecological knowledge poses a challenge, dendrochronology of *Prosopis* genus has demonstrated to have the power to generate knowledge and answer questions regarding regional and global ecological factors and their complex interactions.

Acknowledgments This chapter could not be possible without the encomiastic contribution of all researches interested on *Prosopis* genus whose publications contribute to this review. We also thank the two anonymous referees for improving the quality of the manuscript.

References

Álvarez JA, Villagra PE, Villalba R et al (2011a) Wood productivity of *Prosopis flexuosa* D.C. woodlands in the Central Monte: influence of population structure and tree-growth habit. J Arid Environ 75:7–13. https://doi.org/10.1016/j.jaridenv.2010.09.003

Álvarez JA, Villagra PE, Villalba R (2011b) Factors controlling deadwood availability and branch decay in two Prosopis woodlands in the Central Monte, Argentina. For Ecol Manag 262:637–645. https://doi.org/10.1016/j.foreco.2011.04.032

Archer S (1995) Tree-grass dynamics in a Prosopis-thornscrub savanna parkland: reconstructing the past and predicting the future. Écoscience 2:83–99. https://doi.org/10.1080/11956860.1995.11682272

Beramendi-Orosco LE, Rodríguez-Estrada ML, Morton-Bermea O et al (2013) Correlations between metals in tree-rings of *Prosopis julifora* as indicators of sources of heavy metal contamination. Appl Geochem 39:78–84. https://doi.org/10.1016/j.apgeochem.2013.10.003

Bogino S, Jobbágy E (2011) Climate and groundwater effects on the establishment, growth and death of *Prosopis caldenia* trees in the pampas (Argentina). For Ecol Manag 262:1766–1774. https://doi.org/10.1016/j.foreco.2011.07.032

Bogino S, Villalba R (2008) Radial growth and biological rotation age of *Prosopis caldenia* Burkart in Central Argentina. J Arid Environ 72:16–23. https://doi.org/10.1016/j.jaridenv.2007.04.008

Bogino S, Roa GC, Velasco-Sastre A et al (2015) Synergistic effects of fire, grazing, climate, and management history on *Prosopis caldenia* recruitment in the Argentinean pampas. J Arid Environ 117:59–66. https://doi.org/10.1016/j.jaridenv.2015.02.014

Bogino S, Cangiano ML, Dussart E et al (2019) Dendrochronological studies of indigenous and creole archeological remains in the Argentinean pampas (19th and 20th centuries). Dendrochronologia 55:25–32. https://doi.org/10.1016/j.dendro.2019.03.005

Bovey R (2016) Mesquite: history, growth, biology, uses and management. Texas A&M University Press, College Station

Bravo S, Kunst C, Grau R (2008) Suitability of the native woody species of the Chaco region, Argentina, for use in dendroecological studies of fire regimes. Dendrochronologia 26:43–52. https://doi.org/10.1016/j.dendro.2007.05.004

Buffington LC, Herbel CH (1965) Vegetation changes on a Semidesert grassland range from 1858 to 1963. Ecol Monogr 35:139–164. https://doi.org/10.2307/1948415

Burkart A (1976) A monograph of the genus Prosopis. J Arnold Arbor 57(4):450–525

Carvalho NJF, Alves PM, Vieira AJR et al (2019) The hydrological performance of *Prosopis juliflora* (Sw.) growth as an invasive alien tree species in the semiarid tropics of northeastern Brazil. Biol Invasions 21:2561. https://doi.org/10.1007/s10530-019-01994-y

Contreras S, Santoni C, Jobbágy EG (2012) Abrupt watercourse formation in a semiarid sedimentary landscape of Central Argentina: the roles of forest clearing, rainfall variability and seismic activities. Ecohydrology 6:794–805. https://doi.org/10.1002/eco.1302

Cozzarín GI (2008) Actividad apícola en el caldenal. In: Gabutti E, MJL P, Barbosa OA (eds) El Caldenal Puntano. Caracterización ecológica y utilización sustentable. Editorial El tabaquillo, Villa Mercedes, pp 63–66

Darwin C (1839) Narrative of the surveying voyages of His Majesty's ships Adventure and Beagle, between the years 1826 and 1836, describing their examination of the southern shores of South America, and the Beagle's circumnavigation of the globe, vol III & appendix, 1st edn. Henry Colburn, London

Decuyper M, Chavez RO, Copini P et al (2016) A multi-scale approach to assess the effect of groundwater extraction on *Prosopis tamarugo* in the Atacama Desert. J Arid Environ 131:25–34. https://doi.org/10.1016/j.jaridenv.2016.03.014
Dussart E, Lerner P, Peinetti R (1998) Long-term dynamics of two populations of *Prosopis caldenia* Burkart. J Range Manag 51:685–691
Dussart EG, Chirino CC, Morici E et al (2011) Reconstrucción del paisaje del caldenal pampeano en los últimos 250 años. Quebracho 19(1–2):54–65
Ferrero ME, Coirini RO, Díaz MP (2013) The effect of wood-boring beetles on the radial growth of *Prosopis flexuosa* DC. in the arid Chaco of Argentina. J Arid Environ 88:141–146. https://doi.org/10.1016/j.jaridenv.2012.07.004
Flinn RC, Archer S, Boutton TW et al (1994) Identification of annual rings in an arid-land woody plant, *Prosopis Glandulosa*. Ecology 75:850–853. https://doi.org/10.2307/1941741
Ghezzi I, Rodríguez R (2015) Primera serie dendroarqueológica en los Andes Centrales: resultados preliminares de Chankillo, Casma. Bulletin 'Institut français d'études andines 44:1–21. https://doi.org/10.4000/bifea.7465
Giantomasi MA, Roig-Juñent FA, Villagra PE et al (2009) Annual variation and influence of climate on the ring width and wood hydrosystem of *Prosopis flexuosa* DC trees using image analysis. Trees 23:117–126. https://doi.org/10.1007/s00468-008-0260-5
Giantomasi MA, Roig-Juñent FA, Villagra PE (2013) Use of differential water sources by *Prosopis flexuosa* DC: a dendroecological study. Plant Ecol 214:11–27. https://doi.org/10.1007/s11258-012-0141-2
Giménez AM, Ríos N, Moglia G (1997) Leño y corteza de itín *Prosopis kuntzei* (Harms) en relación a algunas magnitudes dendrométricas. Invest Agr Sist Recur For 6(1–2):163–182
Giménez AM, Ríos NA, Moglia JG (2003) Crecimiento de *Prosopis nigra* (Griseb.) Hieron (algarrobo negro) en Santiago del Estero, Argentina. Foresta Veracruzana 5(2):17–22
Giménez AM, Ríos N, Hernández P et al (2009) Influencia de la edad en el crecimiento de vinal (*Prosopis ruscifolia* Burkart.), en la Provincia de Santiago del Estero, Argentina. Madera Bosques 15(2):45–57
Giménez AM, Ríos N, Moglia G et al (2010) Determinación de la edad del árbol histórico de la casa de Facundo Quiroga en los Llanos de La Rioja. Quebracho 18(1–2):71–78
Giordano CV, Guevara A, Boccalandro HE et al (2011) Water status, drought responses, and growth of *Prosopis flexuosa* trees with different access to the water table in a warm South American desert. Plant Ecol 212:1123–1134. https://doi.org/10.1007/s11258-010-9892-9
Golubov J, Mandujano M, Eguiarte L (2001) The paradox of mesquites (*Prosopis* spp.): invading species or biodiversity enhancers? B Soc Bot Mex 69:23–30. https://doi.org/10.17129/botsci.1644
Holmgren M, López BC, Gutierrez JR et al (2006) Herbivory and plant growth rate determine the success of El Niño Southern Oscillation-driven tree establishment in semiarid South America. Glob Change Biol 12(12):2263–2271. https://doi.org/10.1111/j.1365-2486.2006.01261.x
Juárez de Galíndez M, Giménez AM, Ríos N et al (2005) Modelación de crecimiento en *Prosopis alba* Griseb. empleando dos modelos biológicos. Quebracho 12:34–42
Juárez de Galíndez M, Giménez AM, Ríos N et al (2007) Modelación del crecimiento en diámetro de vinal (*Prosopis ruscifolia*), en Santiago del Estero, Argentina. Foresta Veracruzana 9(2):9–15
Juárez de Galíndez M, Giménez AM, Ríos N et al (2008) Determinación de la edad de aprovechamiento de individuos de *Prosopis alba* mediante un modelo logístico de intercepto aleatorio para incrementos radiales. Ciencia e Investigación Forestal Instituto Forestal Chile 14(2):287–299
Judd BI, Laughlin JM, Guenther HR et al (1971) The lethal decline of mesquite on the Casa Grande National Monument. Great Basin Naturalist 31(3):153–159
Korori SAA, Shirvany A, Mashaei EM et al (2013) Tree utilization as an indicator to evaluate the degradation of Iran's ecosystems in Persian Gulf War. Casp J Appl Sci Res 2(9):85–94

Krebs C, Fischer G (1931) El pluviómetro secular. In: Circular de la Sección de Propaganda e Informes del Ministerio de Agricultura, República Argentina, vol 847, pp 1–24
Leavitt SW, Long A (1991) Seasonal stable-carbon isotope variability in tree rings: possible paleoenvironmental signals. Chem Geol (Isot Geosci Sect) 87:59–70. https://doi.org/10.1016/0168-9622(91)90033-s
López BC, Sabaté S, Gracia CA et al (2005) Wood anatomy, description of annual rings, and responses to ENSO events of *Prosopis pallida* H.B.K., a wide-spread woody plant of arid and semi-arid lands of Latin America. J Arid Environ 61:541–554. https://doi.org/10.1016/j.jaridenv.2004.10.008
López BC, Rodríguez R, Gracia CA et al (2006) Climatic signals in growth and its relation to ENSO events of two *Prosopis* species following a latitudinal gradient in South America. Glob Change Biol 12(5):897–906. https://doi.org/10.1111/j.1365-2486.2006.01138.x
Medina A (2007) Reconstrucción de los regímenes de fuego en un bosque de *Prosopis caldenia*, Provincia de la Pampa, Argentina. Bosque 28(3):234–240. https://doi.org/10.4067/s0717-92002007000300008
Medina A (2008) Cicatrices de fuego en el leño de *Prosopis caldenia* en Luán Toro, provincia de La Pampa. Bosque 29(2):115–119. https://doi.org/10.4067/s0717-92002008000200003
Medina A, Dussart E, Esterich D et al (2000) Reconstrucción de la historia del fuego en un bosque de *Prosopis caldenia* (Burk.) de Arizona, sur de la provincia de San Luis. Multequina 9:91–98
Morales MS, Villalba R (2012) Influence of precipitation pulses on long-term *Prosopis ferox* dynamics in the Argentinean intermontane subtropics. Oecologia 168:381–392. https://doi.org/10.1007/s00442-011-2087-9
Morales MS, Villalba R, Grau HR et al (2001) Potencialidad de *Prosopis ferox* Griseb (Leguminosae, subfamilia: Mimosoideae) para estudios dendrocronológicos en desiertos subtropicales de alta montaña. Rev Chil Hist Nat 74:865–872. https://doi.org/10.4067/S0716-078X2001000400013
Morales MS, Villalba R, Grau HR et al (2004) Rainfall-controlled tree growth in high-elevation subtropical treelines. Ecology 85:3080–3089. https://doi.org/10.1890/04-0139
Morales MS, Villalba R, Boninsegna JA (2005) Climate, land-use and *Prosopis ferox* recruitment in the Quebrada de Humahuaca, Jujuy, Argentina. Dendrochronologia 22:169–174. https://doi.org/10.1016/j.dendro.2005.05.004
Pasiecznik NM, Felker P, Harris PJC et al (2001) The *Prosopis juliflora-Prosopis pallida* complex: a monograph. HDRA, Coventry
Perpiñal E, Balzarini M, Catalán L et al (1995) Edad de culminación del crecimiento en *Prosopis flexuosa* D.C. en el chaco árido argentino. Invest Agr Sist Recur For 4(1):45–55
Piraino S, Roig JFA (2016) Assessing the sensitivity of riparian algarrobo dulce (*Prosopis flexuosa* dc) radial growth to hydrological changes. Geochronometria 43:1–8. https://doi.org/10.1515/geochr-2015-0027
Piraino S, Roig JFA (2017) Diferencias en el hábito de crecimiento como variable explicativa de la influencia climática en *Prosopis flexuosa* en el Desierto del Monte Central (Argentina). Bol Soc Argent Bot 52(3):523–533
Piraino S, Abraham EM, Diblasi A et al (2015) Geomorphological-related heterogeneity as reflected in tree growth and its relationships with climate of Monte Desert *Prosopis flexuosa* DC woodlands. Trees 29:903–916. https://doi.org/10.1007/s00468-015-1173-8
Piraino S, Abraham EM, Hadad MA et al (2017) Anthropogenic disturbance impact on the stem growth of *Prosopis flexuosa* DC forests in the Monte desert of Argentina: a dendroecological approach. Dendrochronologia 42:63–72. https://doi.org/10.1016/j.dendro.2017.01.001
Polley HW, Johnson HB, Mayeux HS (1994) Increasing CO2: comparative responses of the C4 grass Schizachyrium and grassland invader *Prosopis*. Ecology 75:976–988. https://doi.org/10.2307/1939421
R Core Team R, Carey VJ, Deb Roy S, Eglen S, Guha R, Lewin-Koh N, Myatt M, Ptaff B, Warmerdam F, Weigand S (2016) R: a language and environment for statistical computing. Free Software Foundation, Inc. Available at: http://www.R-project.org

Rajput KS, Rao KS, Kim YS (2008) Cambial activity and wood anatomy in *Prosopis spicigera* (Mimosaceae) affected by combined air pollutants. IAWA J 29(2):209–219. https://doi.org/10.1163/22941932-90000180

Ríos NA, Giménez AM, Moglia JG (2001) Crecimiento del itín (*Prosopis kuntzei* Harms) en la región chaqueña argentina. Madera Bosques 7(1):47–56. https://doi.org/10.21829/myb.2001.711318

Risio L, Bogino S, Bravo F (2014) Aboveground and belowground biomass allocation in native *Prosopis caldenia* Burkart woodlands in the semi-arid Argentinean pampas. Biomass Bioenergy 66:249–260. https://doi.org/10.1016/j.biombioe.2014.03.038

Risio L, Lara HW, Bogino S et al (2018) Aridity variations in the semiarid Argentinean Pampas: How they affect *Prosopis caldenia* growth at the edge of the world distribution area. Dendrochronologia 50:126. https://doi.org/10.1016/j.dendro.2018.05.003

Rivera M, Moya J, Shea D (2010) Dendrocronología en la Pampa del Tamarugal, Desierto de Atacama, Norte de Chile. Diálogo Andino 36:33–50

Rodríguez R, Mabres A, Luckman B et al (2005) "El Niño" events recorded in dry-forest species of the lowlands of northwest Peru. Dendrochronologia 22:181–186. https://doi.org/10.1016/j.dendro.2005.05.002

Rogers KE (2000) The magnificent mesquite. University of Texas Press, Austin

Salazar PC, Navarro-Cerrillo RM, Ancajima E et al (2018) Effect of climate and ENSO events on Prosopis pallida forests along a climatic gradient. Forestry: Int J Forest Res 91(5):552–562. https://doi.org/10.1093/forestry/cpy014

Shackleton RT, Le Maitre DC, Pasiecznik NM et al (2014) *Prosopis*: a global assessment of the biogeography, benefits, impacts and management of one of the world's worst woody invasive plant taxa. AoB Plants 6:plu027. doi:https://doi.org/10.1093/aobpla/plu027

Smeins FE (1983) Origin of the brush problem: a geological and ecological perspective of contemporary distribution. In: McDaniel KC (ed) Proceeding of the brush management symposium. Society for Range Management, Albuquerque, pp 5–16

Tapia A, Dussart E (2013) Aportes de la dendrocronología al estudio de la evolución del caldenar pampeano durante la ocupación ranquelina. Revista del Mus de La Plata (Nueva Serie). Sección Antropología 13(87):361–374

Tarhule A, Hughes MK (2002) Tree-ring research in semiarid West Africa: need and potential. Tree-Ring Res 58:31–46

Tello DS, de Prada J, Cristeche ER (2018) Valoración económica del bosque de caldén (*Prosopis caldenia* Burkart) en el sur de Córdoba, Argentina. Rev Chapingo Ser Cie 24(3):297–312. https://doi.org/10.5154/r.rchscfa.2017.03.027

Velasco-Sastre T, Vergarechea M, Tapia A et al (2018) Growth dynamics and disturbances along the last four centuries in the *Prosopis caldenia* woodlands of the Argentinean pampas. Dendrochronologia 47:58–66. https://doi.org/10.1016/j.dendro.2017.12.005

Viglizzo EF, Frank FC (2006) Ecological interactions, feedbacks, thresholds and collapses in the Argentine pampas in response to climate and farming during the last century. Quat Intl 158:122–126. https://doi.org/10.1016/j.quaint.2006.05.022

Villagra PE, Villalba R, Boninsegna JA (2005a) Structure and growth rate of *Prosopis flexuosa* woodlands in two contrasting environments of the Central Monte desert. J Arid Environ 60:187–199. https://doi.org/10.1016/j.jaridenv.2004.03.016

Villagra PE, Boninsegna JA, Alvarez JA et al (2005b) Dendroecology of *Prosopis flexuosa* woodlands in the Monte desert: implications for their management. Dendrochronologia 22:209–213. https://doi.org/10.1016/j.dendro.2005.05.005

Villalba R, Boninsegna JA (1989) Dendrochronological studies on *Prosopis flexuosa* DC. IAWA Bull 10:155–160. https://doi.org/10.1163/22941932-90000483

Villalba R, Villagra PE, Boninsegna JA et al (2000) Dendroecología y dendroclimatología con especies del género *Prosopis* en argentina. Multequina 9(2):1–18

Virginia RA, Jarrell WM (1983) Soil properties in a mesquite-dominated Sonoran Desert ecosystem. Soil Sci Soc Am J 47:138–144. https://doi.org/10.2136/sssaj1983.03615995004700010028x

Chapter 7
How Past and Future Climate and Drought Drive Radial-Growth Variability of Three Tree Species in a Bolivian Tropical Dry Forest

J. Julio Camarero, Hooz A. Mendivelso, and Raúl Sánchez-Salguero

Abstract Seasonally, dry tropical forests (SDTFs) are among the most diverse and threatened biomes in America. Several tree species coexist in these forests but their long-term growth responses to climate are unknown, and this is needed to make growth forecasts as a function of climate scenarios. We investigated the responses to climate, drought and ocean-atmosphere patterns of three tree species (*Acosmium cardenasii* H.S. Irwin & Arroyo, *Centrolobium microchaete* (Mart. ex Benth.) H.C. de Lima ex G.P. Lewis and *Zeyheria tuberculosa* (Vell.) Bureau coexisting in a Bolivian dry tropical forest. Species chronologies of ring-width indices were related to temperature, precipitation, drought indices and sea temperatures. A growth model was also used to forecast growth variability. *C. microchaete* and *A. cardenasii* presented similar year-to-year growth variability. Cool and wet conditions enhanced growth. Shorter droughts constrained more growth of *C. microchaete* and *A. cardenasii*, whilst longer droughts negatively impacted *Z. tuberculosa*. These different growth responses to climate and drought contribute to explain the coexistence of tree species in SDTFs. The growth patterns of the study species are valuable climate proxies for Bolivia. Forecasted warmer conditions after the 2050s will differently affect the growth variability of these species depending on their responses to climate and drought.

Keywords *Acosmium cardenasii* · *Centrolobium microchaete* · Chiquitano forest · Climate change scenarios · Drought · *Zeyheria tuberculosa*

J. J. Camarero (✉)
Instituto Pirenaico de Ecología (IPE-CSIC), Zaragoza, Spain
e-mail: jjcamarero@ipe.csic.es

H. A. Mendivelso
Grupo Ecología de Organismos (GEO-UPTC), Universidad Pedagógica y Tecnológica de Colombia, Sede Tunja, Tunja, Colombia

R. Sánchez-Salguero
Depto. Sistemas Físicos, Químicos y Naturales, Universidad Pablo de Olavide, Sevilla, Spain
e-mail: rsanchez@upo.es

M. Pompa-García, J. J. Camarero (eds.), *Latin American Dendroecology*,
https://doi.org/10.1007/978-3-030-36930-9_7

7.1 Introduction

Among tropical biomes, seasonally dry tropical forests (SDTFs) are among the most threatened and diverse ecosystems due to their intensive use and rapid deforestation (Janzen 1988; Bullock et al. 1995; Dirzo et al. 2011). SDTFs are still understudied despite they face several menaces of Anthropocene global change including climate warming and human exploitation (Malhi et al. 2014). Long-term data on SDTFs radial growth and productivity such as those obtained from tree rings provide retrospective and prospective views on their responses to anthropogenic drivers and global change (Zuidema et al. 2012, 2013). Nevertheless, tropical forests, and particularly Neotropical biomes, are still underrepresented in global tree-ring data compilations (Zhou et al. 2013; Brienen et al. 2016; Mendivelso et al. 2016a).

South America experienced the greatest loss of forest area (ca. 4 million ha.) in the 2000–2010 period (FAO-JRC 2012). This is the case of SDTFs which represent the 22% of the forest area in South America but have experienced high deforestation rates. For instance, it is estimated that in Colombia only 1.5% of the original area occupied by SDTFs remains nowadays. The loss of these forests is critical if we consider that South American forests represent ca. 33% of all the carbon stored by worldwide forests and constitute a major source of biodiversity (FAO-JRC 2012).

The climate conditions in SDTFs are characterized by a seasonal drought lasting from 4 to 7 months and leading to deciduousness (from 2 to 6 months) in many trees, vines and herbs (Murphy and Lugo 1986). In SDTFs the annual rainfall ranges between 250 and 2000 mm (it can be higher than in some rain forests), and the ratio of potential evapotranspiration to precipitation is greater than one (Murphy and Lugo 1986). Despite drought leads to low relative humidity during the day (20–60%) and most woody plants cease their vegetative activities, many of them flower and disperse their seeds during that dry season feeding many animals (Janzen 1988).

Ecologically, SDTFs are unique biomes because they allow the coexistence of many tree species subjected to a marked seasonal drought which constrains growth and controls phenological patterns (Borchert 1999; Brienen and Zuidema 2005; Rozendaal and Zuidema 2011). During the wet season, most of the growth occurs, whereas during the dry season growth arrests and many tree species shed their leaves and present stem shrinkage due to water loss from the sapwood and bark (Mendivelso et al. 2014, 2016b). Therefore, precipitation seasonality and drought cause high resource variability between years. In such pulsed resource environments, different radial-growth patterns and responses to climate and drought can contribute to tree species coexistence through different temporal niche dynamics (Alfaro-Sánchez et al. 2017). Tree-ring data can help to answer if increasing temporal resource heterogeneity due to precipitation and drought seasonality supports the coexistence of tree species (cf. Pulla et al. 2017).

Many growth and phenological responses to climate drought have been described among coexisting tree species in SDTFs (Enquist and Leffler 2001; Mendivelso et al. 2013, 2014, 2016a). In SDTFs, stem radial growth is coupled with leaf development and persistence during the wet season or during the transition from the dry to the wet seasons (Borchert 1994). However, the relationships between climate, drought and radial growth may differ among coexisting tree species depending on their phenology (e.g., leaf flushing, period with maximum growth rate), functional traits (e.g., wood density), site wetness conditions or climate seasonality (e.g., Mendivelso et al. 2016b). In most South American STDFs such climate seasonality is mainly controlled by coupled atmospheric-sea patterns related to El Niño Southern Oscillation (ENSO) and captured by the Southern Oscillation Index (SOI) which has been shown to be correlated with radial-growth variability in dry forests (López et al. 2005; Rodríguez et al. 2005; Brienen et al. 2010; Ramírez and del Valle 2011; Pucha-Cofrep et al. 2015; Espinosa et al. 2018). The ENSO depends on changes in sea temperature across the equatorial Pacific Ocean (Aceituno 1988).

The Chiquitania region, located in the Bolivian lowlands, is one of the major remnants of SDTFs in South America despite being subjected to several pressures including deforestation and climate warming (Pennington et al. 2009). Every summer multiple fires are initiated and used to expand agricultural fields and pastures by destroying SDTFs, and forecasts indicate that climate warming and drying could increase fire risk and severity reducing the extension of Chiquitania SDTFs (Devisscher et al. 2016). The influence of the ENSO on growth of trees in Chiquitania STDFs is understudied. There are a few studies conducted on *Centrolobium microchaete* (Mart. ex Benth.) H.C. de Lima (López and Villalba 2011) and *Machaerium scleroxylon* Tul. (Paredes-Villanueva et al. 2013). Precipitation patterns in eastern Bolivia are greatly controlled by this atmospheric pattern, but in the rest of the country they are influenced by other factors such as the Andean highs and the zones of intertropical convergence and of convergence of the south Atlantic (Ronchail 1995; Seiler et al. 2013). Previous studies in a Chiquitania SDTF found that tree species such as *Zeyheria tuberculosa* (Vell.) Bureau, *Acosmium cardenasii* H.S. Irwin & Arroyo and *Centrolobium microchaete* (Mart. ex Benth.) H.C. de Lima were the most responsive in terms of radial growth, changes in water availability and drought stress (Mendivelso et al. 2013, 2014).

Here we explore the relationships between the year-to-year variability in climate, drought severity and radial growth in *Z. tuberculosa, A. cardenasii* and *C. microchaete*. We also analyze if climate-drought coupling is related to atmospheric patterns as reflected by the SOI. We test the hypothesis that these three coexisting species will show different growth responses to climate and drought depending on their strategies to tolerate water shortage.

7.2 Materials and Methods

7.2.1 Study Area and Tree Species

The study area is located in the private property INPA forest (Fig. 7.1), which covers an area of 30,000 ha and is located in the Ñuflo de Chávez province (61° 43'W, 16° 07' S, 503 m a.s.l.), 40 km northeast of Concepción, Santa Cruz department, south east Bolivia. According to climate data (mean monthly temperature, total monthly precipitation) from the Concepción station (62° 06' W, 16° 15' S, 497 m a.s.l.) the mean annual temperature is 24.3 °C, but it can drop to minimum values of 8 °C during the dry season (May–October), and the annual rainfall is 1,100 mm (Fig. 7.1d). During the dry season, most canopy tree species shed their leaves (Mostacedo 2007).

The study area is located in the Brazilian Precambrian shield forming oxisol and acid soils, poor in nutrients and with a sandy-loam texture. INPA topography is varied from slightly flat plains to wavy hills (Navarro and Maldonado 2002).

The INPA forest is a Chiquitano SDTF where 115 different species of trees with a diameter larger than 10 cm have been identified (Mostacedo 2007). The most abundant tree species is *A. cardenasii* despite it is not the most exploited species (Table 7.1). The canopy is quite open with a mean height of 20 m. The tree density and the basal area are 437 stems ha^{-1} and 19.7 m^2 ha^{-1}, respectively (Mostacedo 2007).

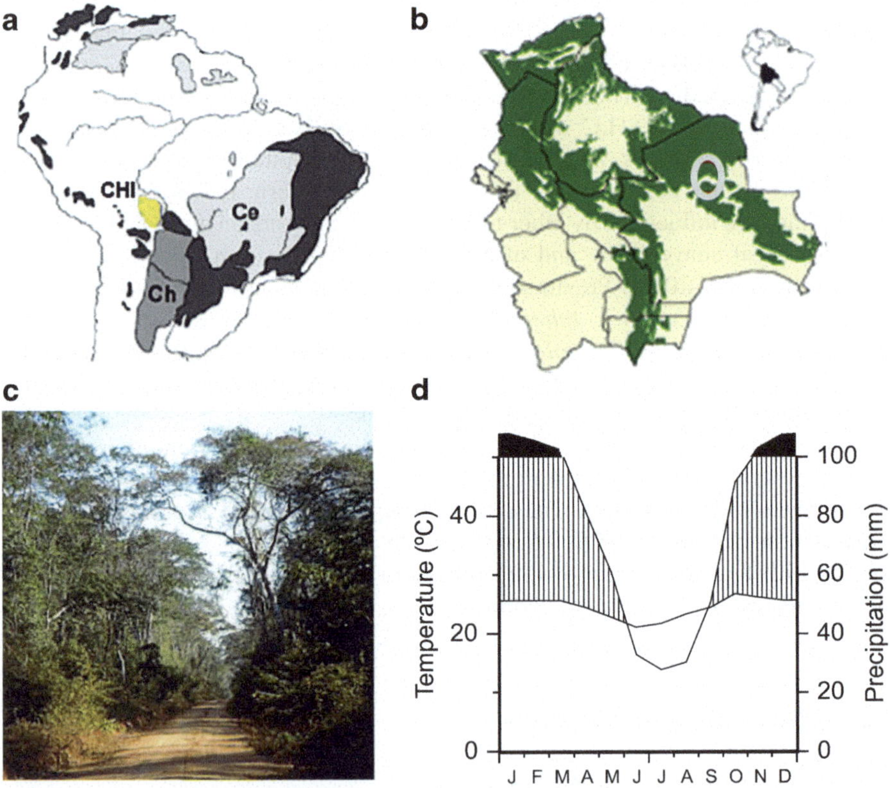

Fig. 7.1 Situation and features of the study site. (**a**) Distribution of South American tropical dry forests and scrublands (dark-grey and black patches) and savanna (grey patches) biomes, (**b**) location of the study area (oval) in the Chiquitania (SE Bolivia; green patches are forests), (**c**) view of the study site (INPA) and (**d**) climate diagram (based on data from Concepción station). In the (**a**) plot the yellow patch indicates the Chiquitania (CHI) forests, and Ch and Ce refer to the Chaco and Brazilian Cerrado formations, respectively. (Modified from Pennington et al. 2009)

Table 7.1 Characteristics of the three studied species

Tree species (common name)	Family	Diameter at 1.3 m (cm)	Basal area (m^2 ha^{-1})	Density (stems ha^{-1})
Zeyheria tuberculosa (Vell.) Bureau (tajibo cabeza de mono)	Bignoniaceae	38.2 ± 3.6	0.04	3
Acosmium cardenasii H.S. Irwin & Arroyo (tasaá)	Fabaceae	29.2 ± 3.9	6.89	159
Centrolobium microchaete (Mart. ex Benth.) H.C. de Lima ex G.P. Lewis (tarara amarilla)	Fabaceae	36.0 ± 3.2	0.55	11

Diameter data refer to the sampled individuals and are means ± standard deviation (SD). Basal area and density were taken from Killeen et al. (1998), Mostacedo (2007) and Mostacedo et al. (2009)

About 60% of trees have some degree of infestation of lianas. The understory is dense with abundant shrubs, lianas and terrestrial bromeliads.

According to Mostacedo et al. (2009) and Villegas et al. (2009), *Z. tuberculosa,* and *C. microchaete* are long-lived pioneers forming the canopy, whereas *A. cardenasii* can be considered a sub-canopy, shade-tolerant species. Leaf expansion starts in October–November in the two Fabaceae species (*C. microchaete* and *A. cardenasii*) and in December in *Z. tuberculosa*. The leafless period may last from two (*A. cardenasii, Z. tuberculosa*) to five (*C. microchaete*) months (Mostacedo 2007). Among the studied species, *A. cardenasii* is restricted to the Bolivian Chiquitano forests, whereas the others two are found in other Bolivian tropical forests.

7.2.2 Sampling and Chronology Building

The exploitation of some of tree species in INPA allowed us to collect wood discs from 7 to 14 living and dominant trees per species with diameters at 1.3 m bigger than 20 cm (Table 7.2), which were felled for timber collection in 2010 and 2011 (see more details in Mendivelso et al. 2013). In tropical dendrochronology it is important to select appropriate species showing annual and conspicuous rings and studying wood discs, given the variability of growth within the stem and the difficulty of recognizing ring borders (Worbes 1995; Schöngart et al. 2017).

In the laboratory, the wood samples were air dried and polished using sandpapers with increasing grit (from 80 to 1000) until the tree rings were clearly visible. The rings were visually cross-dated and measured with a resolution of 0.01 mm along three radii per disc, separated by 120°, using a LINTAB-TSAP system. The dating was checked using the program COFECHA, which compared the series of each tree to a master chronology of each species (Holmes 1983).

To generate the chronologies or means of ring-width indices (TRWi) of each species, standardization and detrending were performed. This process allows eliminating the biological and geometric growth trends that are not associated with climatic

Table 7.2 Characteristics of the sampled trees and analyzed wood samples. Values are means ± standard error (SE)

Tree species (common name)	No. trees (No. radii)	Age (years)	Ring width ± SE (mm)	AR1	Rbar	EPS
Z. tuberculosa	7 (21)	112 ± 11	1.77 ± 0.12	0.51	0.20	0.83
A. cardenasii	11 (38)	92 ± 5	1.93 ± 0.09	0.54	0.24	0.92
C. microchaete	14 (42)	89 ± 3	1.82 ± 0.10	0.47	0.26	0.94

Abbreviations: *AR1* first-order autocorrelation, *Rbar* mean inter-series correlation, *EPS* Expressed Population Signal

variables but with enlargement as the tree ages, making the data comparable among species (Cook and Kairiukstis 1990; Fritts 2001). We fitted a negative exponential function and obtained the TRWi by dividing the observed by fitted ring-width values. Then, the temporal autocorrelation associated with the growth of the previous year was eliminated by fitting autoregressive models, and the resulting TRWi were averaged using robust bi-weight means to obtain a residual chronology of each species. These procedures were developed with the ARSTAN v 4.4. software (Cook 1985; Cook and Krusic 2007).

To characterize the chronologies, a series of basic statistics were calculated (Briffa and Jones 1990): the first-order autocorrelation of ring-width data (AR1), the mean inter-series correlation of the chronologies (Rbar), and the expressed population signal (EPS) which measures how well replicated is a chronology (Wigley et al. 1984).

7.2.3 Climate, Drought, SOI, Sea Surface Temperature and Projected Climate Scenarios

Since local climate data from Concepción station showed gaps and heterogeneities, we obtained long-term climate data from the 0.5°-gridded CRU TS data set v. 4.02 (Harris et al. 2014). Monthly mean temperature and total precipitation data were obtained from the grid with coordinates 61.0–61.5° W and 16.0–16.5° S using the KNMI-Climate explorer webpage (https://climexp.knmi.nl/start.cgi). Sea surface temperatures at 0–100 m depth for the equatorial and South Pacific Ocean was also retrieved from the same webpage for the 1955–2010 period (Levitus et al. 2012).

To quantify drought severity, we used the monthly values of the self-calibrating Palmer drought severity index (hereafter PDSI; Wells et al. 2004) and the standardized precipitation evapotranspiration index (SPEI; see Vicente-Serrano et al. 2010). Both drought indices are calculated using CRU gridded climate data. The PDSI allows assessing long-term droughts by using surface air temperature and a physical water balance model, but it is not calculated at different time scales. The PDSI measures relative dryness and may vary from −10 (dry conditions) to +10 (wet conditions). The SPEI is calculated at different time scales and considers the effect of temperature on the rate of evapotranspiration, with negative and positive SPEI values reflecting dry and humid conditions, respectively. The SPEI was calculated for the same 0.5° grid as climate data and it was obtained for 1- to 24-month long scales from the web page http://sac.csic.es/spei/index.html. SOI and PDSI data were obtained from the webpage https://crudata.uea.ac.uk/cru/data/. The SOI measures the large-scale changes in air pressure between the western and eastern tropical Pacific by calculating the sea level pressure differences between Tahiti and Darwin (Australia).

The climate data projected for the twenty-first century was downloaded from the webpage https://esgf-node.llnl.gov/projects/esgf-llnl/ and downscaled (cf. Moreno and Hasenauer 2015) at a 0.5° spatial resolution from the fifth phase of the Coupled Model Intercomparison Project (ensemble CMIP5; Taylor et al. 2012). We selected two Intergovernmental Panel on Climate Change (IPCC 2013) AR5 (Fifth Assessment Report) scenarios, specifically the scenario (RCP 8.5) that most closely tracked recent historical emissions of greenhouse gases, and one lower-emission scenario (RCP 2.6) in which the increase in annual emissions is more gradual and declines after the mid-twenty-first century. These scenarios result in +4.8° C (RCP 8.5) +0.3° C (RCP 2.6) global warming rates, respectively, by the year 2100, relative to the late-twentieth-century baseline.

7.2.4 Assessing Relationships Between Growth Variability, Climatic Variables and Drought

To assess the similarity between growth variability, Pearson correlations (r) and their associated significance levels (p) were calculated between the three chronologies for the common, best-replicated period (1950–2010). These correlations were also used to relate the three chronologies with monthly climate data (mean temperature, total precipitation) and also with the two drought indices (PDSI, SPEI). In the case of the SPEI correlations were calculated considering all months and 1- to 24-month scales. Correlations between the chronologies and the climate data were calculated focusing on the assumed growing period from November to February (cf. Mendivelso et al. 2016b). Since lagged growth responses to climate are common, we considered a wide window of analyses from June 2 years before up to August.

To analyze the influence of large-scale precipitation and atmospheric patterns (ocean temperatures) on TRWi, spatial correlations were calculated between monthly values of those variables and the species' chronologies. These correlations were calculated using the KNMI Climate Explorer webpage.

We simulated TRWi for each species as a function of climate using the VS-Lite model and a Bayesian parameter estimation approach (Tolwinski-Ward et al. 2011, 2013). For each year, the model simulates standardized tree-ring width anomalies from the minimum of the monthly growth responses to temperature (gT) and moisture (gM), modulated by insolation (gE). This model uses the Leaky Bucket Model of hydrology to estimate monthly soil moisture. Day length is determined from site latitude and does not vary from year-to-year. The growth response functions for temperature (gT) and moisture (gM) in VS-Lite involve two parameters. The first parameter represents the temperature (T_1) or moisture (M_1) thresholds below which growth will not occur, whereas the second parameter is the optimal temperature (T_2) or moisture (M_2) thresholds above which growth is not limited by climate. The growth function parameters were estimated for each species using Bayesian calibration assuming normally distributed errors for the modeled ring-width indices. The posterior median for each parameter was used to obtain the calibrated growth response for the selected species. Finally, the model was run over the entire period 1950–2010 using the calibrated parameters to produce a simulated chronology that represents an estimate of the site climate signal of growth (Tolwinski-Ward et al. 2013).

The model was evaluated 10,000 times for each site using three parallel Markov Chain Monte Carlo chains with uniform prior distribution for each parameter and a white Gaussian noise model error (Tolwinski-Ward et al. 2013). To compute annual TRWi values, we integrated the overall simulated growth rates (gT, gM and gE) over the time window from November of the year prior to growth to February of the year of tree-ring formation (cf. Mendivelso et al. 2016b). To evaluate the temporal stability of the calibrated growth response functions, we divided the period 1950–2010 into two intervals (1950–1980, 1980–2010) and withhold the second half for validation of the parameters estimated in the first half.

7.2.5 *Testing Forecasts of Year-to-Year Growth Variability*

The climate-growth relationships (mean temperature and precipitation) calculated using Pearson correlations were re-examined by applying stepwise multiple linear regressions to identify the effects of climate on the observed tree-ring growth data, and to project ring-width indices as a function of climate for each species (Sánchez-Salguero et al. 2017a, b). Only those climate variables highly correlated with growth ($r > |0.30|$, $p < 0.05$) were considered in these projections. Monthly climate variables were transformed into normalized standard deviations to give them the same weight in the models. Additionally, we evaluated the existence of multicollinearity among explanatory variables by calculating the variance inflation factor (VIF). In all cases VIF was lower than 2, confirming no redundancy problems. The selected models were run to forecast TRWi of each species for the 2011–2050 and 2050–2100 periods under the two emission scenarios. Finally, we ran VS-Lite models on forecasted ring-width series over the same periods to estimate growth responses (gT, gM) and parameters (T_1, T_2, M_1, M_2) under future climate projections. The four parameters were compared between periods using Tukey HSD post-hoc tests.

All analyses were conducted using the R statistical package (R Development Core Team 2019). We used the function dredge (Barton 2012) of the R package *MuMIn* using the lowest Akaike information criterion (AIC) for selecting the most parsimonious regression model (Burnham and Anderson 2002). The models were fitted using the generalized least-squares estimation (GLS) and the R package *nlme* (Pinheiro et al. 2016).

7.3 Results

7.3.1 Climate and Growth Variability and Patterns

Considering their common overlapping period, local (data from Concepción station) and CRU growing-season climate data (Fig. 7.2a) showed a significant correlation regarding precipitation (r = 0.56, p = 0.0003, 1970–2006 period) but not for temperature (r = −0.03, p = 0.89; 1982–2006 period). Precipitation and temperature were inversely related considering local climate data (r = −0.45, p = 0.025), but this correlation was not significant for CRU climate data (r = −0.21, p = 0.20). The growing-season PDSI showed two dry periods in the late 1960s to early 1970s and in the mid-1990s, preceded by wet periods in the late 1950s and early 1980s (Fig. 7.2b). According to the PDSI, the 1994 and 1982 years showed the driest and wettest growing seasons, respectively.

In total, we measured 101 radii and the oldest tree (years) corresponded to *Z. tuberculosa* with 139 years (1879–2010 period) followed by an *A. cardenasii* individual with 103 years, and a *C. microchaete* individual with 95 years (Table 7.2). The mean radial-growth rate of the three species was 1.84 mm yr^{-1}, and the mean AR1 was 0.52. The Rbar and EPS values were high, particularly in *C. microchaete*, and the EPS was above the 0.85 threshold for the 1950–2010 period, excepting in *Z. tuberculosa* (0.83). This period was used in subsequent analyses.

The year-to-year growth variability captured by the species residual chronologies (ring-width indices) showed that *A. cardenasii* and *C. microchaete* presented more similar patterns, whereas *Z. tuberculosa* presented a slightly different high-frequency pattern (Table 7.3, Fig. 7.3).

7.3.2 Associations Between Climate, SOI and Growth Variability

The climate-growth associations were characterized by negative responses to growing-season temperature during the growth and previous years in the three species, but responses during the prior year were more important in *C. microchaete* and *A. cardenasii* (Fig. 7.4a). The three species showed positive correlations with monthly precipitation during the current and prior growing seasons which peaked in November (*C. microchaete*, *A. cardenasii*) or February (*Z. tuberculosa*) (Fig. 7.4b). Growth responses to prior precipitation were notable in the case of *Z. tuberculosa*,

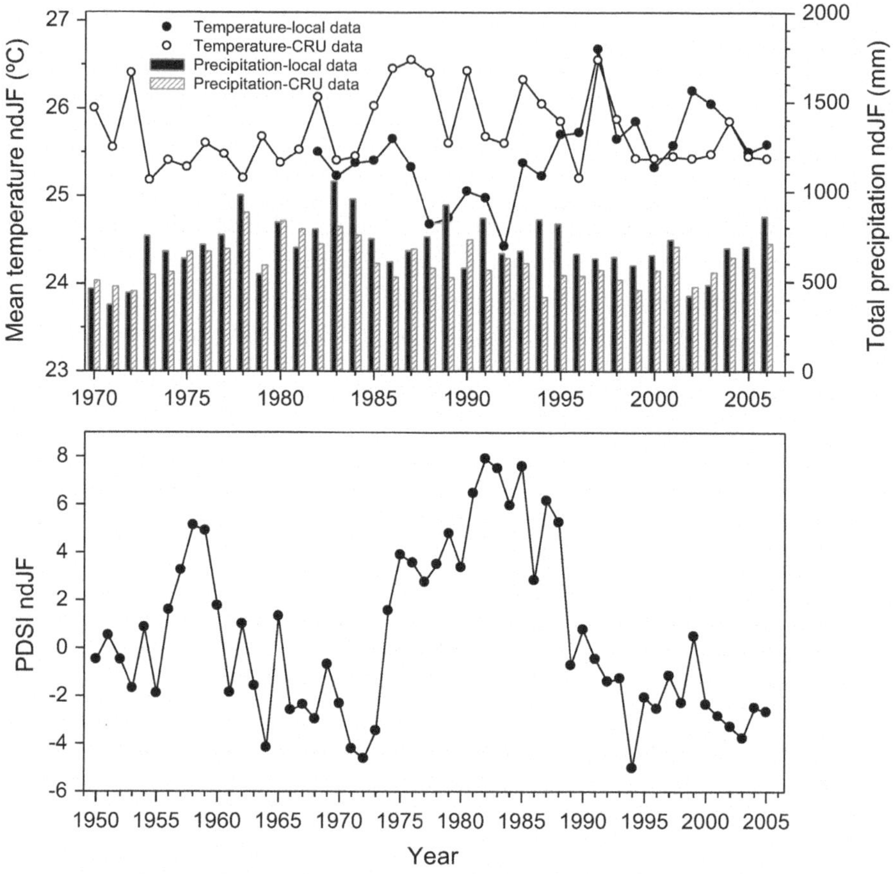

Fig. 7.2 Climate trends and patterns in the study area based on (**a**) temperature and precipitation data for the assumed growing season, i.e. from November to February (abbreviated ndJF) and (**b**) drought severity (PDSI). Local climate data were obtained from the Concepci6n meteorological station

Table 7.3 Pearson correlations and significance levels (between parentheses) calculated among the residual chronologies of the three species and considering the common, best-replicated period 1950–2010

Species	*Z. tuberculosa*	*A. cardenasii*
Z. tuberculosa		
A. cardenasii	0.25 (0.052)	
C. microchaete	0.26 (0.045)	0.58 (0.0001)

which explains why this species showed the highest correlations with the PDSI, peaking in prior June (Fig. 7.4c). The correlation between the growing-season PDSI and the *Z. tuberculosa* chronology was positive and significant (r = 0.48, p = 0.0001). By contrast *C. microchaete* and *A. cardenasii* showed the highest

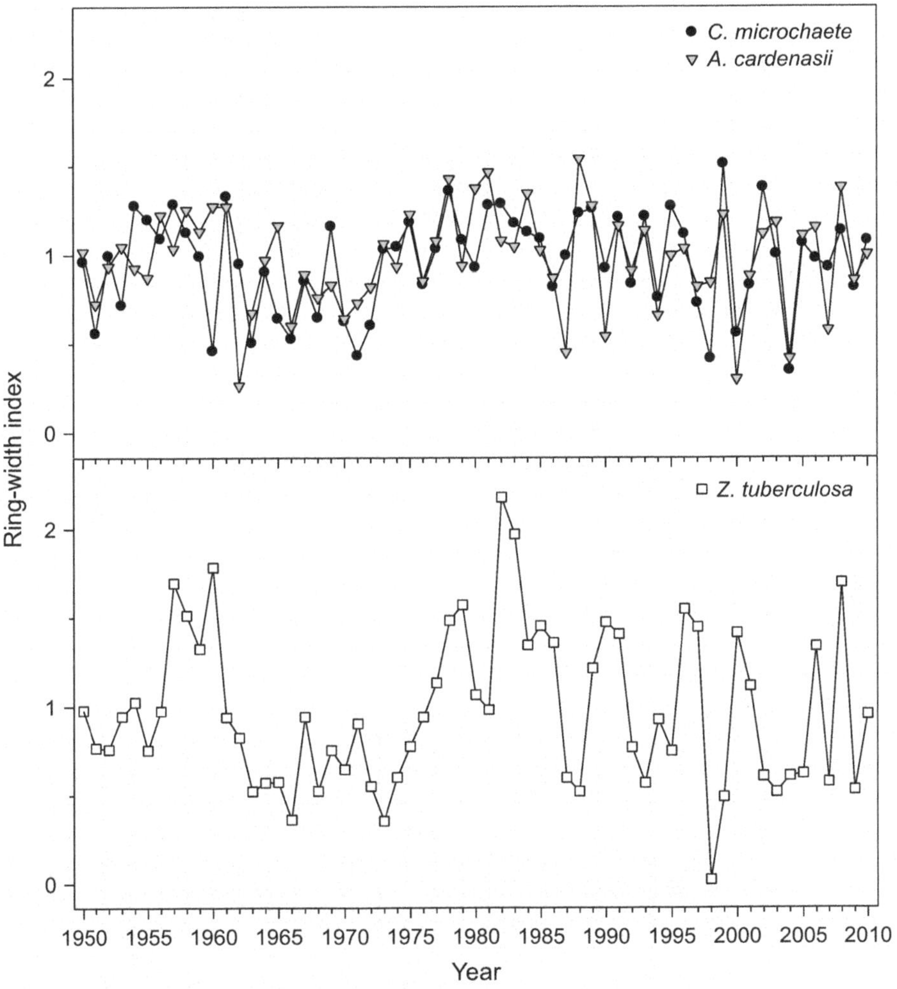

Fig. 7.3 Chronologies of the three study species

correlations with the PDSI in December and October, respectively, i.e. during the mid and early growing season.

Regarding the correlations between the SOI and the chronologies, we found that the November, January, February and March SOI values were negatively related to the *Z. tuberculosa* ring-width indices (Fig. 7.4d). In the other two species, SOI values of the prior growing-season (January, March, April) also showed negative and significant correlations.

The *Z. tuberculosa*chronology was negatively related to the SOI values averaged for the growing season ($r = -0.45$, $p = 0.0002$), but reached a minimum correlation with March SOI values ($r = -0.53$, $p = 0.000009$; Fig. 7.5). The period January–March showed the minimum correlations of the *Z. tuberculosa* chronology with SOI

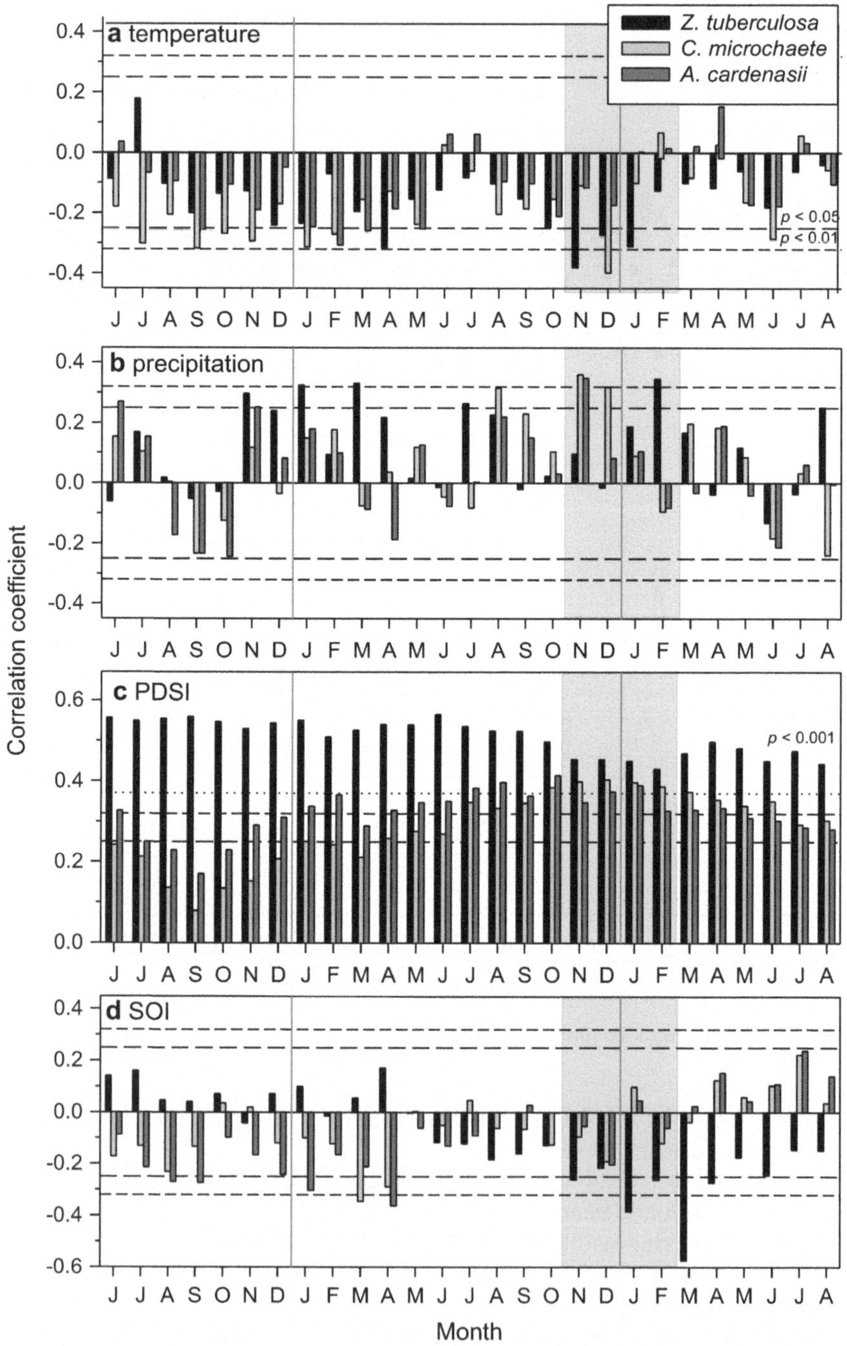

Fig. 7.4 Pearson correlations calculated between the residual chronologies of the three study species and monthly climate data (**a**, mean temperature; **b**, total precipitation; CRU gridded data set), drought severity (**c**, PDSI) and the SOI (**d**). In all plots the horizontal dashed and dotted lines show significance levels as indicated ($p < 0.05$, $p < 0.01$, $p < 0.001$). The grey vertical boxes include the assumed growing period from November to February. The previous period of growth includes the previous growing season and the prior autumn and spring

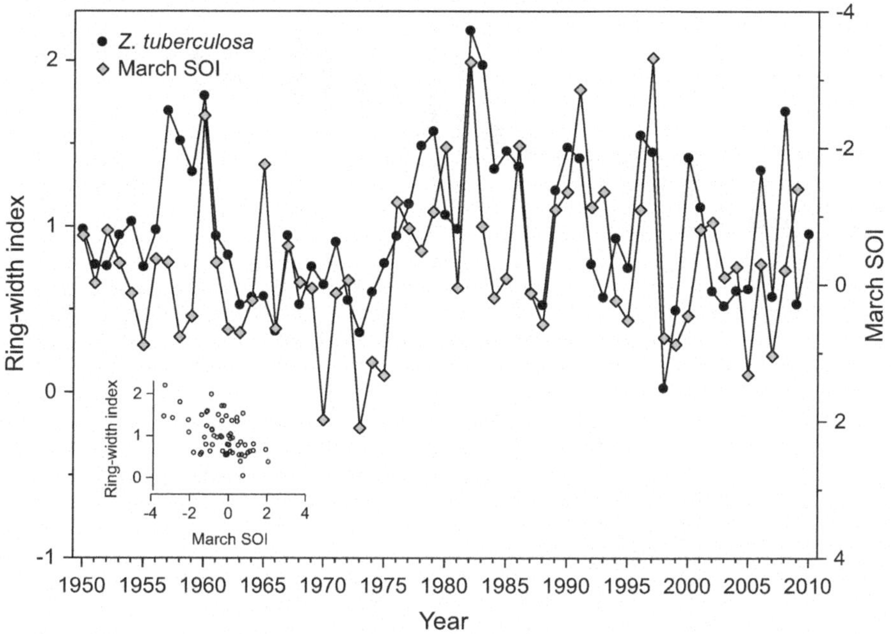

Fig. 7.5 Negative association observed between the *Z. tuberculosa*chronology and the March SOI values. Note the reverse scale of the SOI values (right *y* axis). The inset shows the scatter plot relating the March SOI and the *Z. tuberculosa* ring-width indices

suggesting that this corresponds to the growing season of the species and the period with wettest conditions. The *Z. tuberculosa* chronology was also negatively related to the SOI annual values ($r = -0.41$, $p = 0.001$).

7.3.3 Associations Between the SPEI and Growth Variability

When comparing species, *Z. tuberculosa* growth variability was much more responsive to longer droughts than *C. microchaete* and *A. cardenasii* (Fig. 7.6). The SPEI scales (drought duration) when maximum SPEI-growth correlations (shown between parentheses) were found for each species were 23 ($r = 0.62$), 9 ($r = 0.48$) and 10 ($r = 0.41$) months for *Z. tuberculosa*, *C. microchaete* and *A. cardenasii*, respectively (Fig. 7.6).

The highest correlations of the *C. microchaete* with the SPEI were found for April SPEI values and 9-month scales, i.e. for droughts lasting from prior August to April ($r = 0.47$, $p = 0.00011$; Fig. 7.7). This was the same SPEI month (April) for *A. cardenasii* but considering 10-month long droughts ($r = 0.41$, $p = 0.0014$), but not in *Z. tuberculosa* which showed a maximum response to November SPEI values calculated at 23-month long scales ($r = 0.61$ $p = 1.33\ 10^{-7}$).

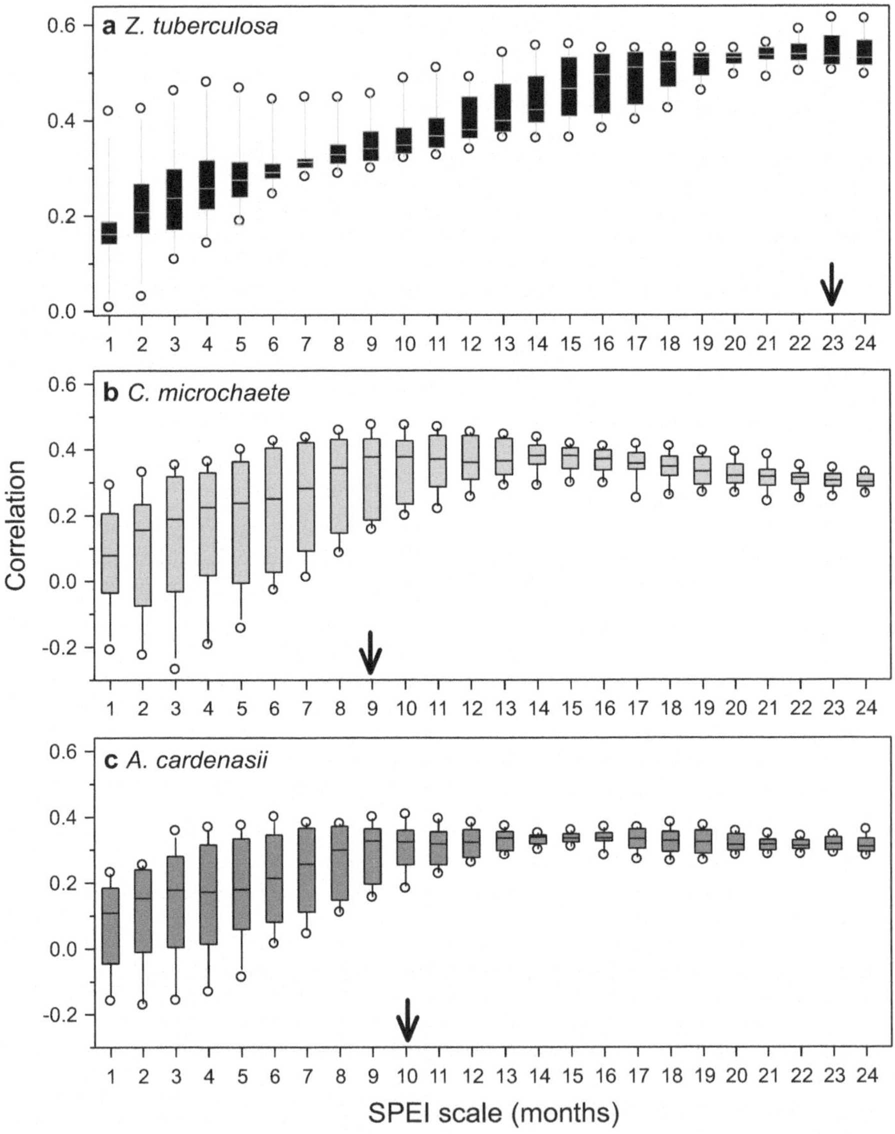

Fig. 7.6 Correlations (Pearson coefficients, r) calculated between droughts of different duration (month, x axis) based on the SPEI values and the chronologies of the three study species. The box plots show all correlations values calculated for the 12 months of a year considering from 1- to 24-month long scales. Significance levels and corresponding correlation thresholds are $p < 0.05$ for $|r| > 0.25$; $p < 0.01$ for $|r| > 0.30$; and $p < 0.001$ for $|r| > 0.40$. The arrows indicate the months when maximum correlations were found for each species

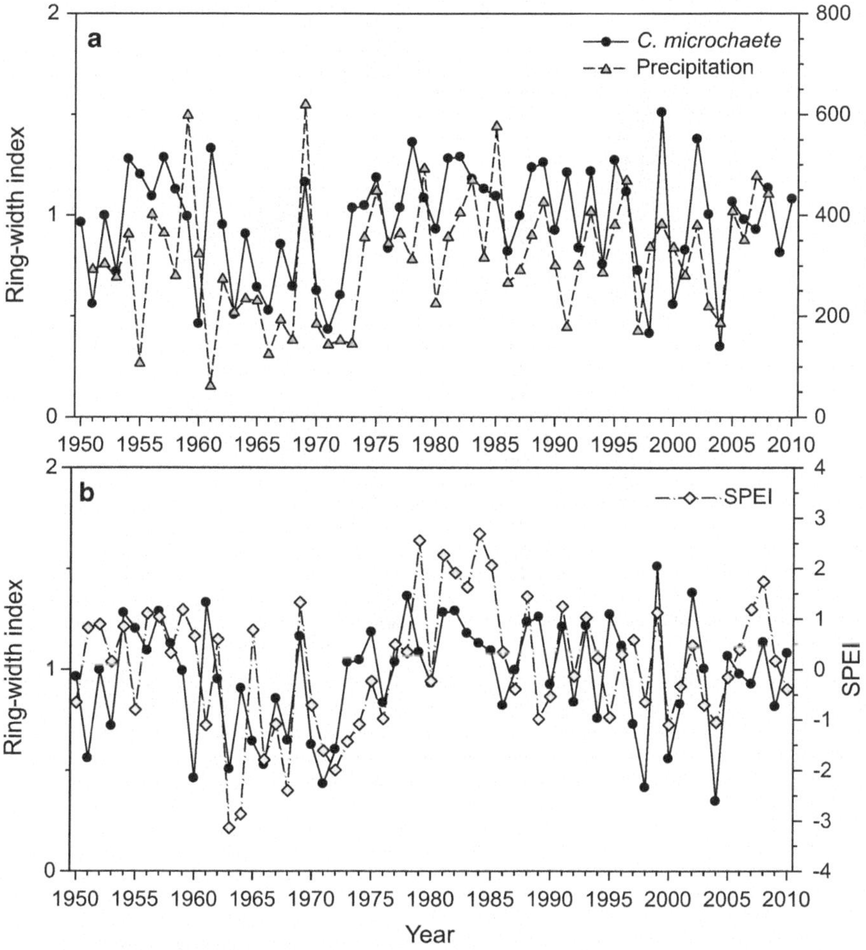

Fig. 7.7 Main climate variable (precipitation) and drought index (SPEI) influencing *C. microchaete* growth variability. The (**a**) precipitation accumulated from November to December and (**b**) the 9-month SPEI April values are positively correlated with the *C. microchaete* chronology

7.3.4 Responses of Growth Variability to Regional Climate

In this analysis, we first compared *C. microchaete* and *A. cardenasii* which showed similar responses to precipitation (Fig. 7.4b) and the two drought indices (PDSI, SPEI; see Figs. 7.4c and 7.6). The spatial correlations showed that their chronologies positively responded to prior November precipitation variability across a region centered in eastern Bolivia (Fig. 7.8a, b). Similar results were observed when using November PDSI or SPEI values (not presented). In the case of *C. microchaete*, the

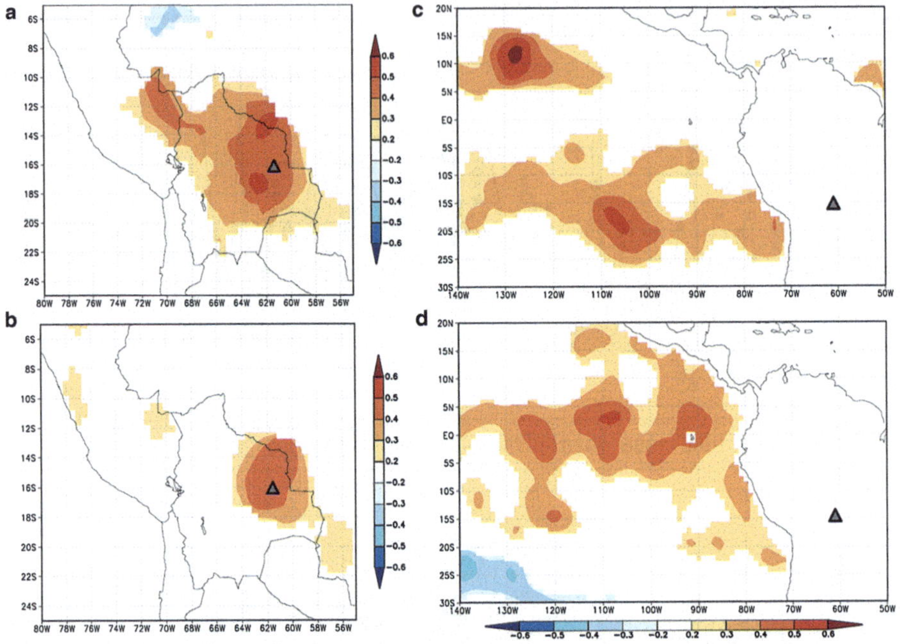

Fig. 7.8 Spatial correlations (color scale) calculated between prior November precipitation across Bolivia and nearby countries and the chronologies of (**a**) *C. microchaete* and (**b**) *A. cardenasii* and between prior April or March sea surface temperatures across the Pacific Ocean and the (**c**) *C. microchaete* and (**d**) *A. cardenasii* chronologies, respectively. The overall significance of the correlation plots is lower than 0.1. The triangles show the approximate location of the study site (INPA forest)

area with positive correlations not only included eastern Bolivia lowlands but also reached south-eastern Peru. The highest correlation of the *C. microchaete* chronology with sea surface temperatures was found for prior April considering an area across the equatorial Pacific Ocean and at longitudes 125–130° W and latitudes 5–15° N and at 100–110° W and 15–25° S (Fig. 7.8c). In the case of the *A. cardenasii* chronology, the strongest correlation was found for prior March sea surface temperatures at longitudes 105–110° W and latitudes 0–5° N (Fig. 7.8d).

In the case of *Z. tuberculosa* the positive spatial correlation between February precipitation and its chronology encompassed central Bolivia (Fig. 7.9a) whilst the highest correlation of the chronology was observed with April sea surface temperatures across the equatorial Pacific Ocean and at longitudes 130–140° W and latitudes 5–10° N and at 70–80° W and 20–25° S near the northern Chile coast (Fig. 7.9b).

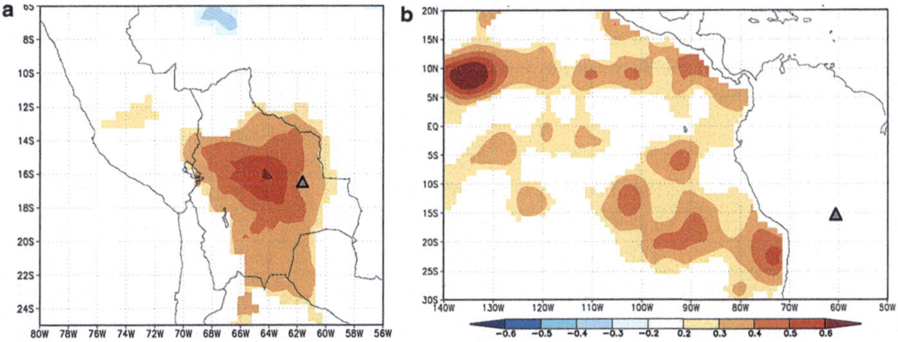

Fig. 7.9 Spatial correlations (color scale) calculated between (**a**) current February precipitation across Bolivia and nearby countries or (**b**) April sea surface temperatures across the Pacific Ocean and the *Z. tuberculosa* chronology. The overall significance of the correlation plots is lower than 0.1. The triangles show the approximate location of the study site (INPA forest)

7.3.5 Process-Based Model of Growth and Climate-Based Forecasts

The VS-Lite model accurately predicted TRWi during the 1950–2010 period for the three species (Table 7.4). The modeled growth response to temperature (gT) and soil moisture (gM) were similar among species (Fig. 7.10). The mean growth response to soil moisture (gM) dropped during late winter to early spring as expected in response to high evapotranspiration rates and soil moisture deficit. The mean growth response to temperature (gT) dropped from May to August only in *C. microchaete*. Tree growth was limited by soil moisture (gM < gT) in the three species during the year with peaks from May to November; see Fig. 7.10).

The selected climate-based models explained on average 62%, 55% and 57% of the TRWi variance in *Z. tuberculosa*, *A. cardenasii* and *C. microchaete*, respectively (Table 7.5). The climate predictors were similar to the most influential monthly climate variables detected in the analyses of climate-growth relationships (Fig. 7.4). The predicted TRWi series based on GLS climate-growth models were within the range of the observed TRWi values for the 1950–2010 period (Fig. 7.11).

The warmer and drier conditions projected by the scenario tracking recent historical emissions (RCP 8.5) forecasted a reduction of TRWi in *A. cardenasii* and *C. microchaete* during the late-twenty-first century, this growth decline being the most pronounced since 2050 (Fig. 7.11). Contrastingly, *Z. tuberculosa* TRWi would increase until 2100.

Under the warmest climate scenario (RCP 8.5), the three species growth is predicted to be increasingly constrained by soil moisture availability (gM < gT), and also by warmer and drier conditions as temperatures rise, thus experiencing more drought stress during the growing season (Fig. 7.10). The estimated minimum and

Table 7.4 Pearson correlation coefficients (r) calculated between observed and fitted series of ring-width indices, and statistics of the Bayesian estimation of the VS-Lite growth response parameters (T_1, T_2, M_1, and M_2) of the three species considering the 1950–2010 period and the second half of the twenty-first century with two IPCC AR5 emission scenarios (RCP 2.6 and RCP 8.5)

Tree species	r	Parameter (units)	Observed 1950–2010	IPCC emission scenarios 2050–2100 r	RCP 2.6	r	RCP 8.5
Z. tuberculosa	0.43	T_1 (°C)	4.13aA	0.32	7.27b	0.45	6.96B
		T_2 (°C)	18.21b		15.27a		18.89
		M_1 (v/v)	0.04bB		0.01a		0.03A
		M_2 (v/v)	0.49b		0.20a		0.49
A. cardenasii	0.25	T_1 (°C)	4.69aA	0.30	5.75b	0.28	6.09B
		T_2 (°C)	14.79aA		15.71b		18.54B
		M_1 (v/v)	0.07bB		0.03a		0.05A
		M_2 (v/v)	0.50b		0.35a		0.49
C. microchaete	0.27	T_1 (°C)	5.76bA	0.26	3.80a	0.25	7.80B
		T_2 (°C)	18.25B		18.20		16.14A
		M_1 (v/v)	0.02bB		0.01a		0.01a
		M_2 (v/v)	0.49b		0.34a		0.50

Different letters indicate significant ($p < 0.05$) differences between periods (lowercase letters for RCP 2.6; uppercase letters for RCP 8.5 scenario) based on Tukey HSD post-hoc tests

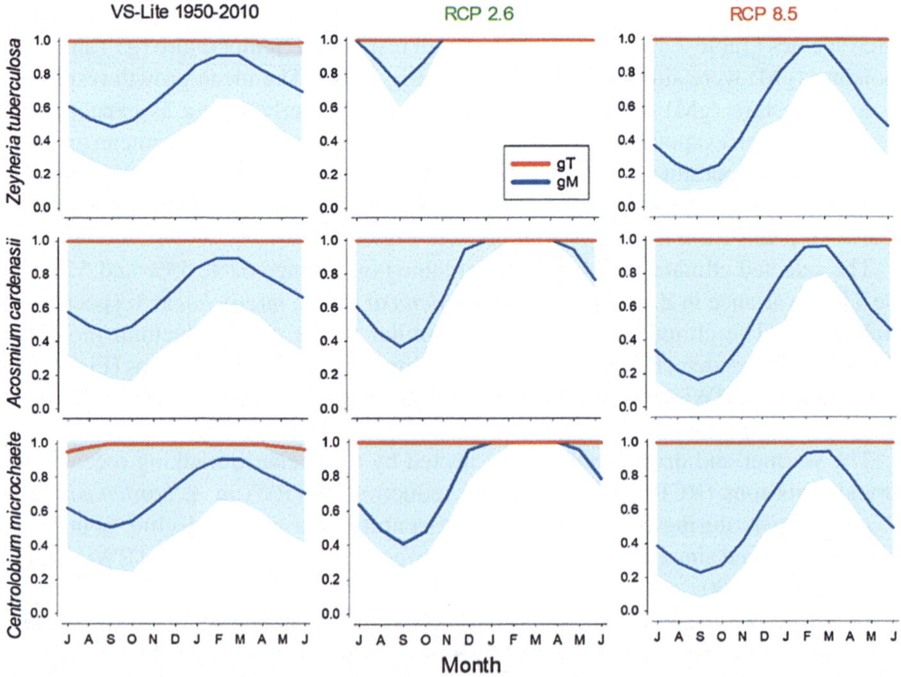

Fig. 7.10 Simulated (first column of plots, 1950–2010 period) and projected (second and third columns of plots) monthly growth response curves considering temperature (gT, red lines and areas) and soil moisture limitations (gM, blue lines and areas) for each tree species. Projected growth response curves correspond to two IPCC AR5 emission scenarios (RCP 2.6 and RCP 8.5, with +0.3° and + 4.8 ° C projected warming, respectively) and to the 2050–2100 period

Table 7.5 Summary of the generalized least squares regression models used to forecast species chronologies of ring-width indices

Tree species	R^2adj (%)	Equation	
Z. tuberculosa	0.62	$0.96 + 0.19\ T_7 - 0.09T_{4p} + 0.12T_{9p} - 0.11T_{11p} - 0.16T_{12p} - 0.13Pr_6 + 0.11Pr_{1p} + 0.10Pr_{3p} + 0.21Pr_{4p}$	(7.1)
A. cardenasii	0.55	$0.96 - 0.05T_{ndJF} - 0.10T_{5p} + 0.05Pr_1 - 0.11Pr_6 + 0.05Pr_{3p} + 0.05Pr_{7p} + 0.08Pr_{8p} + 0.13Pr_{11p}$	(7.2)
C. microchaete	0.57	$0.95 - 0.08T_{ndJF} + 0.10T_{3p} - 0.08T_{5p} + 0.10T_{8p} + 0.07Pr_{11p} - {}_{12p} + 0.13Pr_{8p} + 0.08Pr_{11p}$	(7.3)

The percentage of variance explained by climate-based models is given by the adjusted coefficient of determination (R^2adj). Abbreviations: *Pr* precipitation, *T* mean temperature. Numbers after climate variables indicate months (i.e., from 1 January to 12 December), whereas the subscript "p" indicates the previous year

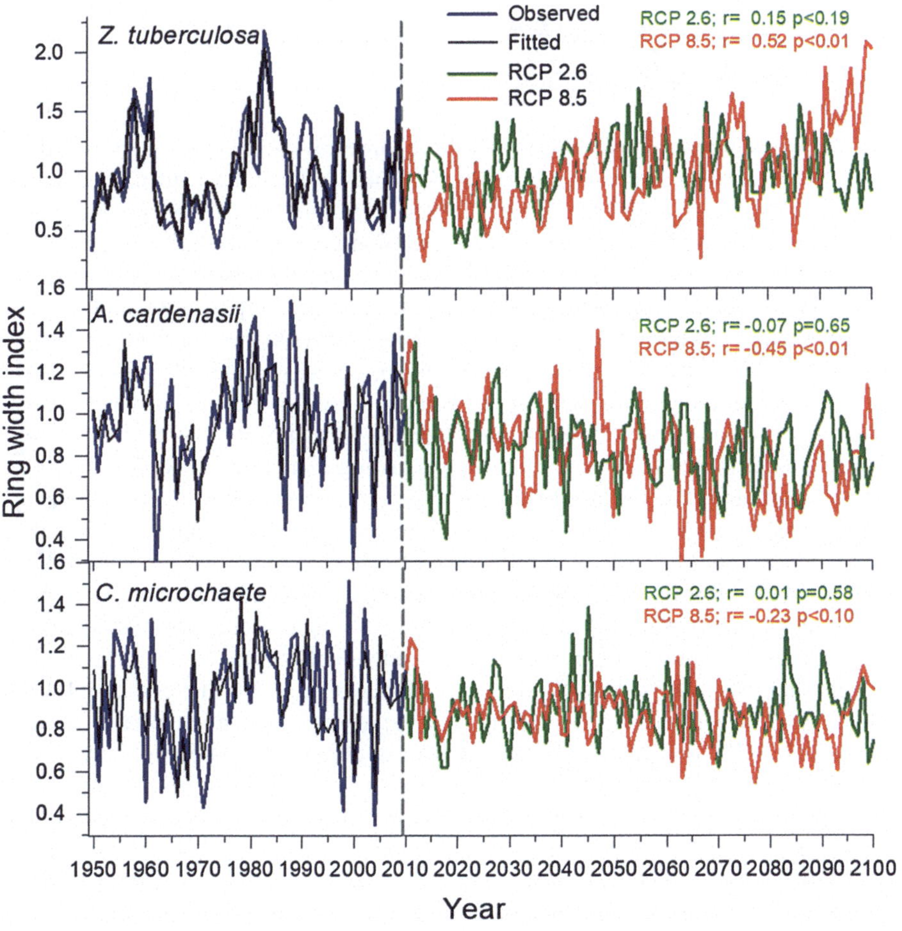

Fig. 7.11 Observed, fitted and projected tree-ring width indices for the three study species. Mean series for each species correspond to the 1950–2010 period (mean observed and fitted values as a function of climate are shown as dark blue and black lines, respectively). Projected indices are based on two IPCC AR5 emission scenarios for the 2011–2100 period (RCP 2.6 -green line- and RCP 8.5 -red line- with +0.3° and +4.8 °C projected warming, respectively). The Pearson correlation (*r*) values quantify the linear trends of the corresponding projected indices (probability levels are also presented)

optimal thresholds (T_1, T_2 and M_1, M_2) for growth confirmed the highest sensitivity of the species to low soil moisture (maximum M_2 value) and that of *Z. tuberculosa* to warm temperatures (minimum T_1 values). We observed an increase of the minimum and optimum temperatures thresholds of growth (T_1 and T_2) in the *A. cardenasii* and *Z. tuberculosa* under the warmer RCP 8.5 scenario (Table 7.4). Thus, the estimated minimum soil moisture (M_1) would decrease in the three species. Consequently, growing conditions would deteriorate because of declining soil water availability, this limitation being more pronounced in *A. cardenasii* and *C. microchaete* than in *Z. tuberculosa* (Table 7.4).

7.4 Discussion

7.4.1 Explaining Different Growth Responses to Climate and Drought

Our findings support the hypothesis that these three coexisting species show different growth responses to climate and drought which indicate they display different functional strategies to tolerate water shortage. On one side, *C. microchaete* and *A. cardenasii* showed a high year-to-year common growth variability (they shared 34% of common variability). Growth of these two species positively responded to cool and wet conditions during the early to mid-growing season and it was constrained by mid-term droughts lasting 9–10 months. On the other side, *Z. tuberculosa* showed a different growth pattern. Its growth was enhanced by cool conditions throughout the growing season and wet conditions in the late growing season, explaining the lowest correlations with January–March SOI. The growth of this species was restricted by long-term droughts lasting 23 months making it the most responsive to drought, followed by *C. microchaete*. In *C. microchaete* a negative relationship between growth and temperature was reported (López and Villalba 2011), which could be explained because higher temperatures rise evapotranspiration rates and constrain wood formation (Clark et al. 2010).

The dendrochronological statistics (Table 7.2) such as the low Rbar values indicate a high variability in growth between individuals of the same species, which is common in tropical studies (Worbes 1995; Brienen et al. 2016), but the high EPS values confirm a reliable cross-dating and appropiate replication of the studied chronologies. The described growth responses to climate do not completely match with the ecological strategies of the study species because *Z. tuberculosa* and *C. microchaete* are shade-intolerant species while *A. cardenasii* is a shade-tolerant species. The similar growth variability and responses to climate and drought of the two Fabaceae species (*C. microchaete* and *A. cardenasii*) may be related to their similar phenological patterns since they show leaf flushing 1–2 months earlier than *Z. tuberculosa*. More long-term data on the phenology of primary and secondary growth would help to answer if aboveground phenology (e.g., bud bursting, leaf flushing and shedding) influences growth responses to climate and drought in these species and in other SDTFs.

According to phenology, xylogenesis and dendrometer data, *C. microchaete* and *A. cardenasii* showed maximum radial-growth rates in the study site from October to March, during the wet season, and their phenological patterns of primary (leaf flushing) and secondary (stem increment) growth were not very coupled (Mendivelso et al. 2016b). Regrettably, no data was available for *Z. tuberculosa*. This uncoupling between primary and secondary growth was explained because leaf phenology responded to water availability at daily to monthly scales, whereas radial increment responded to monthly precipitation (Mendivelso et al. 2016b). Automatic dendrometers recorded high radial increment rates in November (*A. cardenasii*) and in January–March (*C. microchaete*), but such stem growth was also reversed by dry phases leading to reversible stem rapid shrinkage which could account for –40% (*A. cardenasii*) to –25% (*C. microchaete*) reductions in stem radius during the growing season (Mendivelso et al. 2016b). During that season phases of radial contraction and increment were positively and negatively related to air temperature, respectively. This was attributed to the indirect effects of high air temperature on water use by trees through the rise in evapotranspiration rates, vapor pressure deficit and the decrease in soil water availability (Mendivelso et al. 2016b). Hourly to diurnal changes of environmental conditions drive evapotranspiration rates and water loss through the canopy associated to stem shrinkage (Goldstein et al. 1998) which peaked during the afternoon in *C. microchaete* and *A. cardenasii* (Mendivelso et al. 2016b). Stem and bark water stores may be recharged during the night when growth may peak as has been suggested in other tropical forests (Bräuning et al. 2009). Further research on hourly, diurnal, monthly and yearly changes of stem increment as related to environmental variables (air and soil temperatures and humidity, radiation) and xylogenesis studies would help to determine when radial growth occurs (see for instance Volland-Voigt et al. 2011; García-Cervigón et al. 2017), and if this heterogeneity in the temporal niche of growth contributes to explain tree species coexistence in SDTFs (Pulla et al. 2017). In some SDTFs, individualistic and species-specific growth responses to precipitation and soil moisture were observed (Enquist and Leffler 2001) which would partially agree with the two responses detected here, namely *C. microchaete*–*A. cardenasii* vs. *Z. tuberculosa*.

The projections of growth variability (TRWi) suggest that warmer and drier conditions would lead to a more pronounced limitation of growth in *A. cardenasii* and *C. microchaete* because of a decline in soil water availability and increased evaporative demand, whilst *Z. tuberculosa* could show growth enhancement which may be explained by its delayed aboveground phenology. Such late leaf expansion could help to avoid the negative effects of late droughts on *Z. tuberculosa* growth. The high sensitivity of *Z. tuberculosa* and *C. microchaete* to precipitation is related to their sapwood density according to Mendivelso et al. (2013). These authors showed that the wood density of several tree species was negatively related to growth variability driven by water balance. In that study *Z. tuberculosa* growth was very sensitive to changes in the climate water balance, and this species and *C. microchaete* presented the lowest wood density values among seven tree species coexisting in the

INPA forest, whilst *A. cardenasii* presented intermediated wood density values. The climate-driven projections of ring-width indices suggest different responses in these coexisting species with *A. cardenasii* and *Z. tuberculosa* showing the most negative and positive responses, respectively, to the projected increasingly warmer conditions. These differences emphasize the importance of considering functional traits as wood density which could contribute to explain why so different climate-driven forecasts of growth variability are produced by the VS-Lite model. We argue that similar studies based on climate-driven forecasts of TRWi should carried out in diverse SDTFs where drought is the major growth constraint but where coexisting tree species show different responses to the seasonal water shortage. These forecasts could compare projected growth responses to climate warming among coexisting tree species in SDTFs with different drought duration or severity or forecasts for the same species occupying SDTFs with different drought duration or severity. Currently, *C. microchaete* is the most promising candidate for such forecasts in the case of Bolivian SDTFs forming the western margin of the Amazonian tropical forests (López et al. 2017, 2019).

This study has focused on radial growth but this is a sole component of forest dynamics. To better forecast the future of SDTFs subjected to warmer and drier conditions we need more long-term series of other relevant variables characterizing forest dynamics including regeneration and mortality rates. This objective has not been fulfilled, but INPA forest contains valuable species and has provided relevant data (e.g., Markesteijn et al. 2011) to achieve it and to reach conclusions which could be generalized for other SDTFs.

7.4.2 Links Between Climate-Drought-Growth Relationships and Sea-Atmospheric Patterns

In Peruvian and Ecuadorian STDFs, the growth responses to climate and the SOI depended on site conditions with stronger responses in very or moderately dry sites or years (Paredes-Villanueva et al. 2013; García-Cervigón et al. 2017; Espinosa et al. 2018). In other cases, these growth reactions occurred in response to strong or lasting El Niño events (e.g., 1983) associated to warmer Pacific sea temperatures and leading to high pulses of precipitation or strong La Niña events (e.g., 1971) associated to colder Pacific sea temperatures and leading to severe droughts (Rodríguez et al. 2005; Pucha-Cofrep et al. 2015). In extreme environments such as Tumbesian STDFs located in northern Peru and southern Ecuador the high year-to-year rainfall variability is the main factor limiting growth and the available soil moisture mainly depends on precipitation (Lugo et al. 1978). We argue this is not the case of Chiquitania STDFs where growth may be sometimes more responsive to temperature than to precipitation variability (Mendivelso et al. 2013, 2014, 2016b).

In *C. microchaete*, which is widely distributed across Bolivian tropical lowlands (Pirie et al. 2009), the spatial correlations indicated that its growth variability is a reliable proxy of growing-season (November) precipitation across the eastward Andean-Amazonian piedmont at longitudes from 60° to 72° W and latitudes from 12° to 20° S. This agrees with what López et al. (2017, 2019) reported. This species' chronology also captured changes in sea temperature across the equatorial Pacific, albeit *Z. tuberculosa* chronology better recorded those ENSO-related oceanic patterns and similar warmer sea conditions in the northern Chilean coast at ca. 1000 km from the INPA study site. The correlations of the precipitation near INPA show a delay of 1–2 months with respect to the warming of the tropical Pacific which could be explained by hiatus phenomena connected to the cooling trends observed in the study area and in the Chilean coast (Vuille et al. 2015).

Our findings confirm that warmer Pacific sea temperatures were linked to wet conditions and improved growth in the three study species in agreement with the negative associations found with the SOI. Overall, *C. microchaete* and *Z. tuberculosa* seem promising species to reconstruct ENSO-related patterns including sea temperatures in the equatorial Pacific and the northern Chilean coast, and precipitation across eastern Bolivia.

7.5 Conclusion

The year-to-year growth variability and responses to climate were similar in *C. microchaete* and *A. cardenasii*, but *Z. tuberculosa* presented a different pattern. Growth of the three species increased in response to cool and wet conditions during the current and prior growing seasons. However, *Z. tuberculosa* growth was more constrained by longer droughts whereas shorter droughts affected more the other two species. These different growth responses to climate and drought contribute to explain the coexistence of tree species in SDTFs. Forecasts based on a process-based model indicate that *Z. tuberculosa*, a late leaf-flushing, shade-intolerant species with low wood density, will better tolerate the projected warmer and drier conditions. Tree-ring data contribute to a better understanding of the ecology of tree species in SDTFs including assessments of their long-term responses to climate and drought.

Acknowledgments We thank colleagues at the Instituto Boliviano de Investigación Forestal (IBIF) for their support, particularly V. Vroomans and M. Toledo (now director of the "Museo de Historia Natural Noel Kempff Mercado"), and INPA Co. staff (P. Roosenboom, G. Urbano) for allowing the field sampling. This research was funded by the project "Análisis retrospectivos mediante dendrocronología para profundizar en la ecología y mejorar la gestión de los bosques tropicales secos (Dentropicas)" financed by BBVA Foundation.

References

Aceituno P (1988) On the functioning of the southern oscillation in the south American sector part I: surface climate. Mon Weath Rev 116:505–524. https://doi.org/10.1175/1520-0493(1988)116<0505:OTFOTS>2.0.CO;2

Alfaro-Sánchez R, Muller-Landau HC, Wright SJ et al (2017) Growth and reproduction respond differently to climate in three Neotropical tree species. Oecologia 184:531–541. https://doi.org/10.1007/s00442-017-3879-3

Barton K (2012) MuMIn: Multi-model inference R package version 177. Retrieved from https://cran.r-project.org/web/packages/MuMIn/index.html

Borchert R (1994) Soil and stem water storage determine phenology and distribution of tropical dry forest trees. Ecology 75:1437–1449. https://doi.org/10.2307/1937467

Borchert R (1999) Climatic periodicity phenology and cambium activity in tropical dry forest trees. IAWA J 20:239–247

Bräuning A, Volland-Voigt F, Burchardt I et al (2009) Climatic control of radial growth of *Cedrela montana* in a humid mountain rainforest in southern Ecuador. Erdkunde 63:337–345

Brienen RJW, Zuidema PA (2005) Relating tree growth to rainfall in Bolivian rain forests: a test for six species using tree ring analysis. Oecologia 146:1–12. https://doi.org/10.1007/s00442-005-0160-y

Brienen RJW, Lebrija-Trejos E, Zuidema PA et al (2010) Climate-growth analysis for a Mexican dry forest tree shows strong impact of sea surface temperatures and predicts future growth declines. Glob Ch Biol 16:2001–2012. https://doi.org/10.1111/j.1365-2486.2009.02059.x

Brienen RJW, Schöngart J, Zuidema PA (2016) Tree rings in the tropics: insights into the ecology and climate sensitivity of tropical trees. In: Goldstein G, Santiago LS (eds) Tropical tree physiology. Springer, Cham, pp 439–461

Briffa KR, Jones PD (1990) Basic chronology statistics and assessment. In: Cook ER, Kairiukstis LA (eds) Methods of dendrochronology: applications in the environmental sciences. Kluwer, Dordrecht, pp 137–152

Bullock SH, Mooney HA, Medina E (1995) Seasonally dry tropical forest. Cambridge University Press, Cambridge

Burnham KP, Anderson DR (2002) Model selection and multimodel inference: a practical information-theoretic approach. Springer, New York

Clark DB, Clark DA, Oberbauer SF (2010) Annual wood production in a tropical rain forest in NE Costa Rica linked to climatic variation but not to increasing CO_2. Glob Ch Biol 16:747–759. https://doi.org/10.1111/j.1365-2486.2009.02004.x

Cook ER (1985) A time series analysis approach to tree ring standardization [PhD thesis]. Univ. Arizona, Tucson

Cook ER, Kairiukstis LA (1990) Methods of dendrochronology: applications in the environmental science. Kluwer, Dordrecht

Cook ER, Krusic P (2007) ARSTAN 44 A Tree-ring standardization program based on detrending and autoregressive time series modeling with interactive graphics https://www.ldeocolumbiaedu/tree-ring-laboratory/resources/software. Accessed 23-08-19

Devisscher T, Anderson LO, Aragão LEOC et al (2016) Increased wildfire risk driven by climate and development interactions in the Bolivian Chiquitania Southern Amazonia. PLoS One 11:e0161323

Dirzo R, Young HS, Mooney HA et al (2011) Seasonally dry tropical forest: ecology and conservation. Island Press, Washington, DC

Enquist BJ, Leffler AJ (2001) Long-term tree ring chronologies from sympatric tropical dry-forest trees: individualistic responses to climatic variation. J Trop Ecol 17:41–60. https://doi.org/10.1017/S0266467401001031

Espinosa CI, Camarero JJ, Gusmán AA (2018) Site-dependent growth responses to climate in two major tree species from tropical dry forests of Southwest Ecuador. Dendrochronologia 52:11–19. https://doi.org/10.1016/j.dendro.2018.09.004

FAO-JRC (2012) Global forest land-use change 1990–2005. Food and Agriculture Organization (FAO) of the United Nations, European Commission Joint Research Centre (JRC), Rome

Fritts HC (2001) Tree rings and climate. The Blackburn Press, Caldwell

García-Cervigón AI, Camarero JJ, Espinosa CI (2017) Intra-annual stem increment patterns and climatic responses in five tree species from an Ecuadorian tropical dry forest. Trees 31:1057–1067. https://doi.org/10.1007/s00468-017-1530-x

Goldstein G, Andrade JL, Meinzer FC et al (1998) Stem water storage and diurnal patterns of water use in tropical forest canopy trees. Plant Cell Environ 21:397–406. https://doi.org/10.1046/j.1365-3040.1998.00273.x

Harris I, Jones PD, Osborn TJ et al (2014) Updated high-resolution grids of monthly climatic observations – the CRU TS310 dataset. Int J Climatol 34:623–642. https://doi.org/10.1002/joc.3711

Holmes RL (1983) Computer-assisted quality control in tree ring dating and measurement. Tree Ring Bull 43:69–78

IPCC- Intergovernmental Panel on Climate Change (2013) Climate change 2013: the physical science basis. Contribution of working group I to the fifth assessment report of the intergovernmental panel on climate change. Cambridge University Press, Cambridge

Janzen DH (1988) Tropical dry forest: the most endangered major tropical ecosystem. In: Wilson EO (ed) Biodiversity. National Academy Press, Washington, DC, pp 130–137

Killeen TJ, Jardim A, Mamani F et al (1998) Diversity, composition and structure of a tropical semideciduous forest in the Chiquitanía region of Santa Cruz Bolivia. J Trop Ecol 14:803–827. https://doi.org/10.1017/S0266467498000583

Levitus S, Antonov JI, Boyer TP et al (2012) World ocean heat content and thermosteric sea level change (0–2000 m) 1955–2010. Geophys Res Lett 39:10603. https://doi.org/10.1029/2012GL051106

López L, Villalba R (2011) Climate influences on the radial growth of *Centrolobium microchaete* a valuable timber species from the tropical dry forests in Bolivia. Biotropica 43:41–49. https://doi.org/10.1111/j.1744-7429.2010.00653.x

López BC, Sabaté S, Gracia CA et al (2005) Wood anatomy description of annual rings and responses to ENSO events of *Prosopis pallida* HBK a wide-spread woody plant of arid and semi-arid lands of Latin America. J Arid Environ 61:541–554. https://doi.org/10.1016/j.jaridenv.2004.10.008

López L, Stahle D, Villalba R et al (2017) Tree ring reconstructed rainfall over the southern Amazon Basin. Geophys Res Lett 44:7410–7418. https://doi.org/10.1002/2017GL073363

López L, Rodríguez-Catón M, Villalba R (2019) Convergence in growth responses of tropical trees to climate driven by water stress. Ecography 42:1899. https://doi.org/10.1111/ecog.04296

Lugo AE, Gonzalez-Liboy JA, Cintron B et al (1978) Structure productivity and transpiration of a subtropical dry forest in Puerto Rico. Biotropica 10:278–291. https://doi.org/10.2307/2387680

Malhi Y, Gardner TA, Goldsmith GR et al (2014) Tropical forests in the Anthropocene. Ann Rev Env Resour 39:125–159. https://doi.org/10.1146/annurev-environ-030713-155141

Markesteijn L, Poorter L, Bongers F et al (2011) Hydraulics and life history of tropical dry forest tree species: coordination of species' drought and shade tolerance. New Phytol 191:480–495. https://doi.org/10.1111/j.1469-8137.2011.03708.x

Mendivelso HA, Camarero JJ, Royo Obregón O et al (2013) Differential growth responses to water balance of coexisting deciduous tree species are linked to wood density in a Bolivian tropical dry forest. PLoS One 8:e73855. https://doi.org/10.1371/journal.pone.0073855

Mendivelso HA, Camarero JJ, Gutiérrez E et al (2014) Time-dependent effects of climate and drought on tree growth in a Neotropical dry forest: short-term tolerance vs long-term sensitivity. Agric For Meteorol 188:13–23. https://doi.org/10.1016/j.agrformet.2013.12.010

Mendivelso HA, Camarero JJ, Gutiérrez E (2016a) Dendrocronología en bosques neotropicales secos: métodos avances y aplicaciones. Ecosistemas 25:66–75

Mendivelso HA, Camarero JJ, Gutiérrez E et al (2016b) Climatic influences on leaf phenology xylogenesis and radial stem changes at hourly to monthly scales in two tropical dry forests. Agric For Meteorol 216:20–36. https://doi.org/10.1016/j.agrformet.2015.09.014

Moreno A, Hasenauer H (2015) Spatial downscaling of European climate data. Int J Climatol 36:1444–1458. https://doi.org/10.1002/joc.4436
Mostacedo B (2007) Natural regeneration of canopy trees in a tropical dry forest in Bolivia [PhD thesis]. Univ. Florida, Gainesville
Mostacedo B, Villegas Z, Licona JC et al (2009) Ecología y Silvicultura de los Principales Bosques Tropicales de Bolivia. Instituto Boliviano de Investigación Forestal, Santa Cruz
Murphy PG, Lugo AE (1986) Ecology of tropical dry forest. Annu Rev Ecol Syst 17:67–88. https://doi.org/10.1146/annurev.es.17.110186.000435
Navarro G, Maldonado M (2002) Geografía ecológica de Bolivia: vegetación y ambientes acuáticos. Fundación Simón I Patiño, Cochabamba
Paredes-Villanueva K, Sánchez-Salguero R, Manzanedo RD et al (2013) Growth rate and climatic response of *Machaerium scleroxylon* in a dry tropical forest in Southeastern Santa Cruz Bolivia. Tree-Ring Res 69:63–79. https://doi.org/10.3959/1536-1098-69.2.63
Pennington RT, Lavin M, Oliveira-Filho AT (2009) Woody plant diversity evolution and ecology in the tropics: perspectives from seasonally dry tropical forests. Annu Rev Ecol Evol Syst 40:437–457. https://doi.org/10.1146/annurev.ecolsys.110308.120327
Pinheiro J, Bates D, DebRoy S et al (2016) nlme: linear and nonlinear mixed effects models. R package version 3.1-141. http://cran.r-project.org/web/packages/nlme/index.html. Accessed 01.09.19
Pirie MD, Klitgaard BB, Pennington RT (2009) Revision and biogeography of *Centrolobium* (Leguminosae-Papilionoideae). Syst Bot 34:345–359. https://doi.org/10.1600/036364409788606262
Pucha-Cofrep D, Peters T, Bräuning A (2015) Wet season precipitation during the past century reconstructed from tree-rings of a tropical dry forest in Southern Ecuador. Glob Planet Chang 133:65–78. https://doi.org/10.1016/j.gloplacha.2015.08.003
Pulla S, Suresh HS, Dattaraja HS et al (2017) Multidimensional tree niches in a tropical dry forest. Ecology 98:1334–1348. https://doi.org/10.1002/ecy.1788
R Core Team (2019) R: a language and environment for statistical computing, version 3.6.1. R Foundation for Statistical Computing, Vienna. http://www.R-project.org. Accessed 01.09.19
Ramírez JA, del Valle JI (2011) Paleoclima de La Guajira, Colombia según los anillos de crecimiento de *Capparis odoratissima* (Capparidaceae). Rev Biol Trop 59:1389–1405
Rodríguez R, Mabresa A, Luckman B et al (2005) "El Niño" events recorded in dry-forest species of the lowlands of Northwest Peru. Dendrochronologia 22:181–186. https://doi.org/10.1016/j.dendro.2005.05.002
Ronchail J (1995) Variabilidad interanual de las precipitaciones en Bolivia. Bull Inst Fr Études Andin 24:369–378
Rozendaal DMA, Zuidema PA (2011) Dendroecology in the tropics: a review. Trees 25:3–16. https://doi.org/10.1007/s00468-010-0480-3
Sánchez-Salguero R, Camarero JJ, Gutiérrez E et al (2017a) Assessing forest vulnerability to climate warming using a process-based model of tree growth: bad prospects for rear-edges. Glob Chang Biol 23:2705–2719. https://doi.org/10.1111/gcb.13541
Sánchez-Salguero R, Camarero JJ, Carrer M et al (2017b) Climate extremes and predicted warming threaten Mediterranean Holocene firs forests refugia. PNAS 114:E10142–E10150. https://doi.org/10.1073/pnas.1708109114
Schöngart J, Bräuning A, Maioli AC et al (2017) Dendroecological studies in the neotropics: history status and future challenges. In: Amoroso MM, Daniels LD, Baker PJ, Camarero JJ (eds) Dendroecology: tree-ting analyses applied to ecological studies. Springer, Switzerland, pp 35–73
Seiler C, Hutjes RW, Kavat P (2013) Climate variability and trends in Bolivia. J Appl Meteorol Climatol 52:130–146. https://doi.org/10.1175/JAMC-D-12-0105.1
Taylor KE, Stouffer RJ, Meehl GA (2012) An overview of CMIP5 and the experiment design. B Am Meteorol Soc 93:485–498. https://doi.org/10.1175/BAMS-D-11-00094.1

Tolwinski-Ward SE, Evans MN, Hughes MK et al (2011) An efficient forward model of the climate controls on interannual variation in tree-ring width. Clim Dyn 36:2419–2439. https://doi.org/10.1007/s00382-010-0945-5

Tolwinski-Ward SE, Anchukaitis KJ, Evans MN (2013) Bayesian parameter estimation and interpretation for an intermediate model of tree-ring width. Clim Past 9:1481–1493. https://doi.org/10.5194/cp-9-1481-2013

Vicente-Serrano SM, Beguería S, López-Moreno JI (2010) A multiscalar drought index sensitive to global warming: the standardized precipitation evapotranspiration index. J Clim 23:1696–1718. https://doi.org/10.1175/2009JCLI2909.1

Villegas Z, Peña-Claros M, Mostacedo B et al (2009) Silvicultural treatments enhance growth rates of future crop trees in a tropical dry forest. For Ecol Manag 258:971–977. https://doi.org/10.1016/j.foreco.2008.10.031

Volland-Voigt F, Bräuning A, Ganzhi O et al (2011) Radial stem variations of *Tabebuia chrysantha* (Bignoniaceae) in different tropical forest ecosystems of southern Ecuador. Trees 25:39–48. https://doi.org/10.1007/s00468-010-0461-6

Vuille M, Franquist E, Garreaud R et al (2015) Impact of the global warming hiatus on Andean temperature. J Geophys Res Atmos 16:3745–3757. https://doi.org/10.1002/2015JD023126

Wells N, Goddard S, Hayes MJ (2004) A self-calibrating Palmer Drought Severity Index. J Clim 17:2335–2351. https://doi.org/10.1175/1520-0442(2004)017<2335:ASPDSI>2.0.CO;2

Wigley TML, Briffa KR, Jones PD (1984) On the average value of correlated time series with applications in dendroclimatology and hydrometeorology. J Clim Appl Meteorol 23:201–213. https://doi.org/10.1175/1520-0450(1984)023<0201:OTAVOC>2.0.CO;2

Worbes M (1995) How to measure growth dynamics in tropical trees a review. IAWA J 16:337–351. https://doi.org/10.1163/22941932-90001424

Zhou X, Fu Y, Zhou L et al (2013) An imperative need for global change research in tropical forests. Tree Physiol 33:903–912. https://doi.org/10.1093/treephys/tpt064

Zuidema PA, Brienen RJ, Schöngart J (2012) Tropical forest warming: looking backwards for more insights. Trends Ecol Evol 27:193–194. https://doi.org/10.1016/j.tree.2011.12.007

Zuidema PA, Baker PJ, Groenendijk P et al (2013) Tropical forests and global change: filling knowledge gaps. Trends Plant Sci 18:413–419. https://doi.org/10.1016/j.tplants.2013.05.006

Part III
Forest Dynamics, Climate and Disturbances

Chapter 8
Forest Dynamics in the Argentinean Patagonian Andes: Lessons Learned from Dendroecology

Ana M. Srur, Mariano M. Amoroso, Ignacio Mundo, Mariano S. Morales, Milagros Rodríguez-Catón, Valeria Aschero, and Ricardo Villalba

Abstract The study of forest dynamics over large temporal and spatial scales has widely benefited from dendrochronological techniques. Patagonia is home to several long-lived tree species (*Austrocedrus chilensis*, *Araucaria araucana*, *Fitzroya cupressoides*, *Nothofagus dombeyi* and *N. pumilio*) with well-defined tree rings suitable for reconstructing tree establishment, mortality, spatio-temporal growth patterns and disturbance regimes with annual resolution. The first dendrochronological studies in the region date back to the 1950s and had a strong emphasis on hydroclimatology. It was not until the last few decades that studies using dendroecological techniques began to emerge. In this chapter, we review the experience gained by the tree-ring lab at IANIGLA (CONICET, Mendoza, Argentina) and colleagues from other institutions over the past 30 years applying dendroecological techniques to

A. M. Srur (✉) · M. S. Morales · R. Villalba
Laboratorio de Dendrocronología e Historia Ambiental. IANIGLA, Centro Científico Tecnológico CONICET Mendoza, Mendoza, Argentina
e-mail: asrur@mendoza-conicet.gob.ar; mmorales@mendoza-conicet.gob.ar; ricardo@mendoza-conicet.gob.ar

M. M. Amoroso
Centro Científico Tecnológico CONICET Patagonia Norte, San Carlos de Bariloche, Río Negro, Argentina

Instituto de Investigaciones en Recursos Naturales, Agroecología y Desarrollo Rural (IRNAD), Sede Andina, Universidad Nacional de Río Negro, Viedma, Río Negro, Argentina
e-mail: mamoroso@unrm.edu.ar

I. Mundo · V. Aschero
Laboratorio de Dendrocronología e Historia Ambiental. IANIGLA, Centro Científico Tecnológico CONICET Mendoza, Mendoza, Argentina

Facultad de Ciencias Exactas y Naturales, Universidad Nacional de Cuyo, Mendoza, Argentina
e-mail: iamundo@mendoza-conicet.gob.ar; vaschero@mendoza-conicet.gob.ar

M. Rodríguez-Catón
Lamont-Doherty Earth Observatory, Columbia University, Palisades, NY, USA
e-mail: milagros@ldeo.columbia.edu

M. Pompa-García, J. J. Camarero (eds.), *Latin American Dendroecology*, https://doi.org/10.1007/978-3-030-36930-9_8

understand the role of climate and disturbances (insect outbreaks, snow avalanches, windblows, fires and decline) on forest dynamics. For each case, we summarized the process, and the dendrocronological methods used. In this way, it was possible to detect those gaps of knowledge that still can be explored using dendroecological methods in the Patagonian forests of Argentina.

Keywords IANIGLA tree ring lab · Climate variability · Disturbances

8.1 Introduction

Biotic and abiotic factors modulate in the long-term the structure and dynamics of the forests through the regulation of establishment, growth rates and mortality of trees in a community. These factors include climate, soil, interactions between organisms and natural or human-induced disturbances. These factors can influence the dynamic processes of the forest in more than one aspect, even combining between them and amplifying their effects on the dynamics of forests. In this way, the studies of forest dynamics focus on the characterization of changes in the structure and composition of forests. The structure of the forest is defined by the age distribution, sizes and dominance of individuals within a stand, while the composition refers to the species that constitute the stand. The changes in the forest induced by variations in environmental conditions (e.g. temperature increase, droughts, extreme precipitation), in competitive and/or facilitation interactions and in disturbance regimes (e.g. avalanches, fires, blowdown) occur on a very wide range of spatial and temporal scales, making their characterization more difficult (Frelich 2002; Pretzsch 2009).

Natural disturbances shape the structure and function of the ecosystems. As disturbance regimes are changing, new spatial patterns, altered regimes and new trajectories of change will occur concurrent to global changes (Bonan 2008). On the other hand, extreme events must be studied in a special way since their impacts on ecosystems and people are substantial (Turner 2010). Depending on the intensity and magnitude of the disturbances or environmental changes, a sequence with minor variations in the composition and structure of the forest will start or a new succession process that will depend not only on the intensity of the disturbance, but also on the traits of the dominant tree species (White and Pickett 1985). In some cases, the combination of species and types of disturbance form a pattern of stand and landscapes dynamics that persists over several generations of trees. In contrast, large-scale environmental changes or disturbances cause the replacement of the main canopy leading to a process of plant succession where complex interactions between the biological legacy, the attributes of the species and the type of disturbance will determine the trajectory and final state of the succession. These successional processes tend to be slow involving, in most cases, centenary or longer periods.

Changes in the structure of a forest, and consequently in its dynamics, can be examined through dendrochronological techniques that enable us to document the dates of establishment and mortality of individuals, the patterns of radial growth and the frequency and magnitude of disturbances related to forest ecosystems (Fritts and Swetnam 1989, Baker et al. 2005; Amoroso et al. 2017a, b; among others). Overall, the dendrochronological techniques allow inferring patterns and processes for longer periods than those resulting from regular ecological monitoring studies and with a precise resolution on annual scale. Therefore, dendroecology today has become a very useful tool for understanding ecological processes beyond the time scales commonly used by classical ecological studies. Dendroecology represents a valuable and irreplaceable discipline for understanding forest dynamics and for assessing the long-term consequences of environmental changes on forest stands. The application of dendroecological techniques in the Patagonian Andes has increased substantially in recent decades in search of documenting and clarifying the complex processes associated with forest dynamics in the region. In this context, we will conduct in this chapter a concise review of the dendroecological applications and their major contributions to the understanding of the dynamics of the forests and species in the Argentinean Southern Andes.

To facilitate the understanding and organization of this review, we have decided to address the dendroecological studies in Patagonia following those environmental changes or major disturbances that induce variations in the dynamics of Patagonian forests. First, we provide a brief account of how climate variability affects forest dynamics, and then focus on the description and characterization of disturbance regimes at different temporal and spatial scales. For each disturbance, special emphasis is given to the contribution of dendroecology in the understanding of the dynamical processes. Finally, and based on our review, we try to identify the existing gaps of knowledge on the dynamics of the Patagonian forests that could be addressed from a dendroecological perspective.

8.2 Climate Variability and Forest Dynamics

Unprecedented changes in climate experienced in the last decades have altered the magnitude and speed of the environmental changes having direct effect on the biological systems and, therefore, on forest dynamics (Wolkovich et al. 2014). Climate changes impacts on forest dynamics have been recognized on a global scale (Allen et al. 2010, among others). Climate modulates forest dynamics directly by affecting the vital physiological processes of trees (McDowell et al. 2013) and indirectly by altering disturbance regimes such as fire, avalanches, insect defoliations, among others (Veblen et al. 2011).

In this section, we emphasize the direct effects of climate on forest dynamics, since the climatic influence on disturbance regimes will be discussed later, when we refer to each particular disturbance in detail. To this end, it is essential to perceive the climate as a "continuum" of atmospheric processes varying at different temporal and spatial scales (Wolkovich et al. 2014). In this "continuum," millennium-scale climatic changes induced the migrations of the species following the glaciation-interglaciation transitions in the planet. The individualistic responses of each species to climatic changes after the last glaciation (Overpeck et al. 2003) forged the structures, and particularly the current floristic composition, of most forest on the planet. These millennial climate oscillations, largely modulated by changes in the total energy received from the sun, interacted with the pre-existing topography creating environmental gradients along which species migrated to their current distributions.

On shorter spatiotemporal scales, climatic variations have modulated changes in the boundaries of the species distribution or in sectors where growth is strongly constrained by water availability or critical thermal conditions. For example, the expansion of forest masses over the upper limit of their distribution has long been documented in different regions since the mid-twentieth century of the planet in response to global warming. However, the influences on forest dynamics of decadal to centennial climatic changes are not always easy to visualize, since climate influences are mostly concurrent with changes in land use or in disturbance regimes, which in turn could be indirectly associated with these variations in climate. For example, the effect of decadal to multi-decadal climatic variations on establishment or mortality may be very difficult to visualize in forests where cattle have been introduced and/or greater climatic variability has changed the frequency of fires. In turn, interannual variability represents one of the fundamental axes that make up ecological systems and it is fundamental for understanding how spatial and temporal variation in the environment controls species distributions, and how such environmental variation affects the dynamics of diverse populations and community structures (Wolkovich et al. 2014). At this point also, dendroecology can provide fundamental contributions in order to understand the dynamics of the limits of woody species distribution.

Finally, extreme climatic events, associated with short, but severe climatic anomalies, expose forests to environmental conditions very different from the mean climate in which the forests are developing. Severe droughts or extraordinary floods are examples of extreme climatic events with marked impacts on forest worldwide.

8.2.1 Dendroecological Methods and Climate Influences on Forests

Dendrochronology is a very useful tool to identify the influence of climate on forest dynamic processes such as establishment, growth rate and tree mortality. The precise determination, with annual or seasonal resolution, the dates of tree establishment and mortality, allows establishing close relationships between climatic conditions and forest demographic processes (Kitzberger et al. 2000). On the other hand, old trees contain valuable information on both climate and past disturbances

(Schulman 1954, Brown 2006, Pederson 2010). Variations in the rate of growth of trees can also be associated with the climatic changes that caused them.

The precise determination of date of tree establishment requires the application of careful techniques of sampling and dating of the selected dendrochronological cores or cross-sections (Villalba and Veblen 1997a). In order to ensure that the sample pith is included in the core, sampling should be done as close to the root collar as possible. Precise sampling methodologies, combined with techniques to correct age determination in samples taken above the root collar or with missing the pith have been proposed by several authors (Duncan 1989; Villalba and Veblen 1997b). In the case of shade-tolerant species, combining the above methods with crown class models, which, in turn, take into account the diameter of the individuals, helps to determine the date of establishment or total age (Baker 2003).

The determination of the date of mortality is based on the cross-dating of wedges or cross sections from dead trees with chronologies of the same species available in the region. The establishment of precise dates of mortality also depends on precaution during sampling and on the conservation of the woody material. It is advisable to collect samples from dead individuals that still maintain the bark, to avoid losing the outermost rings due to erosion or wood decay. Details on the methods to follow for the precise dating of mortality events are indicated in Villalba (1997) and Villalba and Veblen (1998).

Dendroecological studies intended to establish the climatic influence on long-term processes, such as forest decline, require the precise determination of stand growth variation patterns (Rodríguez-Catón et al. 2015, 2016). Precise comparison of these growth patterns with climatic variations at seasonal or interannual scales can provide a clear retrospective view of the extreme climatic events that triggered changes in growth trends among individuals on the same stand. In most dendroecological studies, the precise dating through adequate tree-ring techniques represent the most valid tool to achieve solid and reliable results (Mast and Veblen 1994; Bigler and Bugmann 2004).

8.3 Climatic Influences on Forest Dynamics in Patagonian Forests

8.3.1 Establishment

The careful and detailed determination of tree establishment dates for ≥800 individuals from *Austrocedrus chilensis* along forest-steppe ecotone in northern Patagonia revealed the occurrence of episodic periods of tree establishment that differ markedly from the background establishment. These episodes of tree establishment extend for a decade or longer periods and are associated with persistent humid and relatively cool conditions that favor the germination of this species, but particularly the survival of seedlings recently established in the xeric environments of the forest-steppe ecotone in northern Patagonia (Villalba and Veblen 1997a).

Suárez and Kitzberger (2010) studied the relationships between climatic variability and the dynamics of the evergreen *Nothofagus dombeyi* forests along the west-east precipitation gradient in northern Patagonia. In the west, in humid sites, climatic events that mechanically damaged the trees crowns and, therefore, opened the canopy, such as snowstorms or strong winds, favored the establishment of new individuals under the canopy. In contrast, droughts in xeric sites produce massive tree crown mortality and the consequent canopy opening. However, the success of tree establishment depends on the occurrence of wet periods after droughts, which demonstrates the dependence of the establishment on the occurrence of wet periods in the most xeric environments of the *N. betuloides* forests.

Similarly, recent dendroecological studies indicate that the expansion of the *Nothofagus pumilio* forest above the upper treeline is episodic, limited to relatively long periods (from one to two decades) with favorable climatic conditions. These favorable climatic periods follow changes in continental or hemispheric atmospheric circulation such as the Pacific Decadal Oscillation or the Southern Annular Mode, which strongly affect the climate in Patagonia (Srur et al. 2016, 2018). Along a precipitation gradient in northern Patagonia, relatively warm temperatures at upper treeline are associated with the establishment of new individuals while humid decade-scale periods favor survival.

8.3.2 Mortality

The earlier studies of Villalba and Veblen (1997a, 1998) pointed out the episodic character of the massive mortality events of *Austrocedrus chilensis* in relation to climatic variability. Mortality episodes are largely determined by extreme climatic events characterized by warm and dry summers. Although these mortality episodes could be caused by a single summer with extreme warm and dry conditions, they are also related to adverse climatic conditions prevailing for two consecutive growing seasons. During the twentieth century, the three massive mortality events coincided with the extremely warm and dry summers that occurred in 1912–13, 1942–43 and, more recently, between 1956 and 1962 (Fig. 8.1, Villalba and Veblen 1997a, 1998).

Recent studies indicate that the massive mortality in compact, mesic-wet forests of *A. chilensis*, known as "mal del ciprés," could also be related to dry and warm climatic conditions during the growing season (Amoroso 2009; Amoroso and Larson 2010; Amoroso et al. 2015). Climate variability acts as one of the environmental stresses that triggers or contributes to the dieback of forests at stand level. The onset of the reduction in radial growth as well as the mortality of individual trees were significantly associated with summer moisture deficits. However, the authors emphasize that the *A. chilensis* mortality process is driven by complex interactions between allogeneic abiotic and autogenic biotic stress factors that predispose, incite and contribute to stand-level decline in these forests (sensu Manion

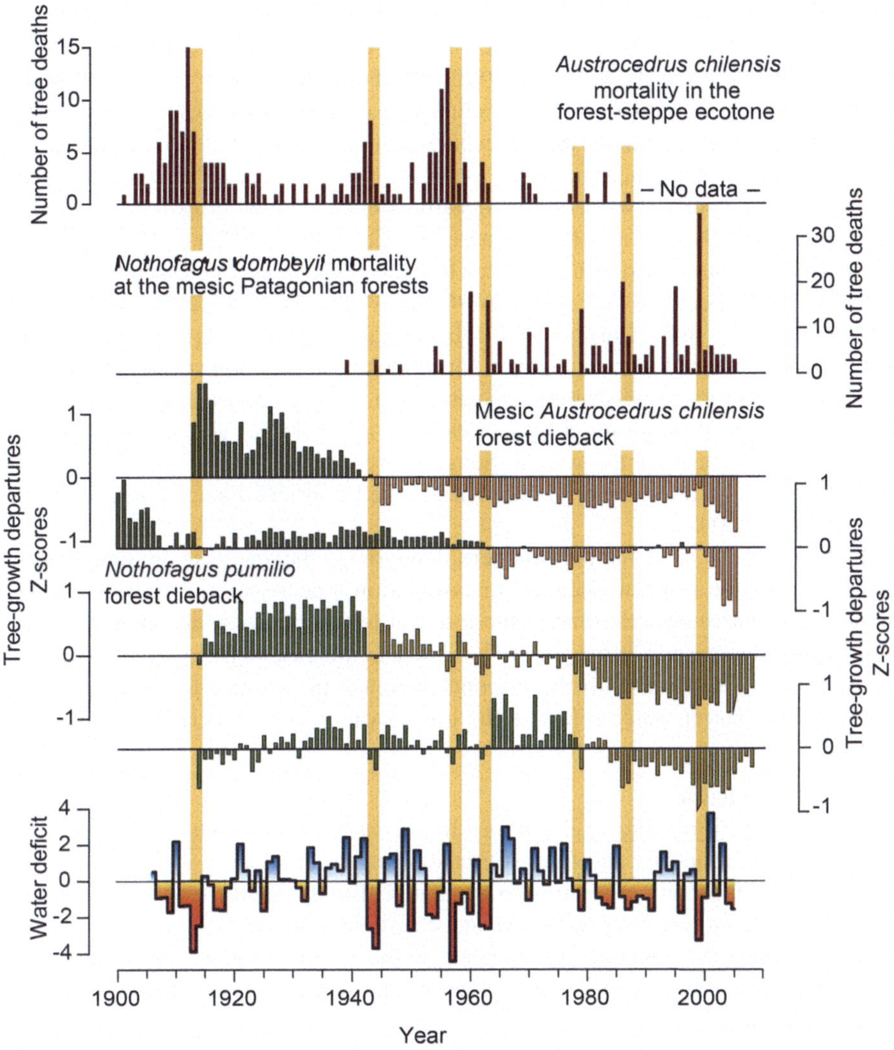

Fig. 8.1 Severe drought influences on forest dynamics in northern Patagonia

1991; Manion and Lachance 1992). Specifically, site conditions, genetic variation, and tree sex are predisposing factors that likely interacts with pathogens such as Phytophthora, in connection with stages of stand development, to cause tree mortality (Amoroso et al. 2015, 2017a, b).

The severe droughts of 1913–14, 1943–44 and 1957 were associated with episodic mortality events in *Austrocedrus chilensis* forests in the steppe-forest ecotone in northern Patagonia (Villalba and Veblen 1998). Likewise, the droughts of 1962–63, 1986 and in particular in 1998 caused episodic mortality in the evergreen mesic forests of *Nothofagus dombeyi* in northern Patagonia (Suarez et al. 2004). The onset of the reduction in the radial growth of *Austrocedrus chilensis*, associated with the dieback of the forests near the locality of El Bolsón, was associated with the summer droughts of 1943–44 and 1962–63. Following these extreme climatic events, a persistent reduction in the radial growth was recorded in trees showing partial crown mortality and signs of dieback (Mundo et al. 2010a,b). Similarly, the patterns of Basal Area Increments in *Nothofagus pumilio* trees with external manifestations of dieback shows persistent negative trends in radial growth that started at the droughts of 1943–44 and 1978 (Rodríguez-Catón et al. 2016).

Severe droughts in northern Patagonia have also been associated with forest mortality in some species of the genus *Nothofagus* (Suarez et al. 2004; Suarez and Kitzberger 2010). In the Bariloche-El Bolsón region, the severe drought of the summer 1998–1999 resulted in the total or partial death of *N. dombeyi* trees in the eastern sector of its range of distribution (Suarez et al. 2004). The mean growth rate was a good predictor of mortality in adult trees, showing that trees with a slower growth rate were more sensitive to drought. Sensitive trees to death may have been negatively affected by previous droughts that affected northern Patagonia in 1956–57 (Suarez et al. 2004).

In the area of El Chaltén, southern Patagonia, the increase in temperature recorded over the last few decades, in association a decrease in precipitation, has induced a negative growth trend in the forest at lower elevation in contact with the steppe. This decrease in radial growth has been an increase in tree mortality starting in the 1970s consistent with the regional increase in summer temperatures and, consequently, larger water deficit in the low elevation forests (Fig. 8.1, Srur et al. 2004, 2008).

Gaps of Knowledge

- Modeling the dynamics of elevational limits based on data on tree establishment, growth rate, mortality and future climate change scenarios
- Model radial growth in relation to simulated climate change for the twenty-first century based on ecophysiological and statistical models
- To determine the effect of CO_2 increase on the radial growth of different species and for the same species in different environments within the range of distribution
- Interactions with climate-modulated disturbances

8.4 Disturbances and Forest Dynamics

Vegetation patterns at landscape scales are the result of two overlapping patterns: that of species distribution determined by environmental gradients and that imposed by disturbances. Disturbance regimes can change rapidly, generating large effects on ecosystems and landscapes in general and on forest dynamics in particular. The study of different types of disturbances can provide a unique insight into ecological patterns and processes (Turner 2010). The importance of natural and anthropogenic disturbances on forest dynamics has been recognized and documented for more than 40 years in the Patagonian Andes (Veblen and Ashton 1978). The framework for analyzing the characteristics and consequences of disturbances is given by the concept of disturbance regime defined as the set of spatial and temporal characteristics of the disturbance in a given landscape (Paine and Levin 1981; White and Pickett 1985). Some of the characteristics of disturbance regimes such as the spatial distribution of the occurrence of events, the frequency of occurrence, the size of the disturbed area and the return interval between events are fundamentally based on the precise dating of the events. It is also important to determine the area and severity of the event, which for many types of disturbances such as fire and insect attacks is feasible thanks to dendroecological techniques (Kitzberger et al. 2000; Veblen et al. 2005; Speer 2010). Tree rings provide information not only on dates of past disturbances but also the anatomical changes in the wood related to these events. For example, disturbance events such as insect outbreaks are identified not only by a decrease in growth but also by the presence of clear or missing rings in the samples (Girardin et al. 2001; Paritsis et al. 2009; Palacios 2013). Below, we describe the most important disturbances recorded in the Patagonian Andes and how dendrochronology has contributed to their characterization and understanding.

8.4.1 Insect Outbreaks

Process description: Very intense defoliations of *Nothofagus pumilio* and *N. antarctica* forests caused by larvae of the genus *Ormiscodes amphimone* Fabricius (Saturniidae) have been reported both in northern (Paritsis et al. 2009; Mundo et al. 2010b; Paritsis and Veblen 2010, 2011) and southern Patagonia (Morales et al. 2004; Morales and Villalba 2009; Palacios 2013). In general, the published works on *Ormiscodes* defoliations refer to the reconstruction of the dates of outbreaks on the different stands studied and their effect on growth.

Description of methods: The use of dendrochronological techniques for the study of this type of disturbance is based on the precise dating of the event and on documenting abrupt reductions in radial growth. One of the major contributions of dendroecology to the study of insect outbreaks is the reconstruction of dates of occurrence of defoliations. Defoliations of *Nothofagus* forests by *Ormiscodes* induce the formation of very thin growth rings during the growing season following the attack. In this way, a chronology of defoliations could be established by individualizing and dating the extremely thin rings. However, since growth reduction may be caused by other factors such as drought, it is important to verify the occurrence of defoliations with historical documents or other complementary techniques (Paritsis et al. 2009; Paritsis and Veblen 2011).

In mixed forests in the Northern Hemisphere, where two or more species cohabit, chronologies from species affected (host) and unaffected (non-host) by insects are usually used. The contrasts in growth between both species allow identifying the old defoliations with greater precision. *N. pumilio* forests are generally monospecific, so these comparative studies between defoliated and undefoliated species are difficult to conduct. Since *Ormiscodes*outbreaks on *N. pumilio* forests are rather localized and restricted to valleys or corridors and limited from lower to middle elevations, comparison with chronologies of the same species in unaffected valleys or at higher elevations might facilitate the identification of past defoliations. However, since the growth of *N. pumilio* varies along altitudinal gradients (Villalba et al. 1997), these comparisons should be made with caution. Another method is to use regional chronologies that strongly record the regional climatic variability, so very narrow rings in the record of interest, not shown in the regional chronology could be related to past defoliations. Considering these limitations to reconstruct the history of *Ormiscodes* outbreaks, it is also advisable to explore the anatomical characteristics of the rings such as changes in density or coloration caused by defoliation (Morales and Villalba 2009; Paritsis et al. 2009).

As indicated above, *Ormiscodes* defoliations produce marked reductions in radial growth of affected trees during the following growth period. It is interesting to quantify the reduction in radial growth, information that could be used to appreciate the intensity of the attack. In defoliated sites, the development of chronologies along environmental gradients would provide an estimate of the degree of affectation with elevation, soil moisture or other environmental parameters.

Contributions of dendroecology: The analysis of interannual variations in the growth of *N. pumilio* allows inferring that *Ormiscodes*defoliations induce a decrease in radial growth during the year following the outbreaks. Mortality due to *Ormiscodes* outbreaks has not been reported for the *Nothofagus* forests in Patagonia. Most studies in southern Patagonia have reported a decrease in growth of up to 80% (Mundo et al. 2010b; Palacios 2013). The *Ormiscodes* attack showing the largest spatial distribution (affecting a total of 10 sites) in northern Patagonia occurred in 1986, followed by the year 1869 (in 4 sites; Paritsis et al. 2009; Paritsis and Veblen 2011). In southern Patagonia, Palacios (2013) reports how the largest outbreak reconstructed in 1880 (with 4 sites), followed by the outbreaks of 1882, 1907, 1999 and 2005 (3 sites). However Paritsis and Veblen (2011) had reported nine outbreaks in the year 1999 in the river basin Las Vueltas.

The frequency of outbreaks in the last few decades has been higher in the south than in the north (Fig. 8.2). The last outbreak that reached several sites was in 1986 in Northern Patagonia, after that attack there were very few and isolated sites. However, in the south since 2000, both the work of Paritsis and Veblen (2011) and that of Palacios (2013) show a greater number of outbreaks that affect a larger area (Fig. 8.2).

The defoliation histories reconstructed for the north and south Patagonian Andes provided a first estimation of the climatic influence on the massive attacks of *Ormiscodes*. In general, the defoliations during the last decades have been concurrent with the increase in temperature registered along Patagonia (Paritsis and Veblen 2011; Palacios 2013). For northern Patagonia, Paritsis and Veblen (2011) attributed the insect attacks to the occurrence of hotter and drier springs. In contrast, no clear relationships between outbreaks and climate have been reported for southern Patagonia. More recently, Palacios (2013) report similar results to those found by

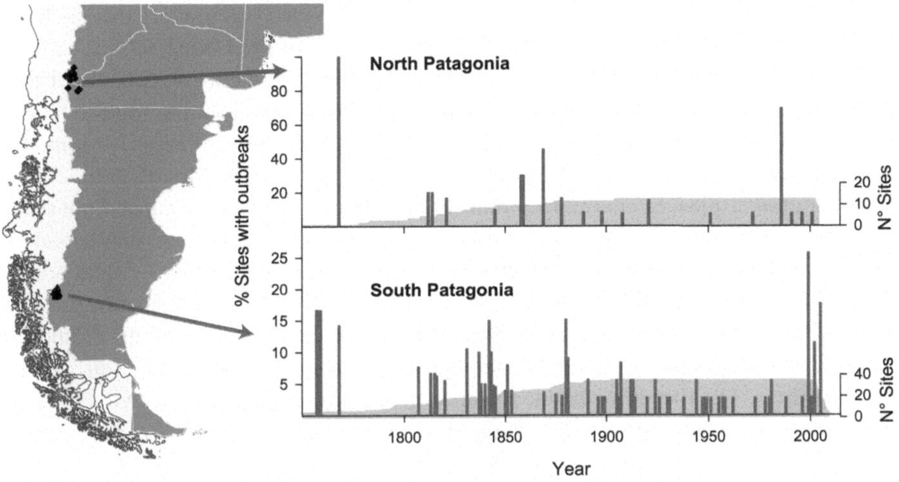

Fig. 8.2 Percentage of area affected by Ormiscodes outbreaks in both north and south Patagonian Andes. Compilation based on Paritsis and Veblen (2011) and Palacios (2013)

Paritsis and Veblen (2011) for northern Patagonia and a relationship between the attacks and the positive phase of the Antarctic oscillation (SAM) during the previous year. For southern Patagonia, the positive phase of the SAM is related to warmer and driers spring and summers.

Gaps of Knowledge

- To establish the relationships between defoliation and the patterns of tree establishment and mortality
- The reconstruction of the spatial distribution of defoliations based on systematic sampling in affected areas
- Given the low number of past defoliation events reconstructed, the determination of outbreak periodicities remains unclear. We have not yet been able to respond, whether or not there is periodicity in the attacks, and whether those cycles have remained constant over time or are accelerating in recent decades

8.4.2 Snow Avalanches

*Process description:*Snow avalanches are a common disturbance in the Patagonian Andes (Mundo et al. 2007; Casteller et al. 2008, 2009, 2011). In most cases, avalanches are recurrent processes producing periodic and abrupt changes in forest structure. Tree rings represent a useful tool for reconstructing avalanche histories at an annual resolution (Burrows and Burrows 1976; Carrara 1979). The evidence left by avalanches in tree ring sequences varies greatly, and depends primarily on (1) the type and size of the avalanche, (2) topography and (3) tree characteristics such as species, degree of inclination, diameter, height and age of the tree.

Description of methods: Dendroecological studies intendent to reconstruct the history of avalanches in the southern Andes are largely based on the identification of several indicators including scars, sudden changes in growth (releases or suppressions), eccentricity changes and the presence of reaction wood (Mundo et al. 2007; Casteller et al. 2008). Scars on trees that survive the avalanches are caused not only by snow impact, but also by debris, broken trees and rocks, which are incorporated into the avalanche (Burrows and Burrows 1976). Growth changes vary depending on the location of individuals within the avalanche zone. The decrease in growth may be the result of a decrease in photosynthetic rate due to branch ruptures and/or canopy loss (Rayback 1998) or by damage to the root system (Gers et al. 2001) among others. On the other hand, releases in general are associated with greater availability of resources due to the death of neighboring trees (Smith 1973). In some cases, the increase in growth extends over time, between 2 and 5 years after the avalanche occurred, with the radius exposed to the avalanche, and consequently close to the dead trees, showing the greatest growth (Mundo et al. 2007; Casteller et al. 2008).

There are also important methodological limitations to reconstruct past avalanche events from tree rings. The loss of evidence from previous events due to the destructive effect of more recent avalanches is one of the main limitations (Carrara 1979; Stoffel and Bollschweiler 2008).

Contributions of dendroecology: Important advances in our understanding of snow avalanches in the Patagonian Andes come from the relatively few dendroecological studies conducted in the region. In several mountainous areas of Patagonia, tree rings have made possible to reconstruct past events, provide an estimate of their frequency, advance in the compression between the occurrence of avalanches and climatic variations, and use dendrochronological methods to validate numerical models that simulate the physical properties (thickness, affected area, speed, transported mass) of avalanches (Casteller et al. 2008, 2009, 2011). In addition, the dynamic responses of the forest to the impact of avalanches have also been analyzed (Mundo et al. 2007).

As mentioned above, in addition to changes in growth rates (releases and suppression), demographic processes, such as mortality and establishment, have been recorded for avalanches in the Patagonian Andes using dendroecological techniques. Mundo et al. (2007) found that *Nothofagus pumilio* forest recovery was faster on the upper edges of the avalanche paths, showing this sector the largest number of individuals established after the year of the event.

These dendroecological studies also provide a first approximation on the relationships between avalanches and climate variations along a precipitation gradient in the area El Chaltén - Lago del Desierto (Fig. 8.3; Casteller et al. 2009, 2011). Although, the occurrence of avalanches is different between wet and dry areas along the gradient, avalanche activity is related to years with above-average winter regional precipitation. These years were concurrent with ENSO cold events (La

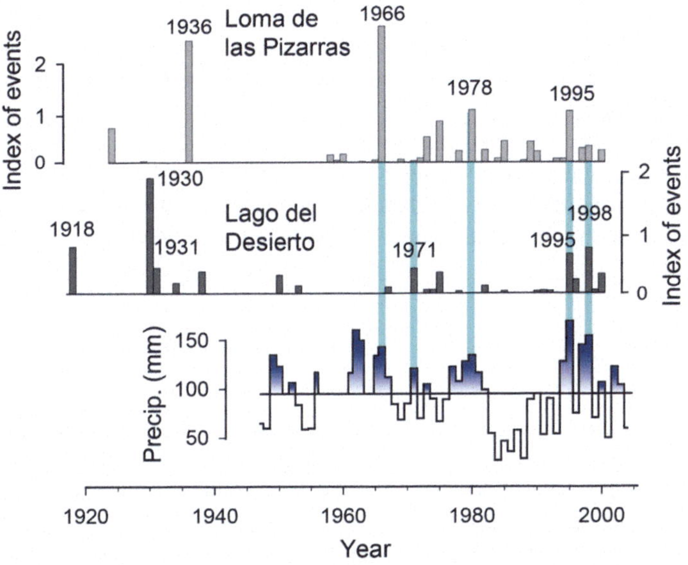

Fig. 8.3 Reconstructions of snow avalanche events in the relatively dry (Lomas de las Pizarras) and humid (Laguna del Desierto) areas in the Río de las Vueltas basin, Santa Cruz, Argentina. The index of events is a weighted measure of the avalanche indicators recorded on all selected avalanche tracks at these two sectors. Interannual variations in winter rainfall in the locality of El Calafate, Santa Cruz are shown below. With the exception of the 1995 event, most events were not simultaneous at both localities; however, outstanding events (years) were concurrent with abundant winter snow precipitation in the region (Casteller et al. 2008, 2011)

Niña), associated with more frequent storms from the South Pacific increasing wind speed and precipitation in the Patagonian Andes by 10–20% (Schneider and Gies 2004).

In Cerro Ventana, in the Patagonian Andes, dendroecological techniques were used to verify the results of numerical avalanche modeling using the AVAL-1D and RAMMS programs (Casteller et al. 2008). The characteristics of the avalanche, such as extension and date, were determined by dendrochronological techniques. The model results for the simulation of an important event in 2002 were compared with the documentary evidence and the evidence provided by the tree rings. The results showed good agreement between the simulated projection of the avalanche and its reconstructed extent using tree ring records. The differences between the observed and the simulated avalanche, mainly related to the height of the snow deposited in the run-out zone, are mainly attributed to the low resolution of the digital elevation model used to represent the topography of the valley (Casteller et al. 2008). In summary, the studies conducted in the Patagonian Andes highlight the potential of tree ring records of *Nothofagus pumilio*, the dominant subalpine species, to reconstruct snow avalanche events in time and space.

Gaps of Knowledge

- Given the limited number of avalanche event reconstructions in Patagonia, it would be interesting to replicate the dendroecological reconstructions along the Patagonian Andes
- To determine the post-disturbance forest dynamics, including patterns of tree-establishment and recovery after recurrent events. Studies of the intensity and areas frequently affected by avalanches need to be conducted
- Interactions with other disturbances in the Patagonian Andes

8.4.3 Wind

Process description: The constant and strong winds originating in the Pacific Ocean have a strong effect on the dynamics of the temperate forests of the region, resulting in one of the main natural disturbances in the *Nothofagus* forests (De Agostini 1941; Eskuche 1973; Mutarelli and Orfila 1973; Rebertus and Veblen 1993; Veblen et al. 1996; Rebertus et al. 1997; Puigdefábregas et al. 1999; Frangi et al. 2005; Veblen et al. 2004). The effect of this disturbance agent varies depending on its magnitude and frequency, allowing to distinguish three major types of patterns in forests, massive treefalls or blowdown, mortality waves and individual treefalls (González et al. 2014). Massive treefalls are one of the most important disturbance affecting *Nothofagus* forests in southern Patagonia (De Agostini 1941; Eskuche 1973; Veblen et al. 1996). In Tierra del Fuego, strong storms can affect large areas where massive blowdown generates discrete patches of mortality (Rebertus et al. 1997). Mortality resulting from strong winds in southern Patagonia can also result in what is known as "mortality waves," parallel bands of disturbed area perpendicular to the predominant direction of the winds (Rebertus and Veblen 1993; Puigdefábregas et al. 1999; González et al. 2014).

Description of methods: The use of dendroecology in this type of disturbance has been widely used not only to study the nature of this agent and its effects, but also to reconstruct the history of the disturbance by dating events and determining intervals of recurrence by dating dead material as well as abrupt changes in radial growth (Rebertus et al. 1997; Amoroso et al. 2013, 2016; Ruiz 2013). To interpret the changes in growth patterns resulting from the wind effect, and not from climate or other types of disturbances, tree-ring chronologies from nearby stands unaffected by windblows have comparatively been used (Amoroso et al. 2013).

Contributions of dendroecology: The effects of winds on the dynamics of the *N. pumilio* forests have been intensively studied at in Tierra del Fuego (Rebertus et al. 1997). The blowdown history was reconstructed since early 1700 to understand the repetitive impacts of this disturbance on the dynamics at the stand and landscape levels. The authors recorded the size of the patches originated by massive treefalls, the event dates and the recurrence intervals. Almost two thirds of the area under study was affected by massive blowdowns, resulting in patches of variable size ranging from 0.1 to 100 has. At the sampling date, the age of the disturbed patches ranged between 19 and 200 years, providing a recurrence interval of 20–30 years between events. However, the mean recurrence interval between 34 selected sites was 145 years, with most intervals longer than 100 years. While 90% of patches were less than 6 ha, the larger patches represented more than 70% of the area turned over. Most of the resulting stands were single-cohort with trees ranging between 20 and 32 cm in diameter. According to Rebertus et al. (1997), the susceptibility of the stands to blowdown would increase with age, although after exceeding 100–200 years it would decrease. These authors determined a vulnerability threshold for trees with diameter greater than 18–20 cm, corresponding to stands of 100–125 years of age. Site characteristics such soil depth would also predispose the occurrence of massive blowdowns. Tree rings have also been used to study the occurrence of "mortality waves" in mature *Nothofagus* forests in the southeast corner of the island of Tierra del Fuego. Although these studies did not provide the dendrochronological dating of the trees, they were based on the counting of rings to determine the ages of the trees. Puigdefábregas et al. (1999) studied the nature and propagation rate of the mortality waves in mature low-elevation forests of *N. betuloides*. The waves resulted in a 50% reduction of the basal area in the disturbed stands, and the speed of propagation depends on the aspect reaching values between 0.85 and 0.4 m year^{-1} in the middle and upper parts of the mountain slopes, respectively. Rebertus and Veblen (1993) reported, on the other hand, the occurrence of these waves of mortality at upper montane and mature *Nothofagus* forests. Unlike the mortality waves occurring at low elevations in the near coastal stands, mortality patterns seem to be more localized and resulting from stochastic wind events. They present propagation rates between 0.5 and1 m year^{-1} (Rebertus and Veblen 1993). Consequently, the forest structure is heterogeneous including remnant trees of different ages emerging from advanced and newly established regeneration in the understory. The authors recorded a temporal coincidence in the occurrence of the wind events affecting the different sampled stands, postulating that regional storms could be an important factor in the development of these cyclical patterns of tree mortality.

Although mortality waves have been mainly recorded in Tierra del Fuego, recent studies reported its occurrence in continental forests at transversal valleys in the Andes in Santa Cruz (Ruiz 2013; Amoroso et al. 2013, 2016). Ruiz (2013) dated and reconstructed the occurrence of wind blowdown at *N. pumilio* mortality waves in two valleys around El Chaltén, Santa Cruz. The dating of the events was carried out through tree-ring analyses combining mortality dates from trees killed by the storms (snapped and uprooted) with abrupt changes in the radial growth. Twelve events were determined in the two valleys, two of which occurred during the same year. Similarly, Amoroso et al. (2013) analyzed the forest structure and the number of small blowdown events in six disturbed patches and their contiguous undisturbed forests, in the Rio Toro valley, Santa Cruz (Fig. 8.4). The authors noted, based on the mean diameter, mean height and mean age of the regeneration patches and the adjacent undisturbed forests, highly variable pre- and post-blowdown structures in the forests. The undisturbed forest structures differed greatly from the affected forests indicating different conditions for tree establishment and mortality before wind

Fig. 8.4 Mortality waves induced by interactions between wind storms and avalanches in *Nothofagus pumilio* forests at the Rio Toro valley, Santa Cruz, Argentina

blowdown events. At least three occurrences resulted in significantly different radial growth in surviving trees compared to unaffected trees; five mortality events were common among waves indicating some degree of synchronicity between windstorms. Waves are dominated by younger trees with some residual living trees and few standing dead trees. More recent, Amoroso et al. (2016) expanded on the previous studies and reconstructed and dated wind blowdown events in five valleys in the area concluding the occurrence of blowdown events was highly variable between valleys, summing up a total of sixteen blowdown events, three of them common between valleys.

Although tree-fall gaps are a very common process in different forests, very few dendrochronological studies have studied the fall of individuals or small group of trees in the Patagonian Andes (Veblen et al. 1996, 2004). The first study in Patagonia using tree ages based on ring counting was conducted by Veblen (1989). Age structures and species composition in gaps originated from the fall of individuals were determined for mixed stands of *A. chilensis, N. dombeyi and N. antartica*. The most successful species regenerating in the gaps was *N. dombeyi* due to the lack of competing species in these relatively xeric forests with less dense bamboo understory. In the *A. chilensis* forests, the fall of individual trees is also very frequent since the root system are weakened by pathogens, although the tree falls also occur as a consequence of strong winds (Relva et al. 2009).

Gaps of Knowledge

- Post-disturbance forest dynamics including the age structure and dominant growth patterns
- Studies on dynamics and dating of events for different wind intensity and magnitude

- History at a regional scale for mortality waves and individual treefalls
- Studies of individual and group treefalls

8.4.4 Fires

Process description: Fire is one of the most important disturbances that affects the Patagonian forests. Over the last decades (2002–2016) Patagonian forests have experienced unprecedented mega fire events unusual in the context of historical records. Large and severe extraordinary events burned vast areas in several national parks and natural reserves in Chile (ca. 36,000 ha burned; González and Lara 2015) and also more recently in mountainous forests in Neuquén, Río Negro and northern Chubut (2013 and 2015) in Argentina (ca. 47,000 ha burned; Mohr-Bell 2015). In this context, information about past fire regimes can be a helpful reference to understand the current and future fire regime characteristics, patterns, and forest structure characteristics. Several fire management activities might be benefited from understanding past fire regimes including prescription of fires, mechanical treatments to reduce fire risk and ecological restoration, among others. Dendroecology has the ability to provide basic information on the different fire regime attributes with annual resolution.

Description of methods: To study fire regimes, dendrochronological methods are based on two types of evidence: fire scars and age structure analyses. Fire scars are injuries formed in tree boles due to heat exposure from wildfire which partially kills a tree's cambium and have a triangular or cat-face pattern. Following the formation of a fire scar, the uninjured portion of a tree's cambium, with the intention of restoring circular continuity, divides and produces new wood over the injury. In this way, fire scars can be used to date exact years of fires. In addition, using the position of a fire scar within an annual growth ring it is sometimes possible to determine the approximate season that a fire occurred. Historical fire frequency can be easily calculated by determining the specific years that fires scarred trees and then averaging the intervals between the scars (Swetnam and Baisan 1996). Fire scars can also be used to reconstruct type and severity of fires. Yocom Kent (2014) stated that fire scars are most often used to reconstruct low-severity surface fire regimes because the presence of a fire scar is an indication of low-severity fire at the scale of a tree; a fire injured but did not kill the tree. In this way, when several trees are scarred by low-severity fire in the same year, it can be inferred that a fire was a low-severity fire throughout the area. However, in the case of mixed forests where species have different bark thicknesses and consequently different sensitivity to fire, special care must be taken when inferring severity as a function of fire scars. In addition, fire scars can also be used to supplement evidence of stand-replacing fires in crown fire regimes. However, fire scars usually cannot be used to reconstruct with certainty the size of fires that burned in the past, because "absence of evidence is not evidence of absence"; fires do not scar every tree within a fire perimeter, and some fires may burn few or no trees (Swetnam and Baisan 1996). Tree-ring data besides fire scars can also be used to obtain information about historical fire regimes (Kipfmueller and Swetnam 2001). In particular, age structures determined by dendrochronological methods are used to look for even aged stands of trees that presumably initiated after a stand-replacing fire. Other tree-ring approaches for determining past fire regime attributes include examining growth releases and determining the death dates of fire-killed trees (Johnson and Gutsell 1994). This information is useful in conjunction with other methods to reconstruct historical fire regime attributes using tree-ring multiple lines approach to reconstruct fire regime (Harley et al. 2018).

Contributions of dendroecology: Since the late 1990s, dendrochronological studies have been conducted on both sides of the Andes Cordillera to reconstruct attributes of historical fire regimes in the Patagonian forests. Fire scars dating was mainly used to reconstruct fire frequency mediated by historical land-use changes and climate variability. Studies on fire regimes have primarily relied on the conifers *Araucaria araucana*, *Austrocedrus chilensis* and *Fitzroya cupressoides* from northern and central Patagonia, which are proven sources for dendroclimatological and dendroecological studies. Detailed information about fire history reconstructions on these conifers is presented in the recent compilation by Mundo et al. (2017a). For example, in the case of *Araucaria* forests, at ca. 37°–39° S on both Andean slopes, fire histories show a significant variation in fire activity over the last 500 years (González et al. 2005; Quezada 2008; Mundo et al. 2013). For years of widespread fires (i.e. 10% recorder trees scarred) in the Argentinean *Araucaria* forests, the mean fire interval, a measure of fire frequency, varied between 8.9 and 32 years among sites (Mundo et al. 2013). In all forests from Northern Patagonia fire frequency over the pre-1880 period was lower than over the Euro-settling period. A regional trend of increasing fire frequency during the latter half of the nineteenth century coincides with the establishment of new Native American groups at the steppe and adjacent open *Austrocedrus* woodlands that burned for guanaco hunting (Kitzberger and Veblen 1997; Veblen et al. 1999). In the *Araucaria* forest, during the Euro-settling period from 1880 ignitions increased due mostly to logging and clearing land for livestock (Mundo et al. 2013). At the end of nineteenth century and beginning of twentieth century, the records show a significant decline in fire frequency following the demise of the Native American population. Since the 1920s fire frequency diminished in both mesic and xeric forests in association with increasingly effective fire suppression policies and the creation of National Parks (Kitzberger and Veblen 1997; Veblen et al. 1999; Mundo et al. 2013).

Recently, dendrochronological techniques were used to reconstruct frequency from the angiosperm *Nothofagus pumilio* at 16 sites distributed from ~44° to 50° S (Mundo et al. 2017b, Fig. 8.5). These fire histories show a pattern of higher fire frequency during the twentieth century, concurrent with the human occupation and colonization processes in southern Patagonia.

Only some specific studies showed age structure analyses to detect cohorts related to fire events in Patagonia. In the case of the xeric *Austrocedrus* ecotone woodlands, Veblen and Lorenz (1988) found that age frequency distributions were indicative of establishment of this tree species following a relatively sudden release of resources caused by disturbance at the beginning of twentieth century. The same

Fig. 8.5 Burned area in a *N. pumilio* forest at Brazo Rico on Lago Argentino, Los Glaciares National Park. (**a**) A big fire event took place in 1959 at the northern slope of Brazo Rico (the building on the lower-left corner corresponds to Los Notros Hotel). (**b**) Standard dendrochronological techniques were used to reconstruct fire history on partial cross-sections. (**c**) However, dendroecological methods have not been used to study the post-fire recruitment pulses

authors also found that mixed *Austrocedrus-Nothofagus dombeyi* forests tend to maintain their mixed composition with post-fire recruitment of *Austrocedrus* occurring slightly after *N. dombeyi* (Veblen and Lorenz 1987). Burns (1993), using age structure analysis, developed a successional vegetation model and stated that fire acts as a medium of species co-existence between *Nothofagus antarctica*, a vigorously sprouting, shade-intolerant species, and *Araucaria,* a species that partly survives fire above-ground and is more shade-tolerant. In *Araucaria* pure stands were the respective fire histories were reconstructed with fire scars, age structure analyses allowed the detection of multiple cohort structures in response to fire recurrence (Mundo 2011). In regional terms, the pattern of increase in both establishment and number of fires up to 1840 indicates a certain synergism between the two processes. However, when a certain number of fires was exceeded and kept constant for several decades, there was a decrease in the number of individuals established. Almost 70% of the frequency of radial growth releases was explained by the fire frequency of all fire dates reconstructed in the corresponding *Araucaria* stands.

Fire histories reconstructed through dendroecological techniques have also allowed us to explore the influence of climate on the occurrence of forest fires. Climate variability has a significant influence on fire occurrence in the Patagonian forests. Past fire events have been strongly associated with low moisture availability. Years of widespread fire are favored by warmer and drier summers, and also by one or two preceding years of dry climatic conditions (Kitzberger et al. 1997; Mundo et al. 2013). The El Niño–Southern Oscillation (ENSO) is an important factor promoting fire. ENSO events have strong influences on the intensity and latitudinal position of the southeast Pacific subtropical anticyclone that blocks the westerly flow of moist air masses into Patagonia causing temperature and precipitation variations. The year of high fire activity coincides with warm and dry summers following El Niño events (Kitzberger and Veblen 1997; Kitzberger et al. 1997; Veblen et al. 1999). Furthermore, on the east side of the Andes, large fires are also associated with warm and dry years during La Niña positive phase (Kitzberger et al. 1997; Veblen et al. 1999; Mundo et al. 2013).

Other broad-scale climate drivers such as the Pacific Decadal Oscillation (PDO) and the Southern Annular Mode (SAM) play an important control on regional fire occurrence (Holz et al. 2017). Interannual positive anomalies of SAM are associated with higher temperatures (Garreaud et al. 2009) and lower precipitation (Aravena and Luckman 2009) in southern South America. High regional fire activity has been strongly associated with the negative and positive signature of the PDO and SAM, respectively (Mundo et al. 2013).

Gaps of Knowledge

- There are scarce studies focused on complementing the reconstruction of fire histories with the measurement of quantitative parameters of forest structure
- Studies in mixed *Araucaria* and *N. pumilio* forests of Argentina to understand the different sensitivity of woody species to fires and the effect different fire severities on the theses stand dynamics
- Dendroecology appears to be an essential tool to achieve a temporal understanding of the *Pinus* invasion process in Patagonia and its relationship with the recent fires of great severity and intensity
- Tierra del Fuego is still a province where the fire histories have not been reconstructed yet nor the forest dynamics have been studied in relation to the fires
- Interactions with other disturbances in the Patagonian Andes

8.4.5 Forest Decline

*Process description:*Forest declines involve complex processes and multiple interactions between disturbing agents and stressors that act on different time scales. This hinders not only its identification but also studying its dynamics and causes (for example, Pedersen 1998; Cherubini et al. 2002; Hartmann and Messier 2008; Amoroso et al. 2015; Camarero et al. 2011). Several species in the Andean-Patagonian forests experience processes of decline and subsequent death of trees, including *Austrocedrus chilensis* and several species of the *Nothofagus* genus.

Description of methods: The set of dendrochronological techniques used in forest decline are more recent compared with other disturbances, principally the study of the complex declining patterns of radial growth. In some cases, such as the *A. chilensis* forests, it has been complemented with traditional methods such as the dating of the establishment and death of individuals, and the assessment of abrupt changes in radial growth.

Contributions of dendroecology: A. chilensis forests at mesic sites in North Patagonia experience stand-level decline (Fig. 8.6). Tree-ring analyses have been used, principally over the past decade, to better understand the nature of the forest decline. Of all studied aspects, the radial growth patterns of trees have received the most attention. Several studies have shown a reduction in radial growth associated with vigor loss and death of trees in declining *A. chilensis* forests (Calí 1996; Mundo et al. 2010a; Amoroso et al. 2012a). Whereas, early studies reported on a limited number of trees and sites or focused on living symptomatic trees only (Calí 1996; Mundo et al. 2010a), more recent studies assessed radial growth decline on a large number of living asymptomatic and symptomatic and dead trees while discriminating the incidence of the growth decline among trees of different crown health conditions and canopy positions at onset of decline (Amoroso et al. 2012a). Furthermore, the latter authors found two common patterns in the rate of decline of radial growth, an abrupt (1–3 years) decline followed by relatively constant narrow rings over time and gradual and constant decline in radial growth over time (Amoroso et al. 2012a). Some dendrochronological studies have shown the radial growth decline actually precedes crown symptoms (Amoroso et al. 2012a, 2015). Tree-ring studies have

Fig. 8.6 Declining *A. chilensis* stand in northern Patagonia (Rio Negro, Argentina) depicting loss of crown vigor characterized by chlorotic foliage and progressing defoliation

also been used to shed light on the role of climate in the decline and death of *A. chilensis* trees. The onset of radial growth decline of individual trees in declining forests was significantly associated with adverse climatic conditions during the previous and current growing season (Mundo et al. 2010a; Amoroso et al. 2015).

The length of the radial growth decline might vary among trees and sites, ranging from a decade up to 80 years (Amoroso et al. 2012a). The temporal patterns of radial growth decline in overstory trees exhibited great variability and contrasting patterns among stands (Amoroso et al. 2012a). While at some stands 30–50 years elapsed from the onset of decline in the first tree until more than half of all declining trees exhibited decline; alternately, 50% of all trees in the stands were declining in two decades or less since the first tree (Amoroso et al. 2012a).

Dendroecological research has compiled knowledge of major aspects of forest stand dynamics in these forests. Dendrochronological studies have helped to precisely date the death of the trees in declining forests (Relva et al. 2009; Amoroso and Larson 2010). Amoroso and Larson (2010) quantified and described long-term patterns of overstory mortality over almost six decades concluding mortality in these forests was erratic, unpredictable and highly variable over time, resulting in different patterns and variable rates among stands. Similarly to the onset of radial decline, high-magnitude mortality events in these forests were concurrent with significant droughts (Amoroso et al. 2015). Mundo et al. (2010a) and Amoroso et al. (2015) foresee a gradual increase in the number of *A. chilensis* forests affected by the decline associated with the expected extreme climatic conditions for the future.

The radial growth response of residual overstory trees to these small-scale disturbances was variable in time and number among stands (Amoroso and Larson 2010). In general, trees in nearly all stands released after the onset of the mortality, but the pattern was not distinctive, varying in time, number and magnitude. Furthermore, growth releases in some stands occurred after single and/or multiple tree deaths suggesting a relationship between processes, while in others not (Amoroso and Larson 2010).

The death of trees in these forests creates gaps in the canopy that in the absence of other disturbances, such as grazing, allows the successful establishment of regeneration in the understory. Amoroso and Larson (2010) reconstructed the temporary pattern of understory establishment concluding that while successful in all stands, it was low in some stands and high in others depending on the density of the overstory. In some stands, there has been an increase in the abundance of *N. dombeyi* regeneration relative to *A. chilensis* in the last years, representing an important shift in composition and increased complexity in stand structure (Amoroso et al. 2012b). These authors also found the establishment of both *A. chilensis* and *N. dombeyi* was synchronous with overstory tree mortality events, but it was more consistent among stands and prolonged for *N. dombeyi*. Establishment of *A. chilensis* was not associated with climatic events but *N. dombeyi* establishment was synchronous with droughts, possibly related to climate-driven mortality, creating canopy gaps or reducing competition within gaps (Amoroso et al. 2012b). Likewise, overstory mortality and resulting canopy changes influence significantly the growth patterns of regeneration. The average radial growth and the average diameter of *A. chilensis* and *N. dombeyi* individuals established in the understory of declining forests differed significantly (Losada et al. 2018). Both species exhibited a positive response when the loss of basal area of the overstory due to mortality exceeded 20%.

Lastly, dendroecological research has also helped understanding the partial death of the cambium associated with the forest decline. On average, the partial death of the cambium may affect more than a third of all living *A. chilensis* trees in declining stands and its duration may vary between 1 and 39 years (Amoroso and Daniels 2010); early research found this can happen up to 75 years prior to the death of individuals (Calí 1996).

Rodríguez-Catón et al. (2016) studied the patterns of radial growth of *N. pumilio* over the last 100 years in declining forests. These authors developed basal area increment (BAI) chronologies in 11 sites in northern Patagonia from 38 °S to 43 °S approximately and used principal component analysis (PCA) in order to identify the dominant growth patterns in these stands and identified different degrees of growth reduction at the stand level. From the PCA, three regional growth patterns were determined. The first pattern was characterized by high growth in the first decades of the twentieth century but marked decrease in growth staring 1940s. The second pattern consisted of individuals who presented an intermediate growth level until the middle of 1960s when growth began to decrease. Finally, the third pattern showed a stable growth until 1960s where it started to increase.

The trees with most crown dieback showed the steepest long-term negative trends in radial growth triggered by extreme drought events in the eastern-dry edge of the *N. pumilio* distribution (Rodríguez-Catón et al. 2016). Indeed, the growth of *N. pumilio* forests with external manifestations of decline is limited by low rainfall and high temperatures during the growing season. The sequence of years with these characteristics, such as 1942 to 1944 and/or particular droughts in periods of high temperatures such as the years 1978 and 1979, is associated with the beginning of decline processes in *N. pumilio* forests in northern Patagonia, Argentina. Although the climate-growth relationship is similar for the different regional growth patterns, their response to climatic variations varies between and within patterns over time. Therefore, to facilitate the study of the effect of climate on declining forests it is recommended to first identify the existing growth patterns in each stand, compare the different levels and growth pattern, to finally determine the differential responses of each growth pattern to the climate (Rodríguez-Catón et al. 2015).

Gaps of Knowledge

- The study of forest dynamics in *Nothofagus* forests including the age structure and dominant growth patterns
- Model prediction of tree mortality using pre-death rates of growth and individual variables (percentage of crown mortality) and population variables (density)
- To determine the decline history for other species and at regional scales

8.5 Concluding Remarks

Although Theodor Hartig laid the foundations of dendroecology in the mid-nineteenth century, it was not until the mid-1970s that it gained attention and became accepted as a subdiscipline in the field of dendrochronology (Hough and Forbes 1943; Lorimer 1980; Schweingruber 1988; Fritts and Swetnam 1989; Speer 2010). In the Patagonian Andes, the first dendroecological studies began in the late 1980s but they were based only on the simple determination of tree ages through the counting of rings (Veblen and Lorenz 1987, 1988; Veblen 1989). It was not until the mid-1990s that more elaborate dendroecological methods began to be applied for the study of the forest dynamics in the Patagonian Andes (Veblen et al. 1995). As indicated above, the initial development of dendroecology in the Patagonian Andes was motivated by previously developed ring-width chronologies. After studying the responses to climate of the main species in the region, an intense era began in order to date the occurrence of the most common disturbances in Patagonia. Finally, progress was made on the analysis of changes in radial growth and the date of establishment and mortality of trees.

What has dendroecology taught us in the Patagonian Andes during the last 30 years? Although it is difficult to completely separate the dendroecological applications from other subdisciplines of dendrochronology, we could synthesize it in the following points:

1. It allowed us to know the range and maximum ages reached by the most common species in the Patagonian Andes of Argentina. The date of the tree rings have, for example, revealed that there are specimens of *Fitzroya cupressoides* that are older than 2000 years in Argentina and that can reach up to 3600 years in Chile (Lara et al. 2000).
2. To establish the nature of episodic establishment and mortality events in relation to climatic variability for different environments and species. Although the first studies focused on *A. chilensis* in the Patagonian steppe-forest ecotone, they soon spread to different species, including *N. dombeyi* in more humid environments and *N. pumilio* in the upper and lower limits of its distribution in forests in the north and south portion of the Patagonian Andes in Argentina. More recently, progress has been made in the study of forest demographic processes and certainly dendrochronology can help us with more knowledge by providing data on forest dynamics to be used to feed models of population dynamics in the face of different climate change scenarios.
3. Although the determination of the frequency of *Ormiscodes* attacks in *N. pumilio* forests, as well as the establishment patterns between these events and the role of climatic variability have not been clearly established, the dendrochronological dating and measurement in sites affected by severe *Ormiscodes* attacks show reductions in radial growth in the year following the attack of more than 50% in relation to unaffected individuals.

4. Pioneering dendrochronological studies in snow avalanches dating show patterns that are not necessarily synchronous along precipitation gradients, suggesting that in addition to the snowpack, other environmental conditions such as temperatures or winds may be affecting the occurrence of these events.
5. Dendrochronology has been extremely useful to develop multi-century fire histories in *Araucaria araucana, Fitzroya cupressoides* and *Austrocedrus chilensis* and mixed *Austrocedrus-Nothofagus dombeyi* forests in Patagonia. In this way, the frequency and its temporal changes have been reconstructed for different environments in relation to human and climate influences.
6. While common to all forest types in the region, the effect of the consequences of wind in these ecosystems received little attention compared to other disturbances. Yet, dendroecology has helped to reconstruct the history of this disturbance at the landscape level by dating events and determining intervals of recurrence mainly in large and small scale blowdowns in *Nothofagus* forests. Along with this, the study of age structures and radial growth patterns at the stand level allowed to distinguish the nature of its magnitude.
7. Understand the complex declining radial growth patterns related to the decline and death of trees in *A. chilensis* and *N. pumilio* forests. The study of the declining patterns of radial growth in these forests revealed the role of climatic variability as a triggering factors many years before the death of individual trees occurs; the severe droughts occurred during the twentieth century triggered individual and stand level decline. In *A. chilensis* forests particularly, dendroecology allowed the reconstruction of the temporal patterns of overstory mortality and understory establishment at the stand level.

Although the dendroecological studies in the region that relate disturbances and forest dynamics using dendrochronological techniques began with simply ring counting, with time they became more complex by the cross-dating of the samples and further analysis of the growth patterns and description of anatomical features in the rings to date disturbance events. To a lesser extent, dendroecological techniques have been used to characterize post-disturbance forest dynamics patterns. Few studies describe the establishment or species replacement after disturbance. Likewise, little is known about the intensity of most disturbances and the development patterns after them. Despite these current limitations, the past contributions of dendroecology to the study of forest dynamics in the Patagonian Andes have allowed us to lay the foundations and evaluate the potential of this technique for the region. New perspectives associated with the study of the interactions between different types of disturbances and the effects that climatic changes predicted for the twenty-first century will have on Patagonian forests are some of the challenges that dendrochronology will have to face in the coming years. Although in recent years the number of studies has increased substantially, there is still much work to be done at the regional and continental levels.

References

Allen CD, Macalady AK, Chenchouni H et al (2010) A global overview of drought and heat-induced tree mortality reveals emerging climate change risks for forests. For Ecol Manag 259:660–684. https://doi.org/10.1016/j.foreco.2009.09.001

Amoroso MM (2009) Stand development patterns as a consequence of the decline in *Austrocedrus chilensis* forests. University of British Columbia, Doctoral Dissertation

Amoroso MM, Daniels LD (2010) Cambial mortality in declining *Austrocedrus chilensis* forests: implications for stand dynamics studies. Can J For Res 40:885–893. https://doi.org/10.1139/X10-042

Amoroso MM, Larson BC (2010) Stand development patterns as a consequence of the mortality in *Austrocedrus chilensis* forests. For Ecol Manag 259:1981–1992. https://doi.org/10.1016/j.foreco.2010.02.009

Amoroso MM, Daniels LD, Larson BC (2012a) Temporal patterns of radial growth in declining *Austrocedrus chilensis* forests in Northern Patagonia: the use of tree-rings as an indicator of forest decline. For Ecol Manag 265:62–70. https://doi.org/10.1016/j.foreco.2011.10.021

Amoroso MM, Suarez ML, Daniels LD (2012b) *Nothofagus dombeyi* regeneration in declining *Austrocedrus chilensis* forests: effects of overstory mortality and climatic events. Dendrochronologia 30:105–112. https://doi.org/10.1016/j.dendro.2010.12.005

Amoroso MM, Villalba R, Ruiz M et al (2013) Wave regeneration after wind blowdowns in *Nothofagus* forests in southern Patagonia. Paper presented at the Second American Dendrochronology Conference – AmeriDendro. Tucson, 13–17 May 2013

Amoroso MM, Daniels LD, Villalba R et al (2015) Does drought incite tree decline and death in Austrocedrus chilensis forests? J Veg Sci 26:1171–1183. https://doi.org/10.1111/jvs.12320

Amoroso MM, Radins M, Villalba R et al (2016) Wind blowdown history in *Nothofagus* forests in southern Patagonia. Paper presented at the Third American Dendrochronology Conference – AmeriDendro. Mendoza, 28–31 March 2016

Amoroso MM, Daniels LD, Baker PJ et al (eds) (2017a) Dendroecology: tree-ring analyses applied to ecological studies, vol 231. Springer, Switzerland

Amoroso MM, Rodríguez-Catón M, Villalba R et al (2017b) Forest decline in Northern Patagonia: the role of climatic variability. In: Amoroso MM, Daniels LD, Baker PJ et al (eds) Dendroecology: tree-ring analyses applied to ecological studies. Springer, Switzerland

Aravena JC, Luckman BH (2009) Spatio-temporal rainfall patterns in southern South America. Int J Climatol 29:2106–2120. https://doi.org/10.1002/joc.1761

Baker PJ (2003) Tree age estimation for the tropics: a test from the southern Appalachians. Ecol Appl 13:1718–1732. https://doi.org/10.1890/02-5025

Baker PJ, Bunyavejchewin S, Oliver C et al (2005) Disturbance history and historical stand dynamics of a seasonal tropical forest in western Thailand. Ecol Monogr 75:317–343. https://doi.org/10.1890/04-0488

Bigler C, Bugmann H (2004) Predicting the time of tree death using dendrochronological data. Ecol Appl 14:902–914. https://doi.org/10.1890/03-5011

Bonan GB (2008) Forests and climate change: forcings, feedbacks, and the climate benefits of forests. Science 320:1444–1449. https://doi.org/10.1126/science.1155121

Brown PM (2006) Climate effects on fire regimes and tree recruitment in Black Hills ponderosa pine forests. Ecology 87:2500–2510. https://doi.org/10.1890/0012-9658(2006)87[2500:ceofra]2.0.co;2

Burns BR (1993) Fire-induced dynamics of *Araucaria araucana-Nothofagus antarctica* forest in the Southern Andes. J Biogeogr 20:669–685. https://doi.org/10.2307/2845522

Burrows CJ, Burrows VL (1976) Procedures for the study of snow avalanche chronology using growth layers (annual rings) of woody plants. Institute of Arctic and Alpine Research. Occas Pap 23(54)

Calí SG (1996) *Austrocedrus chilensis*: estudio de los anillos de crecimiento y su relación con la dinámica del "Mal del Ciprés" en el P.N. Nahuel Huapi, Argentina. Undergraduate Thesis, Universidad Nacional del Comahue

Camarero JJ, Bigler C, Linares JC et al (2011) Synergistic effects of past historical logging and drought on the decline of Pyrenean silver fir forests. For Ecol Manag 262:759–769. https://doi.org/10.1016/j.foreco.2011.05.009

Carrara PE (1979) The determination of snow avalanche frequency through tree-ring analysis and historical Records at Ophir, Colorado. Geol Soc Am Bull 90(8):773–780. https://doi.org/10.1130/0016-7606(1979)90<773:TDOSAF>2.0.CO;2

Casteller A, Christen M, Villalba R et al (2008) Validating numerical simulations of snow avalanches using dendrochronology: the Cerro Ventana event in Northern Patagonia, Argentina. Nat Hazards Earth Syst Sci 8:433–443. https://doi.org/10.3929/ethz-b-000073411

Casteller A, Villalba R, Mayer A et al (2009) Reconstrucción espacial y temporal de la ocurrencia de avalanchas de nieve en los Andes patagónicos utilizando técnicas dendrocronológicas. Rev Chil Hist Nat 82:245–264. https://doi.org/10.4067/s0716-078x2009000200007

Casteller A, Villalba R, Araneo D et al (2011) Reconstructing temporal patterns of snow avalanches at Lago del Desierto, southern Patagonian Andes. Cold Reg Sci Technol 67:68–78. https://doi.org/10.1016/j.coldregions.2011.02.001

Cherubini P, Fontana G, Rigling D et al (2002) Tree-life history prior to death: two fungal root pathogens affect tree-ring growth differently. J Ecol 90:839–850. https://doi.org/10.1046/j.1365-2745.2002.00715.x

De Agostini AM (1941) Andes Patagonicos. In: Viajes De Exploración a La Cordillera Patagónica Austral, Peuser, Buenos Aires

Duncan RP (1989) An evaluation of errors in tree age estimates based on increment cores in kahikatea (*Dacrycarpus dacrydioides*). NZ Nat Sci 16(3):31–37

Eskuche U (1973) Estudios fitosociológicos en el norte de Patagonia. I. Investigación de algunos factores de ambiente en comunidades de bosque y de chaparral. Phytocoenologia 1:64–113

Frangi JL, Barrera MD, Puigdefábregas J et al (2005) Ecología de los bosques de Tierra del Fuego. In: Arturi MF, Frangi JL, Goya JF (eds) Ecología y manejo de bosques nativos de Argentina. Editorial Universidad Nacional de la Plata, La Plata, pp 1–88

Frelich LE (2002) Forest dynamics and disturbance regimes: studies from temperate evergreen-deciduous forests. Cambridge University Press, New York

Fritts HC, Swetnam TW (1989) Dendroecology: a tool for evaluating variations in past and present forest environments. In: Begon M, Fitter AH, Ford ED et al (eds) Advances in ecological research, vol 19. Academic Press, London, pp 111–188

Garreaud RD, Vuille M, Compagnucci R et al (2009) Present-day South American climate. Palaeogeogr Palaeocl 281:180–195. https://doi.org/10.1016/j.palaeo.2007.10.032

Gers E, Florin N, Gartner H (2001) Application of shrubs for dendrogeomorphological analysis to reconstruct spatial and temporal landslide movement patterns-A preliminary study. Zeitschrift fur Geomorphologie Supplementband 125:163–175

Girardin MP, Tardif J, Bergeron Y (2001) Radial growth analysis of *Larix laricina* from the Lake Duparquet area, Quebec, in relation to climate and larch sawfly outbreaks. Ecoscience 8:127–138. https://doi.org/10.1080/11956860.2001.11682638

González ME, Lara A (2015) Large fires in the Andean Araucaria forests: when a natural ecological process becomes a threat. Orix Int J Conserv 49:394. https://doi.org/10.1017/S0030605315000599

González ME, Veblen TT, Sibold JS (2005) Fire history of *Araucaria-Nothofagus* forests in Villarrica National Park, Chile. J Biogeogr 32:1187–1202. https://doi.org/10.1111/j.1365-2699.2005.01262.x

González ME, Amoroso M, Lara A et al (2014) Ecología de disturbios y su influencia en los bosques templados de Chile y Argentina. In: Ecología Forestal: Bases para el Manejo Sustentable y Conservación de los Bosques Nativos de Chile, p 411–502

Harley G, Baisan C, Brown P et al (2018) Advancing dendrochronological studies of fire in the United States. Fire 1:11. https://doi.org/10.3390/fire1010011

Hartmann H, Messier C (2008) The role of forest tent caterpillar defoliations and partial harvest in the decline and death of sugar maple. Ann Bot 102:377–387. https://doi.org/10.1093/aob/mcn104

Holz A, Paritsis J, Mundo IA et al (2017) Southern annular mode drives multicentury wildfire activity in southern South America. Proc Natl Acad Sci 114:9552–9557. https://doi.org/10.1073/pnas.1705168114

Hough AF, Forbes RD (1943) The ecology and silvics of forests in the high plateau of Pennsylvania. Ecol Monogr 13:299–320. https://doi.org/10.2307/1943224

Johnson EA, Gutsell SL (1994) Fire frequency models, methods and interpretations. In: Begon M, Fitter AH (eds) Advances in ecological research. Academic Press, United Kingdom, pp 239–287

Kipfmueller KF, Swetnam TW (2001) Using dendrochronology to reconstruct the history of forest and woodland ecosystems. In: Egan D, Howell EA (eds) The historical ecology handbook: a restorationist's guide to reference ecosystems. Island Press, Washington, D.C., pp 199–228

Kitzberger T, Veblen TT (1997) Influences of humans and ENSO on fire history of *Austrocedrus chilensis* woodlands in northern Patagonia, Argentina. Écoscience 4:508–520. https://doi.org/10.1080/11956860.1997.11682430

Kitzberger T, Veblen TT, Villalba R (1997) Climatic influences on fire regimes along a rain forest-to-xeric woodland gradient in northern Patagonia, Argentina. J Biogeogr 24:35–47. https://doi.org/10.1111/j.1365-2699.1997.tb00048.x

Kitzberger T, Steinaker DF, Veblen TT (2000) Effects of climatic variability on facilitation of tree establishment in northern Patagonia. Ecology 81:1914–1924. https://doi.org/10.1890/0012-9658(2000)081[1914:EOCVOF]2.0.CO;2

Lara A, Villalba R, Aravena JC et al (2000) Desarrollo de una red de cronologías de *Fitzroya cupressoides* (alerce) para Chile y Argentina. In: Roig F (ed) Dendrocronología en América Latina. EDIUNC, Mendoza, pp 217–244

Lorimer CG (1980) Age structure and disturbance history of a southern Appalachian virgin forest. Ecology 61:1169–1184. https://doi.org/10.2307/1936836

Losada S, Amoroso MM, Bogino SM (2018) Regeneration dynamics of *Austrocedrus chilensis* and *Nothofagus dombeyi* in declining forests. Bosque 39:333–345. https://doi.org/10.4067/s0717-92002018000200333

Manion PD (1991) Tree disease concepts. Prentice-Hall, Englewood Cliffs

Manion PD, Lachance D (1992) Forest decline concepts: an overview. In: Manion PD, Lachance D (eds) Forest decline concepts. American Phytopathological Society, St. Paul, pp 181–190

Mast JN, Veblen TT (1994) A dendrochronological method of studying tree mortality patterns. Phys Geogr 15:529–542. https://doi.org/10.1080/02723646.1994.10642533

McDowell NG, Ryan MG, Zeppel MJ et al (2013) Improving our knowledge of drought-induced forest mortality through experiments, observations, and modeling. New Phytol 200:289–293. https://doi.org/10.1111/nph.12502

Mohr-Bell D (2015) Superficies afectadas por incendios en la región Bosque Andino Patagónico durante los veranos de 2013–2014 y 2014–2015. SAyDS-CIEFAP, Nodo Regional Bosque Andino Patagónico

Morales M, Villalba R (2009) Anillos de árboles como evidencias de ataques de insectos defoliadores en la Patagonia. Bol Soc Entomol Argent 20:13–14

Morales MS, Villalba R, Srur A et al (2004) Buscando el anillo perdido: efecto del ataque de cuncuna (*Ormiscodes* sp.) en bosques de lenga (*Nothofagus pumilio*), El Chaltén, Santa Cruz. II Reunión Binacional de Ecología Argentino-Chilena. Mendoza

Mundo IA (2011) Historia de incendios en bosques de *Araucaria araucana* (Molina) K. Koch de Argentina a través de un análisis dendroecológico. Tesis Doctoral, Universidad Nacional de La Plata

Mundo IA, Barrera MD, Roig FA (2007) Testing the utility of *Nothofagus pumilio* for dating a snow avalanche in Tierra del Fuego, Argentina. Dendrochronologia 25(1):19–28. https://doi.org/10.1016/j.dendro.2007.01.001

Mundo IA, El Mujtar VA, Perdomo MH et al (2010a) *Austrocedrus chilensis* growth decline in relation to drought events in northern Patagonia, Argentina. Trees 24:561–570. https://doi.org/10.1007/s00468-010-0427-8

Mundo I, Morales M, Casteller A et al (2010b) Influences of past climate variations on forest disturbance in Patagonia. II International symposium, "Reconstructing climate variations in South America and the Antarctic Peninsula over the last 2000 years", Valdivia

Mundo IA, Kitzberger T, Roig JFA et al (2013) Fire history in the *Araucaria araucana* forests of Argentina: human and climate influences. Int J Wildland Fire 22:194–206. https://doi.org/10.1071/WF11164

Mundo IA, Holz A, González ME et al (2017a) Fire history and fire regimes shifts in Patagonian temperate forests. In: Amoroso MM, Daniels LD, Baker PJ, Camarero JJ (eds) Dendroecology: tree-ring analyses applied to ecological studies. Springer International Publishing, Cham, pp 211–229

Mundo IA, Villalba R, Veblen TT et al (2017b) Fire history in southern Patagonia: human and climate influences on fire activity in *Nothofagus pumilio* forests. Ecosphere. https://doi.org/10.1002/ecs2.1932

Mutarelli E, Orfila E (1973) Algunos resultados de las investigaciones de manejo silvicultural que se realizan en los bosques andino-patagónicos de Argentina. Rev For Argent 17(3):69–75

Overpeck JT, Whitlock C, Huntley B (2003) Terrestrial biosphere dynamics in the climate system: past and future. In: Bradley RS, Pedersen TF, Alverson KD, Bergmann KF (eds) Paleoclimate, global change and the future. Springer, Berlin

Paine RT, Levin SA (1981) Intertidal landscapes: disturbance and the dynamics of pattern. Ecol Monogr 51:145–178. https://doi.org/10.2307/2937261

Palacios M (2013) Reconstrucción dendrocronológica de ataques de cuncuna (*Ormiscodes amphimone*) en bosques de lenga (*Nothofagus pumilio*) en El Chaltén, Santa Cruz. Undergraduate Thesis. Universidad Nacional de Cuyo

Paritsis J, Veblen TT (2010) Temperature and foliage quality affect performance of the outbreak defoliator *Ormiscodes amphimone* (F.)(Lepidoptera: Saturniidae) in northwestern Patagonia, Argentina. Rev Chil Hist Nat 83(4):593–603. https://doi.org/10.4067/S0716-078X2010000400012

Paritsis J, Veblen TT (2011) Dendroecological analysis of defoliator outbreaks on *Nothofagus pumilio* and their relation to climate variability in the Patagonian Andes. Glob Chang Biol 17(1):239–253. https://doi.org/10.1111/j.1365-2486.2010.02255.x

Paritsis J, Veblen TT, Kitzberger T (2009) Assessing dendroecological methods to reconstruct defoliator outbreaks on *Nothofagus pumilio* in northwestern Patagonia, Argentina. Can J For Res 39(9):1617–1629. https://doi.org/10.1139/X09-085

Pedersen BS (1998) Modeling tree mortality in response to short- and long-term environmental stresses. Ecol Model 105(2–3):347–351. https://doi.org/10.1016/s0304-3800(97)00162-2

Pederson N (2010) External characteristics of old trees in the eastern deciduous Forest. Nat Areas J 30(4):396–407. https://doi.org/10.3375/043.030.0405

Pretzsch H (2009) Forest dynamics, growth and yield. From measurement to model. Springer, Berlin/Heidelberg

Puigdefábregas J, Gallart F, Bianciotto O et al (1999) Banded vegetation patterning in a subantarctic forest of Tierra del Fuego, as an outcome on the interaction between wind and tree growth. Acta Oecol 20(3):135–146. https://doi.org/10.1016/S1146-609X(99)80027-7

Quezada JM (2008) Historia de incendios en bosques de *Araucaria araucana* (Mol.) Koch del Parque Nacional Villarrica, a partir de anillos de crecimiento y registros orales. Tesina de grado, Universidad Austral de Chile, Facultad de Ciencias Forestales

Rayback SA (1998) A dendrogeomorphological analysis of snow avalanches in the Colorado Front Range, USA. Phys Geogr 19:502–515. https://doi.org/10.1080/02723646.1998.10642664

Rebertus AJ, Veblen TT (1993) Partial wave formation in old-growth *Nothofagus* forests on Tierra del Fuego, Argentina. Bull Torrey Bot Club 120:461–470. https://doi.org/10.2307/2996751

Rebertus AJ, Kitzberger T, Veblen TT et al (1997) Blowdown history and landscape patterns in the Andes of Tierra del Fuego, Argentina. Ecology 78(3):678–692. https://doi.org/10.1890/0012-9658(1997)078[0678:BHALPI]2.0.CO;2

Relva MA, Westerholm CL, Kitzberger T (2009) Effects of introduced ungulates on forest understory communities in northern Patagonia are modified by timing and severity of stand mortality. Plant Ecol 201:11–22. https://doi.org/10.1007/s11258-008-9528-5

Rodríguez-Catón M, Villalba R, Srur AM et al (2015) Long-term trends in radial growth associated with *Nothofagus pumilio* forest decline in Patagonia: integrating local-into regional-scale patterns. For Ecol Manag 339:44–56. https://doi.org/10.1016/j.foreco.2014.12.004

Rodríguez-Catón M, Villalba R, Morales M et al (2016) Influence of droughts on *Nothofagus pumilio* forest decline across northern Patagonia, Argentina. Ecosphere 7(7):e01390. https://doi.org/10.1002/ecs2.1390

Ruiz M (2013) Reconstrucción de volteos por vientos en los valles del Río Toro, y del Río de las Vueltas (El Chaltén, provincia de Santa Cruz) a través del análisis de los anillos de crecimiento de *Nothofagus pumilio* (Poepp. et Endl.) Krasser (lenga). Undergraduate Thesis, Universidad Nacional de Cuyo

Schneider C, Gies D (2004) Effects of El Niño–southern oscillation on southernmost South America precipitation at 53°S revealed from NCEP–NCAR reanalysis and weather station data. Int J Climatol 24(9):1057–1076. https://doi.org/10.1002/joc.1057

Schulman E (1954) Longevity under adversity in conifers. Science 119(3091):396–399. https://doi.org/10.1126/science.119.3091.396

Schweingruber FH (1988) Tree rings: basics and applications of dendrochronology. D. Reidel Publishing Company, Dordrecht

Smith L (1973) Indication of snow avalanche periodicity through interpretation of vegetation patterns in the north cascades, Washington. In: Methods of avalanche control on Washington mountain highways- third annual report, Washington State Highway Commission Department of Highways, Olympia, Washington, DC, p 187

Speer JH (2010) Fundamentals of tree-ring research. University of Arizona Press, Tucson

Srur AM, Villalba R, Villagra PE et al (2004) Estructura y dinámica del bosque de *Nothofagus pumilio* a lo largo de un gradiente altitudinal en El Chaltén, Santa Cruz. In: II Reunión Binacional de Ecología. XI Reunión de la Sociedad de Ecología de Chile y XXI Reunión Argentina de Ecología (Mendoza)

Srur AM, Villalba R, Villagra PE et al (2008) Influencias de las variaciones en el clima y en la concentración de C02 sobre el crecimiento de *Nothofagus pumilio* en la Patagonia. Rev Chil Hist Nat 81(2):239–256. https://doi.org/10.4067/S0716-078X2008000200007

Srur AM, Villalba R, Rodríguez-Catón M et al (2016) Establishment of *Nothofagus pumilio* at upper treelines across a precipitation gradient in the northern Patagonian Andes. Arct Antarct Alp Res 48(4):755–766. https://doi.org/10.1657/AAAR0016-015

Srur AM, Villalba R, Rodríguez-Catón M et al (2018) Climate and *Nothofagus pumilio* establishment at upper treelines in the Patagonian Andes. Front Earth Sci 6:57. https://doi.org/10.3389/feart.2018.00057

Stoffel M, Bollschweiler M (2008) Tree-ring analysis in natural hazards research? An overview. Nat Hazards Earth Syst Sci 8(2):187–202. https://doi.org/10.5194/nhess-8-187-2008

Suarez ML, Kitzberger T (2010) Differential effects of climate variability on forest dynamics along a precipitation gradient in northern Patagonia. J Ecol 98(5):1023–1034. https://doi.org/10.1111/j.1365-2745.2010.01698.x

Suarez ML, Ghermandi L, Kitzberger T (2004) Factors predisposing episodic drought-induced tree mortality in Nothofagus-site, climatic sensitivity and growth trends. J Ecol 92(6):954–966. https://doi.org/10.1111/j.1365-2745.2004.00941.x

Swetnam TW, Baisan CH (1996) Historical fire regime patterns in the southwestern United States since AD 1700. In: Allen CD (ed) Fire effects in southwestern forest: proceedings of the 2nd La Mesa fire symposium. USDA Forest Service, Rocky Mountain Research Station, pp 11–32

Turner MG (2010) Disturbance and landscape dynamics in a changing world. Ecology 91(10):2833–2849. https://doi.org/10.1890/10-0097.1

Veblen TT (1989) Tree regeneration responses to gaps along a transandean gradient. Ecology 70(3):541–543. https://doi.org/10.2307/1940197

Veblen TT, Ashton DH (1978) Catastrophic influences on the vegetation of the Valdivian Andes, Chile. Vegetatio 36(3):149–167. https://doi.org/10.1007/BF02342598

Veblen TT, Lorenz DC (1987) Post-fire stand development of *Austrocedrus-Nothofagus* forests in northern Patagonia. Vegetatio 71:113–126. https://doi.org/10.1007/BF00044825

Veblen TT, Lorenz DC (1988) Recent vegetation changes along the forest/steppe ecotone of northern Patagonia. Ann Assoc Am Geogr 78:93–111. https://doi.org/10.1111/j.1467-8306.1988.tb00193.x

Veblen TT, Burns BR, Kitzberger TT et al (1995) The ecology of the conifers of southern South America. In: Enright N, Hill R (eds) Ecology of the southern conifers. Melbourne University Press, Carlton, pp 87–126

Veblen TT, Kitzberger T, Burns RB et al (1996) Perturbaciones y regeneración en bosques andinos del sur de Chile y Argentina. In: Armesto JJ, Arroyo MK, Villagrán J (eds) Ecología del Bosque Nativo de Chile. Universidad de Chile Press, Santiago, pp 169–198

Veblen TT, Kitzberger T, Villalba R et al (1999) Fire history in Northern Patagonia: the roles of humans and climatic variation. Ecol Monogr 69:47–67. https://doi.org/10.1890/0012-9615(1999)069[0047:FHINPT]2.0.CO;2

Veblen TT, Kitzberger T, Villalba R (2004) Nuevos paradigmas en ecología y su influencia sobre el conocimiento de la dinámica de bosques del sur de Argentina y Chile. In: Arturi MF, Frangi JL, Goya JF (eds) Ecología y manejo de bosques nativos de Argentina. Editorial Universidad Nacional de la Plata, La Plata, pp 1–48

Veblen TT, Kitzberger T, Villalba R (2005) Nuevos paradigmas en ecología y su influencia sobre el conocimiento de la dinámica de los bosques del sur de Argentina y Chile. In: Arturo M, Frangi J, Goya JF (eds) Ecología y Manejo de los Bosques de Argentina. Editorial de la Universidad Nacional de La Plata, Argentina, Buenos Aires, p 48

Veblen TT, Holz A, Paritsis J et al (2011) Adapting to global environmental change in Patagonia: what role for disturbance ecology? Austral Ecol 36(8):891–903. https://doi.org/10.1111/j.1442-9993.2010.02236.x

Villalba R (1997) Climatic influences on forest dynamics along the forest-steppe ecotone in northern Patagonia, Argentina

Villalba R, Veblen TT (1997a) Regional patterns of tree population age structures in northern Patagonia: climatic and disturbance influences. J Ecol 85(2):113–124. https://doi.org/10.2307/2960643

Villalba R, Veblen TT (1997b) Improving estimates of total tree ages based on increment core samples. Ecoscience 4(4):534–542. https://doi.org/10.1080/11956860.1997.11682433

Villalba R, Veblen TT (1998) Influences of large-scale climatic variability on episodic tree mortality in northern Patagonia. Ecology 79(8):2624–2640. https://doi.org/10.1890/0012-9658(1998)079[2624:IOLSCV]2.0.CO;2

Villalba R, Boninsegna JA, Veblen TT et al (1997) Recent trends in tree-ring records from high elevation sites in the Andes of northern Patagonia. In: Climatic change at high elevation sites. Springer, Dordrecht, pp 193–222

White PS, Pickett STA (1985) Natural disturbance and patch dynamics: an introduction. In: Pickett S, White P (eds) The ecology of natural disturbance and patch dynamics. Academic Press, New York, pp 3–13

Wolkovich EM, Cook BI, McLauchlan KK et al (2014) Temporal ecology in the Anthropocene. Ecol Lett 17(11):1365–1379. https://doi.org/10.1111/ele.12353

Yocom Kent L (2014) An evaluation of fire regime reconstruction methods. Ecological restoration institute and southwest fire science consortium. Northern Arizona University, Flagstaff

Chapter 9
Dendroecological Potential of *Juniperus deppeana* in Northern Mexico

Julián Cerano-Paredes, Iván Molina-Pérez, Gerardo Esquivel-Arriaga, Rosalinda Cervantes-Martínez, José Villanueva-Díaz, Osvaldo Franco-Ramos, Jorge Méndez-González, Víctor Hugo Cambrón-Sandoval, and Gabriel Fernando Cardoza-Martínez

Abstract The species *Juniperus deppeana* is widely distributed throughout the Sierra Madre Occidental, northern Mexico, and no studies have been conducted to determine its dendroclimatic potential. This study in intended to answer the following questions. Is it possible to date annual growth of *J. deppeana* and develop tree-ring series? What is the most important climate variable that limits its growth? Do its growth rings have the potential to reconstruct climatic variables? A total of 31 cross-sections were collected, 84% from standing dead trees and stumps and 16%

J. Cerano-Paredes (✉) · I. Molina-Pérez · G. Esquivel-Arriaga · R. Cervantes-Martínez
J. Villanueva-Díaz
INIFAP CENID-RASPA, km. 6,5 Margen Derecha del Canal Sacramento,
Gómez Palacio, Durango, Mexico
e-mail: cerano.julian@inifap.gob.mx; molinap.ivan@gmail.com;
esquivel.gerardo@inifap.go.mx; rosy_cervantes23@yahoo.com.mx;
villanueva.jose@inifap.gob.mx

O. Franco-Ramos
Instituto de Geografía, Universidad Nacional Autónoma de México, Ciudad Universitaria,
Coyoacán, Mexico, Mexico
e-mail: ofranco@igg.unam.mx

J. Méndez-González
Universidad Autónoma Agraria Antonio Narro, Departamento Forestal,
Calzada Antonio Narro, Buenavista, Saltillo, Coahuila, Mexico
e-mail: jorgemendez.uaaan@gmail.com

V. H. Cambrón-Sandoval
Universidad Autónoma de Querétaro, Facultad de Ciencias Naturales. Av. de las Ciencias s/n,
delegación Santa Rosa Juregui, Juriquilla, Querétaro, Mexico
e-mail: hugo.cambron@gmail.com

G. F. Cardoza-Martínez
Universidad Juárez del Estado de Durango, Facultad de Ciencias Biológicas,
Avenida Universidad S/N, Fraccionamiento Filadelfia,
Gómez Palacio, Durango, Mexico
e-mail: biologo_gabriel@hotmail.com

M. Pompa-García, J. J. Camarero (eds.), *Latin American Dendroecology*,
https://doi.org/10.1007/978-3-030-36930-9_9

from live trees. The samples were dated based on standard dendrochronological techniques. Of the samples collected (25 cross-sections), 80% were dated giving a total of 51 growth series. Significant correlation ($r = 0.51$, $p < 0.01$) among series was determined. Chronologies of the total ring width, earlywood, and latewood were generated for a period of 223 years (1793–2015). The chronologies of total ring width and earlywood constitute an excellent proxy for reconstructing the variability of yearly and seasonal precipitation (January–July, $r = 0.56$; $p < 0.01$). Moreover, high temperatures from January through April showed greater relation with the ring width ($r = -0.38$; $p < 0.01$). Ring growth of *J. deppeana* possesses the quality to be used as an indirect source for reconstruction of climatic variability and ecological studies in northern Mexico.

Keywords Annual rings · Missing rings · False rings · Dendroecological potential · Southwestern Chihuahua

9.1 Introduction

Currently, one of the most relevant topics of environmental sciences is climate variability attributed to global warming, which has caused changes in patterns of precipitation and, consequently, disturbance of ecosystems. It is expected that this trend will become more accelerated in the near future (Irby et al. 2013). Projections of climate change scenarios for the arid subtropical regions (northern Mexico) predict changes in the pattern and intensity of the hydrological cycle, for which a decrease in average precipitation and an increase in temperature are expected (Hoerling et al. 2012; IPCC 2013). Study of historical climate variability is important to understand climate trends (Cerano-Paredes et al. 2011a). In Mexico, the lack of extensive climate records limits reliable analyses of climate variability beyond the last 70 years (IMTA 2009).

Under these circumstances, an alternative option is to use indirect sources of climate information, or "*proxies*," natural elements that provide more extensive climate information than instrumental records (Irby et al. 2013). There are natural sources that record climate variability such as tree rings, ice caps, fossil pollen and lake sediments, among others (Bradley 1999). Tree rings have been used to develop reconstructions of paleoclimate in diverse ecosystems from semiarid to tropical environments. They have the advantage of having annual resolution unlike other indirect sources (Villalba 1990; Villalba et al. 2003; Cleaveland et al. 2003; Arreola-Ortiz et al. 2010; Cook and Kairiukstis 2013; Stahle et al. 2012, 2016).

In the state of Chihuahua, Mexico, dendroclimatic studies have focused mostly on *Pseudotsuga menziesii* (Mirb.) Franco (Cerano-Paredes et al. 2009; Villanueva-Díaz et al. 2009). The diversity of tree species in these ecosystems constitutes an excellent opportunity to explore and define the feasibility of developing dendroclimatic series with them. One of these species is *Juniperus deppeana* Steud, a species that forms small pure forests or mixed with diverse species of conifers and oaks

(Earl and Bash 1996; Valero-Padilla 2004; Adams and Schwarzbach 2006). It is considered a species of little economic importance because it cannot be used for lumber; it is not included in managing programs (Valero-Padilla 2004), guaranteeing the presence of long-lived individuals (Earl and Bash 1996), making them an option for generating extended dendroclimatic series, which in turn would allow analyzing regional hydroclimatic variability and develop dendroecological studies.

There are 12 species and 6 varieties of *Juniperus* in Mexico. It is a genus that, depending on the species, varies from its distribution to its habitat. There are short-size species and tree species that grow up to 20 m tall; some have diameters of just a few centimeters and others of more than a meter (Borja et al. 2010). The species *J. deppeana* is distributed from southern United States to southern Mexico with six varieties. In particular, *J. deppeana* var. *robusta* is widely distributed along the Sierra Madre Occidental (Adams and Schwarzbach 2006). This species is tolerant to poor soils and benefits from periodic forest fires; it is resistant to drought and grows on hillsides with slopes ranging from 10 to 70% (Earl and Bash 1996; Zavala-García 2003). From an ecological perspective, it is highly important given that it represents a primary and secondary species in succession stages (Valero-Padilla 2004).

Unfortunately, in Mexico it has not been given the importance it deserves. Present-day research on this species is minimal to the point that there are only isolated studies of the genus, among which descriptions of some species are outstanding (Zavala-García 2003). In the United States, there are some previous studies that have analyzed its annual growth (Leavitt and Long 1982; Grissino-Mayer 1993; Earl and Bash 1996), but there is not much information about its potential for dating (Earl and Bash 1996). In this way, this study seeks to answer the following questions: (1) Is it possible to date annual growth of *J. deppeana* and develop dendrochronological series? (2) What is the climate variable that most limits its growth? and (3) Does the annual growth of *J. deppeana* have the potential to reconstruct climatic variables?

9.2 Materials and Methods

9.2.1 Study Area and Sampling

The study area is located in the forests of San Juanito, municipality of Bocoyna, Chihuahua, between 28° 0.132′ N and 107° 32.780′ W, at an altitude of 2520 m. The area is part of the mountain massif of the Sierra Madre Occidental (Fig. 9.1). The soil of the area is classified as “Regosol,” which is characterized as shallow, of varying fertility and whose productivity depends on its depth and rockiness (INEGI 2017). Climate is classified as type C (E) (w_2), cool subhumid with a mean annual temperature of 5–12 °C and a minimum temperature of the coldest month of −3 °C; temperature of the hottest month is 24 °C with a mean annual precipitation of 800 mm (García 2004).

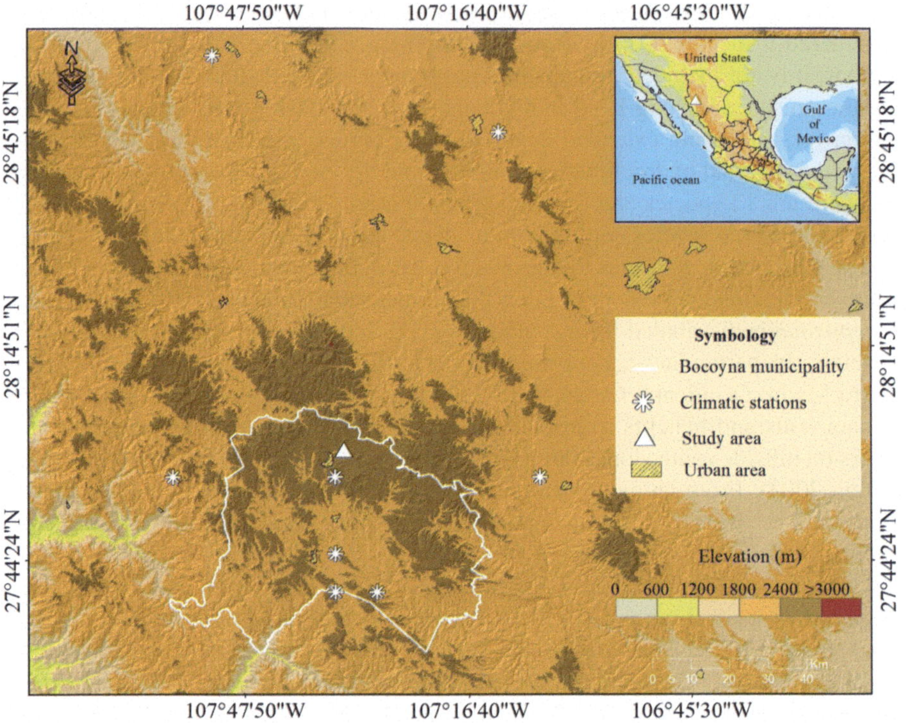

Fig. 9.1 Geographic location of the study area and spatial distribution of the weather stations

Sampling was selective, considering two important aspects: (1) selecting trees with the most concentric growth possible to facilitate dating and (2) identifying the oldest specimens with the objective of extending the growth series back in time. Using a chainsaw, a total of 31 cross-sections were obtained at a height of 0.10–0.20 m from the ground. Of the samples collected, 84% were from dead standing trees and stumps, and 16% were from living trees (Fig. 9.2a, b).

9.2.2 Laboratory Method

The cross-sections were air-dried; the most fragile and broken samples were glued. Later, those with a certain degree of rotten wood were mounted on plywood to provide better support. The samples were polished with different grades of sandpaper (60–600 grains) to highlight the growth rings (Fig. 9.2). Two to three radii were selected from each sample, considering the clearest and most uniform growth as suggested for alpine species (Myers-Smith et al. 2015). For each of the radii,

Fig. 9.2 *J. deppeana* tree (**a**) and cutting of cross-sections using a chainsaw (**b**). Cross-sections of *J. deppeana* indicating the different forms of growth of the species. Some anomalies in growth make dating difficult in different degrees. (**c**, **d**) show a section with damage that caused deformation in growth for several years after the event. (**e**, **f**) indicate non-concentric growth with tension and compression wood, which favor formation of false and missing rings in a given section of the circumference. (**g**, **h**) show concentric growth; these samples with more even growth are ideal and facilitate dating, allowing easier detection of possible problems in growth (false rings, micro-rings and missing rings)

growth graphs, called "*skeleton plots*," were constructed. These graphs allowed comparison of growth patterns of the different samples of a tree and among trees in order to identify the presence of false and missing rings (Stokes and Smiley 1968). Moreover, two reference chronologies generated in the zone, one of *Pinus arizonica* Engelm. (series provided by Rosalinda Cervantes) and another of *P. menziesii* (series provided by Julián Cerano), were used to facilitate dating of *J. deppeana* growth. Once the dating was finalized, total ring widths, earlywood and latewood series were measured with a VELMEX measurement system with a precision of 0.001 mm (Robinson and Evans 1980), using the software Measurament-J2X for Windows.

9.2.3 Data Analyses

The quality of the dating and measurement was verified with the COFECHA program, which is software that statistically compares the dating of each sample by successively correlating 50-year segments and overlaps of 25 years (Holmes 1983). Biological trends unrelated to climate (competition, releases, age, among others) were removed with the ARSTAN program. This program fits the best detrending option (negative exponential, straight line with positive or negative slope, spline of certain rigidity) and then divides each annual measurement value by the value obtained from the curve, thus creating three versions of the chronology (standard, residual, and arstan) for each portion of wood (total ring width, earlywood, and latewood). The resulting time series have a mean around 1.0 and homogeneous variance (Fritts 1976). In our study, the versions used for the analysis were the standard and residual series. The quality of the chronologies was evaluated considering the interseries correlation (IC), mean sensitivity (MS), which is the degree in which one or more common limiting factors are reflected in a series of tree rings (Schweingruber 1996), and the expressed population signal (EPS), which describes the infinite hypothetical population as a finite sample for its estimation (Wigley et al. 1984; Briffa 1995). The chronology is considered reliable with an EPS value over 0.85 (Wigley et al. 1984).

9.2.4 Influence of Climate on Radial Growth

To determine the influence of climate variables in radial growth of *J. deppeana*, a response function analysis was run, for which average monthly data (51 years: 1965–2015) of precipitation and maximum and minimum temperatures from six weather stations were used with the indices of standard and residual chronologies of ring width, earlywood and latewood. The climate records were taken from the software Rapid Extractor of Climate Information (ERIC III) (IMTA 2009), considering the weather stations near the study site and had the most complete climate records (Fig. 9.1).

Pearson correlation analyses between the standard and residual chronologies of total ring width, earlywood, latewood and climate variables were run in Excel. The months that most highly correlated for each variable and the season that most affects *J. deppeana* growth were identified. These analyses were verified and validated statistically with STATISTICA Kernel Release 5.5 software (Stat Soft Inc. 2011). In the same software, regression models were carried on between the climate variables (monthly precipitation and maximum and minimum average temperatures) and the ring indices to define the potential of the species to reconstruct these variables and determine the variable for which radial growth explains the greatest annual or seasonal variability. Finally, the relationship between droughts and radial growth was evaluated by linking the ring-widths index (1800–2012) with the Palmer Drought Severity Index (PDSI, 1800–2012) using The Mexican Drought Atlas (MXDA) (Stahle et al. 2016).

9.3 Results

9.3.1 Description of the Chronology

J. deppeana trees generally have round-shaped stems. However, this does not guarantee concentric growth (Fig. 9.2). The condition of dry, rocky soil where the species grows and presence of anthropogenic disturbance, such as physical damage, blows and fires, cause growth deformations even several years after the damage (Fig. 9.2c, d). Eccentricity is a common feature of the growth of this species, promoting formation of tension and compression wood and giving way to development of false and missing rings in several sections of the stem circumference (Fig. 9.2e, f). Even in individuals with concentric growth, micro-rings and false growth bands are recorded (Fig. 9.2g, h). Nevertheless, problems for dating are comparatively minor in this type of samples. All of the samples had problems for dating; formation of false rings varied in a range of 1–14 rings per sample (1–17%), with an average of 6% per sample. Missing rings varied from 1 to 22 rings (1–23%) with an average per sample of 4% (Fig. 9.3).

Despite this, we were able to date 80% of the collected cross sections (25 sections) and a total of 51 series or radii of growth. The series of ring width exhibit similar patterns of growth among the trees (Fig. 9.4). We determined a significant correlation between series ($r = 0.51$, $p < 0.01$) and a mean sensitivity of 0.728 (Table 9.1).

We were able to generate chronologies of total ring width, earlywood and latewood for a period of 223 years (1793–2015). The series had low values of first order autocorrelation (0.20, $p > 0.05$) and EPS value above 0.85 for most of the chronology length (1870–2014, 144 years) and an EPS above 0.70 as of the year 1844 (172 years) (Fig. 9.5). These series are the first chronologies of *J. deppeana* generated in Mexico.

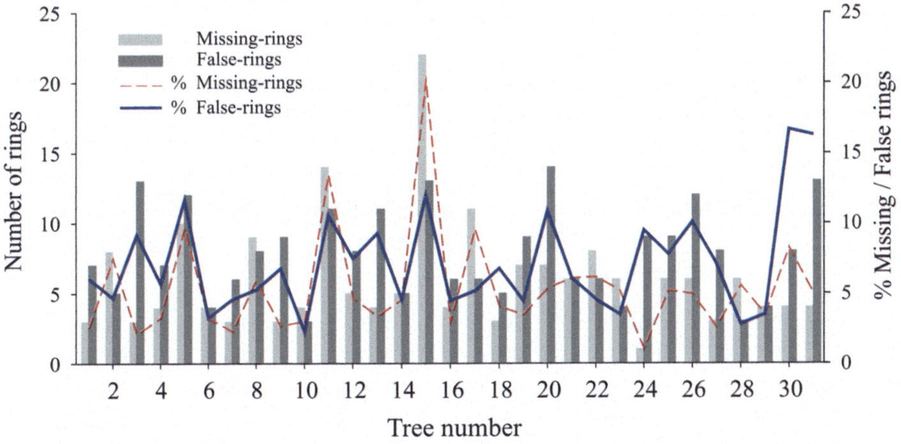

Fig. 9.3 Number and percentage of missing and false rings determined in most of the *J. deppeana* trees. All the trees register problems of growth that makes the dating difficult

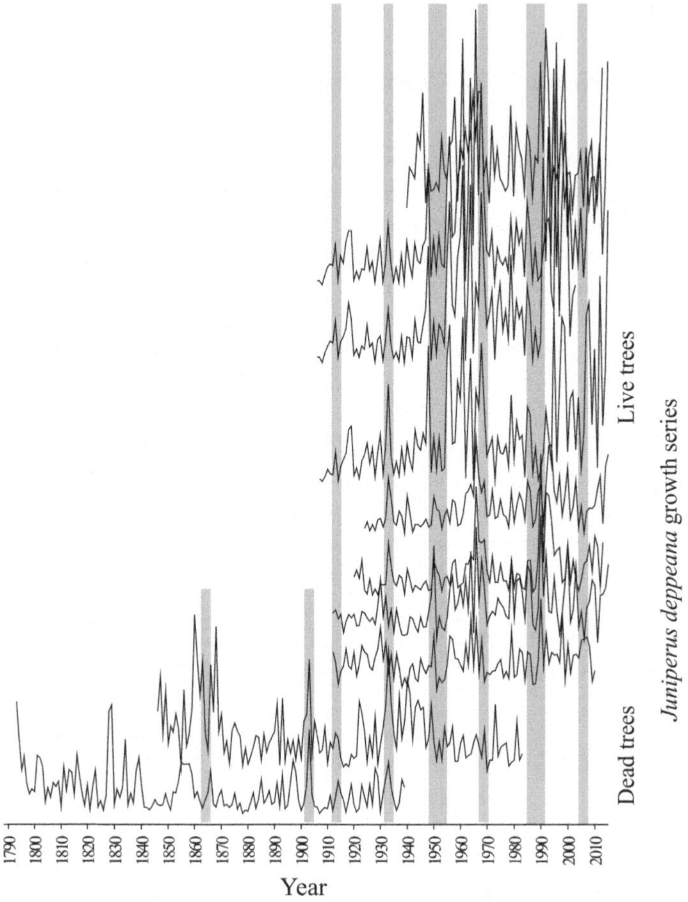

Fig. 9.4 Variability in annual growth common to live and dead *J. deppeana* trees. Gray vertical lines indicate patterns of growth that are similar among trees

Table 9.1 Comparison of dendrochronological statistical parameters between the total ring-width chronology of *J. deppeana* and ring-width chronologies of other species

Species	*Juniperus deppeana*	*Pseudotsuga menziesii*	*Pinus lumholtzii*	*Pinus cembroides*	*Pinus durangensis*	*Juniperus monticola*
State	Chihuahua	Coahuila	Durango	Coahuila	Durango	Mexico State
No. of samples	51	206	41	67	88	33
Total years	223	302	294	410	232	264
Corr.	0.51	0.71	0.60	0.72	0.463	0.55
Sensitivity	0.728	0.35	0.230	0.309	0.309	0.410
Author	Current study	Cerano-Paredes et al. (2011b)	Chávez-Gándara et al. (2017)	Constante-García et al. (2009)	Díaz-Ramírez et al. (2016)	Villanueva-Díaz et al. (2016)

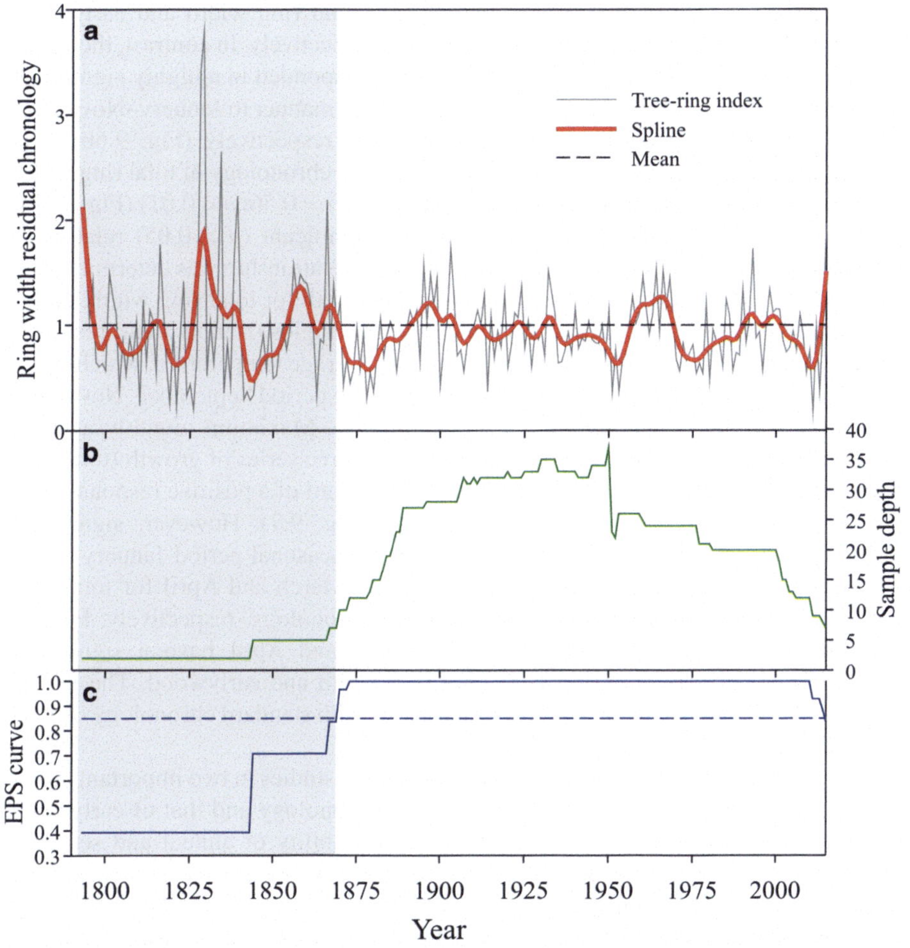

Fig. 9.5 (**a**) Indicates the total ring-width index series of *J. deppeana*, which comprises the period 1793–2015 (223 years). The gray line indicates interannual variability. The red line is a flexible line (spline) of 10 years, which permits observation of low frequency events, and the dashed black line represents the mean of the series. In (**b**) the green line indicates sample size in each of the segments of the series and in (**c**) the blue line represents the EPS curve. The area shaded in gray indicates the portion of the chronology with an EPS value <0.85

9.3.2 Climate-Growth Relationship

By relating the standard and residual chronologies to regional monthly precipitation of the period 1965–2015 (51 years), we determined that the months of January, February and April significantly ($p < 0.05$) affect growth of total ring width and earlywood. In the case of latewood, we determined a positive, but not significant ($p > 0.05$) relationship (Fig. 9.6a). Accumulated precipitation from January to September and January to November had highly significant ($r = 0.36$; $p < 0.01$) and significant ($r = 0.33$; $p < 0.05$) association with total ring width and earlywood growth of the standard version of the chronology, respectively. In contrast, the residual chronology of total ring width and earlywood responded in a highly significant ($r = 0.39$; $p < 0.01$) and significant ($r = 0.34$; $p < 0.05$) manner to January–November and January–December accumulated precipitation, respectively (Fig. 9.6b). The highest correlation was obtained between the residual chronology of total ring width and the period of accumulated rainfall January–July ($r = 0.56$; $p < 0.01$) (Fig. 9.6b).

Response to minimum temperature was not significant ($p > 0.05$) relative to growth of *J. deppeana*. The most important positive relationship was determined for the months of February, September and November both for total ring width, earlywood and latewood and for both versions of chronologies (standard and residual) (Fig. 9.7). The most important monthly minimum average temperature, which influences positively the growth of the species involves the period September–November ($r = 0.16$; $p > 0.05$), and negative response in May. Maximum monthly average temperatures indicate negative correlations for the three series of growth (total ring width, early, and latewood). Only in July is there record of a positive response with growth of total ring width and of earlywood (Fig. 9.7). However, significant ($p < 0.05$) negative correlations were found for the seasonal period January–April and the months of January, March and April, and March and April for total ring width, earlywood and latewood of the standard chronology, respectively. For the residual chronology, only the months of March and April have a significant ($p < 0.05$) negative relationship with total ring width and earlywood. The period January–April shows greater negative relationship with standard chronology of total ring width ($r = -0.38$; $p < 0.01$) (Fig. 9.7).

The generated chronologies will allow developing studies in two important areas. (1) Reconstructing climate, the total ring-width chronology and that of earlywood constitute excellent proxies for reconstructing variability of annual and seasonal precipitation. The total ring-width residual chronology, however, is considered the most appropriate for reconstructing seasonal January–July precipitation variability because it explains the highest variance (Fig. 9.8, Table 9.2). Also, it is feasible to reconstruct variability of high temperatures of the period January–April based on total ring-width standard chronology (Fig. 9.8, Table 9.2). (2) Reconstructing dendroecological studies, it is feasible to date scars registered by disturbances (Fig. 9.2c, d), which represents an alternative to develop studies such as fire regime and their effect on the growth and dynamics of the population.

The interannual variability of growth attributed to the climatic signal (PDSI) is shown in Fig. 9.9. It shows the variability of growth in humid and drought conditions in the last two centuries (1800–2012). It is clear to appreciate rings corresponding to years with favorable humidity conditions (wide rings) as in 1816, 1829, 1834, 1839, 1897–1898, 1903, 1924, 1933, 1959–1960, 1964–1969, 1979, 1985–1986 and 2007, or dry conditions (narrow rings), as in 1819, 1841–1845, 1869–1875, 1893–1894, 1950–1956, 1970–1979, 2001–2003 and 2010–2011 (Fig. 9.9a). Positive and negative climatic conditions of PDSI have significantly ($p < 0.05$) conditioned the greater and lesser growth of *J. deppeana* in this region of northern Mexico (Fig. 9.9b).

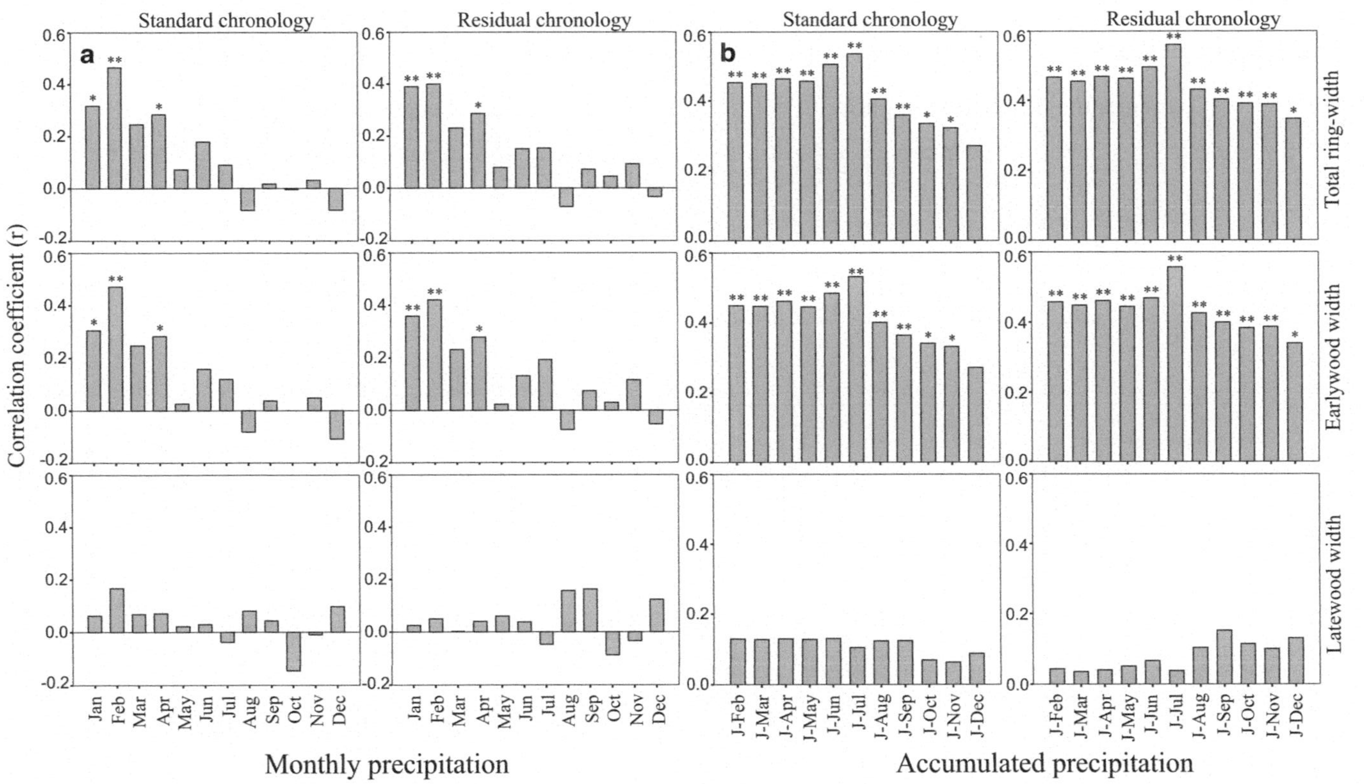

Fig. 9.6 Response function between ring indices (total ring width, earlywood, and latewood) of *J. deppeana* and the monthly (**a**) and accumulated (**b**) precipitation where $* = P < 0.05$ and $** = P < 0.01$ indicate the degree of significance of the correlation between the two variables

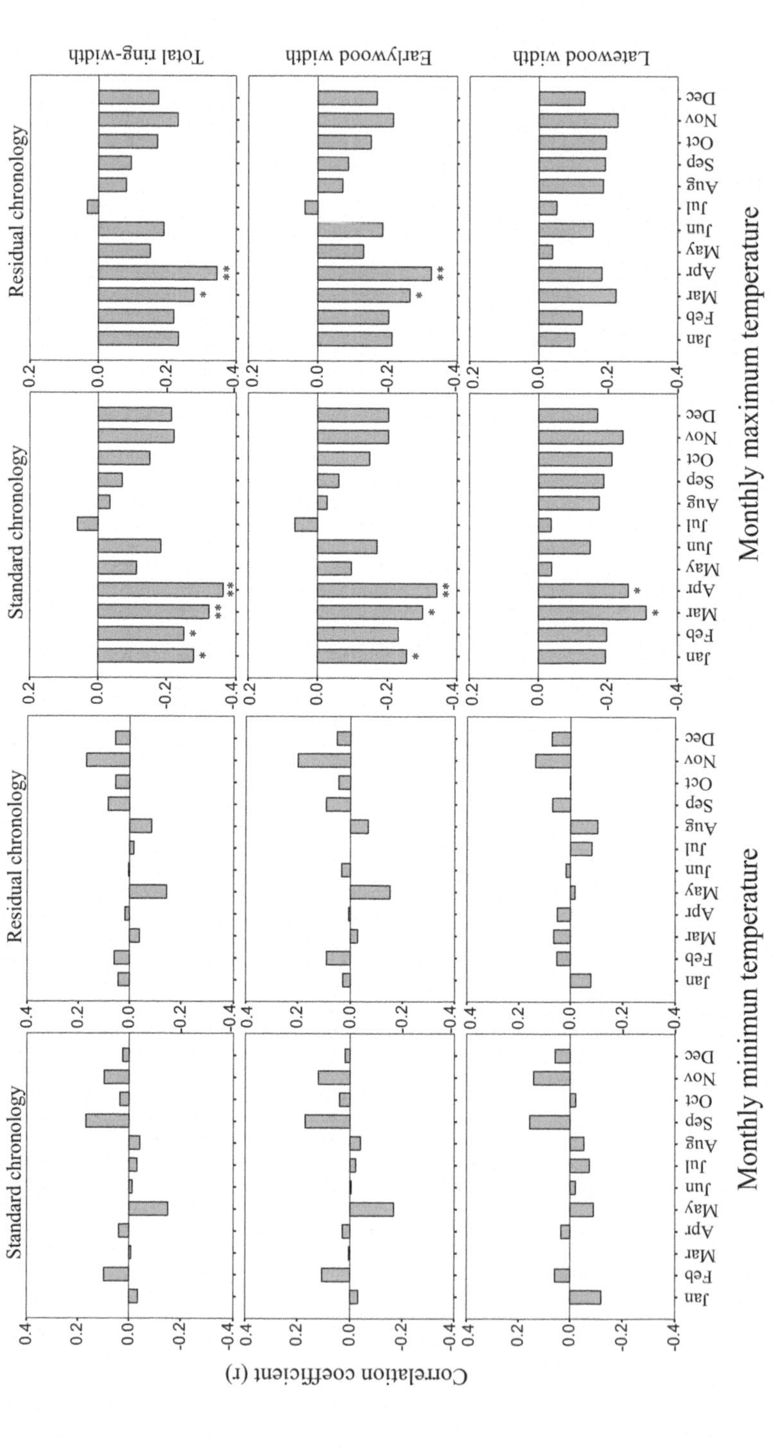

Fig. 9.7 Response function between the ring indices (total ring with, earlywood, and latewood) of *J. deppeana* and average monthly minimum and maximum temperatures where * = $P < 0.05$ and ** = $P < 0.01$ indicate the degree of correlation between the two variables

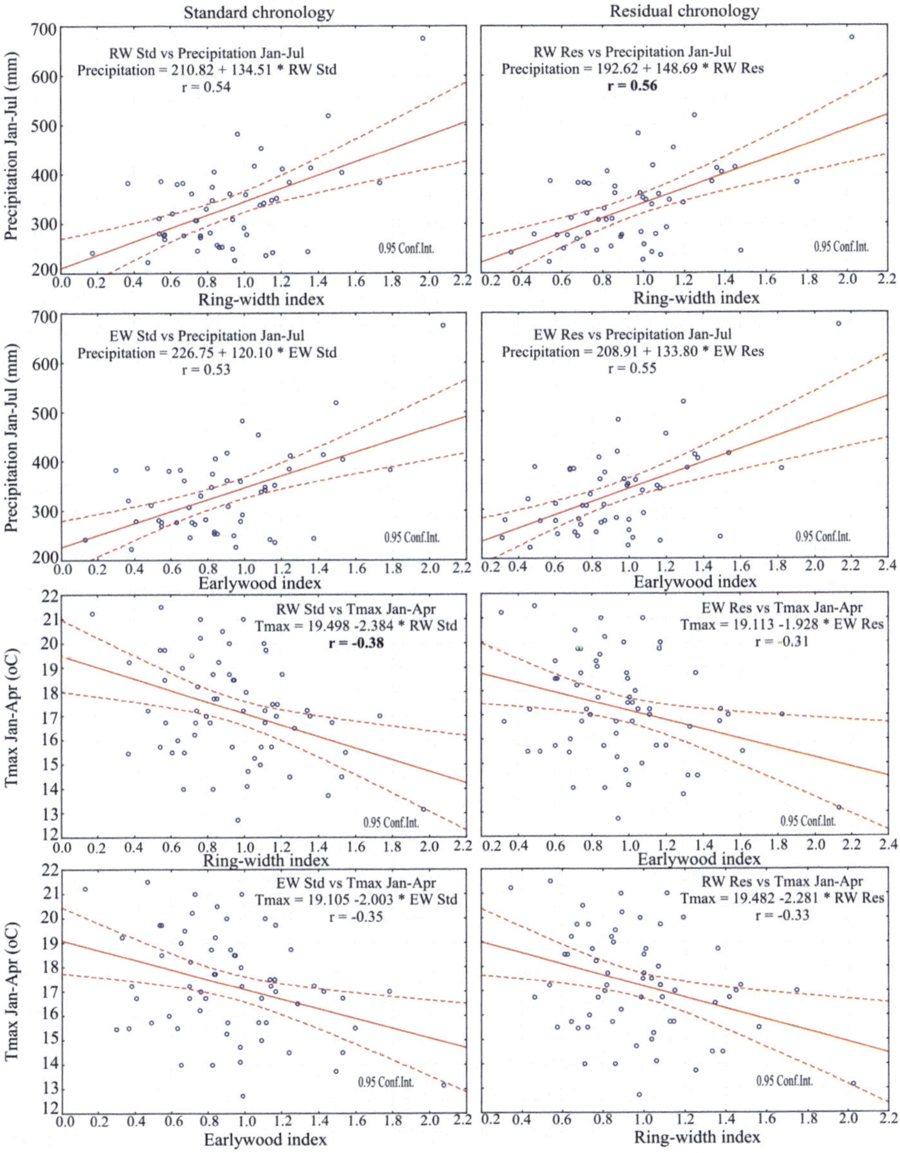

Fig. 9.8 Dispersion diagrams of the regression models between the climate variables (precipitation and maximum temperature) and the dendrochronological series (residual and standard) of total ring width and earlywood

Table 9.2 Linear regression models between the growth indices and climatic variables

Climate variable	Model	Estimated parameters	Standard error	*t*	*r*	*P*
			Standard chronology			
Precip. Jan–Jul	Intercept	210.8172	29.47869	7.15151	**0.54**	**<0.000**
	Ring-width	134.5101	30.29631	4.43981		**<0.000**
	Intercept	226.7478	26.42075	8.58218	**0.53**	**<0.000**
	Earlywood	120.1043	27.37850	4.38681		**<0.000**
	Intercept	258.7520	102.5744	2.52258	0.11	<0.05
	Latewood	77.3168	105.0910	0.73571		0.465
Tmin Sep–Nov	Intercept	2.38837	0.48155	4.95967	0.14	<0.000
	Ring-width	0.52578	0.48422	1.08581		0.281
	Intercept	2.39860	0.43304	5.53894	0.15	<0.000
	Earlywood	0.52482	0.43724	1.20028		0.234
	Intercept	1.53932	1.42337	1.08146	0.12	0.283
	Latewood	1.36688	1.44101	0.94855		0.346
Tmax Jan–Apr	Intercept	19.49777	0.74404	26.20517	**–0.38**	**<0.000**
	Ring-width	–2.38438	0.74816	–3.18696		**<0.01**
	Intercept	19.10465	0.67818	28.17040	**–0.35**	**<0.000**
	Earlywood	–2.00253	0.68475	–2.92446		**<0.01**
	Intercept	22.77411	2.26108	10.07220	**–0.30**	**<0.000**
	Latewood	–5.61379	2.28910	–2.45239		**<0.05**
		Residual chronology				
Precip. Jan–Jul	Intercept	192.6171	31.39739	6.13481	**0.56**	**<0.000**
	Ring-width	148.6924	31.38504	4.73768		**<0.000**
	Intercept	208.9066	28.40748	7.35392	**0.55**	**<0.000**
	Earlywood	133.8043	28.51697	4.69209		**<0.000**
	Intercept	305.5782	106.8891	2.85883	0.04	<0.01
	Latewood	28.8393	108.9083	0.26480		0.792
Tmin Sep–Nov	Intercept	2.278389	0.531979	4.28285	0.15	<0.000
	Ring-width	0.619734	0.521407	1.18858		0.239
	Intercept	2.299355	0.481353	4.77685	0.16	<0.000
	Earlywood	0.606535	0.473057	1.28216		0.204
	Intercept	1.727317	1.528645	1.12996	0.10	0.262
	Latewood	1.169964	1.541853	0.75880		0.450
Tmax Jan–Apr	Intercept	19.48154	0.838033	23.24674	**–0.33**	**<0.000**
	Ring-width	–2.28050	0.821379	–2.77643		**<0.01**
	Intercept	19.11297	0.766139	24.94714	**–0.31**	**<0.000**
	Earlywood	–1.92786	0.752934	–2.56046		**<0.05**
	Intercept	21.17612	2.488077	8.51104	–0.20	<0.000
	Latewood	–3.96718	2.509576	–1.58082		0.119

Note: The parameters in bold type represent significant values and those with gray background highlight the best models

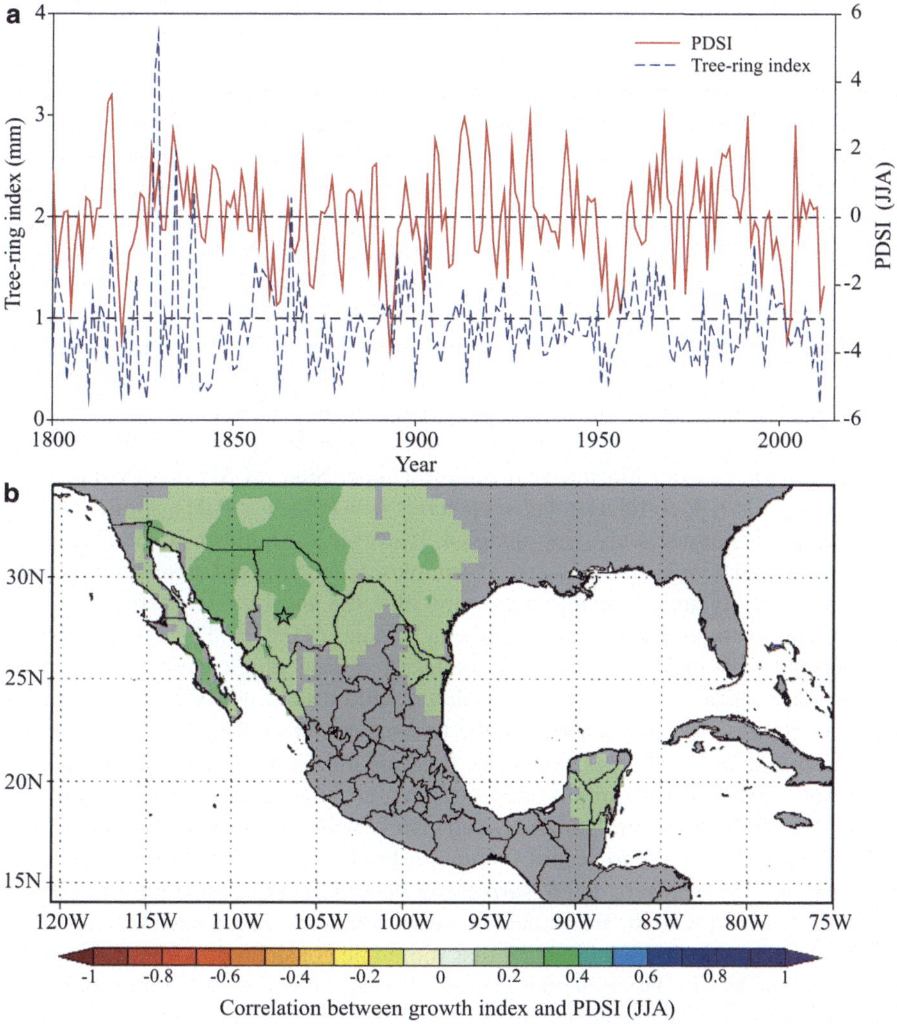

Fig. 9.9 Relationship between the growth index and the Palmer Drought Severity Index (PDSI). (**a**) Shows the variability of growth in humid and drought conditions in the last two centuries (1800–2012). (**b**) Indicates a significant correlation ($p < 0.05$) between the growth of *J. deppeana* and the PDSI of the months June, July and August (JJA) for the period 1800–2012. The PDSI data were extracted from the Mexican Drought Atlas (MXDA) (Stahle et al. 2016). The star indicates the study site

9.4 Discussion

This is the first study to analyze the influence of climatic variables in annual growth of *J. deppeana* in Mexico. We determined that this species records growth patterns that are similar among trees. The result of the intercorrelation between series ($r = 0.51, p < 0.01$) surpassed the parameter established by the COFECHA program ($r = 0.3281, p < 0.01$) to consider the series correctly dated (Holmes 1983). Average mean sensitivity was above 0.2, indicating sufficient interannual variation, which is the ideal condition for dendrochronological studies with the objective of reconstructing climate (Speer 2010). First-order autocorrelation was not significant, indicating that growth of 1 year is not conditioned by the growth of the previous year (Fritts 1976). The statistical parameters of the chronology indicate the potential of the species for developing subsequent dendrochronological studies.

One species of *Juniperus* that has been analyzed for its dendroclimatic potential is *Juniperus monticola* Martínez on Mount Tlaloc, State of Mexico (Villanueva-Díaz et al. 2016). A correlation between series ($r = 0.55$, $p < 0.01$), similar to that with the species in our study, was obtained. However, the number of dated series is lower than with *J. deppeana*, 33 and 51 samples, respectively. *J. deppeana* records a mean sensitivity and a signal to noise ratio of 0.728 and 1.93, respectively, which are higher than those of *J. monticola* for which values of 0.41 and 1.23, respectively, are reported. This may be attributed to the fact that the ecological conditions where they grow are different (Table 9.1). Dendroclimatic studies use between 20 and 30 trees to obtain a chronology, although the International Tree Ring Data Bank established a minimum of 15 trees necessary to obtain a reliable chronology (Cook and Kairiukstis 2013). These sample sizes, however, are subject to the sensitivity of the species and to the intercorrelation between series that depends on the capacity of the individuals of a species to record common climate variability (Fritts 1976). For *J. deppeana*, 25 trees were dated, surpassing the minimum necessary for a reliable chronology in terms of the mentioned dendrochronological parameters.

The main problems identified and that should be considered in future studies to guarantee success in dating annual growth of *J. deppeana* are: (1) deformation of growth caused by mechanical damage and fires, (2) formation of false growth bands, and (3) missing rings due to eccentricity of the trunk, formation of tension and compression wood (Fig. 9.2). Working with cross-sections assures defining more clearly the presence of false and missing rings. Using this type of samples allowed dating of 80% of the trees. Use of a Pressler increment bore to take samples in future studies with this species could cause a major reduction in the number of dated individuals or it might not be possible to develop a chronology. We were not able to date 20% of the trees because of the high degree of growth deformation from damage, development of tension and compression wood, which caused the formation of a large number of suppressed and missing rings making correct dating difficult. Success in dating depends on the conditions of the site where the trees grow, which can cause dating to vary 43–90% (Lara and Villalba 1993; Neira 1995; Villalba et al. 1990). Specimens that do not co-date are regularly derived from trees whose development has been strongly affected by specific ecological conditions of the area (competition, fires and reduction in density) (Lara 2000).

The *J. deppeana* chronology had an interseries correlation higher than that reported for other species with which climatic variability has been reconstructed (Table 9.1). Cerano-Paredes et al. (2014) reports a correlation between series of $r = 0.473$ ($p < 0.01$) for *Abies religiosa* (Kunth) Schltdl. & Cham. Díaz-Ramírez et al. (2016) found a correlation between samples of $r = 0.463$ ($p < 0.01$) for *Pinus durangensis*. Villanueva-Díaz et al. (2015) report that *Hesperocyparis guadalupensis* (S. Watson) Bartel. Var. Guadalupensis had a correlation between samples of $r = 0.389$ ($p < 0.01$). Only species such as *P. menziesii* $r = 0.69$ (Cerano-Paredes et al. 2011b; Villanueva-Díaz et al. 2014) and *Pinus cembroides* Zucc. $r = 0.67$ (Constante-García et al. 2009; Cardoza-Martínez et al. 2014), which have excellent potential for generating dendroclimatic series, had interseries correlation higher than those found for *J. deppeana*. Regarding sensitivity, *J. deppeana* had a higher value than *P. menziezi, P. durangensis, Pinus lumholtzii* Robins & Ferns, *P. cembroides, P. durangensis, H. guadalupensis* and *J. montícola* (Table 9.1). This indicates that the species has high interannual variability (Speer 2010) and thus can be considered as promising species for dendroclimatic and dendroecological reconstructions in northern Mexico.

J. deppeana responded significantly ($r = 0.56$; $p < 0.01$) to seasonal precipitation in the period January–July, like other widely studied species such as *P. cembroides* in the north (Constante-García et al. 2009) and center (Cardoza-Martínez et al. 2014) of the country, *P. menziesii* in the Sierra Madre Occidental (Cleaveland et al. 2003; Cerano-Paredes et al. 2009; Villanueva-Díaz et al. 2014) and Sierra Madre Oriental (Cerano-Paredes et al. 2011b). One explanation of this response is that winter rains tend to be of low intensity, favoring infiltration; low temperatures reduce evapotranspiration, allowing moisture to be stored in the soil profile and be ready available for tree use in the growing season (Fritts 1976; Seiler and Gat 2007). Water deficit decreases size and differentiation of cambium derivate as well as the rate of initial fusiform cell division and mother cells in the xylem and phloem (Kozlowski and Pallardy 1997).

Minimum temperature correlated positively, but not significantly, with growth; the period September–November was the most important. This can be attributed to the beginning of progressive decrease in temperature in September, from 5 °C to −9 °C in December (IMTA 2009), which reduces evaporation and allows more infiltration of winter rains, thus more available moisture accumulates in the soil to be used by the tree in the following growing season. Maximum temperature, in contrast, responds significantly ($p < 0.05$) but negatively to the period January–April; the highest correlation was determined for the months of March and April. The reason for this behavior is that during these months, the greatest increase in temperature is recorded and the most extreme temperatures occur (30 °C) (IMTA 2009). High temperature in the early months of the year when rainfall is scarce causes soil moisture to be readily lost through evapotranspiration, reducing available moisture for the tree at the beginning of the growing season. This situation is reflected in reduced growth. Temperatures above 20 °C drastically reduce stoma aperture and causes a negative response in terms of growth (Leverenz 1981).

An increase in temperature can result in a decrease in production of carbohydrates because the trees in conditions of high evapotranspiration capture less CO_2

when they close their stomata to prevent water loss (Kozlowski and Pallardy 1997). Similar results have been reported for *P. durangensis*, *P. lumholtzii* (Chávez-Gándara et al. 2017) and *Pinus cooperi* Blanco (Pompa-García et al. 2013; Chávez-Gándara et al. 2017) in the Sierra Madre Occidental in the state of Durango and *A. religiosa* in the National Park Pico de Tancitaro in the state of Michoacán (Cerano-Paredes et al. 2014).

In relation to the drought, during the last two centuries, the growth of *J. deppeana* has been significantly influenced ($p < 0.05$) by the variability of the PDSI (Fig. 9.9b), negative conditions (drought) reduce the growth and positive conditions (humidity) influence in more growth (Fig. 9.9a). This species is tolerant to poor soils and benefits from periodic forest fires; it is resistant to drought and grows on hillsides with slopes ranging from 10% to 70% (Earl and Bash 1996; Zavala-García 2003). The topography has an important role because although drought has an impact on the growth, it has a stronger effect on the trees established in sites with irregular topography since the loss of water is probably greater (Martín-Benito et al. 2008).

During the twentieth century, the two most severe droughts were observed during the 1950s and 1970s, results that synchronize with other works that report these same periods of severe drought for the north (Cerano-Paredes et al. 2011a; Seager et al. 2009) and southwest of the United States (Florescano et al. 1980; Stahle et al. 2009). Likewise, one of the few works developed with *J. deppeana* in south-central New Mexico, reports these periods as the driest of the last 1300 years and a direct negative effect on growth (Earl and Bash 1996). These results corroborate the significant relationship of the PDSI in the modulation of growth in the northern region of Mexico and the southwestern United States, which is shown in Fig. 9.9b.

From the point of view of reproductive ecology of *J. deppeana*, drought has an important role in its establishment; Johnsen (1959) indicates that this species has the ability to establish itself in drought conditions. The reduction of competition for pastures due to drought or grazing and the elimination of fire risks benefit the establishment of this species (Johnsen 1962). This survival strategy seems to have a conservative effect in response to climatic variability. During wet periods, *J. deppeana* decreases seed production and increases the competition of grasses; during the drought, there is an increase in seed production, competition from grasses is reduced and there is less destruction caused by fires (Earl and Bash 1996). Droughts play an important role in decreasing the growth of *J. deppeana*; however, it is important to highlight the role they play in the establishment of new individuals.

9.5 Conclusions

Our results highlight that growth of *J. deppeana* is regulated and limited by availability of water. Therefore, a decrease in precipitation and increase in temperature projected by the Intergovernmental Panel on Climate Change (IPCC 2013) suggest a decrease in annual growth, which would lead to a decrease in productivity and carbon assimilation of this and other associated species in ecosystems of northern Mexico. Understanding physiological responses of plants to their changing surroundings can contribute to predicting the destiny of forest ecosystems (Ziaco and Biondi 2016). The potential of *J. deppeana* in reconstructing variability in precipitation and temperature offers an option to increase the network of hydroclimatic studies and begin development of temperature series and ecological studies, reconstructions that will allow analysis of its trends in recent centuries.

Acknowledgments We thank all the personnel that contributed to this study in the field and laboratory. This research was funded by the CONAFOR-CONACYT fund, registration number CONAFOR-2014, C01-234547.

References

Adams RP, Schwarzbach AE (2006) Infraspecific adjustments in *Juniperus deppeana* (cupressaceae). Phytologia 88(3):227–232

Arreola-Ortiz MR, González-Elizondo M, Návar-Cháidez JJ (2010) Dendrocronología de *Pseudotsuga menziesii* (Mirb.) Franco de la Sierra Madre Oriental en Nuevo León, México. Madera Bosques 16(1):71–84

Borja A, Machuca R, Fuentes-Salinas M et al (2010) Caracterización tecnológica de la madera de *Juniperus flaccida* var. Poblana Martínez. Rev Chapingo Ser Cie XVI(2):261–280. https://doi.org/10.5154/r.rchscfa.2010.09.083

Bradley RS (1999) Paleoclimatology. Reconstructin climates of the Quaternary. Second Edition. Academic Press, USA, p 613

Briffa K (1995) Interpreting high-resolution proxy climate data - The example of dendroclimatology. In: Analysis of climate variability, applications of statistical techniques. Springer Nueva York, pp 77–94

Cardoza-Martínez GF, Cerano-Paredes J, Villanueva-Díaz J et al (2014) Reconstrucción de la precipitación anual para la región oriental del estado de Tlaxcala. Revista Mexicana de Ciencias Forestales 5(2):110–127

Cerano-Paredes J, Villanueva-Díaz J, Arreola-Avila JG et al (2009) Reconstrucción de 350 años de precipitación para el suroeste de Chihuahua, México. Madera Bosques 15(2):27–44

Cerano-Paredes J, Villanueva-Díaz J, Valdez-Cepeda RD et al (2011a) Sequías reconstruidas en los útimos 600 años para el noreste de México. Rev Mex Cienc Agríc 2(2):235–249

Cerano-Paredes J, Villanueva-Díaz J, Valdez-Cepeda RD et al (2011b) Variabilidad histórica de la precipitación reconstruida con anillos de árboles para el Sureste de Coahuila. Revista Mexicana Ciencias Forestales 3(4):33–48

Cerano-Paredes J, Villanueva-Díaz J, Cervantes-Martínez R et al (2014) Reconstrucción de precipitación invierno-primavera para el Parque Nacional Pico de Tancítaro, Michoacán. Investigaciones Geograficas 83:41–54. https://doi.org/10.14350/rig.35190

Chávez-Gándara MP, Cerano-Paredes J, Nájera-Luna JA et al (2017) Reconstrucción de la precipitación invierno- primavera con base en anillos de crecimiento de árboles para la región de San Dimas, Durango, México. Bosque 38(2):387–399. https://doi.org/10.4067/S0717-92002017000200016

Cleaveland MK, Stahle DW, Therrell MD et al (2003) Tree-ring reconstructed winter precipitation and tropical teleconnections in Durango, Mexico. Climatic Change 59(1):369–388. https://doi.org/10.1023/A:1024835630188

Constante-García V, Villanueva-Díaz J, Cerano-Paredes J et al (2009) Dendrocronología de *Pinus cembroides* Zucc. y Reconstrucción de Precipitación, estacional para el sureste de Coahuila. Revista Ciencia Forestal en México 34(106):17–39
Cook ER, Kairiukstis LA (2013) Methods of dendrochronology: applications in the environmental sciences. Kluwer Academic Publishers, Dordrecht. The Netherlands.
Díaz-Ramírez B, Villanueva-Díaz J, Cerano-Paredes J (2016) Reconstrucción de la precipitación estacional con anillos de crecimiento para la región hidrológica Presidio-San Pedro. Madera Bosques 22(1):111–123
Earl RA, Bash DL (1996) Response of alligator juniper (*Juniperus deppeana* Pinaceae) to historic environmental variability in south central New Mexico. Southwest Nat 41(3):227–238
Florescano EM, Cervera JS, Arias DPG (1980) Las sequías en México: historia, características y efectos. Comercio Exterior 30(7):747–757
Fritts HC (1976) Tree rings and climate. Academic, San Diego, p 567
García E (2004) Climas (clasificación de Koppen, modificado por García). Instituto de Geografía, Universidad Nacional Autónoma de México 6(1):0
Grissino-Mayer HD (1993) An updated list of species used in tree-ring research. Tree-Ring Bull 53:17–43
Hoerling M, Eischeid J, Perlwitz J et al (2012) On the increased frequency of Mediterranean drought. J Clim 25(6):2146–2161
Holmes RL (1983) Computer-assisted quality control in tree-ring dating and measurement. Tree-Ring Bull 43:69–78
Instituto Mexicano de Tecnología del Agua IMTA (2009) Extractor rápido de información climatológica III (ERIC). Software. Jiutepec, Morelos, México
Instituto Nacional de Estadística, Geografía e Informatica INEGI (2017) Carta edafológica escala 1:20 000. Numero de carta G1301
Intergovernmental Panel on Climate Change IPCC (2013) Climate Change: the physical science basis; Contribution of Working Group I to the IPCC Fifth Assessment Report of the IPCC. In: Stocker TF, Qin D, Plattner GK, Tignor M, Allen SK, Boschung J, Nauels A, Xia Y, Bex V, Midgley PM (Eds.), Cambridge University Press: Cambridge, UK; New York, NY, USA.
Irby CM, Fulé PZ, Yocom LL et al (2013) Dendrochronological reconstruction of long-term precipitation patterns in Basaseachi National Park, Chihuahua, Mexico. Madera Bosques 19(1):93–105
Johnsen TN (1959) Longevity of stored juniper seeds. Ecology 40:487–488
Johnsen TN (1962) One seed juniper invasion of northern Arizona grasslands. Ecol Monogr 32:187–207
Kozlowski TT, Pallardy SG (1997) Growth control in woody plants. Academic Press, San Diego
Lara A (2000) Importancia científica, protección legal y uso destructivo de los bosques de alerce (Fitzroya cupressoides): una contradicción que debe resolverse. Bosque Nativo 27:3–13
Lara A, Villalba R (1993) A 3,620-year temperature record from Fitzroya cuppressoides tree rings in South America. Science 260:1104–1106
Leavitt SW, Long A (1982) Stable carbon isotopes as a potential supplemental tool in dendrochronology. Tree-Ring Bull 42:49–55
Leverenz JW (1981) Photosynthesis and transpiration in large forest-grown Douglas-fir: diurnal variation. Can J Bot 59(3):349–356
Martín-Benito D, Cherubini P, Del Río M (2008) Growth response to climate and drought in *Pinus nigra* Arn. trees of different crown classes. Trees 22:363–373. https://doi.org/10.1007/s00468-007-0191-6
Myers-Smith IH, Hallinger M, Blok D et al (2015) Methods for measuring arctic and alpine shrub growth: A review. Earth-Sci Rev 140:1–13
Neira E (1995) Desarrollo de cronologías para Alerce (*Fitzroya cupressoides* (Molina) Johnston) en las cordilleras de la Costa y de los Andes. Facultad de Ciencias Forestales, Universidad Austral de Chile, p 112
Pompa-García M, Cerano-Paredes J, Fule PZ (2013) Variation in radial growth of *Pinus cooperi* in response to climatic signals across an elevational gradient. Dendrochronologia 31:198–204. https://doi.org/10.1016/j.dendro.2013.05.003

Robinson WJ, Evans R (1980) A microcomputer-based tree-ring measuring system. Tree-Ring Bull 40:59–64

Schweingruber FH (1996) Tree rings and environment: dendroecology. Paul Haupt, Berne. AG Bern

Seager R, Ting M, Davis M et al (2009) Mexican drought: an observational modeling and tree ring study of variability and climate change. Atmósfera 22(1):1–31

Seiler KP, Gat J (2007) Groundwater recharge from run-off, infiltration and percolation. Springer Dordrecht, Paises Bajos

Speer JH (2010) Fundamentals of Tree-Ring Research. University of Arizona Press, Tucson, EUA

Stahle DW, Cook ER, Villanueva-Diaz J et al (2009) Early 21st-century drought in Mexico. Eos 90(11):89–100. https://doi.org/10.1029/2009EO110001

Stahle DW, Burnette DJ, Villanueva-Díaz J et al (2012) Pacific and Atlantic influences in Mesoamerican over the past millennium. Clim Dynam 39:1431–1446

Stahle DW, Cook ER, Burnette DJ et al (2016) The Mexican Drought Atlas: Tree-ring reconstructions of the soil moisture balance during the late pre-Hispanic, colonial, and modern eras. Quat Sci Rev 149:34–60. https://doi.org/10.1016/j.quascirev.2016.06.018

Stat Soft Inc (2011) Statistica for Windows (Computer Program Manual). StatSoft, Inc., Tulsa

Stokes MA, Smiley TL (1968) An introduction to tree-ring dating. University of Chicago Press, Chicago

Valero-Padilla J (2004) Estudio taxonómico y distribucional de los mamíferos del bosque fragmentado de *Juniperus*, del ejido San Juan y Puentes Aramberri, Nuevo León, México. Universidad Autónoma de Nuevo León, p 65

Villalba R (1990) Climatic fluctuations in Northern Patagonia during the last 1000 years as inferred from tree-ring records. Quat Res 34(3):346–360. https://doi.org/10.1016/0033-5894(90)90046-N

Villalba R, Leiva JC, Rubulis S et al (1990) Climate, tree rings and glacier fluctuations in the Frías valley, Río Negro, Argentina. Arct Alp Res 22(3):150–174

Villalba R, Lara A, Boninsegna JA et al (2003) Large-scale temperature changes across the southern Andes: 20th-century variations in the context of the past 400 years. Clim Chang 59:177–232

Villanueva-Díaz J, Fulé PZ, Cerano-Paredes J et al (2009) Reconstrucción de la precipitación estacional para el barlovento de la Sierra Madre Occidental con anillos de crecimiento de *Pseudotsuga menziesii* (Mirb.) Franco. Revista Ciencia Forestal en México 34(105):39–71

Villanueva-Díaz J, Cerano-Paredes J, Fulé PZ et al (2014) Cuatro siglos de variabilidad hidroclimática en el noroeste de Chihuahua, México, reconstruida con anillos de árboles. Investigaciones Geograficas. Boletín del Instituto de Geografía 87:141–153. https://doi.org/10.14350/rig.44485

Villanueva-Díaz J, Cerano-Paredes J, Olivares-Bañuelos NC et al (2015) Respuesta climática del ciprés (*Hesperocyparis guadalupensis*) en Isla Guadalupe, Baja California, México. Madera Bosques 21(3):149–160

Villanueva-Díaz J, Vázquez-Selem L, Gómez-Guerrero A et al (2016) Potencial dendrocronológico de *Juniperus monticola* Martínez en el Monte Tláloc. México Rev Fitotec Mex 39(2):175–185

Wigley TM, Briffa KR, Jones PD (1984) On the average value of correlated time series, with applications in dendroclimatology and hydrometeorology. J Clim Appl Meteorol 23(2):201–213. https://doi.org/10.1175/1520-0450(1984)023<0201:OTAVOC>2.0.CO;2

Zavala-García JA (2003) Caracterización de la roya del enebro (*Juniperus* spp.) en la sierra de Arteaga Coahuila. Departamento de Parasitología, Universidad Autónoma Agraria Antonio Narro, Saltillo, Coahuila, México. No. SD 397. J8. Z38

Ziaco E, Biondi F (2016) Tree growth, cambial phenology, and wood anatomy of limber pine at a Great Basin (USA) mountain observatory. Trees 30(5):1507–1152

Chapter 10
Patterns of Tree Establishment Following Glacier-Induced Floods in Southern Patagonia

Claudia M. Guerrido, Ricardo Villalba, Mariano M. Amoroso, and Milagros Rodríguez-Catón

Abstract Whereas tree establishment after large-scale disturbances such as fires, avalanches, and landslides have been documented for Patagonian forests, information on forest recovery following floods is scarce. Glaciar Perito Moreno has remained stable throughout the twentieth century, reaching the Peninsula de Magallanes several times and embalming parts of Lago Argentino. Following each ice-dam flood, the *Nothofagus* forests along the flooded shores die. We combined dendrochronology and spatial statistics to describe forest recolonization following the episodic flood caused by this glacier in 1988. Four tree species recolonized the lakeshores after the flood. Tree establishment started during spring-summer 1988–1989, the growing season following the break of the ice dam in February 1988. More than 60% of the trees were established between the years 1994–2000, suggesting a rapid colonization of bare shores after the ice-dam break. Using Ripley's spatial analysis, a 100% significant association between species was recorded at sites where two or more species were present. Species dominance was largely modulated by the composition of neighboring non-flooded forests and the climatic conditions along the precipitation gradient. While rates of tree colonization are influenced by species and their seed dispersal capacities, the grouping patterns by site may be modulated by differences in substrate and microclimatic conditions.

C. M. Guerrido (✉)
ICASUR-UART Instituto Ciencias del Ambiente Sustentabilidad y Recursos Naturales, Universidad Nacional de la Patagonia Austral, Río Turbio, Santa Cruz, Argentina

R. Villalba · M. Rodríguez-Catón
Instituto Argentino de Glaciología, Nivología y Ciencias Ambientales IANIGLA-CONICET-Mendoza, Mendoza, Argentina
e-mail: ricardo@mendoza-conicet.gob.ar; mrodriguez@mendoza-conicet.gob.ar

M. M. Amoroso
IRNAD Instituto de Investigaciones en Recursos Naturales, Agroecología y Desarrollo Rural, Sede Andina, Universidad Nacional de Río Negro, Bolsón, Río Negro, Argentina

Consejo de Investigaciones Científicas y Técnicas, CCT – Patagonia Norte, S. C. de Bariloche, Río Negro, Argentina
e-mail: mamoroso@unrn.edu.ar

M. Pompa-García, J. J. Camarero (eds.), *Latin American Dendroecology*,
https://doi.org/10.1007/978-3-030-36930-9_10

Keywords Dendrochronology · Forest dynamics · *Nothofagus* · Spatial analysis

10.1 Introduction

Natural and anthropogenic disturbances are important drivers of forest dynamics shaping species composition and tree structure at different temporal and spatial scales (Picket and White 1985). Following severe and catastrophic disturbances such as fires, glacial retreats, avalanches, and extreme wind-blown events, total growing space becomes available and vegetation establishment starts (Oliver and Larson 1996). The colonization rate is variable and depends on diverse factors such as 1) climatic conditions, 2) post-disturbance nutrients availability and soil characteristics, 3) distance to the undisturbed forest, 4) elevation, aspect, and slope, and 5) remnant propagules, among others (Kitzberger et al. 2000). Seeds, seedlings, roots, basal buds, and surviving individuals make up the biological heritage (Franklin et al. 2000). In addition, disturbance attributes (intensity, scale, and frequency) and species characteristics (tolerance to light intensity, growth rate, and competition) also influence successional patterns (Veblen et al. 1996; Fajardo and Alaback 2005). Indeed, the type and magnitude of disturbance, as well as the micro-climatic conditions, have both proven to be relevant for forest recolonization (Heinemann and Kitzberger 2006).

High-intensity disturbances such as fires, avalanches, ash deposition, earthquakes, landslides, and their impacts have been well documented in the Patagonian forests (Amoroso et al. 2012; González et al. 2014; Rebertus and Veblen 1993; Rebertus et al. 1997; Veblen and Ashton 1978; Veblen et al. 1977, 1981, 1996, 2005). In contrast, the effects of less common disturbances such as large floods on forests have rarely been reported in the southern Andes. This chapter focuses on patterns of tree establishment following the drainage of Glaciar Perito Moreno's ice dam in 1988. The periodic advances of Glaciar Perito Moreno have caused successive ice-dam closures of the southern channels of Lago Argentino and subsequent floods along the shores of Brazo Rico and Brazo Sur. According to historical and dendrochronological records, Glaciar Perito Moreno ice dams have caused 23 floodings since the first historical event recorded in 1936 (Guerrido et al. 2014). These periodic, large-scale disturbances eliminate the vegetation in the inundated areas surrounding the lakes, recurrently affecting forest succession (Oliver 1980; Veblen et al. 2005). In the flooded area, the water level in 1988 rose 21.5 m, close to the historical maximum record of 23.5 m in 1956 (Stuefer 1999). Above this level, forests have remained undisturbed serving as the primary seed source for the recurrently disturbed shores.

Species distribution in the successional process is also modulated by local climatic conditions and species plasticity (Veblen et al. 1977; Cuevas 2000, Srur et al. 2018). Changes in forest species following the steep W-E precipitation gradient in Patagonian Andes provides a natural experiment to study colonization patterns and forest composition after the disturbance. We describe the temporal and spatial patterns of tree establishment in the flooded areas caused by Glaciar Perito Moreno ice dam in 1988 using a combination of dendrochronological techniques and spatial analyses. The specific aims of this chapter are to describe (a) the temporal patterns of tree establishment after the 1988 break event, (b) the spatial patterns of tree establishment in terms of clustering distance from the undisturbed forest, and (c) the changes in forest structure and composition of seedlings and saplings established along the precipitation gradient from the humid western forests in the main Cordillera de los Andes to the xeric woodlands in the transition to the Patagonian steppe. This work represents the first attempt to document the dynamic of Patagonian forests in response to periodic flooding caused by glacier-induced dams.

10.2 Materials and Methods

10.2.1 Study Area

Located in the province of Santa Cruz, Lago Argentino is the largest and southernmost lake in Argentina. The southern arms of Lago Argentino comprises the Canal de Los Témpanos (the Iceberg Channel), and the Brazo Rico and Brazo Sur arms, the latter two located to the south of Glaciar Perito Moreno (50° 30' S, 72° 50' W) in Parque Nacional Los Glaciares, southwestern Patagonia, Argentina (Fig. 10.1).

At a regional scale, the marked W-E precipitation gradient influences vegetation type and distribution in the southern Andes (Burgos 1985). At local scale, changes in forest structure and composition are related to slope, steepness, aspect, altitude,

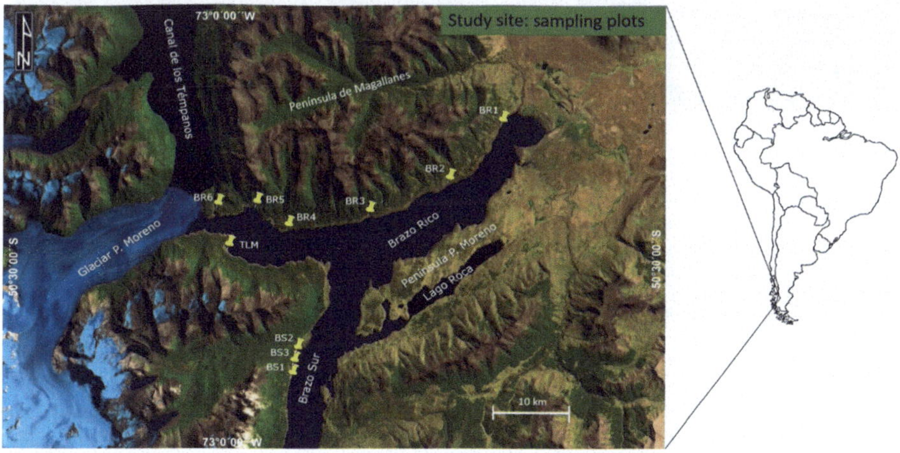

Fig. 10.1 Study area. LANDSAT satellite image of summer 2014 illustrating the southern portion of Lago Argentino and the distribution of sampling sites. The advances of Glaciar Perito Moreno close the natural drainage of the Brazos Rico and Sur producing the increase of the water level and the consequent flooding of their coasts. See Table 10.1 for abbreviations of plots' names

and history of disturbance (Roig et al. 1985; Martínez-Pastur et al. 1994; Dollenz 1995). Forests extend from 180 masl (the water level of Lago Argentino) up to 900–1200 masl, well above the regional lower glacial limit (Perez-Moreau 1959; Armesto et al. 1992, IANIGLA-ING 2018). Three *Nothofagus* species occur in the study area. The evergreen *N. betuloides* Mirb. Blume is the dominant species at wet-low-elevation valleys, whereas the deciduous *N. pumilio* (Poep. et al. Endl) Krasser and *N. antarctica* (G. Foster) Oerst are the dominant trees at the upper elevations and the dry lowlands, respectively (Moore 1983, Dollenz 1995). The conifer *Pilgerodendron uviferum* (D. Don) Florin is limited to a few, small patches in wet environments. Other deciduous, small trees in the area are *Drimys winteri* J. R. Forst and G. Forst and *Embothrium coccineum* J. R. Forst and G. Forst. (Roig et al. 1985). The humid *N. betuloides-N. pumilio* forest is mostly located on the western slope of Brazo Sur and Brazo Rico, whereas the mesic *N. pumilio-N. antactica* forest covers the areas north of Brazo Rico, south of Canal de Los Témpanos and west of Lago Roca. The xeric *N. antarctica* forest occurs in the ecotone with the Patagonian steppe at the eastern sectors of Brazo Rico and Lago Roca (Fig. 10.2).

Fig. 10.2 View of BR6 sampling site (50° 30' S, 72° 50' W), near Glaciar Perito Moreno. Note the seedlings growing along the coast of Brazo Rico, which was previously flooded during the glacier's ice dam. The Brazo Rico is one of the arms of Lago Argentino that floods when Glaciar Perito Moreno reaches the Magellan Peninsula and closes the drainage of the arms of Lake Argentino south of the Glaciar Perito Moreno, Parque Nacional Los Glaciares, Argentina. Photo: Ricardo Villalba

10.2.2 Sampling and Data Analysis

Ten rectangular plots, aligned from the unaffected forest down to the lake, were set at each sampling site, including the humid, mesic, and dry forest types. Plot size was variable and determined by the density of seedlings. We included a minimum of 50 individuals per plot (Table 10.1), resulting in plots ranging between 10 and 145 m in length and 5 and 60 m in width. Tree species, total height, and diameter at the base (DAB) were recorded at each site. For *E. coccineum*, we recorded single and multi-stem individuals. Seedlings with DAB larger than 3 cm were cored using increment borers as close to the ground as possible. When DAB was smaller than 3 cm, seedlings were uprooted for age determination.

We used the ages of the trees to determine the patterns of temporal establishment. Samples were processed following standard dendrochronological methods (Stokes and Smiley 1968). Cross-sections and cores of seedlings and saplings were visually dated using a binocular magnifier (50x; Yamaguchi 1991). For dating the cores, we used the Schulman's convention for the Southern Hemisphere, which assigns to each ring the calendar year when the ring started to form (Schulman 1956). The position along the plot axes were recorded for each of the 780 individuals sampled in the 10 plots. The coordinates of the trees along the plots were used to describe the spatial patterns of establishment. This Y-axis started at the forest undisturbed by the floods and was directed toward the coast of the lake.

We developed maps for tree age, DAB, and height distributions from individual trees for each site based on X-Y location coordinates to characterize the spatial patterns of distribution across the plots (Diggle 2003).

We used Ripley's function to determine the spatial pattern (random, clustered, or regular) of the trees within the sampled plots. We determined the theoretical value of L (Ripley 1981), the theoretical L(r) represents randomly distributed points for a given distance (r) within the area. The L(r) function can be calculated to describe the spatial distribution of one set of points (univariate L(r)) or between two sets of points (bivariate or cross L(r)). For each value of r we generated 99 simulation envelopes of theoretical L(r) specifying the critical points for a Monte Carlo test (Ripley 1981). Consequently, if the estimate of L(r) for a given set of points and a given distance is within the envelope, the spatial pattern is considered as a random distribution. Any time the estimate of L(r) is above (below) the envelope, there is association (rejection) between points at a given distance and the patterns is aggregated (regular) (Wiegand and Moloney 2004; de la Cruz 2006). Since the boundaries of the plots were arbitrarily chosen, an edge effect correction was introduced. The maximum distance that the test allows to evaluate is 1/3 of the smallest side of the plot. We used the free software R and the package spatstats for the spatial analyses (Baddeley et al. 2014).

Tree coordinates on the Y (length) axis of plots provided an indirect way to evaluate the seed dispersal capacity of trees present in the sampling sites. We quantified the relationship between distance to undisturbed forest and age using simple regression. Initially, all data were used for the analysis. In a second step, the data were grouped in 5 m-long sub-plots. The date from the earliest establishment was selected for the analysis at each sub-plot.

Table 10.1 Geographical location, size, and topographical features of the sampling plots shown in Fig. 10.1. The elevation corresponds to the highest point of the plot

	Plot	Latitude	Longitude	Area (m^2)	Length (m)	Width (m)	Elevation (masl)	Slope (°)	Aspect	Sampling date
Brazo Rico	BR1	50°26′10.3	72°45′36.2	8100	135	60	218	12.8	South	Mar-04
Brazo Rico	BR2	50°27′15.0	72°48′40.0	2880	60	20	217	13.2	South	Mar-04
Brazo Rico	BR3	50°29′00.2	72°53′03.6	1450	145	10	219	13.0	South	Dec-05
Brazo Rico	BR4	50°29′23.2	72°57′24.1	420	60	7	203	14.3	South	Jan-06
Brazo Rico	BR5	50°28′28.3	72°59′07.4	1150	115	10	196	14.5	South	Dec-05
Brazo Rico	BR6	50°28′35.2	73°01′20.3	510	10	51	201	35.5	South	Jan-06
Brazo Sur	BS1	50°34′49.9	72°57′35.0	1780	89	20	207	18.0	East	Mar-04
Brazo Sur	BS2	50°33′55.3	72°57′12.9	1400	35	40	194	18.7	East	Mar-04
Brazo Sur	BS3	50°34′21.7	72°57′24.5	230	46	5	218	17.8	East	Mar-04
Monedas	TLM	50°30′31.9	72°59′52.4	890	89	10	186	16.5	North	Jan-12

Averages and standard deviations were calculated for age, height, and DAB at each plot and compared between sites. Richness, density, and percentage of dominance by species were also evaluated. Since *E. coccineum* is heavily browsed by cows, the number of stems per sapling was used as an indicator of grazing pressure at each site. *E. coccineum* is widely distributed on the shores of lakes Brazo Rico and Brazo Sur and was present in 8 of the 10 plots. Since our data was non-normally distributed (Zar 1999), we conducted Kruskal Wallis tests for establishing significant differences in the number of stems per seedling using the Infostat software (Di Rienzo et al. 2011).

Two soil samples were collected at each plot along the Brazo Rico (BR1 to BR6). Soil depths in the field were determined using an incremental soil sampler. Samples were sieved through a 2mm mesh and air dried for 1 week to measure the pH in water and the electrical conductivity (1:2.5 soil–water ratio). Organic matter (OM) was analyzed by the Wakley-Black method (Jackson 1958), while Inorganic N was evaluated by using the Kjeldahl method (Kirk 1950). Phosphorous (P available) was extracted in 0.5M NaHCO3 (1:20, soil–solution ratio) and determined by the Olsen method (Olsen and Sommers 1982).

10.3 Results

10.3.1 Temporal Establishment Patterns

Approximately 60% of the trees were established between 1994 and 2000, 18% between 1988 and 1993, and the remaining 22% after 2001. The glacier-dammed water drained through the ice tunnel in February 1988 and a total of eight trees were established in our sampling plots in the following spring-summer growing season.

Brazo Rico (BR)

During the 5-year period following the 1988 flood, *N. antarctica*, *N. pumilio*, and *N. betuloides* established in the dry, mesic, and humid sites, respectively. In contrast, *E. coccineum* established at most sites along the precipitation gradient (Fig. 10.3). When assessing the establishment over the next five years (i.e., up to 10 years after the flood), *N. antarctica* continued establishing in xeric but also appeared in mesic sites. The *N. pumilio* establishment expanded to intermediate sites in both directions, to drier and more humid sites. *N. betuloides* continued establishing at humid but appeared in the transition to mesic sites. The *E. coccineum* establishment continued at most sites (Fig. 10.3). Considering the establishment of all individuals in BR, recruitment was recorded every year with a steady increase from 1988 until 1994–1995, a secondary peak in the year 2000, and a gradual decrease after that (Fig. 10.5). *N. antarctica* and *N. pumilio* established almost every year. *E. coccineum* and *N. betuloides* started establishing in 1990, two years after the flood. *N. betuloides* recruitment was absent during several years, however, *E. coccineum* established every year (Fig. 10.5).

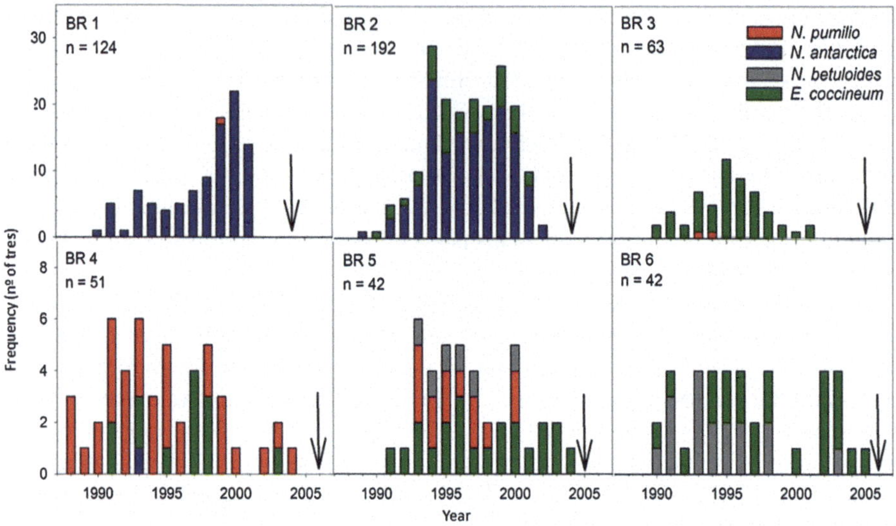

Fig. 10.3 Establishment dates for trees per plot in Brazo Rico following the Glaciar Perito Moreno ice-dam break in February 1988. Black arrows indicate sampling years

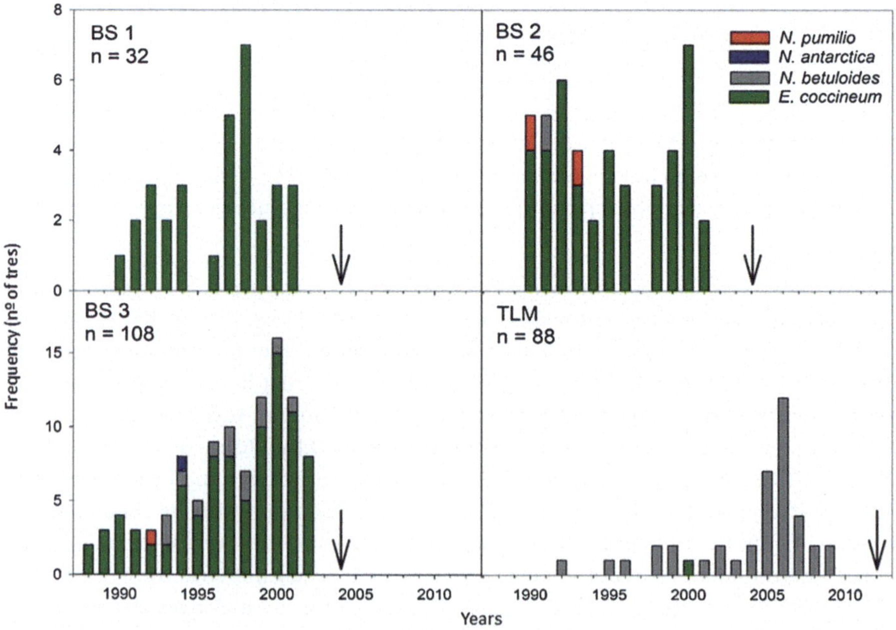

Fig. 10.4 Establishment dates for trees per plot in Brazo Sur and Las Monedas sites. Black arrows indicate sampling years

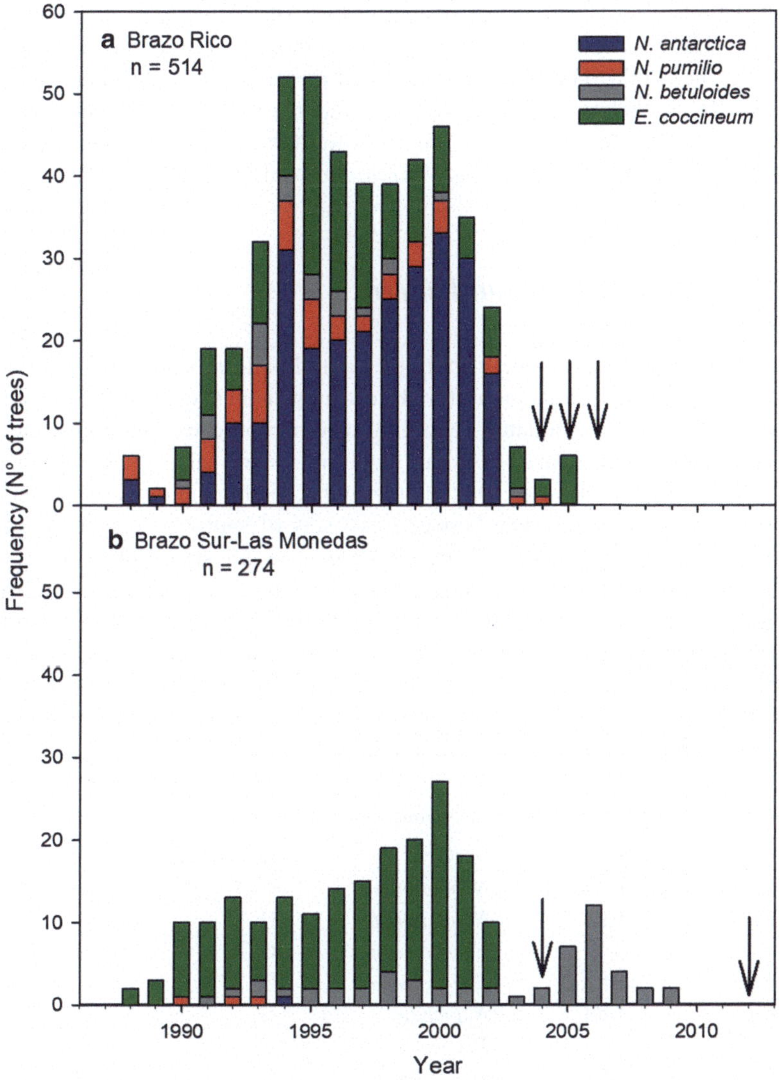

Fig. 10.5 Establishment dates for trees in Brazo Rico and Brazo Sur - Las Monedas sites following the Glaciar Perito Moreno ice-dam break in February 1988. Black arrows indicate sampling years

Brazo Sur (BS) and Las Monedas (TLM)

Since all sites along BS are located in the relatively humid sector of the transect, clear patterns of species establishment did not emerge during the 5 years following the 1988 flood. *N. pumilio* and *N. betuloides* were established in two and *E. coccineum* in three from the four sites selected. A single *N. antarctica* was recorded in BS eight years after the ice-dam brake. No establishment of *N. pumilio* occurred after 1994 in any of the plots (Fig. 10.4). When assessing the establishment of all seedlings in Brazo Sur - Las Monedas, a continuous increase in tree recruitment was registered since 1988 until 2000. The year 2000 recorded the largest establishment in coincidence with the secondary peak in tree establishment at BR. *E. coccineum* established from 1988 to 2003, *N. betuloides* from 1991 to 2009, *N. pumilio* only during the years 1990, 1992, and 1993, whereas *N. antarctica* only in 1994 (Fig. 10.5).

10.3.2 Spatial Distribution Patterns

The univariate L(r) function, describing the spatial pattern of all saplings, shows a significantly grouped spatial pattern for all seedlings in all sites from 0.5 to 8 meters, the maximum distance evaluated according to our plot sizes. The univariate L(r) function, describing the spatial association within species, also shows significantly aggregated patterns for all sites. On the other hand, the bivariate or cross L function comparing grouping occurrence between two sets of saplings (e.g., two species) indicates significant association between species in all sites (Table 10.2). Tree grouping is not related to species, age, DAB, or height. Since the results are similar in all sites, we show a graphical representation of the spatial distribution of saplings classified by age, DAB, and height (Fig. 10.6a) and the function L(r) for site BR2 (Fig. 10.6b–e). Overall, we observed saplings spatially distributed in two or three groups. In most sites, trees were concentrated near the undisturbed forest and in the center of the plot, with few isolated individuals near the lake shores.

Table 10.2 Grouping patterns per plot according to Ripley (1981)

Site	Plot	Species	L-function	L-cross
Brazo Rico	BR1	*N. antarctica*	G	
Brazo Rico	BR2	*N. antarctica*	G	
		E. coccineum	G	G
Brazo Rico	BR3	*E. coccineum*	G	
Brazo Rico	BR4	*N. pumilio*	G	
		E. coccineum	G	G
Brazo Rico	BR5	*E. coccineum*	G	G
		N. pumilio	G	G
		N. betuloides	G	G
Brazo Rico	BR6	*E. coccineum*	G	
		N. betuloides	G	G
Brazo Sur	BS1	*E. coccineum*	G	
Brazo Sur	BS2	*E. coccineum*	G	
Brazo Sur	BS3	*E. coccineum*	G	
		N. betuloides	G	G
Las Monedas	TLM	*N. betuloides*	G	

G grouped

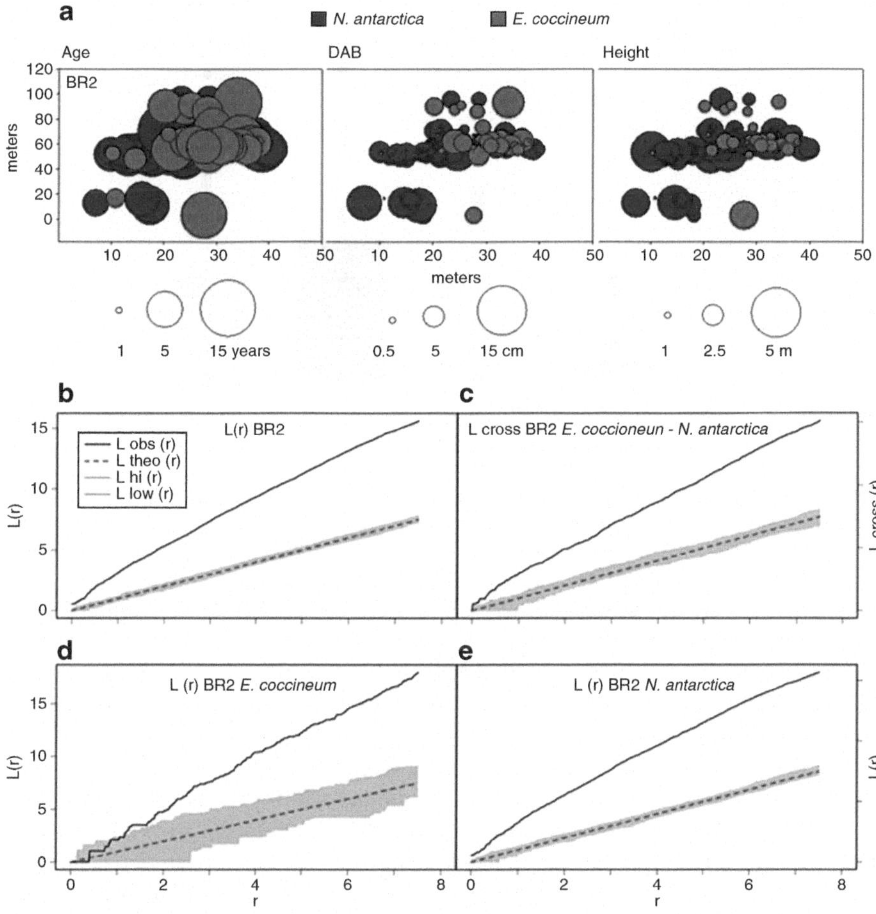

Fig. 10.6 Spatial distribution of saplings and seedlings. (**a**) Saplings and seedlings classified by age, DAB, and height. (**b**) The L univariate function shows that all saplings and seedlings in the plot are arranged in clusters. (**c**) The function L cross shows that *E. coccineum* and *N. antarctica* are 100% significantly associated. (**d**) and (**e**) The L univariate function shows that *E. coccineum* and *N. antarctica* are arranged in clusters, respectively

10.3.3 Distance of Establishment from Undisturbed Forests

We recorded differences in seedling establishment in relation to distances from the seed sources. All four tree species were established at plots larger than 100 m, providing the opportunity to determine wind dispersal distances. *E. coccineum* and *N. antarctica* reached similar maximum distances of 130–135 m from the undisturbed forests. The average dispersal distances were 52 m (ranged between 3 and 60 m) for *N. antarctica* and 24 m (ranged between 9 and 45 m) for *E. coccineum*. *N. betuloides* seedlings established up to 80 m from the forest with a large number

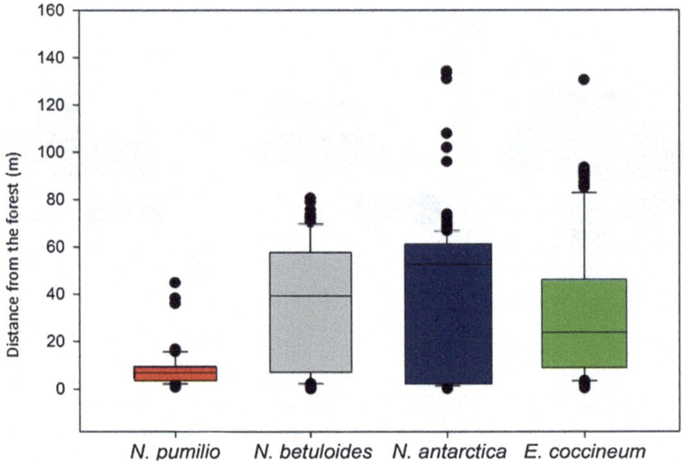

Fig. 10.7 Box plots showing the distance of tree establishment by species from nearby undisturbed forests

of individuals located between 7 and 57 m from the seed source. Comparatively, *N. pumilio* showed the shortest range in distance from seed source. The maximum distance recorded was 45 m, less than half the distance reached by *N. antarctica*. The *N. pumilio* establishment was concentrated between 4 and 8.8 m (mean 6 m) from the forest not affected by floods (Fig. 10.7).

Establishment dates did not always progress gradually from the forest to the lakeshore. At most sites, the age of seedlings and saplings was not significantly related to the distance from seed source ($p < 0.4$). *N. betuloides* showed a positive trend in TLM, no trend in BR6, and negative trend in BS3. *N. antarctica* reported a positive trend in BR2 but negative in BR1. *N. pumilio* showed an inverse trend (older trees at longer distances) in BR5. *E. coccineum* reported a positive trend in BR3, no trend in BR2, and negative trends in BR4, BR5, BR6, BS1, and BS3.

Consistent with our observations, no patterns of trees establishment in relation to distance from undisturbed forests emerged when grouping tress ages by 5-m plots. Positive trends were recorded for *N. pumilio* in BR4, *N. betuloides* in TLM, and *N. antarctica* in BR1 and BR2. *E. coccineum* shows a positive trend in BS3, no trend in BR3, and negative trends in BR2, BR4, BR5, BS1, and BS2.

10.3.4 Structure and Composition

Land use markedly changed throughout recent history in the region. Before the creation of Parque Nacional Los Glaciares in 1937, local inhabitants raised cattle in forested areas. During the process of people relocation following the establishment of the National Park, some cattle were left behind in remote areas (Parques Nacionales 1997;

Guerrido et al. 2014). Nowadays, wild cattle densely graze on the shores of Brazo Sur and the southern part of Brazo Rico. Tourist activities are restricted to the northern sector of Brazo Rico, where the scenic road to Glaciar Perito Moreno runs.

Seedling densities varied between 127 and 4696 trees/ha. On the northern shore of Brazo Rico, the lowest and highest densities of 127 and 1608 trees/ha were recorded at BR1 and BR2, respectively. Along the western shore of Brazo Sur, we also recorded a large range in tree density from 180 trees/ha in BS1 to 4696 trees/ha in BS3 (Table 10.3). Regarding the number of stems per *E. coccineum* individual, the average for the Brazo Sur sites was 7.82, whereas for the Brazo Rico sites was 4.49 stems/tree. The percentages of multi-stem trees were 71% and 46% for Brazo Sur and Brazo Rico, respectively.

When assessing the height of the seedlings, there was a marked difference among sites. The average height was 97.8 cm (SD 32) for Brazo Rico and 31.3 cm (SD 8.5) for Brazo Sur and TLM (Fig. 10.8). A similar pattern was found for the diameter at the base (DAB), where average diameters varied from 3.9 cm (SD 1.2) in Brazo Rico to 1.5 cm (SD 0.4) in Brazo Sur and TLM (Fig. 10.8).

The species composition of the sites reflected the W-E humidity gradient (Figs. 10.3 and 10.4). Two tree species were recorded at xeric and mesic sites in Brazo Rico (BR1, BR2, and BR3); nevertheless, species composition was different in each plot. Three woody species were recorded in the mesic sites of Brazo Rico and Brazo Sur (BR4, BR5 and BS2), whereas four species were present in the humid site of Brazo Sur (BS3) (Figs. 10.2 and 10.3). *N. antarctica* was the dominant species in xeric sites reaching 80% and 90% in BR2 and BR1, respectively, but representing only 10% in the mesic to humid sites BR4 and BS3, and absent in the BS1, BS2, and TLM humid sites. *N. pumilio* abundance ranges from 25% (BR5) to 70% (BR4) in mesic sites reaching only 10% in xeric sites (BR1, BR3). *N. pumilio* was absent at the humid BS3 and TLM sites. *N. betuloides* represented between 10% (BS2) and 90% (TLM) of the trees established in humid sites, but was absent in the xeric and mesic sites. The abundance of *E. coccineum* was variable ranging from 50% (BR6) to 100% (BS1) in humid sites, 20% (BR2) to 90% (BR3) in xeric-mesic sites, and absent in the most xeric site (BR1).

Table 10.3 Characteristics of the forest at the sampling plots

Site	Plot	Density (trees/ha)	Richness (n° of species)	Average age (years)	Age range (years)
Brazo Rico	BR1	127.2	2	5.0	1–15
Brazo Rico	BR2	670.1	2	6.4	1–14
Brazo Rico	BR3	434.5	2	9.0	1–15
Brazo Rico	BR4	1214.3	3	10.1	1–17
Brazo Rico	BR5	365.2	3	8.2	1–14
Brazo Rico	BR6	823.5	2	8.4	1–15
Brazo Sur	BS1	179.8	1	6.6	2–13
Brazo Sur	BS2	328.6	3	8.0	2–13
Brazo Sur	BS3	4695.6	4	6.0	1–15
Las Monedas	TLM	1022.5	2	7.0	2–19

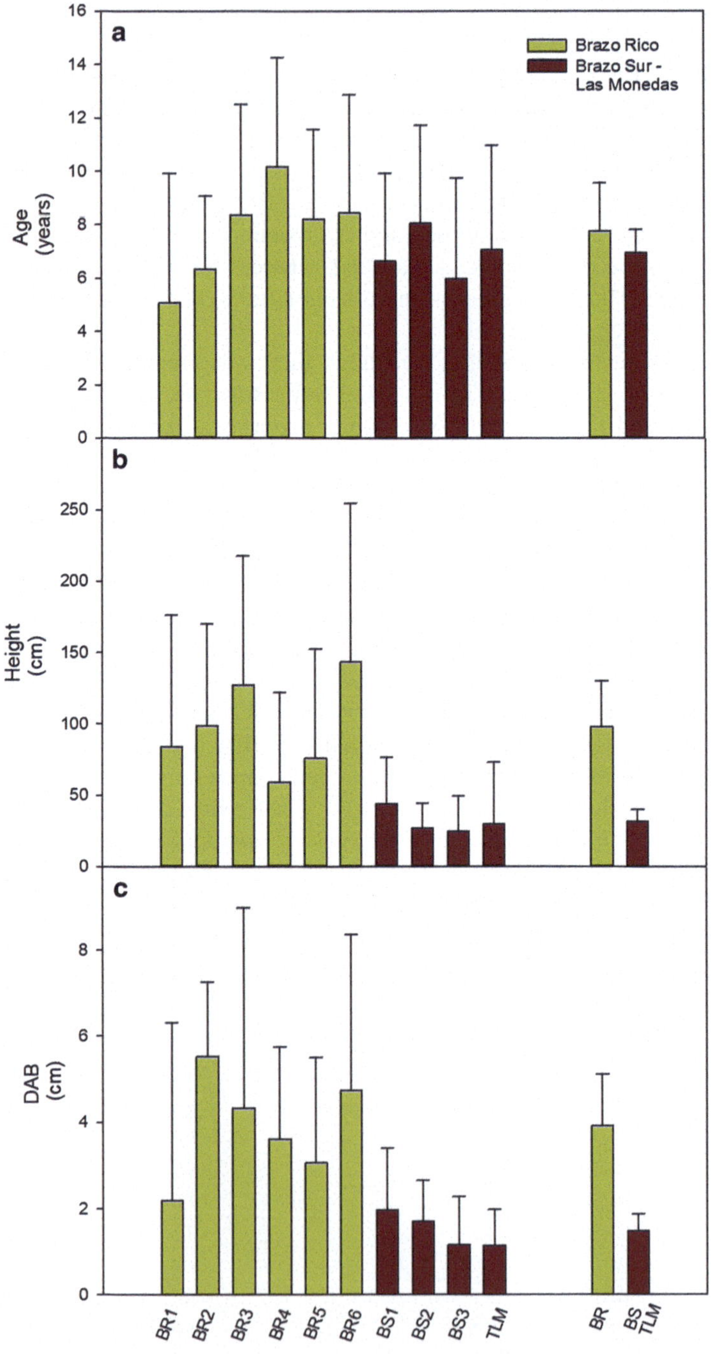

Fig. 10.8 Mean age (**a**), height (**b**), and DAB (**c**) per plot (left) and per sector (right) for Brazo Rico (yellow bars) and for Brazo Sur – Las Monedas (brown bars). Standard deviations are indicated by vertical capped lines

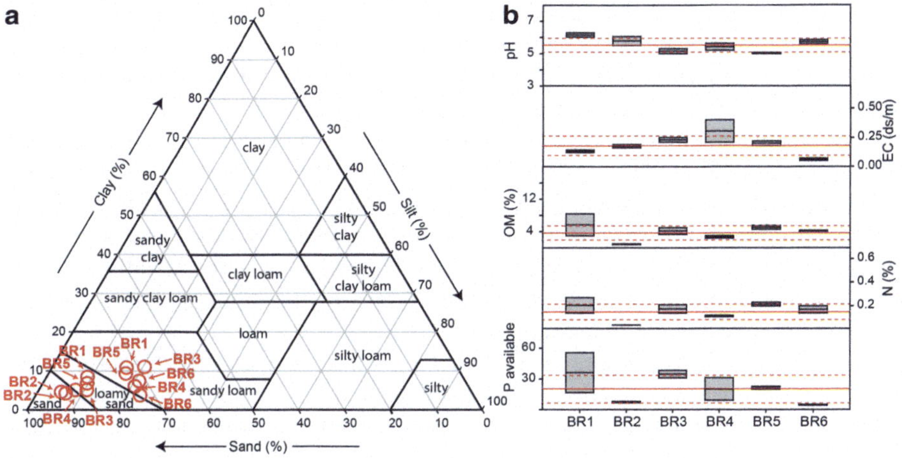

Fig. 10.9 Soil texture diagram showing (**a**) variations in clay, silt, and sand composition and (**b**) variations in pH, electric conductivity, organic matter, N and P contents in soils for sites BR1 to BR6 along the precipitation gradient (Fig. 10.1)

10.3.5 Soil Analysis

The textural analysis of the soils indicates minor differences between plots, grouping all site soils into the categories of sand, loamy sand, and sandy loam (Fig. 10.9a). Most sites have soils with a pH around 5.5 (range 5.03–6.15) and electrical conductivity (EC) of 0.18 ds/m (0.06–0.30). Most soils present medium (3.5–6.0%) organic matter content (mean 3.76%), except BR2 with 0.88% (Fig. 10.9b). Finally, the contents of P and N oscillate around 20.7 mg/kg (7.36–36.1) and 0.15% (0.04–0.21), respectively. No significant relationships were recorded between soil properties and tree species (richness), mean age, mean DAB, or mean height.

10.4 Discussion

The dominant species in the southern Patagonia forests do not survive prolonged floods. The shores of Brazo Rico and Brazo Sur of Lago Argentino were inundated during four years until the ice dam formed by the advance of Glaciar Perito Moreno against Peninsula de Magallanes broke on February 17, 1988 (Guerrido et al. 2014). Even the swamp-adapted *N. antarctica* individuals in the shores of Brazo Rico and Brazo Sur died after the long-term 1984–1988 flooding event. Previous studies showed that floods caused by introduced beavers (*Castor canadensis*) have also caused tree mortality in *Nothofagus* forests in Tierra del Fuego, Argentina. All forest components died in the relatively small ponds created by the beavers (Martínez-Pastur et al. 2006).

Following the 1984–1988 flood, we recorded the establishment of new trees on the bare shores over the 1988–2004 interval (16 years), the most extensive period between two consecutive floods produced by Glaciar Perito Moreno in the last 100 years. Three individuals of *N. pumilio*, three of *N. antarctica*, and two of *E. coccineum* established in the wet sector of the precipitation gradient during the austral growing season (September 1988 to February 1989) that followed the rupture of the ice dam. Individuals from the three *Nothofagus* species and *E. coccineum* established during the following growing season in 1989–1990. By 1994, 60% of all trees recorded at the sampling time had established. The establishment gradually declined after the year 2000, likely due to competition from grasses/shrubs and the reduction of suitable microsites for tree establishment.

In the southern Andes, previous studies on surfaces denudated by large-scale disturbances other than floods have already pointed out the rapid colonization by *Nothofagus* species. Indeed, Rebertus and Veblen (1993) recorded that *N. betuloides* and *N. pumilio* often established during the same year of large-scale tree falls in Tierra del Fuego. The same pattern was observed after debris avalanches, landslides, and mudflows related to seismic activity in the Valdivian rainforests of Chile, where shrubs and trees established on bare surfaces within a year or two following landslides (Veblen and Ashton 1978). Armesto et al. (1992) reported the immediate establishment of *N. betuloides* on recently formed glacier moraines in Torres del Paine, Chile, followed by the deciduous *N. antarctica* in the next 5–7 years. These authors also noted the absence of *N. pumilio* on the moraines although it was abundant in the surrounding hills. They postulated that *N. pumilio* needs well-developed soils to establish successfully. Consistent with these observations, Dollenz et al. (2012) indicated that *N. betuloides*, *N. antarctica*, and *E. coccineum* are the first tree species to colonize glacial deposits in the southern Andes. In similar environments at Glaciar Seco forefields, Lago Argentino, Garibotti et al. (2011) reported a long range (few decades to centuries) in tree establishment time depending on mesotopographic heterogeneity. Differences in species richness and cover in Glaciar Seco forelands were related to decreasing rates of vegetation colonization from the base to the crest of the moraines (Garibotti et al. 2011).

The spatial association between seedlings on the denuded shores of the lake reveals aggregation patterns for all seedlings, both within and between species. The grouping between seedlings of different species is a consequence of the aggregated pattern present in the whole plot. Despite the relatively wide shore exposed for colonization, where a random distribution pattern of trees would have been expected, seedlings and saplings are largely clustered. Previous studies suggest that the dry winds limit the establishment of *Nothofagus* trees in large exposed areas in Patagonia (Kalela 1941; Schmidt and Urzua 1982). We assume that the patterns of tree establishment along the lakeshores are related to local microsite conditions. At microsites protected from severe winds, soil moisture is higher and remains for longer periods. Previous studies indicate that water deficit is a major limitation for the establishment of *N. pumilio* in open areas (Rush 1992). As was previously observed by Armesto et al. (1992), the establishment on moraines recently formed by glacial activity could be related to microsites with different soil conditions.

Differences in exposure to wind and substrate (rocky *versus* soil) along the bare shores could explain the patchy vegetation patterns found in our study area. In some sectors of the lake coasts, outcrops or rocky layers seriously limit the establishment of trees, indicating that substrates also play a dominant role in the grouping patterns recorded in tree establishment. Indeed, the physical-chemical characteristics analyzed in these periodically flooded soils show marked similarities between sites along the precipitation gradient (BR1 to BR6, Fig. 10.9). None of the sites showed consistent differences in soil texture or chemical properties (pH, electrical conductivity, organic matter, and N and P content) that can be related to demographic attributes (age, diameter, and height) of the seedlings established in plots BR1 to BR6. Therefore, micro-climatic conditions and/or substrate type play a larger influence than soil on determining the seedling establishment patterns recorded along the precipitation gradient. It is very likely that the repeated floods during the last century (23 events) have modified the local soil characteristics inducing large similarities in the physical-chemical characteristics of the flooded soils in the coastal sites along the precipitation gradient.

Although the four tree species are anemochorous, seed dispersal distances reflect differences in seed size and weight. *N. antarctica* seeds reached comparatively longer distances from undisturbed forests than those recorded for *N. betuloides* and particularly for *N. pumilio*. *N. antarctica* and *N. dombeyi* (the closest related species to *N. betuloides*) weigh 1.5 and 1.8 gr per 1000 seeds, respectively, whereas *N. pumilio* weighs 16.4 gr per 1000 seeds (Urretavizcaya et al. 2016), that is, approximately 10 times heavier than the other *Nothofagus* species in our study area. Similarly, Veblen et al. (1996) found for northern Patagonia that small and light seeds of *N. dombeyi* are more effectively dispersed than the heavier *N. alpina* seeds. The shorter distances reached by *N. pumilio* are consistent with heavier seeds. Previous studies indicate that *N. pumilio* seeds fall very close to the tree crown (Rush 1993). Current studies at the windy upper tree line in southern Patagonia (49° S) suggest that *N. pumilio* seeds rarely reach distances further than 30 m from the parent trees (Srur et al. 2016, 2018). Seeds are commonly found up to 20 m, rarely up to 40 m, from upper tree lines in Tierra del Fuego (Cuevas 2000).

We recorded an average seed dispersal distance of 25 m with a maximum distance of 140 m for *E. coccineum*. This is five times larger than the distance reported by Rovere and Premoli (2005) for northern Patagonia, where 95% of *E. coccineum* seeds were dispersed within 5 m of the parent trees. Given the weight of the *E. coccineum* seeds (11.9 gr. per 1000 seeds) (Urretavizcaya et al. 2016), we speculate that the ubiquity of *E. coccineum* in our sampling sites relates to the morphology of the one-winged seed specially adapted to the stronger winds in southern Patagonia. Small water streams along the bare lake coasts facilitate seed dispersal and contribute to the longer distances recorded in our sampling sites.

In this chapter we show that after a single large-scale disturbance affecting contrasting environments along the Patagonian forest, colonization of new trees is predominantly determined by local climatic conditions and species plasticity. Indeed, species distribution along the inundated shores appears to be largely regulated by the well-marked W-E precipitation gradient. *N. antarctica* represents 80%

of the seedlings at the stands near the ecotone with the Patagonian steppe. The dominant establishment of *N. antarctica* in the driest sites reflects its ability to withstand extremely dry conditions (Donoso et al. 2006). In contrast to the driest sites, *N. pumilio* dominates the mesic forests in Brazo Rico and Brazo Sur accounting for more than 60% of the seedlings. This species dominates in the mesic-cold areas in the region (González et al. 2006). Finally, *N. betuloides* is the dominant species at low elevation and humid slopes of the wet forests near Glaciar Perito Moreno. According to Donoso and Donoso (2006), *N. betuloides* grows in cold humid oceanic climate along the Pacific Ocean and bordering the low-elevation sectors of the Andean glaciers in southern Patagonia. In contrast to the close association between *Nothofagus* species and site conditions, the *E. coccineum* establishment was recorded at most sampling sites. This generalist species is present at most environments along the Andean precipitation gradient. Due to its polymorphism, *E. coccineum* grows in contrasting habitats (Alberdi and Donoso 2004). Indeed, *E. coccineum* covers a large latitudinal distribution (35° S–56° S) with mean annual temperature ranging from 15°C in the north to 5°C in the southern Patagonian forests (Escobar et al. 2006).

Significant differences in *E. coccineum* morphologies between individuals along shore lakes reflect distinct browsing intensities. Multi-stem trees represent 71% and 46% at Brazo Sur and Brazo Rico shores, respectively. Cattle have been present in the forest before the establishment of Parque Nacional Los Glaciares in 1937 (Parques Nacionales 1997). At present, cattle is abundant on the shores of Brazo Sur and Las Monedas but rare along Brazo Rico. Consistent with our results, Blackhall et al. (2008), who studied the effect of livestock on *Nothofagus – Austrocedrus* establishment after fires in northern Patagonia, found that cattle browsing reduces the heights of both species. Cattle grazing produces changes in soil structure and affects negatively the natural regeneration of *N. pumilio* forests (Peri and Martínez-Pastur 2003). In addition, cattle in the Patagonian steppe induce shorter, multi-stemmed individuals (Bisigato and Bertiller 1997; Srur et al. 2013).

10.5 Concluding Remarks

The structure and composition of forests in the southern Patagonian Andes reflects the interactions between steep environmental gradients and natural- or human-induced disturbances. Our contribution provides the first detailed description of tree establishment following large-scale floods in the Patagonian forests. Consistent with previous studies on forest disturbance in Patagonia, we noted that trees rapidly colonize the exposed areas left by the inundated lakeshores following the rupture of the ice dam of Glaciar Perito Moreno. The rapid colonization of the exposed shores by the trees is consistent with the ecology of the Patagonian *Nothofagus* species adapted to extreme dynamical environments, where natural disturbances have been present throughout the long natural history of the region.

Continued browsing by cattle along exposed shores has a damaging impact on seedling establishment. Seedlings and saplings were less numerous, shorter, and with a greater number of stems at sites intensively browsed. This information could be useful when implementing plans for cattle eradication in protected areas and National Parks in the Patagonian Andes.

The exposure of the lake shore to periodic disturbances such as the floods caused by the dynamics of Glaciar Perito Moreno provide a unique opportunity to conduct in situ long-term experiments, intended to develop a comprehensive view of forest responses to high-intensity severe disturbances in the Patagonian Andes. In addition, our studies show the usefulness of tree-ring methods to date the occurrence of floods (Guerrido et al. 2014), as well as the dates of lake drainage events. For example, individuals of *N. pumilio*, *N. antarctica*, and *E. coccineum* established on denudated shores during the spring following the ice-dam rupture event, providing a precise date for the break of the Glaciar Perito Moreno ice dam in 1988.

Acknowledgments The authors would like to thank to the Administración de Parques Nacionales for granting us the permits to conduct our research and to Hielo y Aventura S.A. for invaluable field assistance. This work was financially supported by a scholarship for the training of doctoral student (PRH 13 PFDT ANPCyT – UNPA – UART) and the following research projects: PICT 00215 ANPCyT and CONICET (PIP 112-2011010-0809 to R.V.). R.V. acknowledges the support of the BNP-PARIBAS Foundation through the THEMES Project.

References

Alberdi M, Donoso C (2004) Variación en *Embothrium coccineum* JR et G. Forster (notro o ciruelillo). In: Donoso C, Premoli A, Gallo L et al (eds) Variación intraespecífica en especies arbóreas de los bosques templados de Chile y Argentina. Editorial Universitaria, Santiago

Amoroso MM, Suarez ML, Daniels LD (2012) *Nothofagus dombeyi* regeneration in declining *Austrocedrus chilensis* forests: effects of overstory mortality and climatic events. Dendrochronologia 30:105–112. https://doi.org/10.1016/j.dendro.2010.12.005

Armesto JJ, Casassa I, Dollenz O (1992) Age structure and dynamics of Patagonian beech forests in Torres del Paine National Park, Chile. Vegetatio 98:13–22. https://doi.org/10.1007/BF00031633

Baddeley A, Diggle PJ, Hardegen A et al (2014) On tests of spatial pattern based on simulation envelopes. Ecol Monogr 84:477–489. https://doi.org/10.1890/13-2042.1

Bisigato AJ, Bertiller MB (1997) Grazing effects on patchy dryland vegetation in northern Patagonia. J Arid Environ 36:639–653. https://doi.org/10.1006/jare.1996.0247

Blackhall M, Raffaele E, Veblen TT (2008) Cattle affect early post-fire regeneration in a *Nothofagus dombeyi–Austrocedrus chilensis* mixed forest in northern Patagonia, Argentina. Biol Conserv 141:2251–2261. https://doi.org/10.1016/j.biocon.2008.06.016

Burgos JJ (1985) Clima del extremo sur de Sudamérica. In: Boelcke O, Moore DM, Roig FA (eds) Transecta Botánica de la Patagonia Austral. Buenos Aires, Santiago, Londres, pp 10–40

Cuevas JG (2000) Tree recruitment at the *Nothofagus pumilio* alpine timberline in Tierra del Fuego, Chile. J Ecol 88:840–855. https://doi.org/10.1046/j.1365-2745.2000.00497.x

Di Rienzo JA, Casanoves F, Balzarini MG et al (2011) InfoStat Versión 2011. Grupo InfoStat, FCA, Universidad Nacional de Córdoba, Argentina

de la Cruz M (2006) Introducción al análisis de datos mapeados o algunas de las (muchas) cosas que puedo hacer si tengo coordenadas. Ecosistemas 15(3):19–39

Diggle PJ (2003) Statistical analysis of spatial point patterns, 2nd edn. Arnold, London
Dollenz O (1995) Los árboles y bosques de Magallanes. Ediciones Universidad de Magallanes, Punta Arenas
Dollenz O, Henríquez J, Domínguez E (2012) La vegetación de las geoformas proglaciares en los glaciares Balmaceda, Tyndall Taraba y Ema, Magallanes, Chile. An Inst Patagon 40:7–17. https://doi.org/10.4067/S0718-686X2012000200001
Donoso C, Steinke L, Premoli A (2006) *Nothofagus antarctica* (G. Forster) Oerst. Familia: Fagaceae. In: Donoso C (ed) Especies arbóreas de los bosques templados de Chile y Argentina. Autoecología Marisa Cúneo Ediciones, Santiago, pp 448–461
Donoso D, Donoso P (2006) *Nothofagus betuloides* (Mirb.) Oerst. Familia: Fagaceae. In: Donoso C (ed) Especies arbóreas de los bosques templados de Chile y Argentina. Autoecología Marisa Cúneo Ediciones, Santiago, pp 448–461
Escobar B, Donoso C, Souto C et al (2006) *Embothrium coccineum* J.R. et. G. Forster Familia: Proteaceae. In: Donoso C (ed) Especies arbóreas de los bosques templados de Chile y Argentina. Autoecología Marisa Cúneo Ediciones, Santiago, pp 448–461
Fajardo A, Alaback P (2005) Effects of natural and human disturbances on the dynamics and spatial structure of *Nothofagus glauca* in south-central Chile. J Biogeogr 32:1811–1825. https://doi.org/10.1111/j.1365-2699.2005.01331.x
Franklin J, Lindenmayer DB, MacMahon JA et al (2000) Threads of continuity: ecosystems disturbance, recovery, and the theory of biological legacies. Conserv Biol Pract 1(1):8–16
Garibotti IA, Pissolito CI, Villalba R (2011) Spatiotemporal pattern of primary succession in relation to Meso-topographic gradients on recently Deglaciated terrains in the Patagonian Andes. Arct Antarct Alp Res 43:555–567. https://doi.org/10.1657/1938-4246-43.4.555
González ME, Donoso C, Ovalle P et al (2006) *Nothofaus pumilio* (Poep. Et Endl) Krasser Familia: Fagaceae. In: Donoso C (ed) Especies arbóreas de los bosques templados de Chile y Argentina. Autoecología Marisa Cúneo Ed, Santiago, pp 448–461
González ME, Amoroso MM, Lara A et al (2014) Ecología de disturbios y su influencia en los bosques templados de Argentina y Chile. In: Donoso C, González ME, Lara A (eds) Ecología Forestal, Bases para el Manejo Sustentable y Conservación de los bosques nativos de Chile. Ed. Universidad Austral de Chile, Valdivia, Chile, pp 411–502
Guerrido CM, Villalba R, Rojas F (2014) Documentary and tree-ring evidence for a long-term interval without ice impoundments from Glaciar Perito Moreno, Patagonia, Argentina. The Holocene 24:1686–1693. https://doi.org/10.1177/0959683614551215
Heinemann K, Kitzberger T (2006) Effects of position, understory vegetation and coarse woody debris on tree regeneration in two environmentally contrasting forests of North-Western Patagonia: a manipulative approach. J Biogeogr 33:1357–1367. https://doi.org/10.1111/j.1365-2699.2006.01511.x
Inventario Nacional de Glaciares IANIGLA (2018) Informe de las subcuencas Brazo Sur del Lago Argentino y río Bote. Cuenca del río Santa Cruz. Parque Nacional Los Glaciares. IANIGLA-CONICET, Secretaría de Ambiente y Desarrollo Sustentable de la Nación
Jackson ML (1958) Soil chemical analysis. Prentice-Hall, Inc., Englewood Cliffs
Kalela EK (1941) Über die Holzarten und die durch die klimatischen Verhältnisse verursachten Holzartenwechsel in den Wäldern Ostpatagoniens. Annales Academiae Scientarium Fennicae, Series A 2: 5–151. In: Veblen TT, Donoso C, Kitzberger T, Rebertus A (1996) Ecology of Southern Chilean and Argentinean Nothofagus forest. In: Veblen TT, Hall RS, Read J (eds) The ecology and biogeography of Nothofagus forest. Yale University Press, New Haven/London
Kirk PL (1950) Kjeldahl method for organic nitrogen. Anal Chem 22:354–358. https://doi.org/10.1021/ac60038a038
Kitzberger T, Veblen TT, Villalba R (2000) Métodos dendroecológicos y sus aplicaciones en estudios de dinámica de bosques templados de Sudamérica. In: Roig F (ed) Manual Latinoamericano de Dendrocronología. Editorial de la Universidad Nacional de Cuyo, Mendoza
Martín G (ed) (1997) Plan Preliminar de Manejo Parque Nacional Los Glaciares. Administración de Parques Nacionales, El Calafate, Argentina, pp 174

Martínez-Pastur G, Lencinas M, Peri P (1994) Variación de parámetros estructurales y de composición del sotobosque de *Nothofagus pumilio* en relación a gradientes ambientales indirectos. Ciencias Forestales 9(1–2):11–22

Martínez-Pastur G, Lencinas M, Escobar J et al (2006) Understory succession in *Nothofagus* forest in Tierra del Fuego, (Argentina) affected by Castor canadensis. Appl Veg Sci 9:143–154. https://doi.org/10.1658/1402-2001(2006)9[143:USINFI]2.0.CO;2

Moore DM (1983) Flora of Tierra del Fuego. Anthony Nelson England and Missouri Botanical Garden, New York

Oliver CD (1980) Forest development in North America following major disturbances. For Ecol Manag 3:153–168. https://doi.org/10.1016/0378-1127(80)90013-4

Oliver CD, Larson BC (1996) Forest stand dynamics. Wiley, New York

Olsen SR, Sommers LE (1982) Phosphorus. In: Page AL (ed) Methods of soil analysis part 2 chemical and microbiological properties. American Society of Agronomy, Soil Science Society of America, Madison, pp 403–430

Pérez-Moreau R (1959) Reseña Botánica sobre el Lago Argentino. Instituto Nacional del Hielo Continental Patagónico, Buenos Aires, Argentina, pp 35

Peri PL, Martínez-Pastur G (2003) Denominación de Origen'Madera Fueguina. Publicación Técnica Forestal. Gobierno de Tierra del Fuego – Consejo Federal de Inversiones, Ushuaia, Argentina, pp 29

Picket ST, White PS (1985) The ecology of natural disturbance and patch dynamics. Academic Press, New York

Rebertus AJ, Veblen TT (1993) Structure and tree-fall gap dynamic of old-growth *Nothofagus* forest in Tierra del Fuego, Argentina. J Veg Sci 4:641–654. https://doi.org/10.2307/3236129

Rebertus AJ, Kitzberger T, Veblen TV et al (1997) Blowdown history and landscape patterns in the Andes of Tierra del Fuego, Argentina. Ecology 78:678–692. https://doi.org/10.1890/0012-9658(1997)078[0678:BHALPI]2.0.CO;2

Ripley BD (1981) Spatial statistics. Wiley, New York

Roig FA, Anchorena J, Dollenz O, Faggi AM, Méndez E (1985) Las comunidades vegetales de la Transecta Botánica de la Patagonia Austral. Primera parte: La vegetación del área continental. In: Boelcke O, Moore DM, Roig FA (eds), Transecta Botánica de la Patagonia Austral (pp. 350–456). CONICET, (Argentina), Royal Society (Gran Bretaña) e Instituto de la Patagonia (Chile)

Rovere AE, Premoli AC (2005) Dispersión asimétrica de semillas de *Embothrium coccineum* (Proteaceae) en el bosque templado de Chiloé, Chile. Ecol Austral 15(1):1–7

Rush V (1992) Principales limitantes para la regeneración de la lenga en la zona NE de su área de distribución. Variables ambientales en claros del bosque. In: Bava J, Schmalz J (eds) Actas Seminario de Manejo Forestal de la Lenga y Aspectos Ecológicos Relacionados, vol 8. CIEFAP, Publicación Técnica, Esquel, pp 61–73

Rush V (1993) Altitudinal variation in the phenology of *Nothofagus pumilio* in Argentina. Rev Chil Hist Nat 66:131–141

Schulman E (1956) Dendroclimatic change in semiarid America. University of Arizona Press, Tucson

Schmidt H, Urzua A (1982) Transformación y manejo de los bosques de lenga en Magallanes. Ciencias Agrícolas (Universidad de Chile, Santiago) 11:1–62. In: Arturi MF, Frangi JL, Goya JF (eds) Ecología y Manejo de bosques de Argentina. Editorial Universidad de La Plata, La Plata

Srur AM, Golluscio RA, Villalba R et al (2013) Grazing-induced morphological and growth rate changes in *Anarthrophyllum rigidum*, a Patagonian leguminous shrub. Dendrochronologia 31:223–227. https://doi.org/10.1016/j.dendro.2013.02.002

Srur AM, Villalba R, Rodríguez-Catón M et al (2016) Establishment of *Nothofagus pumilio* at upper treelines across a precipitation gradient in the northern Patagonian Andes. Arct Antarct Alp Res 48:755–766. https://doi.org/10.1657/AAAR0016-015

Srur AM, Villalba R, Rodríguez-Catón M et al (2018) Climate and *Nothofagus pumilio* establishment at upper treelines in the Patagonian Andes. Front Earth Sci 6:57. https://doi.org/10.3389/feart.2018.00057
Stokes MA, Smiley TL (1968) An introduction to tree-ring dating. University of Chicago Press, Chicago
Stuefer M (1999) Investigations on mass balance and dynamics of Moreno Glacier base on field Measurments and satellite imagery. PhD thesis. University of Innsbruck, Innsbruck
Urretavizcaya MF, Contardi L, Oyharçabal MF et al (2016) Calidad de semillas de especies nativas del bosque andino patagónico de la provincia de Chubut y su importancia para la producción de plantines. Rev Fac Agron 115(1):9–18
Veblen TT, Ashton DH, Schlegel FM et al (1977) Plant succession in a timberline depressed by volcanism in south-Central Chile. J Biogeogr 1:275–294. https://doi.org/10.2307/3038061
Veblen TT, Ashton DH (1978) Catastrophic influences on the vegetation of the Valdivian Andes, Chile. Vegetatio 36:149–167. https://doi.org/10.1007/BF02342598
Veblen TT, Donoso C, Schlegel ZFM et al (1981) Forest dynamics in southern-central Chile. J Biogeogr 8:211–247
Veblen TT, Donoso C, Kitzberger T et al (1996) Ecology of Southern Chilean and Argentinean *Nothofagus* forest. In: Veblen TT, Hall RS, Read J (eds) The ecology and biogeography of *Nothofagus* forest. Yale University Press, New Haven/London
Veblen TT, Kitzberger T, Villalba R (2005) Nuevos paradigmas en ecología y su influencia sobre el conocimiento de la dinámica de los bosques del sur de Argentina y Chile. In: Arturi MF, Frangi JL, Goya JF (eds) Ecología y Manejo de bosques de Argentina. Editorial Universidad de La Plata, La Plata
Wiegand T, Moloney KA (2004) Rings, circles, and null-models for point pattern analysis in ecology. Oikos 104:209–229. https://doi.org/10.1111/j.0030-1299.2004.12497.x
Yamaguchi DK (1991) A simple method for cross dating increment cores from living trees. Can J For Res 21:414–146. https://doi.org/10.1139/x91-053
Zar JH (1999) Biostatistical analysis, 4th edn. Prentice Hall, Upper Saddle River

Chapter 11
A Dendro-Spatial Analysis in Tree Growth Provides Insights into Forest Productivity

Marín Pompa-García, José Manuel Zúñiga-Vásquez, and Eduardo Treviño-Garza

Abstract Climatic disturbance is an increasingly important factor that modifies forest structure and species composition. These modifications strongly affect competition among trees, size and distribution variability, growth and mortality. This leads to declining stocks with serious implications in the provision of various ecosystem services, particularly in natural protected forests. This study combines dendroecological data for interpreting the spatial variability existing in growth of *Picea chihuahuana*, *Abies durangensis* and *Cupressus lusitanica* of northern Mexico along a period of 54 years (from 1960 to 2014). The research was carried out through a retrospective analysis of ring width and basal area increment (BAI) using Moran's I index. We tested whether tree growth responses were spatially autocorrelated and used this as a proxy to evaluate if BAI values are declining under adverse climatic conditions. Such conditions may be affected by temperature, evaporation and drought. The results revealed non-random growth patterns in all three species, with high competition in terms of BAI. The values of BAI showed some variations in productivity with two opposing trends in *Picea chihuahuana* and *Abies durangensis*. Both species exhibited signs of a decline attributable, presumably, to drought, while *Cupressus lusitanica* was less sensitive to this factor. Spatial autocorrelation analysis, along with dendrochronological data, represents a valuable tool to study the productivity of forests.

Keywords Climatic disturbance · Forest structure · Tree-rings · Tree growth · Moran's I index

---

M. Pompa-García · J. M. Zúñiga-Vásquez (✉)
Facultad de Ciencias Forestales, Universidad Juárez del Estado de Durango, Durango, Mexico
e-mail: josema.zuva@gmail.com

E. Treviño-Garza
Facultad de Ciencias Forestales, Universidad Autónoma de Nuevo León, Nuevo León, Mexico

M. Pompa-García, J. J. Camarero (eds.), *Latin American Dendroecology*,
https://doi.org/10.1007/978-3-030-36930-9_11

11.1 Introduction

The rate of loss of global biodiversity is accelerating (Camarero et al. 2012) and could lead to the further extinction of species in the near future. It has been predicted that drought will cause forest decline (Sturrock et al. 2011) and a natural megaclimatic disturbance (Sánchez-Salguero et al. 2011). Such climatic disturbance constitutes an intrinsic element of the dynamics of forest ecosystems (González et al. 2011) and has serious implications for species diversity (Bellard et al. 2012), structural complexity (Sturrock et al. 2011) and ecological successional routes (Rocha-Gil et al. 2017).

Climate-induced changes are even more important in protected areas, which are excluded from timber production (Araújo et al. 2011) and focused on maintaining or increasing biodiversity (Laurance et al. 2012). In northern Mexico, some relict forests of high value for conservation have been segregated from traditional forest use (e.g., timber) (Aguirre et al. 2003) to prioritize their conservation. The presence of threatened species, such as *Picea chihuahuana* Martinez, *Abies durangensis* Martinez and *Cupressus lusitanica* Mill. is an essential ecological attribute of interest to ecologists and forest managers alike. For this reason, these forest stands are an ideal site to determine the implications of disturbance on the dynamics of these habitats.

Firstly to interpret those implications of disturbance, it is essential to differentiate between its intensity (e.g., magnitude) and severity (e.g., tree damage), which are in turn, species-dependent factors (Cohen et al. 2016). Each stand represents a unique expression of the interactions that occur among environmental conditions and plant species. The stand provides a snapshot of the impact of a combination of biotic and abiotic factors, including the effects of human expansion. These factors are governed by their temporal occurrence (Prates et al. 2015). Therefore, it is necessary to characterize their dynamics using a temporal analysis that quantifies the different responses of a species to disturbance factors through time (e.g., drought) (Fortin et al. 2012). One such analysis consists of using basal area increment (BAI) as an indicator of productivity. The BAI is a great indicator of wood production due to the ever-increasing diameter of a growing tree (Biondi and Qeadan 2008). It reflects changes in the conductive area of the stem better than tree-ring width and allows a comparison of productivity among tree species (Martin-Benito et al. 2011; Mohammadi et al. 2017). By using tree-rings data, it is possible to reconstruct BAI values, providing further information of historical climate processes at specific location (Biondi and Qeadan 2008).

Another important aspect to consider is the spatial pattern of tree growth, since the growth of the individual tree is subject to complex, interacting influences, notably competition and micro-site variability (Fox et al. 2007). For this reason a statistical examination of spatio-temporal trends of tree growth is necessary, this could provide a more complete understanding of ecological phenomena over time (Fritts and Shao 1992). With these trends, it is possible to examine whether the tree growth is influenced by the competition from its neighbors and identify signs of temporal decline for any given year. In addition, environmental variables that induce competition (e.g., drought) can be explored.

Based on the above, the spatial analysis through spatial autocorrelation is an appropriate indicator of tree competition (Shi and Zhang 2003). Some authors suggest that when two trees are growing next to each other, competition will tend to create a negative spatial autocorrelation among neighboring trees (Yoda et al. 1957; Fox et al. 2001; Fajardo and McIntire 2007).

Some of the methods for measuring spatial autocorrelation include aggregation and segregation indices (Brown and Chung 2006), Gibbs point process (Mateu et al. 1998), analysis of the nearest neighbor (Kenkel 1988) and the Ripley's K function (Ripley 1979). The purpose of these methods is to determine if the geographic distribution of the data is random, regular or aggregated (grouped), however, only analyze the geographical data (e.g., location of trees) and do not distinguish the data points by their attributes (e.g., diameter, height, basal area) (Shi and Zhang 2003). When an attribute variable is involved in the analysis of the point pattern, the stochastic process is known as a marked points process, in which the "points" are tree positions and the "marks" are tree characteristics (e.g., tree growth) (Baddeley and Turner 2005).

In this sense the Moran's I index is flexible in marked point process and helps to investigate the effects of neighboring trees of different sizes as well as the effects of separation distances on the subject tree's performance (Shi and Zhang 2003).

In Mexico, remarkable efforts have been made to explain the climatic vulnerability of tree species in some conservation status (Pompa-García et al. 2017a, b; Cabral-Alemán et al. 2017; González-Cásares et al. 2018). However, and despite of that, there is a deficit of dendro-spatial approaches to characterize geographical synchrony patterns in tree ring records (Shestakova et al. 2018). The combination of dendroecological data, is an efficient tool to assess the disturbance regime and species sensitivity to environmental change (Carrer and Urbinati 2001). In this work, we propose a methodology for interpreting the variability and common signal strength existing in tree-ring based on Moran's I index principles (Moran 1950). This methodology could be suitable to test the presence of contrasting tree-ring patterns compared to classic chronologies, regarded as the classic approach. So, the present study is the first of its kind in Mexican forests.

The main objective was to analyze the degree to which tree growth responses are spatially autocorrelated in three tree species representatives of a protected area in northern Mexico. We used annual tree growth (i.e., BAI) as a proxy to determine whether *Picea chihuahuana* Martinez, *Abies durangensis*Martinez and *Cupressus lusitanica* Mill. follow spatial synchrony to drought as a disturbance factor. We hypothesized that a negative spatial autocorrelation could reveal non-random patterns in radial growth that would be significantly attributed to adverse climatic conditions.

11.2 Materials and Methods

11.2.1 Study Area

The Sierra Madre Occidental represents one of the most important pine-oak reserves of Mexico and is of considerable value for both commercial timber production and forest conservation (Aguirre-Díaz and Labarthe-Hernández 2003). This region has a temperate, subhumid climate with a cool humid summer because of the monsoon season. Spring and winter are both characteristically dry seasons. The maximum monthly values of evaporation are seen in April and May (Pompa-García et al. 2013). The complexity of its physiographic and climatic characteristics has produced a range of ecosystems dominated by pine and oak forests (González-Elizondo et al. 2012). The soils in the study area include mainly Cambisols, which are characterized by the absence of a layer of accumulated clay, humus and soluble salts (Aguirre-Díaz and Labarthe-Hernández 2003).

The study area preserves some relict forests, where four representative tree species are found: *Picea chihuahuana* Martínez, *Abies durangensis* Martínez, *Cupressus lusitanica* Mill. and *Pseudotsuga menziesii* Mirb. Franco. The study area is located in a site known as Santa Bárbara, within the Ejido El Brillante, in the municipality of Pueblo Nuevo, Durango (Fig. 11.1a). Although the area has been protected since 2011, the forest management plans indicate that timber production has been prohibited for the last 30 years as part of a conservation strategy to promote environmental benefits and biodiversity. This protection status was the main reason for which we conducted the study because the species have not been disturbed for a while. Unlike other commercial trees such as pines and oaks, which are the main source of timber harvesting, tree growth patterns would be the results of natural factors (Fig. 11.1b).

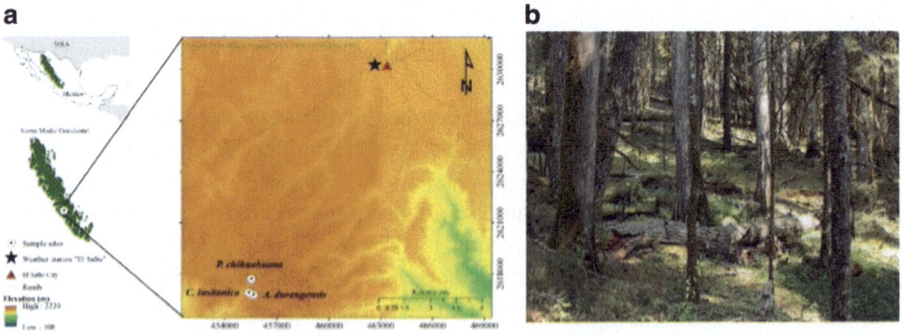

Fig. 11.1 (**a**) Location of the study area in northern Mexico and (**b**) view of sampled stands

11.2.2 Study Species and Data Processing

Since the species under study are naturally grouped by their ecological characteristics (Aguirre-Díaz and Labarthe-Hernández 2003; González-Elizondo et al. 2012), three sites were chosen at random with 15 neighboring trees of the following species: *A. durangensis*, *C. lusitanica* and *P. chihuahuana* (Table 11.1). Given the protection status of the area, an official permit was required to collect the unit samples. This sampling strategy strengthened the temporal approach of growth-climate relationships and its comparability between species (Nehrbass-Ahles et al. 2014). Two radial cores were extracted from diameter breast height (DBH) from each tree using a Pressler increment borer (diameter = 5.1 mm). The wood samples were air-dried, glued onto wooden supports, and polished with a fine sandpaper to improve tree ring identification and cross-dating. The ring widths were measured at 0.01 mm resolution using the Velmex device and dated rigorously from the last ring to the core.

11.2.3 Annual Reconstruction from BAI

To understand the temporal variability of the radial growth of the species, the DBH from each tree was reconstructed based on the ring widths for the 1960–2014 period, which was common for all tree species. The basal area (BA) from each tree was calculated using the following expression:

$$\mathrm{BA} = \pi * r^2 \tag{11.1}$$

where BA is the basal area, π is a constant value equivalent to 3.1416 and r^2 is the distance to the outermost growth ring from the center of the tree. Then, the BAI was calculated as follows (Pompa-García et al. 2017b):

Table 11.1 Characteristics of the sample sites/tree species for the 1960–2014 period

Characteristic	Variable	*A. durangensis*	*P. chihuahuana*	*C. lusitanica*
Tree species	Number of trees	15	15	15
	DBH (cm)	36.3 ± 2.1	40.6 ± 3.03	31.5 ± 1.9
	Height (m)	18.1 ± 0.9	21.9 ± 0.8	16.4 ± 0.7
	BAI (cm^2)	11.2 ± 2.8	13 ± 3.1	6 ± 2.5
	Tree-ring width (mm)	2.05 ± 0.21	2.28 ± 0.26	1.35 ± 0.12
Sample sites	Density (No. trees/ha)	298	433	534
	Elevation (m)	2692	2686	2664
	Slope	18%	12%	19%
	Aspect	North	South	North

Values are means ± SD

$$BAI_i = \mathrm{BA}_n - \mathrm{BA}_{n-1} \tag{11.2}$$

where BAI_i is the increment in basal area of species *i*, BA_n and BA_{n-1} are the basal area values in years *n* and $n - 1$.The BAI can be used in forest modeling to provide a precise quantification of timber production and compare the productivity between species (Biondi and Qeadan 2008).

11.2.4 Spatial Autocorrelation Analysis

Since the spatial pattern of tree growth varies over time, the annual spatial autocorrelation of the BAI for the period 1960–2014 was estimated. Spatial autocorrelation represents the correlation between the value of the BAI at a location *i* and the BAI values at neighboring locations *j* (i.e., measures the similarity or dissimilarity between a BAI value and its neighbors) (Rossi et al. 1992).

Moran's I index (hereafter MI) (Moran 1950) was used to estimate the spatial autocorrelation for BAI of trees of each species, using the software Geoda (Anselin 2004). The value of MI varies from +1 to −1 although some authors have stated that both limits can be exceeded (Cliff and Ord 1981; Upton and Fingleton 1985; Chen 2013). The null hypothesis for spatial autocorrelation assumes complete spatial randomness. The former value indicates a perfect positive correlation while the latter indicates a perfect negative correlation. Zero values represent a completely random spatial pattern. The MI index is estimated as:

$$MI = \frac{N\sum_{i=1}^{N}\sum_{j=1}^{N} W_{ij}\left(BAI_i - \overline{BAI}\right)}{\sum_{i=1}^{N}\sum_{j=1}^{N} W_{ij}\sum_{i=1}^{N}\left(BAI_i - \overline{BAI}\right)^2} \tag{11.3}$$

$$W_{ij} = \begin{bmatrix} 1\,if\ d_{ij} \le d \\ 0\,if\ d_{ij} > d \end{bmatrix} \tag{11.4}$$

where W_{ij} is the binary matrix of contiguity, such that $W_{ij} = 1$ if sites *i* and *j* share a common border, and $W_{ij} = 0$ where this is not the case, and *n* is the number of trees.

Following Elobaid et al. (2009), the vicinity among trees was determined using the *Queen* criterion to estimate proximity of contact in all cardinal directions. In order to contrast the hypothesis of absence of a spatial pattern, MI was placed within a normal curve of probabilities *Z* and the randomness of the spatial distribution of the values among the possible distributions was tested (Vilalta 2005).

11.2.5 BAI Correlation and Climate Relationship

Pearson correlation coefficients were calculated to determine any relationship between MI and climate data. We associated the annual MI (1960–2014) of each species with the monthly climatic data and drought, including maximum temperature (°C), precipitation (mm), evaporation (mm) and, as a proxy for drought, the Standardized Precipitation and Evaporation Index (SPEI). The latter considers the effect of temperature on the evapotranspiration rate where negative SPEI values reflect dry conditions while positive values represent wet conditions (Vicente-Serrano et al. 2010). The maximum temperature and evaporation data were obtained from "El Salto" climatological station (located at 14 km from the study area). The SPEI was calculated using the SPEI library of R software, version 3.5.3 (R Development Core Team 2019), with time scales from 1 to 6 months (i.e., accumulated drought) utilizing precipitation and temperature data (Beguería and Vicente-Serrano 2013). Because tree growth depends on previous climatic conditions, the analysis considered the period prior to growth from January ($t - 1$) to July of the current year (t).

11.3 Results

11.3.1 BAI Dynamics for the Species

Results indicated significant BAI changes along the period of study. As expected, the species grew differently in terms of their genetics and competition with neighboring trees. The average BAI values for *Abies durangensis* and *Cupressus lusitanica* showed a positive trend in the study period (Fig. 11.2a, b, respectively). In contrast, *Picea chihuahuana* showed a negative tendency of average BAI and a remarkable reduction in its growth over time (Fig. 11.2c). Regarding the spatial autocorrelation (MI), all species had negative values (i.e., ungrouped spatial distribution). However, the trends of these values were different among species. Two patterns can be seen from the MI values: the first indicates a decreasing negative spatial autocorrelation for *P. chihuahuana* and *A. durangensis* (Fig. 11.2a, c), while the second an increasing, but still negative, spatial correlation for *C. lusitanica*, which suggests a tendency to random or clustered distribution (Fig. 11.2b).

11.3.2 Spatial Correlation of BAI and Its Association with Climate Data

Significant correlations between the MI values and monthly climatic data ($p \leq 0.05$) are shown in Fig. 11.3. Both *A. durangensis* and *P. chihuahuana* showed negative correlations between maximum temperature and MI (Fig. 11.3a). The negative correlation between MI and maximum temperature of *A. durangensis* was statistically significant in July, prior to the growth season ($t - 1$), and March of the current season (t). Negative correlations between MI and maximum temperature of *P. chihuahuana* were also significant in the months of March, April, October and November of the prior year as well as for March of the current season. This suggests that high temperatures appear to be associated with MI negative values, which, in turn, may

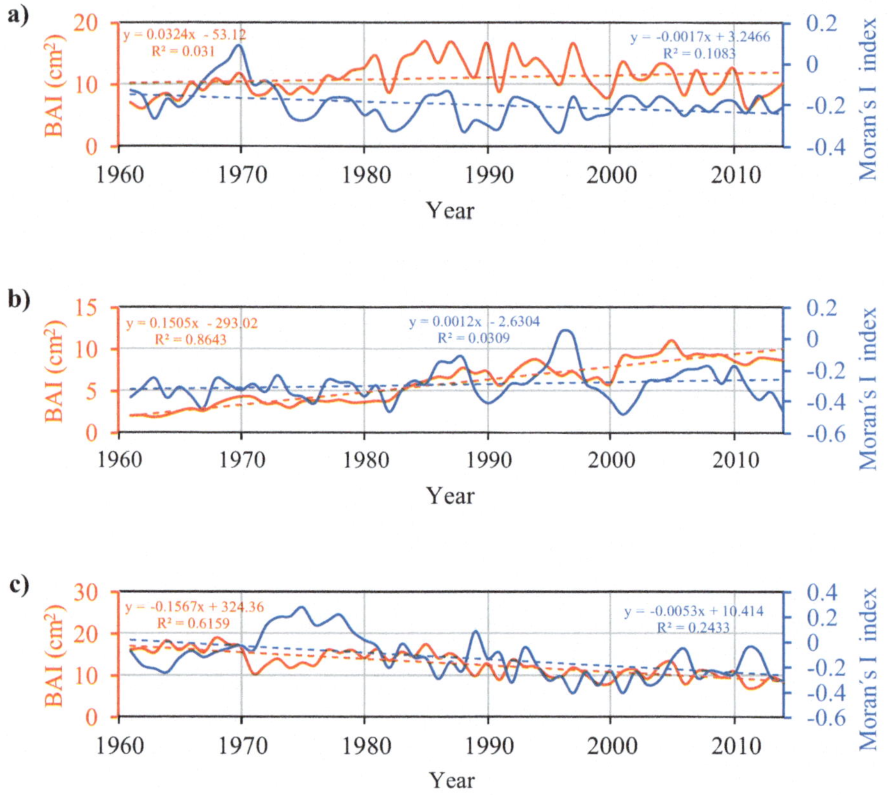

Fig. 11.2 Evolution of the Moran's I index (blue) and the average data of BAI (red) for (**a**) *A. durangensis*, (**b**) *C. lusitanica* and (**c**) *P. chihuahuana*

suggest more competition between individuals. *C. lusitanica* showed no association at all.

In terms of evaporation (Fig. 11.3b), *A. durangensis* had a negative relationship with the MI, specifically in May, August, September, October, November and December. The correlation values for *P. chihuahuana* were negative in December of the previous year and in April of the current year. This indicates that evaporation in April benefits the growth of *P. chihuahuana*, while the December's evaporation in the previous year delays it.

Cupressus lusitanica had a positive correlation value for May and negative correlation values for September of the previous year. Apparently, the evaporation rates in the month of May boosted the growth of *C. lusitanica* as it has a comparative advantage over the other species because it thrives in habitats close to bodies of water.

The coefficients of correlations between MI and the SPEI drought index for *A. durangensis*, *C. lusitanica* and *P. chihuahuana* are shown in Fig. 11.4. *A. durangensis* and *P. chihuahuana* showed a positive association with the accumulated

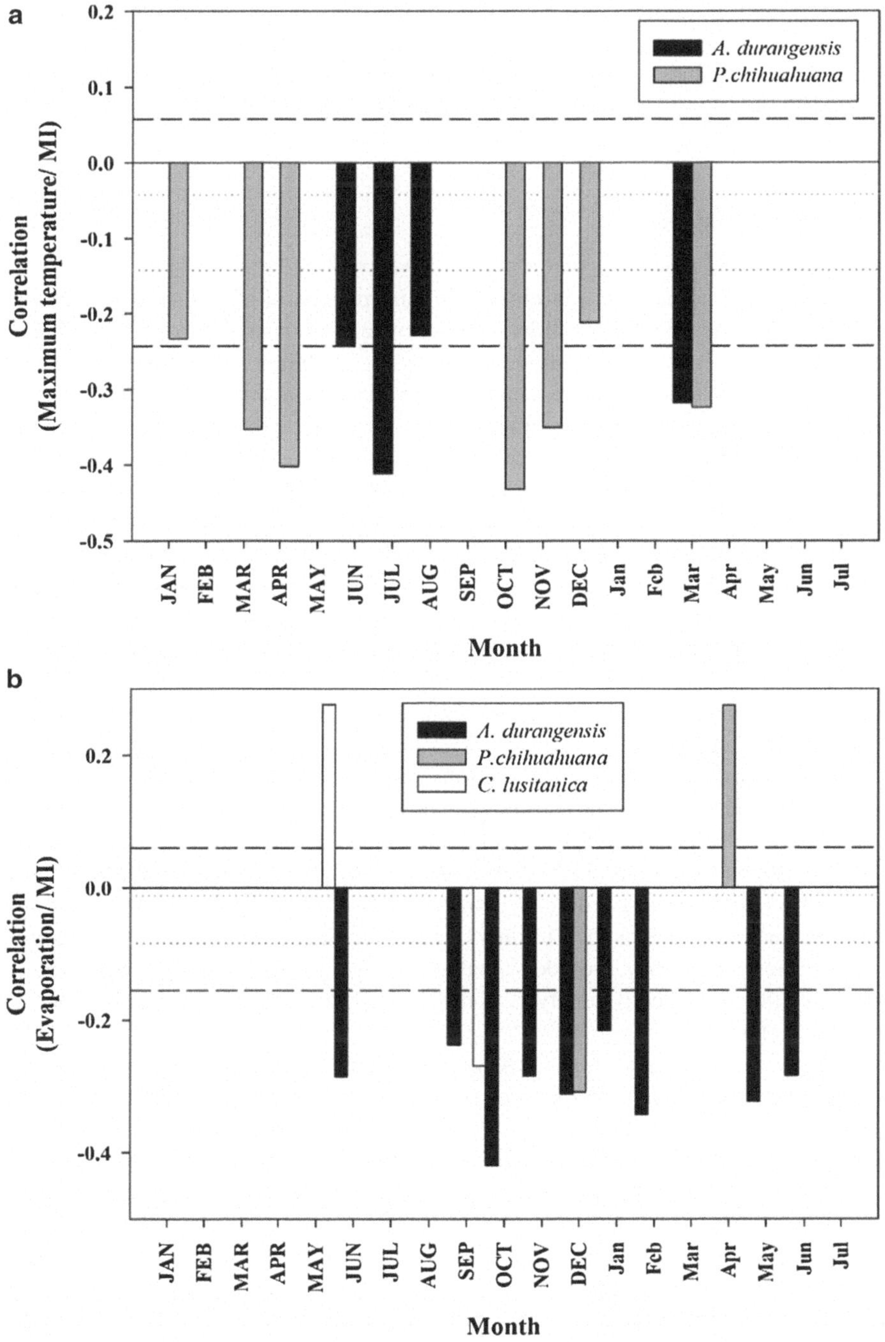

Fig. 11.3 Correlations coefficients between MI from *A. durangensis*, *C. lusitanica* and *P. chihuahuana* and climatic data. The correlations are shown for (**a**) monthly maximum temperature and (**b**) monthly evaporation. Months shown in upper case letters indicate the period prior to the growth season ($t - 1$), while those in lower case letters correspond to current growth periods (t). The dotted lines represent the limit of statistically significant correlation values at $p \leq 0.05$

drought. The high values of positive correlation (in red color) are highlighted, which indicate the influence of the temporal drought on spatial growth pattern (competition). The high values of negative correlation in *C. lusitanica* indicate that the spatial growth pattern of this species is not affected by drought.

11.4 Discussion

The MI results indicated a non-random spatial pattern of tree growth for all tree species. Thus, we reject the null hypothesis of random spatial autocorrelation. All species presented values of negative spatial autocorrelation, suggesting ungrouped spatial pattern. If there is positive spatial autocorrelation amongst the BAI of trees this would indicates that they are more or less alike but on the other hand a negative spatial autocorrelation would indicate that they are unlike (Elobaid et al. 2009). The negative effects of neighbor trees on a subject tree resulting from their consumption, or regulation of access to, limited resources such as light, water and nutrients (Keddy 1989), this results in a heterogeneous distribution of tree growth (i.e., trees tend to have neighbors with different BAI). Based on the above these results suggest that all tree species have grown under conditions of competition over time.

This finding coincide with other authors that found that, when two trees are growing next to each other, competition will tend to create a negative spatial autocorrelation among neighboring trees (Yoda et al. 1957; Fox et al. 2001; Shi and Zhang 2003; Fajardo and McIntire 2007).

The use of dendrochronological datasets combined with the MI, allowed to characterize the spatial nature of disturbances (e.g., drought) influencing tree competition. This result helps improve our understanding of the changes that have taken place at the stand and individual levels due to the effects of drought and competition (Figs. 11.3 and 11.4). According to Schneuwly-Bollschweiler et al. (2013), the use of spatial indexes (such as the Moran's I index) facilitates the quantitative construction of time series. They provide valuable data regarding the spatial clustering of trees and allow an objective analysis of the spatial clustering of affected trees during specific events.

The spatial growth patterns of BAI obtained here suggest in almost all cases that competition occurred among the trees (Fig. 11.2). These results are consistent with those reported by Elobaid et al. (2009), who analyzed the spatial autocorrelation of average tree diameters in a Malaysian forest and found a similar trend over a period of 11 years. However, our approach of basal area reconstruction allowed a wider period of analysis. The inclusion of BAI and its spatial autocorrelation provide a more detailed analysis because it allows the identification of temporal factors as possible causes of disturbance.

While different factors are associated with the growth rates of trees over time (e.g., species, size, age, and environmental factors) (Gea-Izquierdo et al. 2011; Williams et al. 2013; Stoffel and Corona 2014; Bar-Hen et al. 2015), our results suggest that the specific response to drought appears to be an important factor that

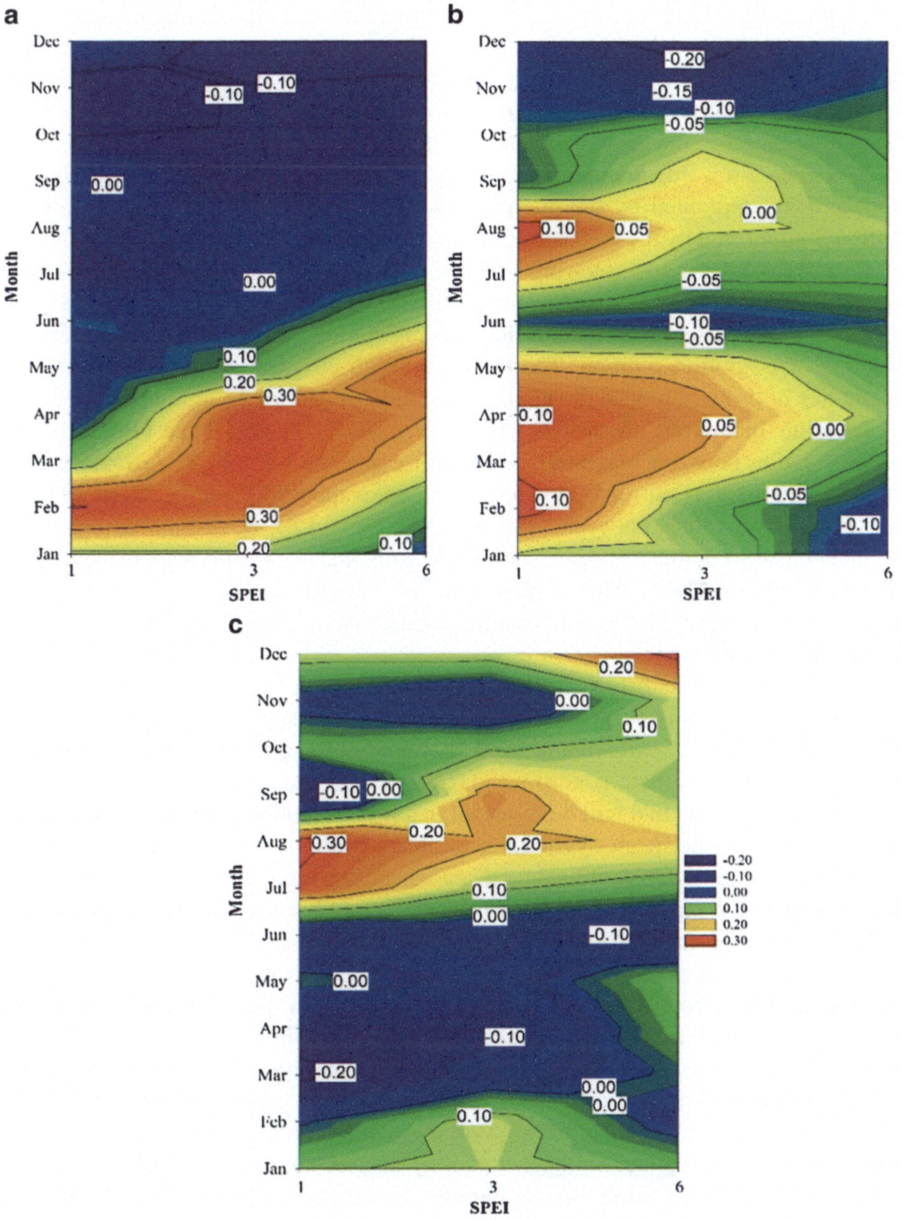

Fig. 11.4 Pearson correlation coefficients between SPEI drought index and MI for (**a**) *A. durangensis*, (**b**) *C. lusitanica* and (**c**) *P. chihuahuana*. The SPEI was calculated for accumulated drought using lags of 1, 3 and 6 months (*x*-axis) from January–December (*y*-axis) of the growth year ($p \leq 0.05$)

determines the differences in susceptibility to decline. This is in agreement with Nelson et al. (2011), who found that historical drought episodes are associated with small annual growth. Climate records from the study area (1960–2014) indicate that in 1971, 1991 and 2011, extreme drought occurred. Thus, drought seems to be a regulating factor in the growth dynamic of the species studied.

Tendencies towards decline (Fig. 11.2) in *A. durangensis* and *P. chihuahuana* appear to be explained by the intensity and severity of drought (Fig. 11.4). This notion is supported by the results of Cabral-Alemán et al. (2017) and Pompa-García et al. (2017a), who used a tree-ring analysis. They found that high temperatures and accumulated periods of droughts caused stress limiting the growth of these tree species. Warm winters usually extend the growing season, but at the same time affect the rate of evapotranspiration. Therefore, the high costs of respiration due to the high rates of drought reduce the net accumulation of carbon, bringing these species to a decline. In contrast, the results from *C. lusitanica* (Figs. 11.3 and 11.4) indicate no correlation between maximum temperature and the MI. For *C. lusitanica*, the study found positive correlation values between evaporation and the MI as well as low correlation values between SPEI and MI (Fig. 11.4). *C. lusitanica* thrives in habitats close to bodies of water and humid soils and with little responsiveness to hydroclimatic variability. Therefore, despite the increase in vapor pressure, the trees do not experience problems of hydraulic conductivity and the long-term photosynthesis rate is maintained.

Given that climatic disturbances (e.g. drought) are factors governing the species and structural diversity (Cohen et al. 2016), our results suggest that at least two of the species studied (*P. chihuahuana* and *A. durangensis*) could reduce their productivity due to drier climate conditions. That is, stress caused by drought implies a high metabolic effort that will be reflected in a loss of biodiversity in the long-term. This would leave the species vulnerable to attack from insects and other pathogens, as well as an increased risk of fire and diseases, among other disturbances. However, such a dynamic condition should not be seen as an obstacle for sustainability, but rather as an opportunity to consider the natural processes of the ecosystem into conservation strategies. While these disturbances have been explained in other countries, in Mexico this kind of knowledge is limited in spite of its high diversity.

Our results are based on observational tree-ring data alone, which may be insufficient to establish cause and effect relationships. Consequently, it is recommended to broaden the base of physical and dendrometric data in order to support our results. Czaplewski et al. (1994) state that abnormal growth reductions are sometimes associated with forest decline, although this may also be caused by a variety of other natural factors. There are numerous hypotheses for the causes of variability forest growth. These combined factors drive the spatial distribution of changes in tree radial-growth. Spatio-temporal dynamics of tree growth provided evidence that our hypotheses was consistent. However, the results of our spatial analysis might not often get categorical conclusions. Yet, some challenges remain that deserve special attention before the analysis of synchrony patterns in tree-ring records may reach its full potential (Czaplewski et al. 1994).

The indirect nature of the effect allows choices to be made from a wide range of management options: selection of better species adapted to warm dry climates, assisted migration of species, and adapted colonization (Sáenz-Romero et al. 2016). It has been stated that disturbances impede the forest from reaching a final equilibrium by killing the trees and providing spaces and habitats for new individuals (Burgeon et al. 2015). However, we should note that many forest ecosystem dynamics depend on the presence of disturbances, and a final equilibrium or climax is almost never reached. Further research should be developed to better understand the ecological mechanisms that these species face.

11.5 Conclusion

This study demonstrated that spatial autocorrelation analysis, together with dendrochronological data, represents a valuable tool to study the dynamics of forests. The correlation of radial growth between the current and past year of the species did not follow a random pattern, suggesting the occurrence of strong competition among individuals. Furthermore, the magnitude of the autocorrelation is species dependent. Analyses of tree growth patterns indicate that *P. chihuahuana* and *A. durangensis* are declining, while *C. lusitanica* is increasing. Drought, as measured in terms of the SPEI, is suggested to be the cause of the observed decline and driver of tree competition in the first two species. More research of other mechanisms that affect variations in the temporal productivity of these protected forests is necessary.

Acknowledgments Funding was provided by CONACYT (Consejo Nacional de Ciencia y Tecnología, CB-2013/222522 project). We thank the Dirección General de Vida Silvestre, SEMARNAT (Secretaría de Medio Ambiente y Recursos Naturales, Mexico) for facilitating field sampling. We acknowledge the comments given by anonymous reviewers. Gustavo Pérez-Verdín and Citlalli Cabral-Alemán helped by commenting on a previous version of this manuscript.

References

Aguirre O, Hui G, von Gadow K et al (2003) An analysis of spatial forest structure using neighbourhood-based variables. For Ecol Manag 183:137–145. https://doi.org/10.1016/S0378-1127(03)00102-6

Aguirre-Díaz GJ, Labarthe-Hernández G (2003) Fissure ignimbrites: fissure source origin for voluminous ignimbrites of the Sierra Madre Occidental and its relationship with Basin and Range faulting. Geology 31:773–776. https://doi.org/10.1130/G19665

Anselin L (2004) GeoDa 0.95i Release Notes. Spatial Analysis Laboratory (SAL). Department of Agricultural and Consumer Economics, University of Illinois, Urbana-Champaign

Araújo MB, Alagador D, Cabeza M et al (2011) Climate change threatens European conservation areas. Ecol Lett 14:484–492. https://doi.org/10.1111/j.1461-0248.2011.01610.x

Baddeley A, Turner R (2005) spatstat: an R package for analyzing spatial point patterns. J Stat Softw 12:1–42

Bar-Hen A, Emily M, Picard N (2015) Spatial cluster detection using nearest neighbor distance. Spat Stat 14:400–411. https://doi.org/10.1016/j.spasta.2015.07.006

Beguería S, Vicente-Serrano SM (2013) SPEI: calculation of the standardized precipitation-evapotranspiration index, R Package Version 1.6. http://CRAN.R-project.org/package=SPEI. Accessed on 13 March 2018

Bellard C, Bertelsmeier C, Leadley P et al (2012) Impacts of climate change on the future of biodiversity. Ecol Lett 15:365–377. https://doi.org/10.1111/j.1461-0248.2011.01736.x

Biondi F, Qeadan F (2008) A theory-driven approach to tree-ring standardization: de-fining the biological trend from expected basal area increment. Tree-Ring Res 64:91–96

Brown LA, Chung CY (2006) Spatial segregation, segregation indices and the geographical perspective. Popul Space Place 12:125–143. https://doi.org/10.1002/psp.403

Burgeon D, Hofer T, Van Lierop P et al (2015) Árboles y bosques, salvavidas que aseguran la resiliencia. Unasylva 243:86–89

Cabral-Alemán C, Pompa-García M, Acosta-Hernández AC et al (2017) Earlywood and latewood widths of Picea chihuahuana show contrasting sensitivity to seasonal climate. Forests 8:173. https://doi.org/10.3390/f8050173

Camarero JJ, Barreda GS, Alla AQ et al (2012) Los precedentes y las respuestas de los árboles a sequías extremas revelan los procesos involucrados en el decaimiento de bosques mediterráneos de coníferas. Ecosistemas 21:22–30. https://doi.org/10.7818/ECOS.2012.21-3.04

Carrer M, Urbinati C (2001) Spatial analysis of structural and tree-ring related parameters in a timberline forest in the Italian Alps. J Veg Sci 12:643–652

Chen Y (2013) New approaches for calculating Moran's index of spatial autocorrelation. PLoS One 8:e68336. https://doi.org/10.1371/journal.pone.0068336

Cliff AD, Ord JK (1981) Spatial processes: models and applications, 1st edn. Taylor & Francis Pion, London

Cohen WB, Yang Z, Stehman SV et al (2016) Forest disturbance across the conterminous United States from 1985–2012: the emerging dominance of forest decline. For Ecol Manag 360:242–252. https://doi.org/10.1016/j.foreco.2015.10.042

Czaplewski RL, Reich RM, Bechtold WA (1994) Spatial autocorrelation in growth of undisturbed natural pine stands across Georgia. For Sci 40:314–328

Elobaid RF, Shitan M, Ibrahim NA et al (2009) Evolution of spatial correlation of mean diameter: a case study of trees in natural dipterocarp forest. J Math Stat 5:267–269. https://doi.org/10.3844/jmssp.2009.267.269

Fajardo A, McIntire EJB (2007) Distinguishing microsite and competition processes in tree growth dynamics: an a priori spatial modeling approach. Am Nat 169:647–661

Fortin MJ, James PM, MacKenzie A et al (2012) Spatial statistics, spatial regression, and graph theory in ecology. Spat Stat 1:100–109. https://doi.org/10.1016/j.spasta.2012.02.004

Fox JC, Ades PK, Bi H (2001) Stochastic structure and individual-tree growth models. For Ecol Manag 154:261–276

Fox JC, Bi H, Ades PK (2007) Spatial dependence and individual-tree growth models: I. Characterising spatial dependence. For Ecol Manag 245:10–19

Fritts HC, Shao XM (1992) Mapping climate using tree-rings from western North America. In: Bradley RS, Jones PD (eds) Climate since A.D. 1500. Routledge Press, London, pp 269–295

Gea-Izquierdo G, Cherubini P, Cañellas I (2011) Tree-rings reflect the impact of climate change on *Quercus ilex* L. along a temperature gradient in Spain over the last 100 years. For Ecol Manag 262:1807–1816. https://doi.org/10.1016/j.foreco.2011.07.025

González WY, Llambí LD, Smith JK et al (2011) Dinámica sucesional del componente arbóreo en la zona de transición bosque-Páramo en Los Andes Tropicales. Ecotropicos 24:60–79

González-Cásares M, Pompa-García M, Venegas-González A (2018) Climate signals from intra-annual wood density fluctuations in *Abies durangensis*. IAWA J 1(aop):1–12

González-Elizondo MS, González-Elizondo M, Tena-Flores JA et al (2012) Vegetación de la sierra madre occidental, México: Una síntesis. Acta Botánica Mexicana 100:351–403

Keddy PA (1989) Competition. Chapman and Hall, London

Kenkel NC (1988) Pattern of self-thinning in jack pine: testing the random mortality hypothesis. Ecology 69:1017–1024

Laurance WF, Useche DC, Rendeiro J et al (2012) Averting biodiversity collapse in tropical forest protected areas. Nature 489:290–294. https://doi.org/10.1038/nature11318

Martin-Benito D, Kint V, Del Rio M et al (2011) Growth responses of West-Mediterranean *Pinus nigra* to climate change are modulated by competition and productivity: past trends and future perspectives. For Ecol Manag 262:1030–1040. https://doi.org/10.1016/j.foreco.2011.05.038

Mateu J, Uso JL, Montes F (1998) The spatial pattern of a forest ecosystem. Ecol Model 108:163–174

Mohammadi Z, Limaei SM, Lohmander P et al (2017) Estimation of a basal area growth model for individual trees in uneven-aged Caspian mixed species forests. J For Res 29:1–10. https://doi.org/10.1007/s11676-017-0556-7

Moran PA (1950) Notes on continuous stochastic phenomena. Biometrika 37:17–23

Nehrbass-Ahles C, Babst F, Klesse S et al (2014) The influence of sampling design on tree-ring-based quantification of forest growth. Glob Chang Biol 20:2867–2885. https://doi.org/10.1111/gcb.12599

Nelson TA, Laroque CP, Smith DJ (2011) Detecting spatial connections within a dendrochronological network on Vancouver Island, British Columbia. Dendrochronologia 29:49–54. https://doi.org/10.1016/j.dendro.2010.08.002

Pompa-García M, Cerano-Paredes J, Fulé PZ (2013) Variation in radial growth of *Pinus cooperi* in response to climatic signals across an elevational gradient. Dendrochronologia 31:198–204. https://doi.org/10.1016/j.dendro.2013.05.003

Pompa-García M, González-Cásares M, Acosta-Hernández AC et al (2017a) Drought influence over radial growth of Mexican conifers inhabiting mesic and xeric sites. Forests 8:175. https://doi.org/10.3390/f8050175

Pompa-García M, Sánchez-Salguero R, Camarero JJ (2017b) Observed and projected impacts of climate on radial growth of three endangered conifers in northern Mexico indicate high vulnerability of drought-sensitive species from mesic habitats. Dendrochronologia 45:145–155

Prates MO, Dey DK, Willig MR et al (2015) Transformed Gaussian Markov random fields and spatial modeling of species abundance. Spat Stat 14:382–399. https://doi.org/10.1016/j.spasta.2015.07.004

R Development Core Team (2019) R: a language and environment for statistical computing. R Foundation for Statistical Computing, Vienna. Available online https://www.r-project.org/

Ripley BD (1979) Tests of 'randomness' for spatial point patterns. J R Stat Soc B 41:368–374

Rocha-Gil ZE, Ramírez ES, Rodríguez LAC (2017) Influencia de la restauración ecológica sobre la calidad fisicoquímica y biológica del agua, caso quebrada La Colorada. Cuaderno Activa 9:77–91

Rossi RE, Mulla DJ, Journel AG et al (1992) Geostatistical tools for modeling and interpreting ecological spatial dependence. Ecol Monogr 62:277–314

Sáenz-Romero C, Lindig-Cisneros RA, Joyce DG et al (2016) Assisted migration of forest populations for adapting trees to climate change. Revista Chapingo Serie Ciencias Forestales y del Ambiente 22:303–323. https://doi.org/10.5154/r.rchscfa.2014.10.052

Sánchez-Salguero R, Navarro-Cerrillo RM, Camarero JJ (2011) Selective drought-induced decline of pine species in southeastern Spain. Clim Chang 113:767–785. https://doi.org/10.1007/s10584-011-0372-6

Schneuwly-Bollschweiler M, Corona C, Stoffel M (2013) How to improve dating quality and reduce noise in tree-ring based debris-flow reconstructions. Quat Geochronol 18:110–118. https://doi.org/10.1016/j.quageo.2013.05.001

Shestakova TA, Gutiérrez E, Voltas J (2018) A roadmap to disentangling ecogeographical patterns of spatial synchrony in dendrosciences. Trees 32:359–370

Shi H, Zhang L (2003) Local analysis of tree competition and growth. For Sci 49:938–955

Stoffel M, Corona C (2014) Dendroecological dating of geomorphic disturbance in trees. Tree-Ring Res 70:3–20. https://doi.org/10.3959/1536-1098-70.1.3

Sturrock RN, Frankel SJ, Brown AV et al (2011) Climate change and forest diseases. Plant Pathol 60:133–149. https://doi.org/10.1111/j.1365-3059.2010.02406.x

Upton G, Fingleton B (1985) Spatial data analysis by example. Point pattern and quantitative data. Wiley, Chichester, p 410

Vicente-Serrano SM, Beguería S, López-Moreno JI (2010) A multiscalar drought index sensitive to global warming: the standardized precipitation evapotranspiration index. J Clim 23:1696–1718. https://doi.org/10.1175/2009JCLI2909.1

Vilalta CJ (2005) Cómo enseñar autocorrelación espacial. Economía Sociedad y Territorio 5:323–333

Williams AP, Allen CD, Macalady AK et al (2013) Temperature as a potent driver of regional forest drought stress and tree mortality. Nat Clim Chang 3:292–297. https://doi.org/10.1038/nclimate1693

Yoda K, Kira T, Hozumi K (1957) Intraspecific competition among higher plants. IX. Further analysis of the competitive interaction between adjacent individuals. J Inst Polytechnics 8:161–178

Chapter 12
Temporal Growth Variation in High-Elevation Forests: Case Study of *Polylepis* Forests in Central Andes

Edilson J. Requena-Rojas, Doris B. Crispín-DelaCruz, Ginette Ticse-Otarola, Harold Rusbelth Quispe-Melgar, Janet G. Inga Guillen, Vladimir Camel Paucar, Anthony Guerra, Fressia Nathalie Ames-Martinez, and Mariano Morales

Abstract *Polylepis* species form the dominant high-altitude forests in the tropical Andes, one of the most vulnerable regions to future climate change scenarios. The study of the growth of these forests provides useful information about their ontogeny and the environmental conditions where they develop. The identification of growth patterns is relevant for understanding the dynamics of the forests in response to climatic variables. In this chapter, we present a brief review of dendroecological studies on *Polylepis* species. We also developed for the first time in the central Andes of Peru three new Polylepis ring-width chronologies together with a diameter growth modeling for the following species: *Polylepis rodolfo-vasquezii*, *Polylepis rugulosa*, and *Polylepis tarapacana*. Dendrochronological techniques together with

E. J. Requena-Rojas (✉) · D. B. Crispín-DelaCruz · G. Ticse-Otarola · J. G. Inga Guillen
Laboratorio de Dendrocronología, Universidad Continental, Huancayo, Peru
e-mail: erequena@continental.edu.pe

H. R. Quispe-Melgar
Programa de Investigación de Ecología y Biodiversidad, Asociación ANDINUS, Huancayo, Peru

V. Camel Paucar
Grupo de Investigación de Fisiología Vegetal, Facultad de Ciencias, Universidad Nacional Agraria La Molina, Lima, Peru

A. Guerra
Facultad de Ciencias Forestales y del Medio Ambiente, Universidad Nacional del Centro del Perú, Huancayo, Peru

F. N. Ames-Martinez
Laboratorio de Biotecnología y Biología Molecular, Universidad Continental, Huancayo, Peru

M. Morales
Instituto Argentino de Nivología, Glaciología y Ciencias Ambientales (IANIGLA), CCT-CONICET, Mendoza, Argentina
e-mail: mmorales@mendoza-conicet.gob.ar

M. Pompa-García, J. J. Camarero (eds.), *Latin American Dendroecology*,
https://doi.org/10.1007/978-3-030-36930-9_12

a biologically based model help us to obtain information on forestry traits of *Polylepis* species. *P. rodolfo-vasquezii* to growth response to summer temperature of the current growth period determined radial growth, whereas spring and summer precipitation from the previous growth period determined the radial growth in *P. rugulosa* and *P. tarapacana,* respectively. The radial growth models indicated differences in the growth of the three *Polylepis* species with *P. rodolfo-vasquezii* reaching the highest rate (0.11 cm/yr), while *P. tarapacana* showed the lowest (0.08 cm/yr). Due to the low growth rates of these *Polylepis* species, long periods (>100 years) are required to establish and provide ecosystem services. As these forests face the challenge of climate change and anthropogenic pressure, there is a clear need to obtain precise information in order to formulate guidelines for the conservation of these forests, and the application of dendroecology is indispensable in this context.

Keywords Relative growth rate · von Bertalanffy · Tropical Andes · *Polylepis* · Dendroecology · Radial growth

12.1 Introduction

In the tropical Andes, the genus *Polylepis* accounts for the predominant arboreal vegetation above the treeline, extending from Venezuela to Argentina, reaching almost 5200 m altitude in the Bolivian Altiplano, whereas its lowest forests are found at 900 m in the mountains of central Argentina (Simpson 1979; Kessler 2006; Marcora et al. 2008; Segovia-Salcedo et al. 2018).

These forests are characterized by a fragmented spatial distribution (Coblentz and Keating 2008; Toivonen et al. 2018), growing in areas with rugged topography and steep slopes, which often makes them inaccessible (Sylvester et al. 2017). The spatial distribution patterns of different *Polylepis* species have been studied extensively and are believed to have been shaped by changes in the historical conditions during their natural evolution (Hooghiemstra and Van der Hammen 2004; Valencia et al. 2016, 2018), microhabitat conditions (Toivonen et al. 2018), and human activities such as logging, burning of pastures, cattle ranching, and firewood extraction (Ellenberg 1979; Kessler and Driesch 1993; Chepstow-Lusty and Winfield 2000; Kessler 2002; Sylvester et al. 2017).

Polylepis forests represent one of the most vulnerable ecosystems in the world (i.e., human activities and climate change) and provide multiple ecosystem services, including water regulation, carbon sequestration, erosion control, firewood production, scenic beauty, and habitat for endemic flora and fauna species (Fjeldså and Kessler 2004). From an ecological and forest conservation perspective, dendroecology is fundamental to understanding how climatic factors and disturbances (e.g., fires, landslides, drought, etc.) regulate tree growth. Previous studies show that the development of some species such as *P. pepei*, *P. besseri*, and *P. reticulata* are controlled by the summer temperature. (Roig et al. 2001; Gareca et al. 2010; Alvites et al. 2019). In contrast, the precipitation of the previous summer favors the growth of *P. tarapacana* (Argollo et al. 2004; Moya and Lara 2011). Other species such as *P. subsericans* do not show a relationship to variations in temperature and/or precipitation (Jomelli et al. 2012). Thus, there are no clear patterns regarding the effect of climatic conditions on the growth of these forests. Given the high diversity of the genus *Polylepis* and its wide geographical distribution (900–5200 m), it offers a unique possibility to understand the climatological and ecological processes that drive these ecosystems in the Andes, from the local to the subcontinental scale. In this chapter, we present a brief review of dendroecological studies on *Polylepis* species. We also developed three new tree-ring chronologies of *P. rodolfo-vasquezii*, *P. rugulosa*, and *P. tarapacana* from the central Andes of Peru. Finally, we evaluated the growth response of each species to local climate variables and we modeled the growth.

12.2 Dendroecology in *Polylepis*

Studies in dendroclimatological on genus *Polylepis* increased during the mid-1990s (Boninsegna et al. 2009). In studies on the dendroecology of *Polylepis*, seven species have been evaluated to date, namely, *P. australis*, *P. besseri*, *P. pepei*, *P. reticulata*, *P. rugulosa*, *P. subsericans*, and *P. tarapacana* (Roig et al. 2001; Argollo et al. 2004; Morales et al. 2004; Suarez et al. 2008; Domic and Capriles 2009; Gareca et al. 2010; Lanza et al. 2018; Alvites et al. 2019). These studies indicate that despite the ability of these trees to grow in varied climatic conditions, they fulfill the necessary conditions for dendroecological studies.

From the many factors that affect the dynamics of the forest, climatic and topographical conditions are the primary ones that limit tree growth and development. Each *Polylepis* species shows a particular climate growth response. For instance, the growth of *P. pepei* in the Bolivian Yungas depends on the summer temperatures (Roig et al. 2001), whereas that of *P. tarapacana* in the Altiplano is highly dependent on water availability, a scarce resource in this region (Morales et al. 2004, 2015; Solíz et al. 2009; García-Plazaola et al. 2015). The growth of *P. subsericans*, *P. rugulosa*, and *P. pepei* in Peru and Bolivia is modulated by the temperature, precipitation, and terrain conditions (Jomelli et al. 2012). Other studies on *P. reticulata* have indicated that its growth is determined mainly by temperature changes, and to a lesser extent, by precipitation (Kessler et al. 2007, 2014), the latter study having been carried out with dasometric variables. The temporal growth variation in *P. besseri* is subject to various limiting factors such as local disturbances (fires, grazing, nutrient variation), nutrient availability, and climate. Variation in rainfall across years seems to have a slight effect on growth, whereas the high temperature before the growth season positively affects radial growth (Gareca et al. 2010).

Little research has been carried out using growth rings to explain the ecological patterns and processes and forest dynamics of *Polylepis* species. Domic and Capriles (2009) identified a positive relationship between the total height and diameter of trees, but neither variable was related to the age of the trees. On the contrary, Suarez et al. (2008) demonstrated a certain level of correlation between the characteristics of the microsites and the establishment of individual *P. australis* trees. Finally, growth rings and the characteristics of the vascular elements (diameter and frequency of vessels mm^2) of exposed *P. australis* roots have been used for estimating date erosion events. This was possible owing to the variation in the reduction in the mean vessel area, percentage of the vessel area, and increase in the number of vessels after root exposure (Chartier et al. 2016).

Numerous studies have evaluated the relationship among dasometric variables such as total height and diameter of trees, for multiple purposes, including carbon sequestration estimation, forest structure analysis, and growth dynamics (Henry and Aarssen 1999; Fehse et al. 2002; Chave et al. 2005; Feldpausch et al. 2011; Kearsley et al. 2013; Sumida et al. 2013). Some *Polylepis* studies have analyzed allometric relationships to estimate carbon sequestration, dasometric structure patterns, and spatial distribution (Rada et al. 1996; Vásquez et al. 2014; Camel et al. 2019).

However, they did not make use of tree-ring data, likely because of the lack of information on the presence of tree species with clearly identifiable rings.

12.3 Growth Modeling

12.3.1 Modeling Based on Relative Growth Rate

Tree growth is a complex process, and growth models represent a frequently used tool for understanding this process. Most tree growth models are empirical and tend to concentrate on mathematical analysis based on data; few models are based on biological principles where growth is considered to represent the difference between anabolic rates (photosynthesis) and catabolic rates (respiration), that is, the principle underlying von Bertalanffy's model (von Bertalanffy 1957; Vanclay 1995; West et al. 2001; Pretzsch 2010).

In studies of tropical tree growth using growth rings, most authors represent only the trajectory of a smoothed mean curve without an apparent use of growth models (Detienne et al. 1998; Mattos 1999; Worbes 1999, 2004; Worbes et al. 2003; Brienen 2005; Wood et al. 2010; Rozendaal et al. 2010; Rozendaal and Zuidema 2011; Szejner 2011; López 2011; López et al. 2013; Groenendijk et al. 2014; Costa et al. 2015). While some other studies use growth models, they do not adequately report the statistical validation of their models in terms of self-relation, heteroscedasticity, and waste normality, nor do they use independent validation (Mushove et al. 1995; Zuidema and Boot 2002; Schöngart 2008; Schöngart et al. 2008; Castilho 2013; Andrade 2015). Consequently, several studies recommend using the von Bertalanffy model owing to its theoretical origin, versatility, and wide use in forestry (Vanclay 1995; Giraldo and Del Valle 2011; Guarín and Del Valle 2014; Inga and Del Valle 2017).

12.4 Case Study in the Central Andes of Peru

Of the many *Polylepis* species found in the central Andes, some have clearly delimited growth rings which can be used for dendroecological studies that can help understand tree growth responses to local climatic variations.

12.4.1 Climate Characteristics of the Study Forests

In the central Andes of Peru, the *Polylepis* forests are distributed in four terrestrial ecoregions: the Puna, the Sechura Desert, the Peruvian Yungas, and the dry Puna (Olson et al. 2001). *Polylepis rodolfo-vasquezii*, *Polylepis rugulosa*, and *Polylepis tarapacana* are found in these regions (Fig. 12.1). The present study was conducted in two of these ecoregions from central and southern Peru: Toldopampa (*P. rodolfo-vasquezii*) in the Yungas, Tarata (*P. rugulosa*) and Paucarani (*P. tarapacana*) in the Puna (Fig. 12.1). The greatest height and diameter were recorded for *P. rodolfo-vasquezii* trees, whereas *P. rugulosa* and *P. tarapacana* trees were small (Table 12.1).

The endemic species *P. rodolfo-vasquezii* was recently described; it inhabits stony sites with high levels of humidity, which is associated with other types of tree vegetation such as *Gynoxys*, *Lupinus*, *Hesperosmeles,* and *Clethra* (Valenzuela and Villalba 2015). These forests experience a humid period from December to March and a dry period from June to August, with a mean annual temperature of 14.3 °C and mean precipitation of 824 mm (Fig. 12.2a). *P. rugulosa* is distributed on the western slopes of northern Chile and southern Peru. The area is predominantly dry (April to December), with a mean temperature of 15.8 °C and a mean precipitation of 212 mm (Fig. 12.2b). *P. tarapacana* distribution reaches relatively high altitudes (up to 5200 m), being one of the tree species with high tolerance to extreme climatic conditions; its distribution region has a wet period from January to March and a dry period from April to October, with a mean temperature of 10.9 °C and a mean precipitation of 535 mm (Fig. 12.2c).

12.4.2 Tree-Ring Chronologies and Climate-Growth Response

Dendrochronology allows the evaluation of tree-ring width with respect to the environmental conditions that modulate the growth of the species (Stokes and Smiley 1968). Samples from the trees were collected between March and April 2017. In all three sites, the samples were taken using an increment borer, obtaining cylindrical sections of 5 mm diameter and variable length, depending on the radius of each sample. The samples were complemented with cross-sections from dead trees. The samples were polished with sandpapers of gradually finer grains (from 120 to 1500) to allow a clear visualization of tree-rings. The tree-rings were visually dated using

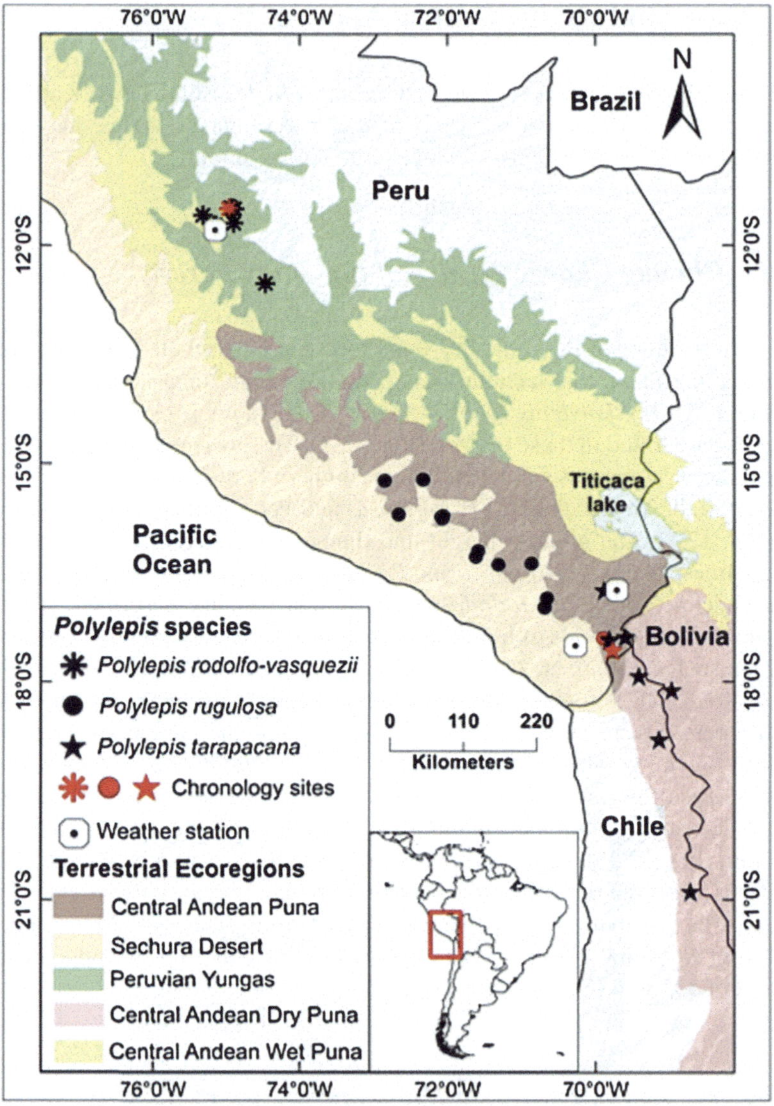

Fig. 12.1 Distribution map of *P. rodolfo-vasquezii* (asterisk), *P. rugulosa* (cross), and *P. tarapacana* (star) in the Andes. Location of the tree-ring sites and the local meteorological stations are indicated with red symbols and squares, respectively

a stereoscope, taking into account marker rings. Following the Southern Hemisphere convention; annual rings were assigned to the year in which their formation began (Schulman 1956). The ring widths were measured with a precision of 0.001 mm using a Velmex equipment. The quality of the visual dating and measurement was cross-checked using the CofechaXP software (Grissino-Mayer 2001). All correctly

Table 12.1 Geographical location and characteristics of the *Polylepis* spp. study sites in the Peruvian Andes

Site	Latitude (°S)	Longitude (°W)	Elevation (m)	Aspect	*H* ± SD (m)	*D* ± SD (cm)
Toldopampa	11°30′	74°53′	4097	SE	5.2 ± 1.2	16.7 ± 5.1
Tarata	17°34′	69°45′	4696	NE	2.8 ± 1.4	5 ± 1
Paucarani	17°31′	69°33′	4175	NW	1.5 ± 2.8	4.5 ± 0.8

H mean height, *D* mean diameter, *SD* standard deviation

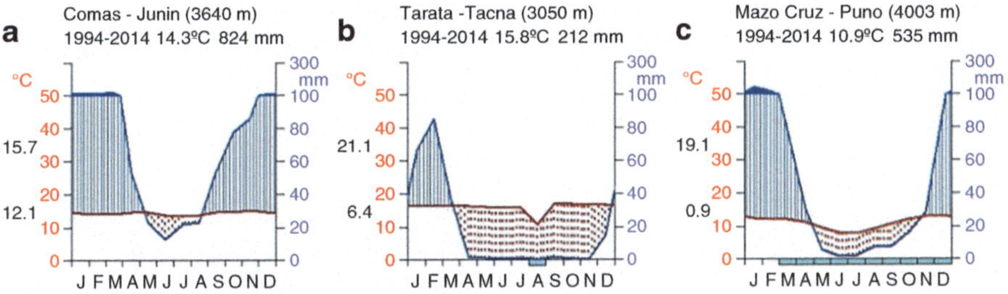

Fig. 12.2 Climate diagram of representative weather stations for the forests of (**a**) *P. rodolfo-vasquezii*, (**b**) *P. rugulosa*, and (**c**) *P. tarapacana*

dated series were standardized by fitting exponential negative curves using the ARSTAN44XP software (Cook and Krusic 2005). The standardization process eliminates the biological tendencies of growth, reduces most of the effects of endogenous stand disturbances, and enhances the common signal present in the tree-ring series (Fritts 1976). The chronologies of *P. rodolfo-vasquezii*, *P. rugulosa*, and *P. tarapacana* were developed in this way (Fig. 12.3). The quality of the chronologies was assessed using the RBAR and EPS statistics (Table 12.2). The RBAR is the common signal measure in the tree-ring variations included in the chronology and the EPS is the measure of the total signal present in the chronology compared with a hypothetical infinitely replicated chronology (Cook et al. 1990). To calculate the EPS, we used a 30-year window with an overlap of 15-year between adjacent windows. While there is no level of significance for EPS, values close to 0.85 are generally accepted as a good level of common signal fidelity between trees (Wigley et al. 1984). Relatively high RBAR and EPS values (>0.8) were recorded in each chronology, indicating a common growth signal between trees and a good population representation of the tree-ring chronologies (Fig. 12.3). The recorded RBAR and EPS values for each chronology were consistent with the previously documented data for *Polylepis* species from other sites in Bolivia, Peru, Chile, and Argentina (Roig et al. 2001; Christie et al. 2009; Solíz et al. 2009; Gareca et al. 2010; Moya and Lara 2011; Jomelli et al. 2012).

Relationships between radial growth and climatic variation are studied through correlation functions (Blasing et al. 1984). The radial growth of a given year can be influenced by climatic conditions of the previous year, and, therefore, we evaluated

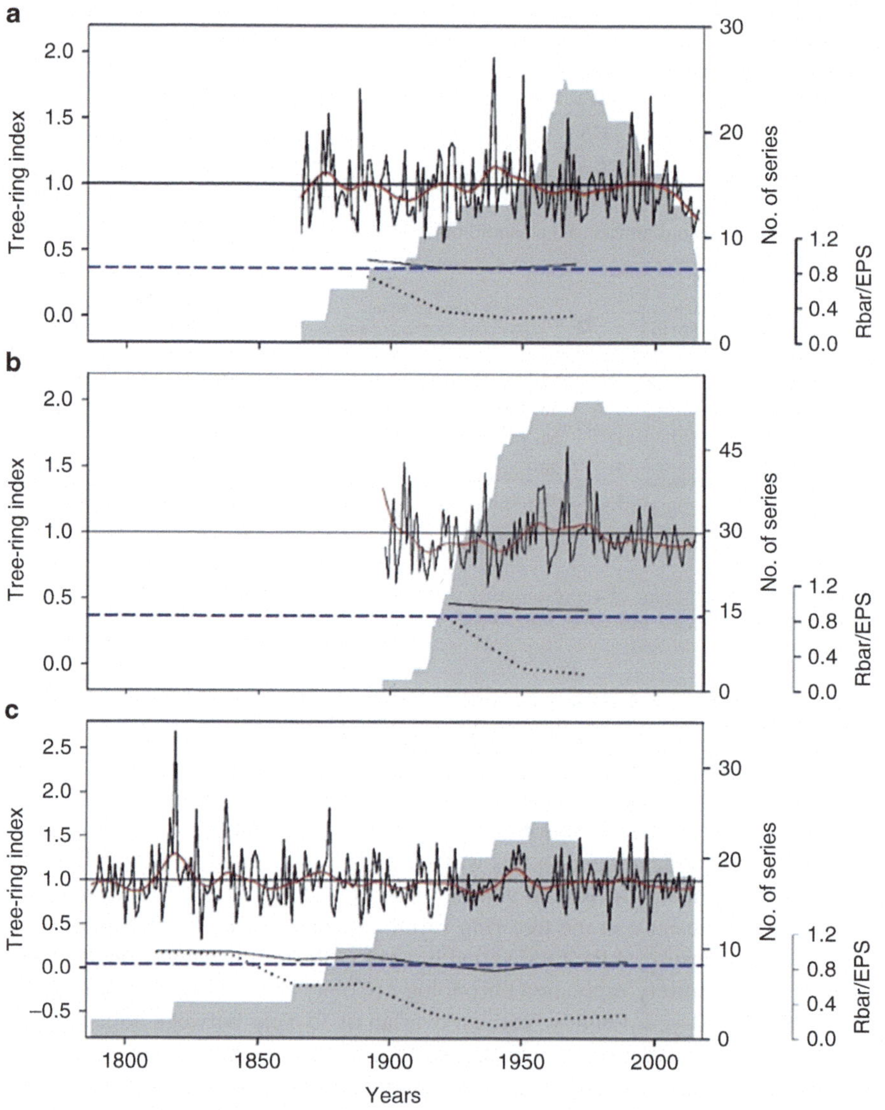

Fig. 12.3 Tree-ring chronologies for the three species: (**a**) *P. rodolfo-vasquezii*, (**b**) *P. rugulosa*, and (**c**) *P. tarapacana*. The time-varying EPS (black solid line) and RBAR (dotted line) values are represented below each chronology. The short-dashed blue line indicates the EPS critical value of 0.85; values close to or higher than 0.85 are considered acceptable. The gray area represents the number of series in the chronology

the relationship of *Polylepis* radial growth with the temperature and precipitation during a period of 21 months.

Variations in the radial growth of *P. rodolfo-vasquezii* were related to changes in monthly temperature during the current growing season. In contrast, low relationships

Table 12.2 Descriptive statistics of the chronologies for the three *Polylepis* species

Statistics	*P. rodolfo-vasquezii*	*P. rugulosa*	*P. tarapacana*
No. of series	27	54	24
No. of trees	19	26	11
Range	1865–2015	1897–2015	1787–2015
Series intercorrelation	0.49	0.41	0.59
Mean sensitivity	0.37	0.36	0.32
Mean EPS	0.89	0.95	0.90
Mean RBAR	0.42	0.42	0.51

were found with monthly precipitation, except during August previous to the current season (Fig. 12.4a). The radial growth of *P. rugulosa* was also related to variations in temperature for the months of December and January of the previous growth period, with significance observed only for January (Fig. 12.4b). Precipitation of November of the previous growth period was significantly correlated with tree growth. Also, winter-spring precipitation was positive correlated with tree growth, being significant during August (Fig. 12.4b). Similar growth response to temperature has been previously reported for *P. pepei* (Roig et al. 2001; Jomelli et al. 2012), *P. besseri* (Gareca et al. 2010), *P. subsericans* (Jomelli et al. 2012), and *P. australis* (Lanza et al. 2018). On the contrary, variations in radial growth of *P. tarapacana* were positively related to changes in monthly precipitation during the previous growth period, with significance observed only for February. In contrast, the relationship with the precipitation of the current growth period was negative (Fig. 12.4c), which coincides with the wet season (Fig. 12.2c). Other studies on *P. tarapacana* also reported a positive response to precipitation during the previous growing season and a negative response during the current growing season (Argollo et al. 2004; Morales et al. 2004; Moya and Lara 2011). Moreover, it is important to note a common pattern for the three species growing in different biogeographical regions, that is, the positive influence of precipitation on the tree growth during the dry season (August–September).

The most striking finding of the positive relationship between the monthly variations of temperature and the radial growth of *P. rodolfo-vasquezii* could be attributed to the fact that this species grows in the ecoregion of the Yungas, with a season wet marked which coincides with the time of tree growth (Fig. 12.2a), and, therefore, temperature is likely the limiting factor for photosynthetic processes. However, precipitation during the winter dry season is also very important for the tree growth.

These results indicate the differences in the tree growth climate drivers of the *Polylepis* species along central Andes. Therefore, further dendrochronological research is needed to understand the complex variety in climate – tree growth relationship in these high-altitude ecosystems.

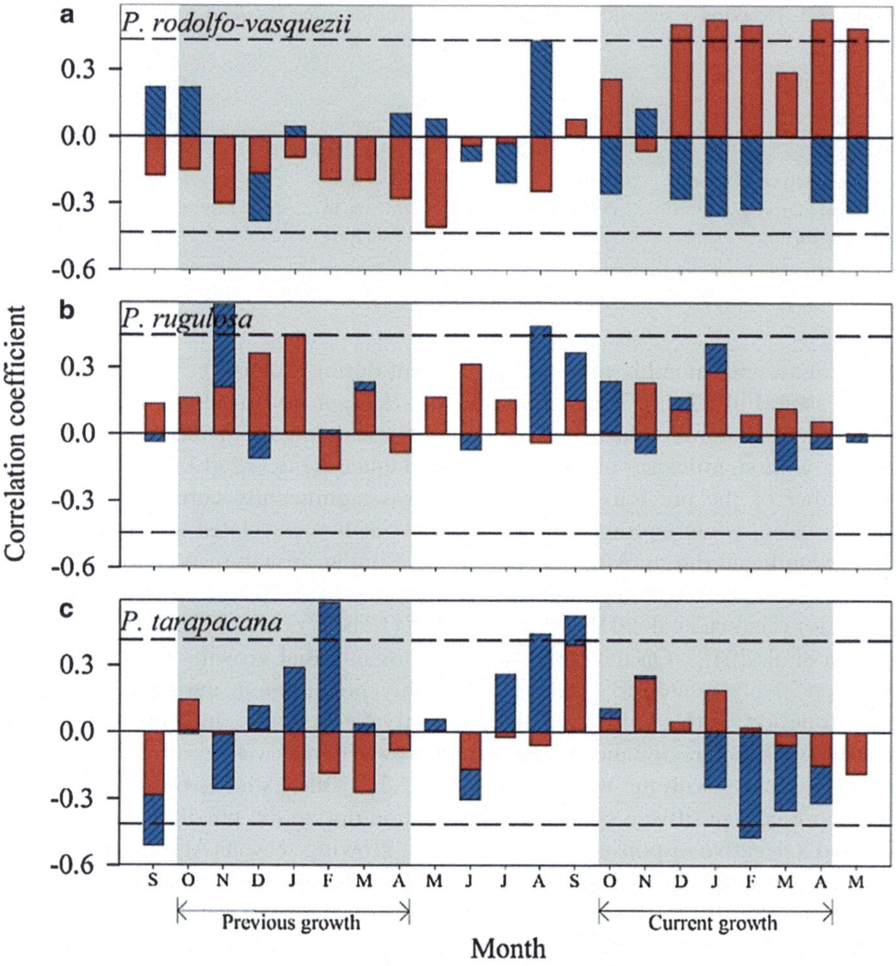

Fig. 12.4 Correlation coefficients between tree-ring indexes, temperature (red bars), and precipitation (hatched blue bars). The length of the comparison period between the tree-ring index and climate variables correspond to 1994–2014. The gray-vertical areas represent the growing seasons. Dashed-black lines represent 95% significant confidence levels

12.4.3 Polylepis Growth Modeling

To develop the growth modeling based on the relative growth model (Inga and Del Valle 2017), we selected those samples that presented the pith and have mean series intercorrelation values higher than 0.4. A total of 18, 20, and 16 series measurements of *P. rodolfo-vasquezii*, *P. rugulosa*, and *P. tarapacana*, respectively, were used.

The variability in growth rates of tropical trees reflects the changes in size relative to age. It is often assumed that trees with larger diameters are older, which may not necessarily be true. Absolute growth rates (*dD*/*dt*) express growth in diameter without considering differences in tree size. The relative growth rates of diameter (1/*D*) (*dD*/*dt*), obtained from the quotient of absolute growth rates (*dD*/*dt*) and diameter (*D*), represent a standardized measure of growth that avoids the inherent differences in scale between organisms (Hunt 1990). Therefore, several authors argue that relative growth rates are the most appropriate option for studying the growth of organisms (Hunt 1990; Pommerening and Muszta 2016).

Likewise, Inga and Del Valle (2017) reported that the logarithmic transformation of the relative growth rate of *D*, using the von Bertalanffy model (Eq. 12.2), reached high statistical adjustments and satisfies all the assumptions of the least-square regression. From the differential equation of von Bertalanffy (Eq. 12.1), the parameters are estimated and then replaced in the integrated equation (Eq. 12.3). This demonstrates the versatility of the model.

$$dD / dt = \eta D^{m} - \gamma D \tag{12.1}$$

$$\ln\left(1/D\right)\left(dD/dt\right) = \ln\eta + \ln\left(D^{m-1} - A^{m-1}\right) \tag{12.2}$$

$$D = A\left(1 - \mathrm{be}^{-k(t)}\right)^{1/(1-m)} \tag{12.3}$$

The growth equation for the trees, $D = f(t)$, obtained by integrating the differential equation (Eq. 12.1) allows the calculation of various features. Therefore, novel and accurate growth information was obtained for three species of genus *Polylepis* (Table 12.1).

The radial growth model indicated that the *P. rodolfo-vasquezii* forest presents individuals with 286 years, older than those from *P. rugulosa* and *P. tarapacana*. A previous dendrochronological study in *P. tarapacana* forests reported living trees with an age higher than 500 years (Moya and Lara 2011). On the other hand, *P. rugulosa* showed 112 years average age, higher to the reported in previous studies (Jomelli et al. 2012). The growth rate found for the three *Polylepis* species (Table 12.3) are similar to those reported by other studies and are relatively low in comparison with the growth rates of tree species from other regions such as the Amazonian forests (Table 12.4).

The weighted absolute growth rate (*WAGR*) expresses the total performance of a species during its lifespan, and, therefore, it does not depend on age (Del Valle and

Table 12.3 Growth traits of the forestry for the three study species in the Peruvian Andes

Features	*Polylepis rodolfo-vasquezii*	*Polylepis rugulosa*	*Polylepis tarapacana*
$t_{0.5}$	286 years	112 years	114 years
WAGR	0.11 cm yr^{-1}	0.09 cm yr^{-1}	0.08 cm yr^{-1}
WRGR	2% annual	2.9% annual	2.5% annual
$D_{\text{of the 100 years}}$	15 cm	12 cm	10 cm

$t_{0.5}$ mean age, *WAGR* weighted absolute growth rate, *WRGR* weighted relative growth rate, *D* of the 100 years: diameter reached in 100 years

Table 12.4 Mean radial growth and geographic location for tree species growing in the central Andes, Amazon and central Argentine

Species	Mrw (cm/yr^{-1})	Lat. (°S)	Long. (°W)	Elev. (m)	Source
P. besseri	0.07	17°44′	65°34′	3600	Gareca et al. (2010)
P. pepei	0.23	17°0′	65°39′	4100	Roig et al. (2001)
P. australis	0.15	31°37′	64°40′	2100	Suarez et al. (2008)
P. reticulata	0.10	2°47′	79°12′	4060	Alvites et al. (2019)
P. tarapacana	0.03	17°26′	69°20′	4429	Solíz et al. (2009)
P. tarapacana	0.08	22°32′	66°35′	4500	Solíz et al. (2009)
Cedrela nebulosa	0.32[a]	11°18	75°19	2150	Layme-Huaman et al. (2018)
Cedrela odorata	0.28[a]	10°25′	74°15′	1248	Inga and Del Valle (2017)
Juglans neotropica	0.25[a]	11°45′	74°14′	1250	Inga and Del Valle (2017)

Mrw mean ring width
[a]Species of the Amazon

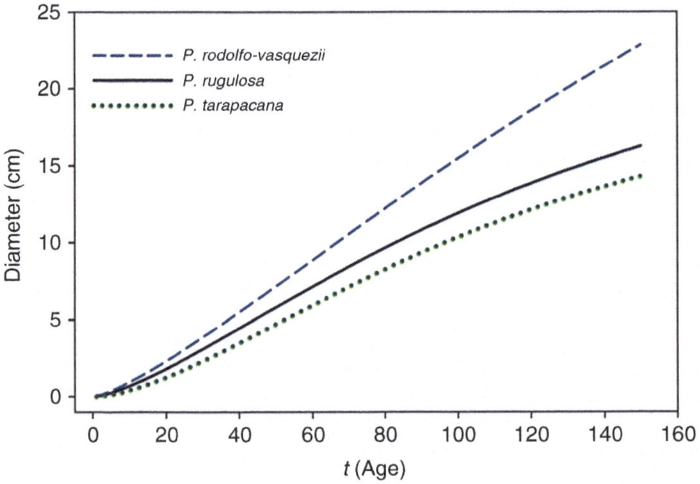

Fig. 12.5 Growth diameter (*D*), as a function of age (*t*), using the von Bertalanffy model in *P. rodolfo-vasquezii* (blue dashed line), *P. rugulosa* (brown solid line), and *P. tarapacana* (green dotted line)

Lema 1999). This criterion is considered objective as it includes the growth of the species throughout its life; *P. rodolfo-vasquezii* recorded a value of 0.11 cm/yr, *P. rugulosa* had a value of 0.09 cm/yr, and *P. tarapacana* had a value of 0.08 cm/yr. The relative weighted growth rate (*WRGR*) allows evaluation of the growth efficiency in percentage (Table 12.3). The diametric growth of the *Polylepis* species (Fig. 12.5) as a function of age indicated that it maintains an upward growth curve up to 150 years, despite the difference in growth between the three species. For example, *P. rodolfo-vasquezii* presents a higher absolute growth rate, reaching 15 cm in diameter in 100 years. All three species have relatively low growth rates.

12.5 Concluding Remarks

The few studies that have used tree-rings from an applied ecology viewpoint in the central Andes and adjacent regions are mainly focused on deciphering the association of climatic variables with the radial growth of trees. A variety of tree growth responses, mainly summer temperature and precipitation, were recorded in those previous studies. Here, we present three new tree-ring chronologies of *Polylepis* species. The reported chronology of *P. rodolfo-vasquezii* is the first one developed for this species, where its radial growth was found to be affected by the summer temperatures of the current growing season, showing a high dendrochronological potential. In addition, the other two *Polylepis* ring width chronologies (*P. rugulosa* and *P. tarapacana*) represented the first dendrochronological records for Peru. The annual growth of *P. rugulosa* is positively influenced by the precipitation in November of the previous growing.

Season, while *P. tarapacana* recorded a positive growth response to precipitation during summer of the previous growth period. Despite the different climate responses between the three *Polylepis* species, they share a common growth response pattern to winter dry season precipitation. It is likely that the precipitation before the current growing season is extremely important when the dry conditions became stressful for the *Polylepis* growth in the different ecoregions. The von Bertalanffy model was applied to the three *Polylepis* species, which indicated that the *Polylepis* species present slow radial growth, not exceeding 0.11 cm/yr. In addition, they provide evidence that up to 150 years continue with an ascending growth. The low growth rates of *Polylepis* species indicate that these forests need long periods to establish and provide ecosystem services. The current challenges of climate change and anthropogenic pressure represent threats to their permanence. The loss of these forests would require long-term restoration processes (> than 100 years) and would be associated with high costs and even social conflicts related to land use.

Acknowledgments The authors are grateful for the collaboration of Luis Salome, Diana Rivera, Christian Marmanillo, and Andre Llanos in the field work. We are grateful to 147-2015 FONDECYT, ANPCyT (PICT 2013-1880), CONICET, PIP 11220130100584, partially funded by the BNP Paribas Foundation in the frame of its "Climate Initiative" program, NSF_P2C2, and PIRE NSF. We also are grateful to the editors and anonymous reviewers for their useful comments and suggestions.

References

Alvites C, Battipaglia G, Santopuoli G et al (2019) Dendrochronological analysis and growth patterns of *Polylepis reticulata* (Rosaceae) in the Ecuadorian Andes. IAWA J 40(2):1–21. https://doi.org/10.1163/22941932-40190240

Andrade VHF (2015) Modelos de crecimiento para *Hymenaea courbaril* L. e *Handroanthus serratifolius* (Vahl) S.O.Grose em floresta de terra firme utilizando análise de anéis de crescimento. Disertation, Universidade Federal do Paraná

Argollo J, Solíz C, Villalba R (2004) Potencialidad dendrocronológica de *Polylepis tarapacana* en los Andes Centrales de Bolivia. Ecología en Bolivia 39(1):5–24

Blasing TJ, Solomon A, Duvick D (1984) Response functions revisited. Tree-Ring Bull 44:1–15

Boninsegna JA, Argollo J, Aravena JC et al (2009) Dendroclimatological reconstructions in South America: a review. Palaeogeogr Palaeoclimatol Palaeoecol 281:210–228. https://doi.org/10.1016/j.palaeo.2009.07.020

Brienen RJW (2005) Tree rings in the tropics: a study on growth and ages of Bolivian rain forest trees, PROMAB Scientific Series:10. Riberalta, Beni

Camel PV, Quispe-Melgar HR, Ames-Martínez FN et al (2019) Forest structure of three endemic species of the genus *Polylepis* (Rosaceae) in central Peru. Ecología Austral 29:285–295. https://doi.org/10.5167/uzh-173473

Castilho NTF (2013) Manejo da regeneração natural e produção de madeira de pau mulato em floresta de várzea do estuário amazônico. Disertation, Universidade Federal do Amapá (UNIFAP)

Chartier MP, Giantomasi MA, Renison D et al (2016) Exposed roots as indicators of geomorphic processes: a case-study from *Polylepis* mountain woodlands of Central Argentina. Dendrochronologia 37:57–63. https://doi.org/10.1016/j.dendro.2015.11.003

Chave J, Andalo C, Brown S et al (2005) Tree allometry and improved estimation of carbon stocks and balance in tropical forests. Oecologia 145(1):87–99. https://doi.org/10.1007/s00442-005-0100-x

Chepstow-Lusty A, Winfield M (2000) Inca agroforestry: lessons from the past. AMBIO J Hum Environ 29:322–328. https://doi.org/10.1579/0044-7447-29.6.322

Christie DA, Lara A, Barichivich J et al (2009) El Niño-Southern Oscillation signal in the world's highest-elevation tree-ring chronologies from the Altiplano, Central Andes. Palaeogeogr Palaeoclimatol Palaeoecol 281(3–4):309–319. https://doi.org/10.1016/j.palaeo.2007.11.013

Coblentz D, Keating PL (2008) Topographic controls on the distribution of tree islands in the high Andes of south-western Ecuador. J Biogeogr 35:2026–2038. https://doi.org/10.1111/j.1365-2699.2008.01956.x

Cook E, Krusic PJ (2005) Program ARSTAN: a tree-ring standardization program based on detrending and autoregressive time series modeling, with interactive graphics. Lamont-Doherty Earth Observatory, Columbia University, Palisades

Cook E, Briffa K, Shiyatov S et al (1990) Tree ring standardization and growth-trend estimation. In: Cook ER, Kairiukstis LA (eds) Methods of dendrochronology: applications in the environmental sciences. Kluwer Academic Publishers, Dordrecht, pp 104–123

Costa MS, Ferreira KEB, Botosso PC et al (2015) Growth analysis of five Leguminosae native tree species from a seasonal semidecidual lowland forest in Brazil. Dendrochronologia 36:23–32. https://doi.org/10.1016/j.dendro.2015.08.004

Del Valle JI, Lema A (1999) Crecimiento de cohortes de árboles coetáneos en rodales espacialmente dispersos: el caso de *Campnosperma panamensis*. Ecología 23:249–260

Detienne P, Oyono F, Durrieu de Madron L et al (1998) L'analyse de cernes: Applications aux études de croissance de quelques essences en peuplements naturels de forét dense africaine. CIRAD-Foret, France, p 36

Domic AI, Capriles JM (2009) Allometry and effects of extreme elevation on growth velocity of the Andean tree *Polylepis tarapacana* Philippi (Rosaceae). Plant Ecol 205:223–234. https://doi.org/10.1007/s11258-009-9612-5

Ellenberg H (1979) Man's influence on tropical mountain ecosystems in South America. J Ecol 67:401–416. https://doi.org/10.2307/2259105

Fehse J, Hofstede R, Aguirre N et al (2002) High altitude tropical secondary forests: a competitive carbon sink? For Ecol Manag 163(1):9–25. https://doi.org/10.1016/S0378-1127(01)00535-7

Feldpausch TR, Banin L, Phillips OL et al (2011) Height-diameter allometry of tropical forest trees. Biogeosciences 8:1081–1106. https://doi.org/10.5194/bg-8-1081-2011

Fjeldså J, Kessler M (2004) Conservación de la biodiversidad de los bosques de *Polylepis* de las tierras altas de Bolivia. Una contribución al manejo sustentable en los Andes. DIVA Technical Report 11. Editorial FAN, Santa Cruz de la Sierra

Fritts HC (1976) Tree rings and climate. Academic Press, New York
García-Plazaola JI, Rojas R, Christie DA et al (2015) Photosynthetic responses of trees in high-elevation forests: comparing evergreen species along an elevation gradient in the Central Andes. AoB Plants 7:1–13. https://doi.org/10.1093/aobpla/plv058
Gareca EE, Fernández M, Stanton S (2010) Dendrochronological investigation of the high Andean tree species *Polylepis besseri* and implications for management and conservation. Biodivers Conserv 19:1839–1851. https://doi.org/10.1007/s10531-010-9807-z
Giraldo JA, Del Valle JI (2011) Estudio del crecimiento de *Prioria copaifera* (Caesalpinaceae) mediante técnicas dendrocronológicas. Rev Biol Trop 59:1813–1831
Grissino-Mayer H (2001) Evaluation crossdating accuracy: a manual and tutorial for the computer program COFECHA. Tree-Ring Res 57:205–221
Groenendijk P, Sass-Klaassen U, Bongers F et al (2014) Potential of tree-ring analysis in a wet tropical forest: a case study on 22 commercial tree species in Central Africa. For Ecol Manag 323:65–68. https://doi.org/10.1016/j.foreco.2014.03.037
Guarín JR, Del Valle JI (2014) Modeling the stipe growth of the *Oenocarpus bataua* palm in the Central Cordillera of the Andes, Colombia. For Ecol Manag 314:141–149. https://doi.org/10.1016/j.foreco.2013.12.005
Henry HAL, Aarssen LW (1999) The interpretation of stem diameter-height allometry in trees: biomechanical constraints, neighbor effects, or biased regressions? Ecol Lett 2(2):89–97. https://doi.org/10.1046/j.1461-0248.1999.22054.x
Hooghiemstra H, Van der Hammen T (2004) Quaternary Ice-Age dynamics in the Colombian Andes: developing an understanding of our legacy. Philos Trans R Soc Lond Ser B Biol Sci 359:173–181. https://doi.org/10.1098/rstb.2003.1420
Hunt R (1990) Basic growth analysis: plant growth analysis for beginners. Academic Division of Unwin Hyman Ltd, London
Inga JG, Del Valle JI (2017) Log-relative growth: a new dendrochronological approach to study diameter growth in *Cedrela odorata* and *Juglans neotropica*, Central Forest, Peru. Dendrochronologia 44:117–129. https://doi.org/10.1016/j.dendro.2017.03.009
Jomelli V, Pavlova I, Guin O et al (2012) Analysis of the dendroclimatic potential of *Polylepis pepei*, *P. subsericans* and *P. rugulosa* in the tropical Andes (Peru-Bolivia). Tree-Ring Res 68(2):91–104. https://doi.org/10.3959/2011-10.1
Kearsley E, de Haulleville T, Hufkens K et al (2013) Conventional tree height–diameter relationships significantly overestimate aboveground carbon stocks in the Central Congo Basin. Nat Commun 4. https://doi.org/10.1038/ncomms3269
Kessler M (2002) The "*Polylepis* problem": where do we stand? Ecotropica 8:97–110
Kessler M (2006) Bosques de *Polylepis*. Botánica económica los Andes Centrales 11:110–120
Kessler M, Driesch P (1993) Causas e historia de la destrucción de bosques altoandinos en Bolivia. Ecol Boliv 21:1–18
Kessler M, Bohner J, Klugea J (2007) Modelling tree height to assess climatic conditions at tree lines in the Bolivian Andes. Ecol Model 207:223–233. https://doi.org/10.1016/j.ecolmodel.2007.05.001
Kessler M, Toivonen JM, Sylvester SP et al (2014) Elevational patterns of *Polylepis* tree height (Rosaceae) in the high Andes of Peru: role of human impact and climatic conditions. Front Plant Sci 5. https://doi.org/10.3389/fpls.2014.00194
Lanza MG, Chartier MP, Marcora PI (2018) Relación clima-crecimiento radial de *Polylepis australis* en un gradiente altitudinal en las Sierras Grandes de Córdoba, Argentina. Ecol Austral 28:278–290. https://doi.org/10.25260/EA.18.28.1.1.620
Layme-Huaman ET, Ferrero ME, Palacios-Lazaro KS et al (2018) *Cedrela nebulosa*: a novel species for dendroclimatological studies in the Montane tropics of South America. Dendrochronologia 50:105–112. https://doi.org/10.1016/j.dendro.2018.06.004
López L (2011) Una aproximación dendrocronológica a la ecología y el manejo de los bosques tropicales secos del cerrado boliviano. Universidad Nacional del Comahue

López L, Villalba R, Bravo F (2013) Cumulative diameter growth and biological rotation age for seven tree species in the Cerrado biogeographical province of Bolivia. For Ecol Manag 292:49–55. https://doi.org/10.1016/j.foreco.2012.12.011

Marcora P, Hensen I, Renison D et al (2008) The performance of *Polylepis australis* trees along their entire altitudinal range: implications of climate change for their conservation. Divers Distrib 14(4):630–636. https://doi.org/10.1111/j.1472-4642.2007.00455.x

Mattos PP (1999) Identificação de aneis anuais de crescimento e estimativa de idade e incremento anual em diametro de especies nativas do pantanal da Nhecolandia, MS. Disertation, Universidade Federal do Paraná

Morales MS, Villalba R, Grau HR et al (2004) Rainfall-controlled tree growth in high-elevation subtropical treelines. Ecology 85(11):3080–3089. https://doi.org/10.1890/04-0139

Morales MS, Carilla J, Grau HR et al (2015) Multi-century lake area changes in the Southern Altiplano: a tree-ring-based reconstruction. Clim Past 11(9):1139–1152. https://doi.org/10.5194/cp-11-1139-2015

Moya J, Lara A (2011) Cronologías de ancho de anillos de queñoa (*Polylepis tarapacana*) para los últimos 500 años en el Altiplano de la región de Arica y Parinacota, Chile. Bosque 32(2):165–173. https://doi.org/10.4067/S0717-92002011000200007

Mushove PT, Prior JAB, Gumbie C et al (1995) The effects of different environments on diameter growth increments of *Colophospermum mopane* and *Combretum apiculatum*. For Ecol Manag 72:287–292. https://doi.org/10.1016/0378-1127(94)03468-C

Olson DM, Dinerstein E, Wikramanayake ED (2001) Terrestrial ecoregions of the world: a new map of life on earth. Bioscience 51(11):933–938. https://doi.org/10.1641/0006-3568(2001)051[0933:TEOTWA]2.0.CO;2

Pommerening A, Muszta A (2016) Relative plant growth revisited: towards a mathematical standardisation of separate approaches. Ecol Model 320:383–392. https://doi.org/10.1016/j.ecolmodel.2015.10.015

Pretzsch H (2010) Forest dynamics, growth, and yield. In: Forest dynamics, growth and yield. Springer Berlin Heidelberg, Germany, p 1–39

Rada F, Azócar A, Briceño B et al (1996) Carbon and water balance in *Polylepis sericea*, a tropical treeline species. Trees 10(4):218–222. https://doi.org/10.1007/BF02185672

Roig F, Fernández M, Gareca E et al (2001) Estudios dendrocronológicos en los ambientes húmedos de la puna boliviana. Rev Bol Ecol 9(1):3–13

Rozendaal DMA, Zuidema PA (2011) Dendroecology in the tropics: a review. Trees 25:3–16. https://doi.org/10.1007/s00468-010-0480-3

Rozendaal DMA, Brienen RJW, Soliz-Gamboa CC et al (2010) Tropical tree rings reveal preferential survival of fast-growing juveniles and increased juvenile growth rates over time. New Phytol 185:759–769. https://doi.org/10.1111/j.1469-8137.2009.03109.x

Schöngart J (2008) Growth-Oriented Logging (GOL): a new concept towards sustainable forest management in Central Amazonian várzea floodplains. For Ecol Manag 256:46–58. https://doi.org/10.1016/j.foreco.2008.03.037

Schöngart J, Felfili Fortes C, Nunes da Cunha C et al (2008) Carbon dynamics in aboveground coarse wood biomass of wetland forests in the northern Pantanal, Brazil. Biogeosci Discuss 5(3):2103–2130

Schulman E (1956) Dendroclimatic change in semiarid America. University of Arizona Press, Tucson, p 142

Segovia-Salcedo MC, Domic A, Boza TE et al (2018) Situación taxonómica de las especies del género *Polylepis*. Implicancias para los estudios ecológicos, la conservación y la restauración de sus bosques. Ecol Austral 28:188–201. https://doi.org/10.25260/EA.18.28.1.1.527

Simpson BB (1979) A revision of the genus *Polylepis* (Rosaceae: Sanguisorbeae). Smithson Contrib Bot 43:1–62. https://doi.org/10.5479/si.0081024X.43.1

Solíz C, Villalba R, Argollo J et al (2009) Spatio-temporal variations in *Polylepis tarapacana* radial growth across the Bolivian Altiplano during the 20th century. Palaeogeogr Palaeoclimatol Palaeoecol 281(3–4):296–308. https://doi.org/10.1016/j.palaeo.2008.07.025

Stokes M, Smiley T (1968) An introduction to tree-ring dating. University of Chicago Press, Chicago, p 73

Suarez ML, Renison D, Marcora P et al (2008) Age–size–habitat relationships for *Polylepis australis*: dealing with endangered forest ecosystems. Biodivers Conserv 17(11):2617–2625. https://doi.org/10.1007/s10531-008-9336-1

Sumida A, Miyaura T, Torii H (2013) Relationships of tree height and diameter at breast height revisited: analyses of stem growth using 20-year data of an even-aged *Chamaecyparis obtusa* stand. Tree Physiol 33(1):106–118. https://doi.org/10.1093/treephys/tps127

Sylvester SP, Heitkamp F, Sylvester MDPV et al (2017) Relict high-Andean ecosystems challenge our concepts of naturalness and human impact. Sci Rep 7:1–13. https://doi.org/10.1038/s41598-017-03500-7

Szejner P (2011) Tropical dendrochronology: exploring tree-ring of *Pinus oocarpa* in eastern Guatemala. Georg-August-Universität Göttingen

Toivonen JM, Gonzales-Inca CA, Bader MY et al (2018) Elevational shifts in the topographic position of *Polylepis* forest stands in the Andes of Southern Peru. Forests 9:1–10. https://doi.org/10.3390/f9010007

Valencia BG, Matthews-Bird F, Urrego DH et al (2016) Andean microrefugia: testing the Holocene to predict the Anthropocene. New Phytol 212:510–522. https://doi.org/10.1111/nph.14042

Valencia BG, Bush MB, Coe AL et al (2018) *Polylepis* woodland dynamics during the last 20,000 years. J Biogeogr 45:1019–1030. https://doi.org/10.1111/jbi.13209

Valenzuela LV, Villalba MI (2015) A new species of *Polylepis* (Rosaceae) from Peru. Arnaldoa 22(2):329–338

Vanclay JK (1995) Growth models for tropical forests: a synthesis of models and methods. For Sci 41(1):4–42. https://doi.org/10.1093/forestscience/41.1.7

Vásquez E, Ladd B, Borchard N (2014) Carbon storage in a high-altitude *Polylepis* woodland in the Peruvian Andes. Alp Bot 124(1):71–75. https://doi.org/10.1007/s00035-014-0126-y

von Bertalanffy L (1957) Quantitative laws in metabolism and growth. Q Rev Biol 32(3):217–231

West GB, Brown JH, Enquist BJ (2001) A general model for ontogenetic growth. Nature 413:628–631

Wigley TM, Briffa L, Jones PD (1984) On the average value of correlated time series, with applications in dendroclimatology and hydrometeorology. J Clim Appl Meteorol 23:201–213. https://doi.org/10.1175/1520-0450(1984)023<0201:OTAVOC>2.0.CO;2

Wood SW, Hua Q, Allen K et al (2010) Age and growth of a fire prone Tasmanian temperate old-growth forest stand dominated by *Eucalyptus regnans*, the world's tallest angiosperm. For Ecol Manag 260:438–447. https://doi.org/10.1016/j.foreco.2010.04.037

Worbes M (1999) Annual growth rings, rainfall-dependent growth and long-term growth patterns of tropical trees from the Caparo forest reserve in Venezuela. J Ecol 87:391–403. https://doi.org/10.1046/j.1365-2745.1999.00361.x

Worbes M (2004) Mensuration Tree ring analysis. In: Encypledia of forest sciences. Elsevier, Amsterdam, Holanda

Worbes M, Staschel R, Roloff A et al (2003) Tree ring analysis reveals age structure, dynamics and wood production of a natural forest stand in Cameroon. For Ecol Manag 173:105–123. https://doi.org/10.1016/S0378-1127(01)00814-3

Zuidema PA, Boot RGA (2002) Demography of the Brazil nut tree (*Bertholletia excelsa*) in the Bolivian Amazon: impact of seed extraction on recruitment and population dynamics. J Trop Ecol 18:1–31. https://doi.org/10.1017/S0266467402002018

Part IV
Forest Management and Conservation

Chapter 13
Dendrochronological Study of the Xeric and Mesic *Araucaria araucana* Forests of Northern Patagonia: Implications for Ecology and Conservation

Martín Ariel Hadad, Julieta Gabriela Arco Molina, and Fidel Alejandro Roig

Abstract The identification of the forest structure could give insights of the processes that determine certain aspects of the population dynamics over time. To address this concern in the *Araucaria araucana* forests of northwestern Patagonia in Argentina, 14 forest stands distributed along a strong precipitation gradient were analyzed. A suite of components of the forest structure were considered in order to determine (1) the relationship between climate and tree growth variability, (2) the age structure of each study site by defining age classes through the tree-ring analysis, (3) the relationship between tree age, tree height, and stem diameter, and (4) sex ratio. Xeric forests were affected significantly and negatively by the temperature of the current growing season, while mesic forests were more affected during winter months. Precipitation showed a positive effect on growth at the beginning of the growing season in all forests. Age distribution in the xeric and mesic environments showed a preponderance of trees between 51 and 300, and 51 and 250 years old, respectively. Broadly, male and female trees were more abundant in xeric and mesic

M. A. Hadad (✉)
Laboratorio de Dendrocronología de Zonas Áridas. CIGEOBIO (CONICET-UNSJ), Gabinete de Geología Ambiental (INGEO-UNSJ), San Juan, Argentina
e-mail: mhadad@unsj-cuim.edu.ar

J. G. Arco Molina
Laboratorio de Dendrocronología e Historia Ambiental, IANIGLA-CCT CONICET-Mendoza, Mendoza, Argentina
e-mail: jarco@mendoza-conicet.gob.ar

F. A. Roig
Laboratorio de Dendrocronología e Historia Ambiental, IANIGLA-CCT CONICET-Mendoza, Mendoza, Argentina

Facultad de Ciencias Agrarias, Universidad Nacional de Cuyo, Mendoza, Argentina

Hémera Centro de Observación de la Tierra, Facultad de Ciencias, Universidad Mayor, Santiago, Chile

ESALQ, Universidade de São Paulo, Piracicaba, Brazil
e-mail: froig@mendoza-conicet.gob.ar

M. Pompa-García, J. J. Camarero (eds.), *Latin American Dendroecology*,
https://doi.org/10.1007/978-3-030-36930-9_13

forests, respectively. The tree age-diameter and tree height-diameter relationships were significantly positive in both mesic and xeric environments. Through these elements, some aspects of the *A. araucana* trees and forests in their natural distribution area were revealed, which consolidate a better understanding of the ecology and dynamics of these endemic forests and provide tools for designing proper conservation strategies.

Keywords Forest structure · Precipitation gradient · Sex ratio · Tree age · Tree diameter

13.1 Introduction

13.1.1 Araucaria araucana *Forests*

Araucaria araucana (Molina) K. Koch (Araucariaceae), commonly named "pehuén" or "araucaria," is a dioecious evergreen conifer endemic of the temperate forests of south-central Argentina and Chile that spreads between 37° 20′ and 40° 20′ SL and through an altitudinal range between 900 and 1800 m asl (Roig and Villalba 2008). The pehuén is known to be a long-lived tree species, reaching ages surpassing 900 years, particularly in forest patches placed on rocky outcrops at the forest-steppe boundary (LaMarche et al. 1979; Mundo et al. 2012; Hadad et al. 2015; Aguilera-Betti et al. 2017). The presence of a strong precipitation gradient from the Andes Mountain to the Patagonian steppe produces mesic forests that are dense and mixed mainly with *Nothofagus pumilio* (Poepp. & Endl.) Krasser (lenga) and *N. antarctica* (G. Forst.) Oerst. (ñire) on the eastern slopes of the cordillera, while in the transition to the dry steppe produces open, monospecific xeric forest patches strongly associated with rocky outcrops (Veblen 1982; Donoso 1993; Gandullo 2003). While the mesic araucaria forests show abundant regeneration of sexual origin, the xeric forests have a strong incidence of asexual reproduction (Gandullo 2003).

A. araucana has been declared as a protected species according to the N° 1890/91 and N° 2780/11 laws of the Province of Neuquén in Argentina; it is considered as an endangered species (Category: endangered (EN), Criteria: B2ab(ii,iii,v)) by the Red List of the Threatened Species of the IUCN (International Union for Conservation of Nature) 2017–3 (http://www.iucnredlist.org/) and it has been included in the Appendix I of CITES (Convention on International Trade in Endangered Species of Wild Fauna and Flora) due to its critical status of conservation (https://www.cites.org/). Therefore, *A. araucana* is considered to be facing a very high risk of extinction in the wild.

13.1.2 Dendroecology

The identification, dating, and measuring of annual growth rings are widely used to determine the age of trees and their growth, variables that are useful to study the forest ecology and productivity. Thus, through measuring growth rings it can be established the relationship between climatic factors and tree growth (Fritts 1976) as well as the reconstruction of forest disturbance histories, past climatic variations, and different features associated with the individuals. For example, in the case of *Pinus ponderosa* (Dougl. ex Laws) from North America, various factors affecting seedling establishment have been mentioned, such as reduction of the livestock pressure and years of high seed production following by years free of heat and drought (Mast et al. 1998). These dendrochronological studies also reveal the importance of regional historical processes in the forest composition and structure (Mast et al. 1998; Ehle and Baker 2003; Brown and Wu 2005). Although the use of the dendrochronology for the study of forest ecosystems, climate, and their relationships is relatively new in South America, with the first systematic tree-ring studies developed in the temperate regions of Patagonia (e.g., Schulman 1956; LaMarche et al. 1979), its development has been growing during the last decades (Roig and Villalba 2008). Particularly, for the northern region of Patagonia, factors associated with climatic variations and changes in disturbance regimes have been suggested to be involved in the age structure of *Austrocedrus* and *Nothofagus* forests (Villalba and Veblen 1997; Kitzberger and Veblen 2003; Molowny-Horas et al. 2017).

13.1.3 Structure and Dynamics of Forests

At the landscape scale, patterns of forest regeneration are the result of biotic interactions in relation to two superimposed patterns: the distribution of species controlled by environmental gradients and disturbance regime (Romme and Knight 1982). Since the forest structure plays major roles in ecosystem functions and diversity, its understanding can help to interpret the history, function, and dynamics of forest ecosystems. In this sense, researches on the age structure could give insights into the processes that determine certain aspects of the population structure over time (Svensson and Jeglum 2001), being one of the prerequisites for understanding ecological processes and restoration of natural forests (Wang et al. 2004). The structure, composition, and current pattern of a forest can be influenced by many factors throughout its development history, including competitive interactions between trees or disturbances and differences in available resources (North et al. 2004). These factors affect, in turn, the growth of trees in particular and, consequently, the dynamics of forests in general (Lebourgeois et al. 2014). Trees of natural populations tend to exhibit a great variation in growth, and the analysis of this variation is fundamental for understanding the forest ecology (Coomes and Allen 2007). Forest communities experience changes not only in time but also in space, often being influenced by changes in environmental conditions due to different types of natural and/or anthropogenic disturbances (Donoso 1993).

Dynamics of forest, in particular of those composed by long-lived trees, could be considered as an indicator of vegetation succession influenced by climate variability and change (Wang et al. 2004). Thus, climate variability is an important factor affecting the regeneration dynamics and successional processes in forest communities. It has been argued that the increased sensitivity in the development of tree populations toward their limits of natural distribution (ecotone) is due to climate stress where it constitutes one of the main obstacles to the recruitment of trees (Fritts 1976; Camarero and Gutiérrez 2007). Therefore, quantitative reconstructions of the conditions that affect the age structure (distribution and age range of the trees) in areas of strong climatic gradients, such as those considered here, could also serve as a reference point for conservation, restoration, and management of forest ecosystems.

13.1.4 Sex Ratio

A large number of factors can contribute to the proportion of sexes in a population of a dioecious tree species. These include genetic (Ainsworth 2000) and environmental factors that can influence the fertilization, germination, flowering, and mortality processes or the combination of all or some of the above processes (Sinclair et al. 2012). In many dioecious species, the sex ratio does not differ from 1:1, which is a progeny ratio that optimizes parental reproductive success, according to theoretical models (Charnov 1982), but this ratio can also be modified by environmental factors.

Furthermore, the expression of dioecious plants leads to sex-related differences in the hormones at the whole plant level, which can affect vegetative growth and stress tolerance differently in females and males (Munné-Bosch 2015). In this sense, in context of a warming world, the rate of mortality in female plants may outpace the rate of mortality in co-occurring males, particularly in species where females display a greater physiological sensitivity than males to a warmer climate (Hultine et al. 2016). In the same way, long-lived dioecious species may have sex ratios that are highly susceptible to rapid shifts in weather owing to their long generation turnover times that restrict the rate of adaptation to rapid changes of the environmental conditions (Hultine et al. 2016). Even though the dioecious character is rare in Patagonian trees, there are already some observations about it for the north of this region. Thus, Gallo et al. (2004) reported higher proportion of female trees for mesic araucaria forests while Nuñez et al. (2008) reported a 1:1 sex ratio for *A. chilensis* forests at a regional level.

13.1.5 Relationship Between Tree Growth and Climate in **Araucaria araucana**

Previous dendroclimatological studies reported that the inter-annual variability in the tree rings of *A. araucana* trees reveals a strong dependence to the spring-summer climate conditions during the current and prior period of growth (Mundo et al. 2012; Muñoz et al. 2014; Hadad et al. 2015; Hadad and Roig 2016). At the hemispheric scale, it has been observed that growth is negatively affected by higher spring-summer values of the Southern Annular index (Mundo et al. 2012; Villalba et al. 2012; Hadad et al. 2015). When working at the individual scale, the tree-ring sensitivity to climate appeared to be age and sex dependent (Hadad et al. 2015; Hadad and Roig 2016; Arco Molina 2018). Thus, in general, younger trees appear to be more sensitive to precipitation during the current growing period while male trees are more sensitive to precipitation during part of the current growing season. Female trees seemed to be more sensitive to the air surface temperature during the prior period of growth.

13.1.6 Disturbance Factors in **Araucaria araucana** *Forests*

Araucaria forests are subjected to large-scale regime of disturbances linked to different origins (Armesto et al. 1996). Fire is the main disturbance factor modeling the dynamics of the araucaria woodlands, impacting both at stand and landscape scale (Burns 1993; Aagesen 2004; González et al. 2005; González and Veblen 2006; González et al. 2010; Assal et al. 2018). From an historical perspective, an increase in the fire frequency after 1850 in the *A. araucaria* forests of Argentina has been recognized, period that coincides with the increase in the human-induced fires in the region (Mundo et al. 2013).

The ecotone environment, as the one between the Andean forest and the Patagonian steppe, is often affected by changes in the disturbance regimes of both natural and anthropic origin (League and Veblen 2006). The *A. araucana* forests without or with little evidence of anthropogenic disturbances tend to present a good representation of all age and diameter classes, multi-stratified structure, and good regeneration (Roig 1998; Rechene 2000). One of the most important anthropic factors that negatively influences the tree and forest structures, mainly at the forest-steppe ecotone of Patagonia, is the presence of cattle. Evidence of livestock impacts in the region since the early 1890s shows records of strong intensity during the period between 1930 and 1940 (Eriksen 1971). During this time, these areas were colonized by European settlers who presented a strong culture of forest clearings using fire in order to favor the production of pastures for feeding the domestic livestock (Veblen et al. 2003). According to the local indigenous people, pigs, goats, and rabbits are recognized as factors that negatively affect araucaria forests and their regeneration, since they are important seed consumers (Herrmann 2006).

Although araucaria is highly tolerant to different disturbances (Burns 1993), the increasing levels of cattle ranching have had a strong impact on its regeneration by reducing the number of seedlings and saplings and by decreasing the ratio of sexual/asexual regeneration (Zamorano-Elgueta et al. 2012).

13.1.7 Objective

Due to the strong precipitation gradient along the *A. araucana* distribution area in Argentina, the many ontogenetic and environmental factors that can affect the forest ecology, and the critical state of conservation of the araucaria forests, it is relevant to determine their current state considering dynamics and structure aspects along time and space. In this sense, the aims of this study were to determine (1) the relationship between the climate and tree growth variability, (2) the age structure of the forests, (3) the relationship between tree age, height, and diameter, and (4) the sex ratio of different sites along the precipitation gradient through the analysis of the tree rings in both mesic and xeric conditions. The results may help to better know the current state of these emblematic forests in order to encourage proper conservation, preservation, and management strategies.

13.2 Material and Methods

13.2.1 Study Area, Field, and Laboratory Work

Fourteen *A. araucana* forests in the north of Patagonia in Argentina (corresponding to six xeric and eight mesic sites; Table 13.1; Fig. 13.1 and 13.2) were analyzed. The xeric sites are distributed in an area whose mean annual temperature is 11.11 °C (period 1912–2005) and the total annual rainfall around 573 mm (period 1929–2001; Fig. 13.3a), while the mesic sites comprise other values with a mean annual temperature of 8.71 °C (period 1912–2008) and a total annual rainfall of 1081 mm (period 1931–2010; Fig. 13.3b). At each study site, random plots were selected and all trees included in the plots that presented diameters at breast height (DBH) of 8 cm or more were considered for this study. For each tree, the DBH and the total height from the ground were measured. No tree height data were available for the RK site. The sex of the trees was identified observing the presence and type of the strobiles, and the trees were classified as (1) male, (2) female, and (3) undifferentiated (Fig. 13.4). The "undifferentiated" trees corresponded to individuals that had not yet reached their sexual maturity at the sampling time. The sex ratio (number of male/number of female, M/F) was determined for each study site. Furthermore, at least two wood samples per tree were taken at breast height (1.30 approximately) using a Pressler increment borer.

Table 13.1 Characteristics of the study sites from Neuquén Province in the northwestern Patagonia of Argentina

Forest type	Sites (Code)	Latitude S	Longitude W	Altitude	N° of trees
Xeric	Río Agrio (RA)	37° 50′ 02"	70° 58′ 38"	1594	78
	Chenque-Pehuén (CH)	38° 05′ 59"	70° 52′ 36"	1653	82
	Primeros Pinos (PP)	38° 52′ 09"	70° 34′ 26"	1628	44
	Río Kilca (RK)	38° 53′ 53"	70° 50′ 31"	1442	36
	Carreri Malal (CM)	38° 55′ 36"	70° 32′ 59"	1510	57
	Catan Lil (CL)	39° 14′ 3"	70° 40′ 55"	1290	36
Mesic	Caviahue (CV)	37° 51′ 11"	71° 02′ 23"	1676	37
	Las Lecheras (LL)	37° 58′ 36"	71° 02′ 26"	1605	42
	Batea Mahuida (BM)	38° 50′ 01"	71° 10′ 47"	1598	24
	Valle El Salvo (VS)	38° 56′ 03"	71° 25′ 14"	1294	12
	Verde Chico (VC)	38° 56′ 19"	71° 23′ 33"	1267	15
	Rucachoroi (RC)	39° 13′ 24"	71° 15′ 47"	1214	39
	Pinalada Redonda (PR)	39° 18′ 36"	71° 17′ 05"	1100	52
	Senda a Rucachoroi (SR)	39° 19′ 07"	71° 12′ 44"	1507	16

In the laboratory, wood samples were mounted on wooden supports and delicately polished with sandpaper of increasing granulometry to attain a perfect anatomical identification of the boundary between growth rings (Stokes and Smiley 1968). Ring widths were measured using a measuring device (Velmex, USA) with a precision of 0.001 mm. Statistical quality control of the ring width measurements was performed using the COFECHA program (Holmes 1983). This program converts the original measurement value into an index, eliminating the low frequency of the series and maximizing the inter-annual or high-frequency variability. Then, these series of indexes are compared simultaneously to verify the verisimilitude of the growth ring dates.

13.2.2 Chronologies Development

Due to the variability of the tree-ring width is not only affected by climatic factors, but also by the age and processes of forest dynamics (Fritts 1976), the series of measurements were processed statistically. In order to study the relationship between tree growth and climate, it is necessary to remove those influences that cannot be explained by climate incidence (such as those derived from competition between trees or from the biological age of the trees). To achieve this, the original tree-ring width series were standardized and converted into indexes (Cook 1985). Tree-ring width index chronologies were developed for each forest type using the ARSTAN program (Cook 1985). The ring width series were standardized by applying a smoothed spline with a cutoff frequency of 50% and a response period of 50 years (Cook 1985).

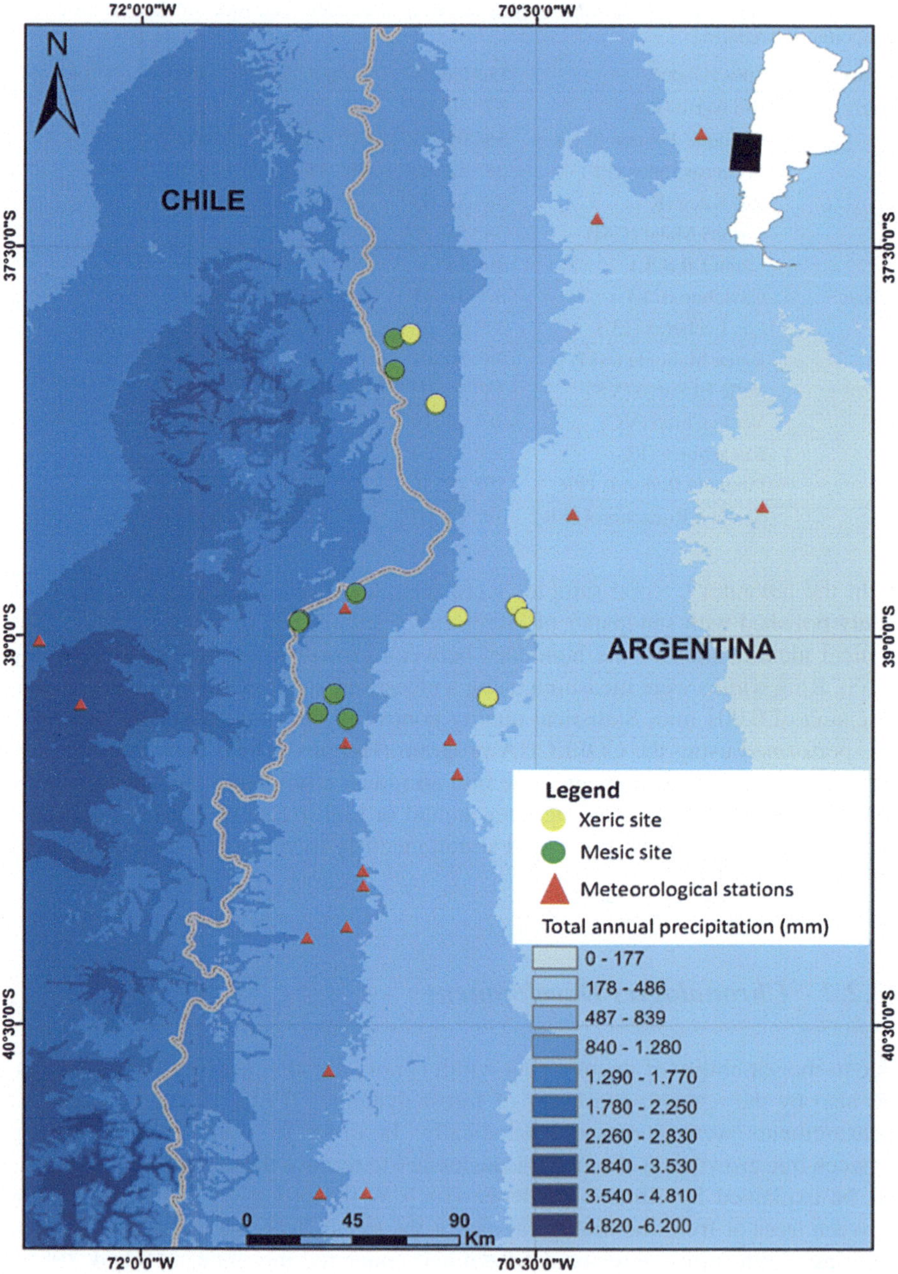

Fig. 13.1 Location of the study sites in the NW Patagonia, Argentina. Yellow and green circles indicate xeric and mesic sites, respectively. Red triangles indicate meteorological stations

Fig. 13.2 Pictures showing the (**a**) *Araucaria araucana-Nothofagus pumilio* forest association characteristic of the mesic environments (Rucachoroi site) and (**b**) open and monospecific *Araucaria araucana* xeric forest in contact with the Patagonian steppe (Primeros Pinos site)

13.2.3 Relationship Between Tree Growth and Climate

The residual versions of the chronologies were compared with climatic variables (mean monthly air temperature and total monthly precipitation) using Pearson correlation functions. The correlation analyses make it possible to identify the main climatic factors that influence the tree growth as well as the seasonality and the strength of the tree growth responses (Fritts 1976). Because climatic conditions from the prior growing season can influence the current tree growth, the correlations

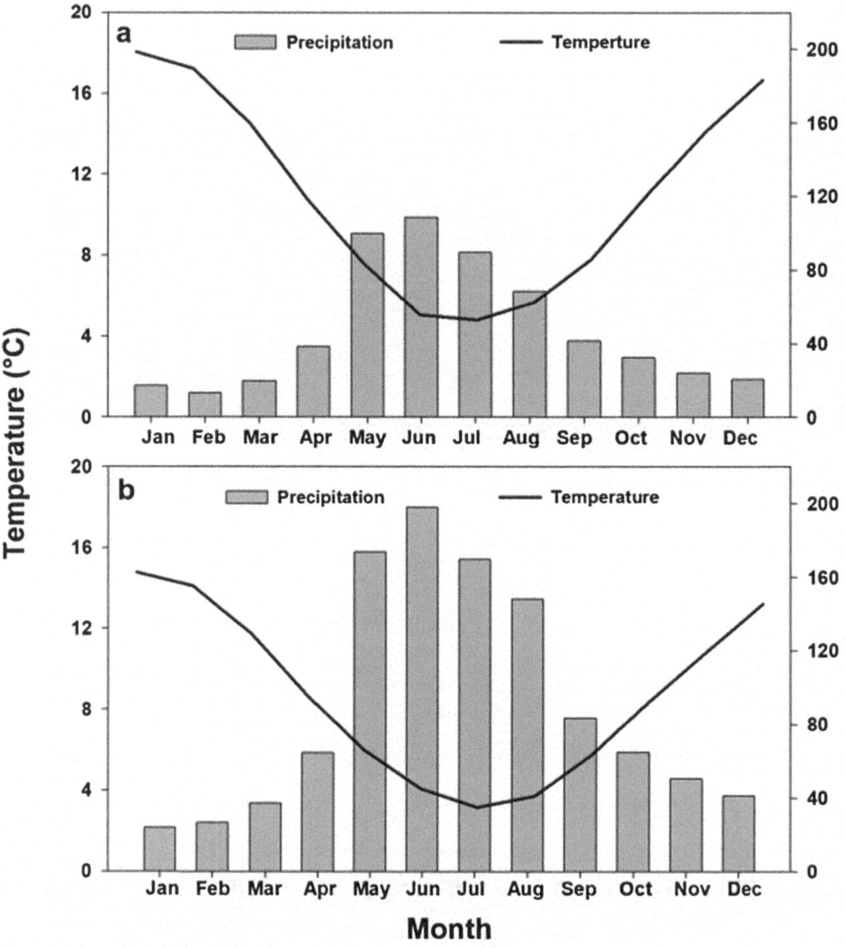

Fig. 13.3 Climate diagram for the (**a**) xeric and (**b**) mesic *Araucaria araucana* forests (period 1901–2018). Solid lines represent the temperature and bars, the precipitation

were carried out for a period of 21 months, comprising from September of the year before the year of formation of the growth ring until April of the current growing period, thus encompassing two growing seasons.

For the radial growth and climate relationship analysis, four regional monthly climate records were performed, one per each variable and per environment. The regional series were built using the MET program (http://www.ldeo.columbia.edu/tree-ring-laboratory/resources/software). Table 13.2 and Fig. 13.1 shows the meteorological stations used. The regional series of mean monthly temperature comprised the period 1912–2008 and 1912–2005 for the mesic and xeric environments, respectively, while total monthly precipitation covered the period 1931–2010 and 1929–2001 for the mesic and xeric forests, respectively. These series were selected because they were the most extensive, complete, and closest to the study sites.

Fig. 13.4 Evidence of the dioecious trait in *Araucaria araucana*: (**a**) male cones and (**b**) female cones in separate trees

13.2.4 Age Structure

The cambial age of the trees (rings from the pith) was determined counting the number of tree rings between the bark and the pith. For those series not reaching the pith, but with the inner arch close to it, the number of missing rings was estimated following the geometric method proposed by Duncan (1989). This minimized the difference between the sample cambial age and the real cambial age. The forest age structure was analyzed in 50-year intervals (1–50 years, 51–100 years, 101–150 years, etc.).

13.2.5 Data Analyses

In each plot of the mesic sites, a transect of 30 m × 5 m was determined and the number of *A. araucana* seedlings within it was counted. The seedlings were classified according to their height in four classes: (1) 0–20 cm, (2) 21–50 cm, (3) 51–100 cm, and (4) more than 101 cm. This allowed characterizing the forest understory composed by araucaria plants, differentiating it by strata and estimating the regeneration status of each site. The relative presence of the dominant or codominant tree species *A. araucana* and *N. pumilio* and the relative presence of the araucaria seedlings of the understory at each mesic site were analyzed through contingency tables. These analyses were carried out for the mesic sites since the seedling counting was performed only in this environment and the xeric sites presented monospecific forests. The statistical significance of the difference between proportions was evaluated using Pearson's chi-squared test with a level of significance of 0.05. The relationship between the variables was measured using the Pearson's contingency coefficient (C). This coefficient is not affected by the sample size and takes values between 0 and 1. If the variables are independent, C is equal to 0, while the greater the association between them the higher the value of C.

Table 13.2 Data of the meteorological stations used for comparison with tree-ring chronologies

Station	Environment	Latitude S	Longitude W	Altitude (m)	Period (*years*)	Parameter	Source
San Martín	Mesic/Xeric	40.166	71.333	426	1936–1972 (*37*)	P	SMN
Rahue	Mesic	39.369	70.931	845	1975–2010 (*36*)	P	AIC
Est. Mamuil Malal	Mesic/Xeric	39.650	71.233	990	1935–1976 (*42*)	P	HIDRONOR
Lago Traful	Mesic	40.700	71.233	474	1936–1976 (*41*)	P	SMN
Lago Aluminé	Mesic	38.966	71.050	533	1933–1969 (*37*)	P	SMN
Flor del Lago	Mesic/Xeric	39.200	72.100	300	1931–1960 (*30*)	P	KNMI
Est. Colluncó	Mesic/Xeric	39.964	71.199	873	1933–2010 (*78*)	P	AIC
Est. Casa de Lata	Mesic	39.846	71.177	848	1933–2010 (*78*)	P	AIC
Campo Grande	Mesic	39.500	70.630	960	1948–1998 (*51*)	P	AIC
Allipen Chiduco	Mesic	39.000	72.400	220	1931–1960 (*30*)	P	KNMI
Bariloche	Mesic	41.150	71.166	700	1956–2009 (*54*)	T	SMN
Buta Ranquil	Mesic/Xeric	37.080	69.750	850	1966–2005 (*40*)	T	SRH
Chos Malal	Mesic	37.380	70.270	900	1976–2005 (*30*)	T	SRH
Colluncó	Mesic/Xeric	40.100	71.200	875	1912–1989 (*78*)	T	AIC
Las Lajas	Xeric	38.516	70.366	713	1916–1973 (*58*)	T	SRH

(continued)

Table 13.2 (continued)

Station	Environment	Latitude S	Longitude W	Altitude (m)	Period (*years*)	Parameter	Source
Paso de los Indios	Mesic	38.530	69.410	498	1963–2005 (*43*)	T	SRH
Bariloche 2	Xeric	41.150	71.266	825	1951–2009 (*59*)	T,P	SRH

P total precipitation, *T* mean temperature, *SMN* Servicio Meteorológico Nacional, *AIC* Autoridad Interjurisdiccional de las Cuencas de los ríos Limay, Neuquén y Negro, *SRH* Subsecretaría de Recursos Hídricos de la Nación, *HIDRONOR* Hidroeléctrica Norpatagónica S.A., *KNMI* Koninklijk Nederlands Meteorologisch Instituut

The relationships between the stem diameter, tree age, and height at the regional level for both mesic and xeric environments were analyzed using linear mixed-effects (LME) models. A generalized functions or models of height diameter and age diameter at a regional level estimate the specific relationship between these variables considering the differences between sites and environments. The LME models represent a valuable methodology for the analysis of different responses that involve many sources of variability, since they incorporate variables with fixed effects (values are independent between observations) and random effects (a certain relationship between the observations within the same sampled unit is assumed). Thus, the incorporation of the hierarchical structure of the data (random effects) in the analysis allows reducing the interdependence between the observations within the same sampling unit (Seoane 2014). The LME models allow, in turn, identifying the real effect of the factors that affect a process or variable avoiding incorporating irrelevant sources of variability (Venables and Ripley 2002). In this sense, the conceptual models for tree age and total height were the following:

$$\text{Tree age} \sim \text{fixed effects}(\text{diameter}) + \text{random effects}(\text{forest type / sites})$$
$$\text{Tree height} \sim \text{fixed effects}(\text{diameter}) + \text{random effects}(\text{forest type / sites})$$

The variables were previously centered and normalized to facilitate their comparison independently of their measurement unit. Heteroscedasticity was corrected in both models. The models were performed using the “nlme” package (Pinheiro et al. 2017) under R environment (R Core Team 2016). The best models were selected using the Akaike information criterion (AIC).

To compare the age, height, and DBH between sites, environments, and sexes, the non-parametric Kruskal–Wallis test (H-value is the test statistic for the K-W test) was applied, and the differences were tested using the post-hoc Dunn’s test.

13.3 Results

13.3.1 Tree Growth Responses to Climate

After the process of dating control, samples with dating problems were rejected, reaching a total of 322 series for the xeric sites and 429 for the mesic ones. The tree-ring index chronology of the xeric forest covered the period AD 1190 to AD 2011 (822 years of extension) and the one of the mesic forest covered the period AD 1271 to AD 2013 (743 years of extension) (Fig. 13.5).

Correlations between regional climate data and residual regional chronologies from both environments showed that radial growth was significantly and negatively affected by mean air temperature during January to March of the prior growing period (Fig. 13.6a). Mean temperature of June and July of the current growing season presented a significant and negative relationship with tree growth in the mesic forests, while tree growth was significant and negatively affected during November and December in the xeric forests (Fig. 13.6a). In general, the response pattern obtained between tree growth and total precipitation was similar for the two forest types (Fig. 13.6b). Xeric and mesic forests had significant and positive correlations during January of the prior growing season and September of the current period of growth.

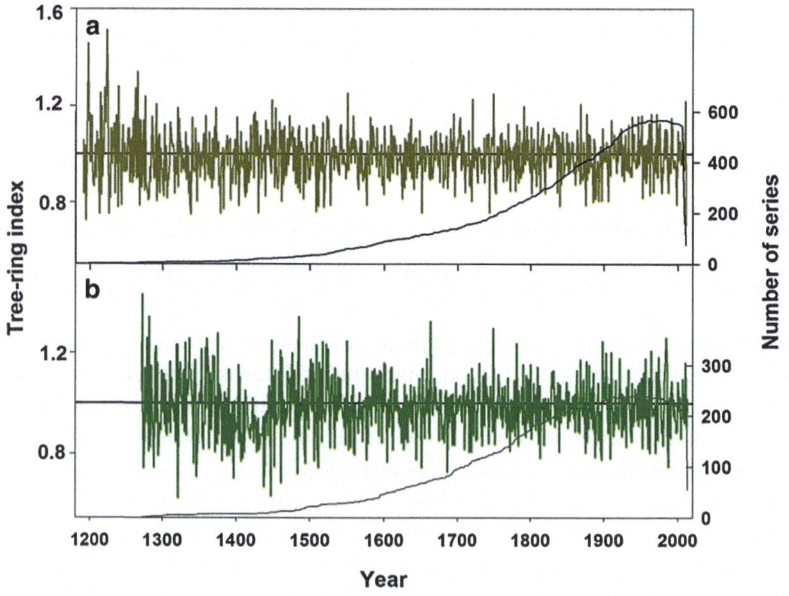

Fig. 13.5 Residual tree-ring index chronologies of the (**a**) xeric and (**b**) mesic *Araucaria araucana* forests

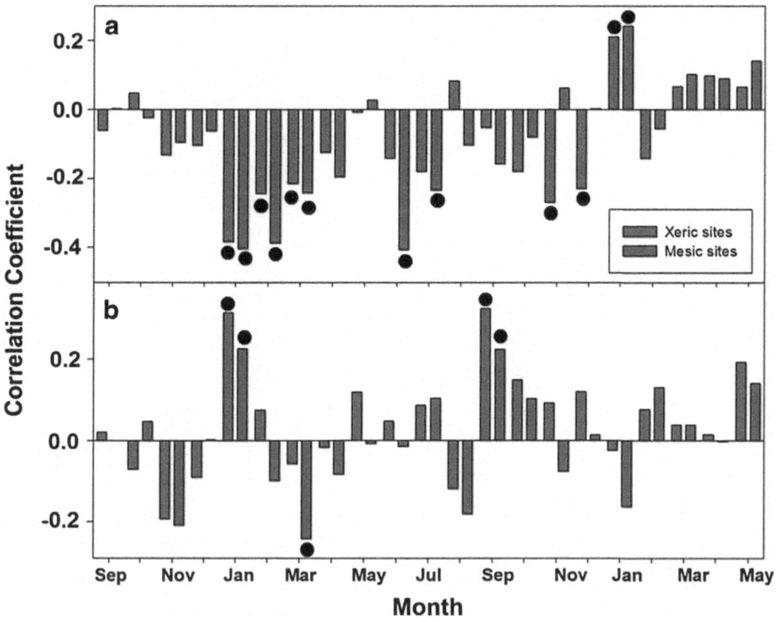

Fig. 13.6 Correlation coefficients between *Araucaria araucana* residual chronologies from the mesic sites (green bars) and from the xeric sites (brown bars) and (**a**) mean monthly regional temperature and (**b**) total monthly regional precipitation. Black circles indicate the significant level at $p < 0.05$

13.3.2 Characterization of the Xeric and Mesic Sites

Tables 13.3 and 13.4 show the number of sampled trees, the total sampled area, and the tree density per plot and per site for the xeric and mesic *A. araucana* forests. It was observed that the density within each site was very variable. In the xeric environment, RA presented the lowest tree density (99.05 tree/ha) and CL the highest (207.12 trees/ha) (Table 13.3). The mesic sites presented, in general, a considerably higher tree density than the xeric ones, varying from 433 to 589 tree/ha in SR and LL, respectively, except PR that presented a considerably higher tree density (1050 tree/ha) (Table 13.4).

Regarding the diversity of tree species, all the mesic sites belonged to the forest subtype "Pehuén-Lenga mixed forest" (*Araucaria araucana-Nothofagus pumilio*) (Burns 1991) although only 80% of the sampled plots presented both species (Table 13.5). The contingency table (Table 13.5) showed that although there was a relationship between the presence of the species and the sites (Pearson's chi squared <0.0001), araucaria was the dominant tree in all sites. The C coefficient was 0.33, indicating an intermediate association between the variables "species" and "sites."

Moreover, the understory of araucaria plants was characterized according to different tree strata. In RC and LL sites, three plots were selected but only in two of

Table 13.3 Tree density per sampled plot and total density of the xeric study sites (see site codes in Table 13.1)

Xeric sites	N° plots	N° trees	Total N° of trees	Area (ha)	Density (tree/ha)	Total density (tree/ha)
RA	1	5	64	0.05	90.77	99.05
	2	4		0.04	98.91	
	3	4		0.05	76.39	
	4	4		0.04	96.80	
	5	2		0.08	225.28	
	6	4		0.06	63.56	
	7	3		0.04	74.58	
	8	22		0.2	94.94	
	9	12		0.2	45.45	
	10	4		0.03	123.08	
CH	1	17	82	0.04	388.237	203.93
	2	9		0.03	135.17	
	3	6		0.1	51.94	
	4	10		0.04	240.38	
PP	1	9	44	0.07	121.31	179.46
	2	4		0.01	201.773	
	3	9		0.1	67.54	
	4	6		0.04	137.06	
	5	11		0.03	326.29	
	6	5		0.02	222.81	
RK	1	3	29	0.02	132.63	149.85
	2	4		0.05	79.71	
	3	3		0.03	76.13	
	4	4		0.07	53.33	
	5	6		0.01	301.32	
	6	9		0.03	256.01	
CM	1	22	58	0.07	311.26	207.12
	2	22		0.1	141.02	
	3	14		0.08	169.08	

them the transect was determined, while in PR and SR only one plot was sampled; therefore, the data corresponded to that single plot. The results showed that, in general, as the height of the seedlings increased, their abundance decreased, as can be observed in the total column of the heights in Table 13.6. An exception was observed in the RC and VS sites where seedling between 21 and 50 cm were more abundant than those among 0 and 20 cm in height. On the other hand, RC and VS presented the highest and lowest number of seedlings, respectively. Pinalada Redonda was the second site with the highest abundance of seedlings despite data corresponded to a single sampling transect. The statistical analysis of the Table 13.6 indicated that there was a relationship between the seedling height and the study sites (Pearson's chi squared <0.0001) while the coefficient C was 0.44, indicating a moderate

Table 13.4 Tree density per sampled plot and total density of the mesic study sites (see site codes in Table 13.1)

Mesic sites	N° plots	N° trees	Total N° of trees	Area (ha)	Density (tree/ha)	Total density (tree/ha)
CV	1	34	57	0.06	566.67	475
	2	23		0.06	383.33	
LL	1	28	106	0.06	466.67	588.89
	2	28		0.06	466.67	
	3	50		0.06	833.33	
BM	1	29	58	0.06	483.33	483.33
	2	29		0.06	483.33	
VS	1	21	54	0.06	3.0	450
	2	33		0.06	550	
VC	1	43	67	0.06	716,67	558.33
	2	24		0.06	400	
RC	1	28	81	0.06	466.67	540
	2	19		0.03	633.33	
	3	34		0.06	566.67	
PR	1	63	63	0.06	1050	1050
SR	1	26	26	0.06	433.33	433.33

relationship between these variables. The BM site was not included in the analysis due to the height categories determined did not match those from the rest of the sites and therefore they could not be compared.

13.3.3 Age Structure of A. araucana *Forests*

From the six xeric sites, a total of 333 trees were considered. In RA, PP, CL, and CM sites, tree establishment was more or less continuous during the last 500 years. Although the age classes of the xeric forests ranged from 51 to 820 years, the highest number of individuals was observed between 51 and 300 years of age (Fig. 13.7). All sites presented trees older than 500 years except CL. In RA it was observed the highest number of old trees while in RK it was found the oldest tree with an age of 820 years. In the age classes younger than 100 years old, there was a decrease in the number of trees, except in CH site where a larger number of individuals was observed in the age classes between 51 and 100 years (Fig. 13.7). In CH was found the youngest tree with 28 years old.

From the eight mesic sites, 237 trees were considered. As in the xeric sites, the age distributions in the mesic forests were variable between sites. PR was the site with the highest presence of young trees, while in VS, SR, and BM it was observed a higher abundance of old trees (Fig. 13.8). In RC and LL, the number of trees was similar in almost all the age classes. In VC the highest number of trees was registered between 100 and 300 years old, and no trees older than 300 years were observed.

Table 13.5 Absolute and relative frequencies of the presence of *Araucaria araucana* and *Nothofagus pumilio* with diameters at breast height (DBH) ≥ 8 cm at each mesic site (see site codes in Table 13.1)

	Mesic sites																Total	
	BM		CV		LL		PR		RC		SR		VC		VS			
Species	*n*	%	*n*	%	*n*	%	*n*	%	*n*	%	*n*	%	*n*	%	*n*	%	*n*	%
Araucaria	49	84.48	57	100	86	81.13	63	100	66	80.49	20	76.92	67	100	34	62.96	442	86.16
Lenga	9	15.52	0	0	20	18.87	0	0	16	19.51	6	23.08	0	0	20	37.04	71	13.84
Total	58	100	5	100	106	100	63	100	82	100	26	100	67	100	54	100	513	100

Table 13.6 Absolute and relative frequencies of the presence of *Araucaria araucana* seedlings according to different height categories at each mesic site (see site codes in Table 13.1)

	Mesic sites														Total	
Height	CV		LL		PR		RC		SR		VC		VS			
(cm)	*n*	%	*n*	%	*n*	%	*n*	%	*n*	%	*n*	%	*n*	%	*n*	%
0–20	27	48.21	52	40.63	117	68.82	42	22.46	66	88.00	63	44.68	13	25.49	380	47.03
21–50	19	33.93	35	27.34	45	26.47	78	41.71	8	10.67	64	45.39	27	52.94	276	34.16
51–100	4	7.14	16	12.50	4	2.35	41	21.93	0	0	9	6.38	5	9.80	79	9.78
> *101*	6	10.71	25	19.53	4	2.35	26	13.90	1	1.33	5	3.55	6	11.76	73	9.03
Total	56	100	128	100	170	100	187	100	75	100	141	100	51	100	808	100

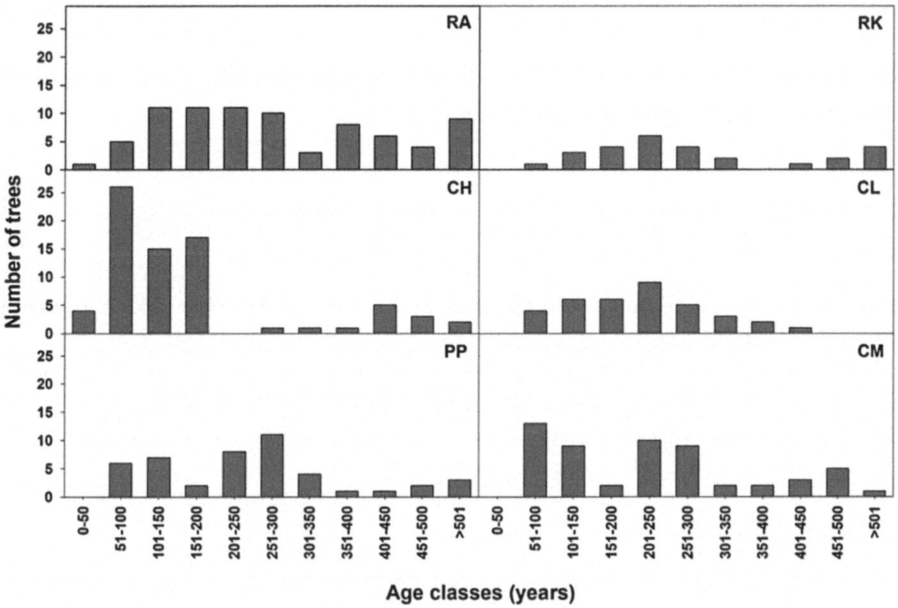

Fig. 13.7 Distribution of the age classes of the *Araucaria araucana* trees of the different sites from the xeric environment (see site codes in Table 13.1)

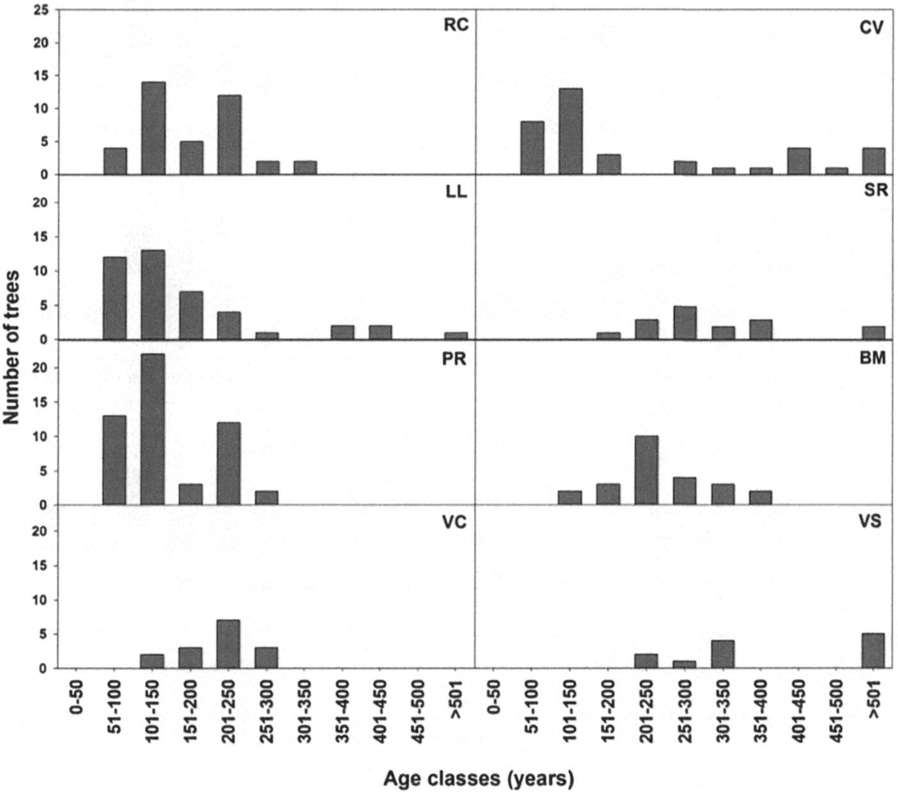

Fig. 13.8 Distribution of the age classes of the *Araucaria araucana* trees of the different sites from the mesic environment (see site codes in Table 13.1)

The oldest trees were found in VS (742 and 779 years old) (Fig. 13.8), while the youngest one was found in LL with 59 years old.

13.3.4 Relationships Between Tree Age, Height, and DBH

In the xeric forests, significant differences in the cambial age ($H = 36.01$, p-value <0.0001), the DBH ($H = 52.87$, p-value <0.0001), and height ($H = 35.91$; p-value <0.0001) between the different sites were observed. The highest mean age was recorded in RK (296 years) while the highest mean DBH was found in RA (66 cm) (Fig. 13.9). The lowest mean age and DBH corresponded to CH (175 years and 39 cm, respectively). The oldest araucaria tree was identified in RK reaching 820 years old. The highest and lowest mean height was shown in RA (8.98 m) and CM (5.56 m), respectively.

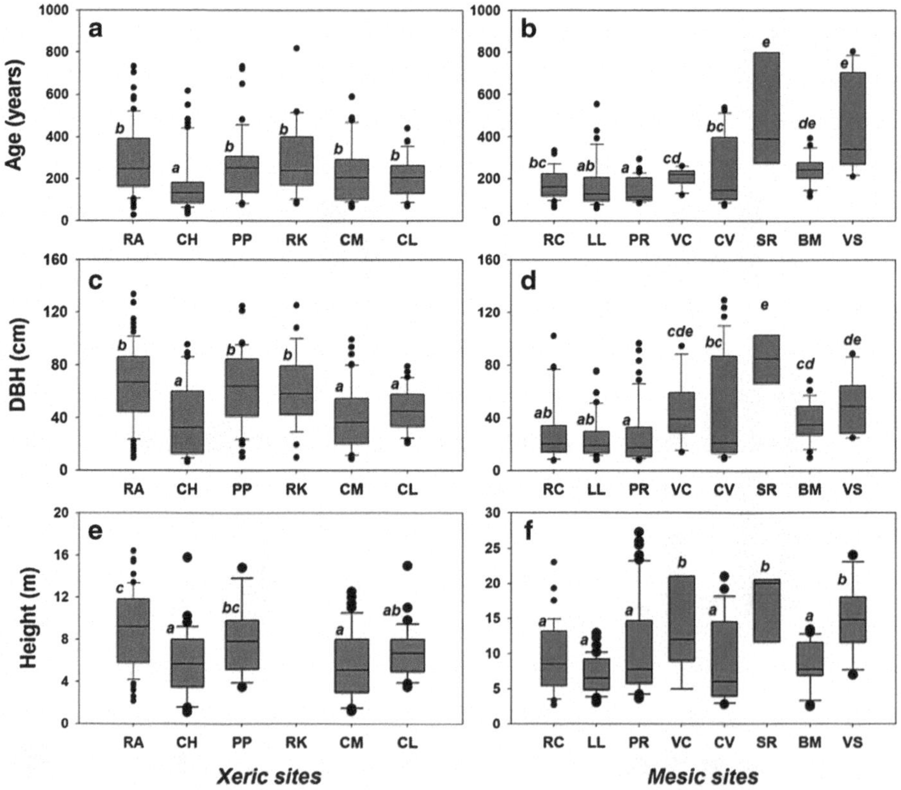

Fig. 13.9 Mean values (± standard deviation) of the tree age (**a** and **b**), DBH (**c** and **d**), and height (**e** and **f**) of *Araucaria araucana* trees from xeric (left panels) and mesic (right panels) forests. Different letters indicate significant differences at $p < 0.05$

In the mesic forests, it was also observed that there were significant differences in the cambial age ($H = 71.76$, *p*-value <0.0001), in the DBH ($H = 38.46$, *p*-value <0.0001) and height ($H = 36.88$, *p*-value <0.0001) between the different sites. The highest mean age and mean DBH were observed in SR (454 years and 85 cm, respectively). The lowest mean age corresponded to PR (142 years) and the lowest DBH to LL (24 cm) (Fig. 13.9). The oldest araucaria tree was 779 years old and it was found in VS. The highest mean heights were observed in SR and VS (16.22 m and 16.01 m, respectively) and the lowest mean height was recorded in LL (7.46 m) (Fig. 13.9).

When comparing the mean age between the two *A. araucana* forest types significantly differences ($H = 4.03$, *p*-value <0.0447) were found, showing the xeric forests a mean age (± Standard Error, SE) of 238 years (± 150.13) and the mesic ones a mean age of 211 years (± 137.37). The mean DBH between the two forest conditions was also significantly different ($H = 58.40$, *p*-value <0.0001). The mean DBH was 0.52 m (± 0.29) and 0.35 m (± 0.27) for the xeric and mesic forests, respectively. The regional mean height between the forest types was significantly higher in the mesic sites (9.95 m) than in the xeric ones (7.02 m) ($H = 0.23$; *p*-value <0.0001).

Table 13.7 shows the coefficients of the fixed effects of the LME models applied to model the tree age and the tree height. The best model selected for both variables showed that the diameter was the variable with the greatest influence on them, showing a positive and significant relationship in both cases. The tree sex did not show any significant influence and because of that it was not considered in the models.

13.3.5 Sex Ratio

Considering 322 trees from xeric sites, the regional sex ratio (male/female, M/F) was 1.13 (38% for male trees and 34% for female trees, Fig. 13.10). The percentage of undifferentiated trees was 28%. Moreover, the number of male and female trees was different depending on the sites. Female trees were more abundant in RA, PP, and RK (M/F = 0.91, 0.85, and 0.43, respectively), while in CH, CM, and CL, the male trees predominated (M/F = 1.5, 1.57, and 2.45, respectively) (Table 13.8). In

Table 13.7 Results of the linear mixed-effect model that relates the age and height of *Araucaria araucana* trees with the diameter at breast height as fixed effect and the site variability within forest type variability as random effects

Fixed effect		Coefficient	SE	*t*-value
Age	Intercept	0.019	0.140	0.138 ns
	Diameter	0.668	0.034	19.543***
Height	Intercept	0.001	0.551	0.002 ns
	Diameter	0.725	0.020	35.636***

SE standard error
Significance levels: *, $p < 0.05$; **, $p < 0.01$; ***, $p < 0.001$; *ns*, no significant

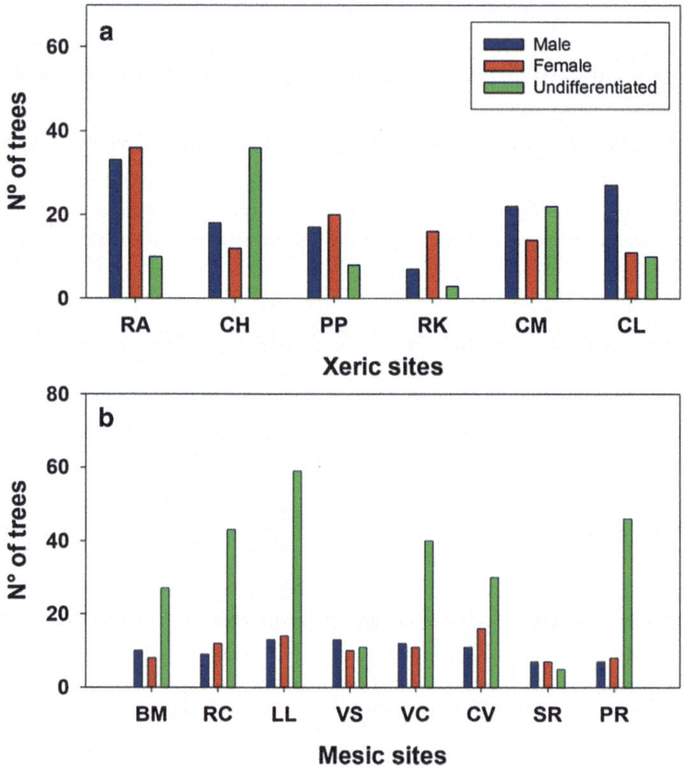

Fig. 13.10 Number of male (blue bars), female (red bars), and undifferentiated (green bars) *Araucaria araucana* sampled trees from (**a**) xeric and (**b**) mesic sites

the mesic forests, 429 trees were considered to calculate the regional sex ratio that was 0.9 (19% for male and 20% for female trees). The undifferentiated trees reached 61%. The sex ratio was variable between the sites. In CV, RC, PR, and LL, the presence of female trees was higher (M/F = 0.7, 0.7, 0.9 and 0.9, respectively), while in BM, VC, and VS, male trees were more abundant (M/F = 1.2, 1.1, and 1.3, respectively), and only in SR the M/F ratio was 1.

The mean age and mean DBH between the sexes were significantly different ($H = 86.95$, p-value <0.0001; $H = 153.09$, p-value <0.0001, respectively, Fig. 13.11a and c) in the xeric forests. In the mesic environments, significant differences were found in the mean age between the sexes ($H = -10.49$, p-value >0.0001, Fig. 13.11b) while the undifferentiated trees showed lower mean DBH than males and females ($H = 113.90$, p-value <0.0001) (Fig. 13.11d).

Table 13.8 Sex ratio of different *Araucaria araucana* forest sites from xeric and mesic environments

Forest type	Sites (code)	Sex ratio (M/F)
Xeric	Río Agrio (RA)	0.91
	Chenque-Pehuen (CH)	1.5
	Primeros Pinos (PP)	0.85
	Río Kilca (RK)	0.43
	Carriel Malal (CM)	1.57
	Catan Lil (CL)	2.45
	Regional	*1.13*
Mesic	Batea Mahuida (BM)	1.2
	Rucachoroi (RC)	0.7
	Las Lecheras (LL)	0.9
	Valle El Salvo (VS)	1.3
	Verde Chico (VC)	1.1
	Senda Rucachori (SR)	1.0
	Pinalada Redonda (PR)	0.9
	Caviahue (CV)	0.7
	Regional	*0.9*

M males, *F* females

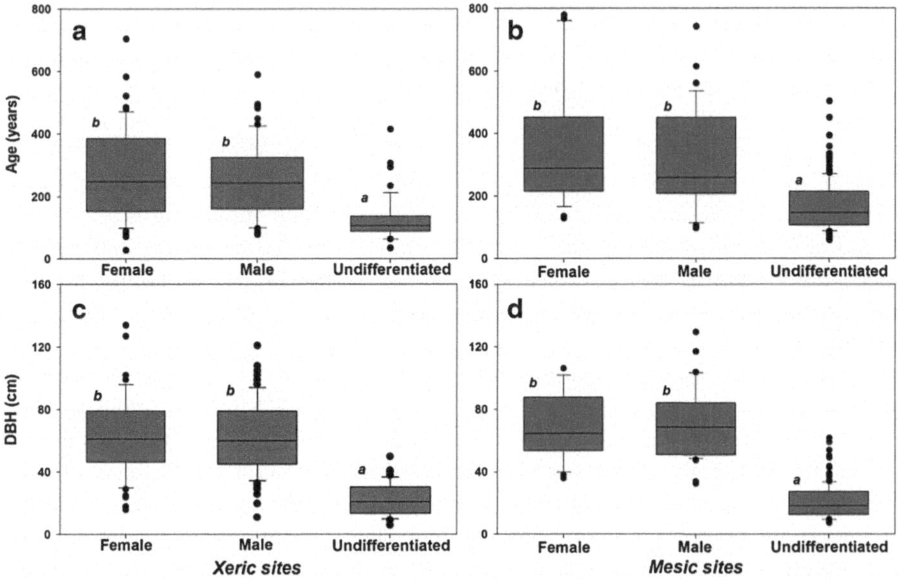

Fig. 13.11 Mean values (± standard deviation) of the age (**a** and **b**) and DBH (**c** and **d**) of *Araucaria araucana* trees of different sexes from xeric (left panels) and mesic (right panels) forests. Different letters indicate significant differences at $p < 0.05$

13.4 Discussion

13.4.1 Relationship Between Radial Tree Growth and Climate in Araucaria araucana

Tree-ring width variability in *A. araucana* trees has been found to be mainly related to surface air temperatures during summer and autumn of the previous growing season and to spring temperatures of the current growth period (Mundo et al. 2012; Muñoz et al. 2014; Hadad et al. 2015; Hadad and Roig 2016). In the present study, similar growth responses to climate were found. In general, the correlations between radial tree growth and regional climate were similar between the two forest types. However, it was observed that trees from the xeric sites were more sensitive to mean temperature during November and December of the current period of growth than those from the mesic forests. Muñoz et al. (2014) also analyzed a network of 21 tree-ring width chronologies to study the growth-climate relationships of *A. araucana* trees along the precipitation gradient and they observed that all sites presented a similar pattern of growth responses throughout the gradient.

Mesic sites showed a negative and significant correlation with winter months (June and July). The influence of winter temperatures on the radial growth of trees has been suggested in other studies (e.g., Jacoby and D'Arrigo 1989; Grace and Norton 1990). Thus, high temperatures during winter months may cause less snowy precipitation and/or may induce early loss of snow cover causing water deficit in the soil at the beginning of the growing season. On the other hand, they can induce increases in respiration rates and as a consequence losses of the stored carbohydrates necessary for the growth restart in the following growing season (Gao et al. 2010). These processes could be the cause of the negative correlations observed between the growth of *A. araucana* and winter temperatures, with greater influence at the mesic sites. Similar results have been observed by Mundo et al. (2012) for those *A. araucana* sites located at higher altitude, but not in those located at low elevations and with more xeric conditions. Moreover, high temperatures during the period of active growth (November and December) affected negatively and significantly the growth of the trees from xeric environments (Fig. 13.6). High temperatures during these months would induce higher water stress conditions and therefore less growth.

On the other hand, precipitations at the beginning of the growing period seemed to significantly favor the tree growth in both forest types. Similar responses were reported by Mundo et al. (2012) for *A. araucana* forests along an altitudinal gradient and by Hadad et al. (2015) and Hadad and Roig (2016) for xeric forests. On the other hand, the positive relationship between tree growth and precipitation during January of the previous growing season in both environments could be related to the accumulation of reserves rather than to a direct effect on the water balance of the soil in the current season of growth (Suárez et al. 2015).

13.4.2 Site Characteristics of the Xeric and Mesic **A. araucana** *Forests*

In the present study, eight mesic sites were analyzed, all belonging to the Araucaria-Lenga forest type (*Araucaria araucana-Nothofagus pumilio*) (Burns 1991). This association is the most extensive type of forest that includes the species *A. araucana* in Patagonia (Schmidt 1977; Veblen 1982). This type of species association would seem to occur at high altitudes where araucaria's tolerance to wetter conditions overlaps with lenga's tolerance to drier conditions (Schmidt 1977; Veblen 1982). The *A. araucana-N. pumilio* mixed forest in Argentina is mainly confined to the southern exposure slopes while northern exposure ones in general present pure forests of *A. araucana* (Veblen 1982). This could partially explain the absence of *N. pumilio* trees in the CV, VC, and PR plots, located on slopes with southeastern exposure in CV and northeastern exposure in VC and PR. On the other hand, the greater resistance of *A. araucana* to fires of low to moderate intensity, the continuous recruitment, and the greater tolerance to shade make this species a better competitor against *N. pumilio* (Fajardo and González 2009). This could also explain the greater relative presence of araucaria trees in all the mesic study sites (Fig. 13.8).

The tree density within each site and between sites was very variable. The tree densities of all the mesic sites were within the forest density range reported by Burns (1991), which varied between 268 and 919 trees/ha in araucaria-lenga mixed forests. Only in PR, the forest density was higher (Table 13.3). It should be noted that the density reported by Burns (1991) included trees with DBH > 4 cm, whereas in the present study only trees with DBH > 8 cm were considered. This could underestimate the site densities of this work compared to that reported by Burns (1991). Similar mesic forest densities were also reported by Mundo et al. (2012). Density-dependent effects have been widely studied in forest ecology, and in many cases they have been mentioned to play an important role in the dynamics of forest populations (He and Duncan 2000). On the other hand, the presence of *A. araucana* seedlings in the understory was analyzed in order to characterize the different araucaria strata. In general, it was observed that as the height of the seedlings increased their abundance decreased, indicating the existence of some stress factors that possible limit and prevent the development of the seedlings as they grow. The RC and VS sites presented more abundance of seedlings with heights between 21 and 50 cm than among 0–20 cm. This could be due to the influence of some disturbance factors affecting mainly the earliest stages of the tree life, or the presence of some factors that limit the production, availability and/or germination of the seeds during the years prior to sampling. It has been suggested that the establishment of araucaria seedlings is controlled by the occurrence of large seed production events (masting), the pre- and post-dispersion predation of seeds, and the understory characteristics (Sanguinetti and Kitzberger 2009; Duplancic 2011). Particularly, in mesic forests seed production is the limiting factor for regeneration during intermasting periods while seed predation regulates the magnitude of the establishment of seedlings. However, it has been also argued that regeneration is limited to microsites favorable for germination (Sanguinetti and Kitzberger 2009), and different levels of regeneration at a local scale could be linked to different types and magnitude of disturbances. In this sense, due to the variety of seedling strata found in the mesic sites, more detailed studies are necessary to better understand these ecological differences.

Along the precipitation gradient, the frequency of araucaria regeneration decreases from west to east, which would be related to the greater dryness conditions to which seeds and seedlings are exposed toward the steppe (Burns 1993). Burns (1991) conducted studies on the dynamics of *A. araucana* stands associated with *Austrocedrus chilensis* in xeric sites, and it was observed a complete absence of seedlings and trees under 80 years old in the only pure *A. araucana* stand analyzed. This may be indicative of a high frequency and/or magnitude of disturbances, mainly fires and grazing, in the more xeric forests. Moreover, it was suggested a continuous but low recruitment rate during stand development in this environment (Burns 1991). In this sense, low tree densities (Table 13.3), low recruitment and almost complete lack of seedlings were observed in the xeric forest sites (data not shown). According to Burns (1991) and to the observations made during the field work of this study, when the cattle, mainly goats, migrate from the steppe to the high forests in the Andes, they travel through the araucaria forests consuming the new shoots and seedlings. This grazing process occurs between spring and early summer. Toward the autumn, the cattle return to the steppe and in their passage through the araucaria forests consume the seeds that have already detached from the cone. These processes complete the seasonal cycle of the cattle production in which the regenerative resources of the araucaria forests are negatively impacted. Moreover, water and wind erosion are triggered as a result of overgrazing and shrub clearing to obtain firewood, processes that deepen the system's desertification of the xeric environments (Roig et al. 2014).

13.4.3 Age Structure of **Araucaria araucana** *Forests*

The age structure of a forest is a reflection of the tree's life history, resulting from the combination of birth and mortality over time (Brown and Wu 2005) as a consequence of the magnitude and frequency of disturbances (Oliver and Larson 1996). Therefore, studies of the age classes' structure offer a better insight of the processes that determine the structure of a population over time while these processes could be considered as drivers of the forest dynamics and as indicators of environmental changes (Dang et al. 2010). Furthermore, the age structure could indicate the environmental factors influencing the reproduction and mortality rate of individuals, e.g., extreme temperature, natural disasters, among others (Wang et al. 2004). The age class distribution analysis of the fragments of *A. araucana* forests from xeric environments showed that they reached a greater representativeness of the age classes between 51 and 300 years old. However, it was generally observed that these forests have produced recruitments during the last 500 years, with greater or lesser intensity over time. On the other hand, it was observed that trees with cambial ages less than 50 years were scarcely represented in these study sites.

On the other hand, the mesic sites presented variable age class distributions indicating differences in the forest dynamics between the sites. In general, the oldest conifers have been recorded in sites of extreme environments, such as those located at high altitudes (Schulman 1954; Rebertus et al. 1991; Wang et al. 2004). In this sense, the BM, CV, LL, and VS, sites located at higher altitudes than the rest of the mesic sites, except VS, presented the greatest number of trees over 300 years of age. It has been suggested that the longevity of conifers could have evolved as an adaptation to persist within the mixed angiosperm forests, allowing them to have greater opportunities for regeneration within a highly competitive system (Burns 1991).

13.4.4 Age, Height, and DBH Relationships

Generally in forestry science, the age of the trees is indirectly estimated from the diameter of their principal stems, based on an empirical relationship between the age of the tree and the diameter (Xing et al. 2012). However, this relationship is not usually linear and strongly depends on the competitive interactions between the trees growing in the same site. Thus, the estimation of the age-DBH relationship and its variations could be linked to aspects related to tree and forest dynamics. Furthermore, the relationship between diameter and total tree height is commonly used to estimate tree volume, and it is, therefore, a fundamental component of many growth and yield forest planning models (Saunders and Wagner 2008, Mehtätalo et al. 2015). Thus, height-diameter functions can be used to indirectly predict the tree height when height measurements are not possible (Larsen and Hann 1987). These functions make it possible to replace tedious measurements in the field, such as height, with another variable of greater practicality, such as diameter, reducing sampling efforts (Mehtätalo et al. 2015). Two types of height-diameter models have been reported in the literature: local models and regional or generalized models. Local models normally depend on the diameter of the tree and have application only at the site from which the data were extracted, while the generalized ones are functions of the diameter, age, sex, or other site variables, such as density, altitude, latitude, and longitude, among others, and, therefore, can be applied at the regional level (Soares and Tomé 2002), avoiding to establish individual height-diameter relationships for each study site (Temesgen and Gadow 2004). Thus, a single equation or model allows estimating quite accurately the height of any tree that develops within the scope of application of the model (Barrio-Anta et al. 2004). Based on this, a generalized height-diameter model was estimated for the *A. araucana* forests. A linear mixed-effect model (LME) was applied and it was observed that total height significantly depended on the DBH. Linear and nonlinear effect models have been applied for other tree species to model the height-diameter relationship at the regional level (e.g., Saunders and Wagner 2008; Mehtätalo et al. 2015). On the other hand, the tree age was also positively and significantly related to DBH and this relationship was variable at the different sites within each forest type (Table 13.7). These results indicated that there is a wide variety of araucaria growth responses along its distribution area in Argentina as a result of the site quality characteristics, the environmental conditions where trees develop, and genetic aspects of the species, among others.

13.4.5 Sex Ratio

As it was mentioned before, the most suitable sex ratio that optimizes parental reproductive success in dioecious species is 1:1 (Charnov 1982); however, a large number of genetic and environmental factors can modify this optimal sex proportion (Ainsworth 2000; Sinclair et al. 2012). According to this model, it was observed in both araucaria forest types that the sex ratio was close to 1:1 (1.3 in xeric and 0.9 in mesic forests). However, it was observed that the sex ratio was different between sites. Gallo et al. (2004) observed a sex ratio of 0.71 for mesic *A. araucana* sites, evidencing a higher proportion of female trees in those areas. Nuñez et al. (2008) reported a 1:1 sex ratio at the regional level in forests of *A. chilensis* from northwest Patagonia although they found spatial differences in the distribution of sexes along environmental gradients present in the area. Ortiz et al. (2002) also reported differences in the sex ratio along an altitudinal gradient in *Juniperus communis* subsp. *alpine*, observing that the number of male trees increased with elevation. These studies reaffirmed the hypothesis that most of the dioecious species are spatially segregated along environmental gradients (Bierzychudek and Eckhart 1988).

In the present study, different sex ratios between the two forest environments were observed, male trees being more abundant than female ones in the xeric forests (M/F = 1.13) while in the mesic environment the abundance of male and female trees was the same (M/F = 1:1). In general, studies of the spatial segregation of the sexes indicate that female trees are more abundant in those sites with less stressful environmental conditions (Bierzychudek and Eckhart 1988). In a context of climate warming and change, Hultine et al. (2016) suggested that in dioecious species, male trees may be less sensitive to increased aridity than co-occurring females, and, consequently, extreme male-biased sex ratios could be possible in a significant number of populations. In this sense, the araucaria xeric forests that are generally subjected to more stressful environmental conditions showed less proportion of female than male trees. Moreover, it was indicated that female araucaria trees from xeric forests are more sensitive to temperature (Hadad and Roig 2016) and have a higher reproduction cost than male trees (Hadad 2013). Thus, prediction of future intensification of warming conditions in Patagonia (IPCC 2013) allows hypothesizing a scenario with more stressful environmental conditions which may influence the physiology of trees, probably inducing a female/male-biased sex ratio and, ultimately, affecting the forest dynamics and structures.

13.5 Conclusion

Knowledge of the structure of temperate forests is relevant to understand the forest diversity and its essential processes in the forest ecosystem stability and services. In this sense, the study of the forest structure becomes necessary to monitor successional processes toward maintaining species and habitat diversity. In the *A. araucana* forests, the analysis of age structure evidenced variable age classes at both xeric and mesic environments in NW Patagonia, a fact that may be linked to the biological cycle of the species, the resources availability, and the history of disturbance processes. In this sense, the environmental and anthropic factors, separately or interacting, might play an important role in the ecology and dynamic of these forests. From both types of araucaria forests analyzed, those distributed in xeric environments showed more sensitivity to summer temperature variability during the current period of growth. However, the observed relationship between the tree growth and climate variability in Patagonia gives arguments to establish hypotheses about the potential different levels of forest vulnerability to future changes in the climate for both mesic and xeric forests. In this sense, Patagonia has experienced an increase in temperature and a decrease in precipitation during the last century, trends that could be intensified in the next decades. Under this scenario, the xeric *A. araucana* forests may be probably more strongly influenced by a decrease in water resources availability and by an increase in drought conditions, with potential modifications in tree recruitment, tree growth, and sex ratios. Furthermore, the current increase in livestock in northern Patagonia may impact the age structure of *A. araucana*, especially in the xeric environments due to the greater pressure on the consumption of seeds and newly installed plants. Therefore, it would be convenient to consider protection and conservation strategies in order to preserve these emblematic forests. Considering that *A. araucaria* was declared as an endangered tree species, results of this study provide arguments to establish programs for the conservation and recovery of degraded areas. Finally, due to all the above stated, more detailed studies about the age classes, sex ratio, tree development, and forest recruitment, among others, together with studies about the forest disturbance factors, will be necessary to better understand how these forests could respond under future climate change scenarios.

Acknowledgments This research was supported by the Consejo Nacional de Investigaciones Científicas y Técnicas (CONICET) through a doctoral fellowship granted to MAH and JGAM, by the Agencia Nacional de Promoción Científica y Tecnológica (PICT-Bicentenario 2010-2679 and PICT-2014-0915 to FAR and MAH, respectively) and by the Programa Bosques Andinos-HELVETAS Swiss Intercooperation (RUC N° 20549449931 to FAR). We would like to sincerely acknowledge C. Moreno, S. Piraino, A. Duplancic, E. Barrio, and S. Papú for the field and laboratory assistance. We thank Dr. Daniel Flores for the design of Fig. 13.1. We also thank the Administración de Parques Nacionales of Argentina for allowing access to the sites.

References

Aagesen D (2004) Burning monkey-puzzle: native fire ecology and forest management in northern Patagonia. Agri Hum Values 21(2–3):233–242. https://doi.org/10.1023/B:AHUM.0000029402.85972.6c

Aguilera-Betti I, Muñoz A, Stahle D et al (2017) The first millennium-age *Araucaria araucana* in Patagonia. Tree-Ring Res 73:53–56. https://doi.org/10.3959/1536-1098-73.1.53

Ainsworth C (2000) Boys and girls come out to play: the molecular biology of dioecious plants. Ann Bot 86(2):211–221. https://doi.org/10.1006/anbo.2000.1201

Arco Molina JG (2018) Intensidad de la señal climática y eficiencia del uso del agua expresada por los anillos de crecimiento de *Araucaria araucana* en relación con la edad de los árboles en bosques mésicos del norte de Patagonia. Tesis Doctoral. Universidad Nacional de Córdoba, p 198

Armesto JJ, León LP, Kalin AM (1996) Los bosques templados del sur de Chile y Argentina: Una isla biogeográfica. In: Armesto JJ, Villagran C, Kalin Arroyo M (eds) Ecología de los bosques nativos de Chile. Editorial Universitaria, Santiago de Chile

Assal T, González M, Sibold J (2018) Burn severity controls on postfire *Araucaria-Nothofagus* regeneration in the Andean Cordillera. J Biogeogr 45:2483–2494. https://doi.org/10.1111/jbi.13428

Barrio-Anta M, Álvarez GJG, Díaz-Maroto IJ et al (2004) Relación altura-diámetro generalizada para *Quercus robur* L. en Galicia. Cuadernos de la Soc Española Cienc Forest 18:141–146. https://doi.org/10.31167/csef.v0i18.9451

Bierzychudek P, Eckhart V (1988) Spatial segregation of the sexes of dioecious plants. Am Nat 132:34–43. https://doi.org/10.1086/284836

Brown PM, Wu R (2005) Climate and disturbance forcing of episodic tree recruitment in a southwestern Ponderosa Pine Landscape. Ecology 86:3030–3068. https://doi.org/10.1890/05-0034

Burns BR (1991) Regeneration dynamics of *Araucaria araucana*. Dissertation. University of Colorado, Boulder

Burns BR (1993) Fire-induced dynamics of *Araucaria araucana-Nothofagus antarctica* forest in the southern Andes. J Biogeogr 20(6):669–685. https://doi.org/10.2307/2845522

Camarero JJ, Gutiérrez E (2007) Response of *Pinus uncinata* recruitment to climate warming and changes in grazing pressure in an isolated population of the Iberian System (NE Spain). Arct Antarct Alp Res 39(2):210–217. https://doi.org/10.1657/1523-0430(2007)39[210:ROPURT]2.0.CO;2

Charnov EL (1982) The theory of sex allocation. Princeton University Press, Princeton

Cook E (1985) A time series analysis approach to tree ring standardization. Dissertation. School of Renewable Natural Resources, University of Arizona, Tucson (AZ), p 171

Coomes DA, Allen RB (2007) Effects of size, competition and altitude on tree growth. J Ecol 95(5):1084–1097. https://doi.org/10.1111/j.1365-2745.2007.01280.x

Dang HS, Zhang YJ, Zhang KR et al (2010) Age structure and regeneration of subalpine fir (*Abies fargesii*) forests across an altitudinal range in the Qinling Mountains, China. For Ecol Manag 259(3):547–554. https://doi.org/10.1016/j.foreco.2009.11.011

Donoso C (1993) Bosques templados de Chile y Argentina. Variación, estructura y dinámica. Editorial Universitaria, Santiago, Chile

Duncan RP (1989) An evaluation of errors in tree age estimates based on increments cores in Kahukatea (*Dacrycarpus dacrydioides*). NZ J Nat Sci 16:31–37

Duplancic MA (2011) Germinación de semillas de *Araucaria araucana* provenientes del bosque xérico nor-patagónico. Multequina 20:113–115

Ehle DS, Baker WL (2003) Disturbance and stand dynamics in Ponderosa Pine forests in Rocky Mountain National Park, USA. Ecol Monogr 73(4):543–566. https://doi.org/10.1890/03-4014

Eriksen W (1971) Betriebsformen und Probleme der Viehwirtschaft am Rande der Argentinischen kordillere. Zeitschriftfuir Ausl Landwirstschaft 10:24–27

Fajardo A, González ME (2009) Replacement patterns and species coexistence in an Andean *Araucaria-Nothofagus* forest. J Veg Sci 20(6):1176–1190. https://doi.org/10.1111/j.1654-1103.2009.01117.x

Fritts H (1976) Tree-rings and climate. Academic Press, London

Gallo L, Izquierdo F, Sanguinetti LJ et al (2004) *Araucaria araucana* forest genetic resources in Argentina. In: Vinceti B, Amaral W, Meilleur B (eds) Challenges in managing forest genetic resources for livelihoods: examples from Argentina and Brazil. International Plant Genetic Resources Institute, Rome, pp 115–143

Gandullo R (2003) Fitosociología de los matorrales de “Ñire” (*Nothofagus antarctica*), con bosque de “Pehuén” (*Araucaria araucana*), del Parque Provincial Copahue. I. Interpretación Sintaxonómica. Candollea 58:163–181

Gao L, Zhang C, Zhao X et al (2010) Gender-related climate response of radial growth in dioecious *Fraxinus mandshurica* trees. Tree-Ring Res 66(2):105–112. https://doi.org/10.3959/2009-5.1

González M, Veblen TT (2006) Climatic influences on fire in *Araucaria araucana-Nothofagus* forests in the Andean cordillera of south-central Chile. Écoscience 13(3):342–350. https://doi.org/10.2980/i1195-6860-13-3-342.1

González ME, Veblen TT, Sibold JS (2005) Fire history of *Araucaria-Nothofagus* forests in Villarrica National Park, Chile. J Biogeogr 32(7):1187–1202. https://doi.org/10.1111/j.1365-2699.2005.01262.x

González M, Veblen TT, Sibold JS (2010) Influence of fire severity on stand development of *Araucaria araucana-Nothofagus pumilio* stands in the Andean cordillera of south-central Chile. Austral Ecol 35(6):597–615. https://doi.org/10.1111/j.1442-9993.2009.02064.x

Grace J, Norton DA (1990) Climate and growth of *Pinus sylvestris* at its upper altitudinal limit in Scotland: evidence from tree growth-rings. J Ecol 78(3):601–610. https://doi.org/10.2307/2260887

Hadad MA (2013) Análisis florístico-dendrocronológico de los bosques xéricos de *Araucaria araucana* (Araucariaceae) en el contacto con la estepa patagónica en Neuquén, Argentina. Tesis Doctoral. Universidad Nacional de Córdoba, p 155

Hadad MA, Roig FA (2016) Sex-related climate sensitivity of *Araucaria araucana* Patagonian forest-steppe ecotone. For Ecol Manag 362:130–141. https://doi.org/10.1016/j.foreco.2015.11.049

Hadad MA, Roig JFA, Boninsegna J et al (2015) Age effects on the climatic signal in *Araucaria araucana* from xeric sites in Patagonia, Argentina. Plant Ecol Divers 8(3):343–351. https://doi.org/10.1080/17550874.2014.980350

He F, Duncan RP (2000) Density-dependent effects on tree survival in an old-growth Douglas fir forest. J Ecol 88(4):676–688. https://doi.org/10.1046/j.1365-2745.2000.00482.x

Herrmann TM (2006) Indigenous knowledge and management of *Araucaria araucana* forest in the Chilean Andes: implications for native forest conservation. Biodivers Conserv 15(2):647–662. https://doi.org/10.1007/s10531-005-2092-6

Holmes RL (1983) Computer-assisted quality control in tree-ring dating and measurement. Tree-Ring Bull 43:69–78

Hultine KR, Grady KC, Wood TE et al (2016) Climate change perils for dioecious plant species. Nat Plants 2:16109. https://doi.org/10.1038/nplants.2016.109

IPCC (2013) Climate Change 2013: The Physical Science Basis. Contribution of Working Group I to the Fifth Assessment Report of the Intergovernmental Panel on Climate Change. In: Stocker TF, Qin D, Plattner GK, Tignor M, Allen SK, Boschung J, Nauels A, Xia Y, Bex V, Midgley PM (eds) . Cambridge University Press, Cambridge, UK

Jacoby GC, D'Arrigo R (1989) Reconstructed Northern Hemisphere annual temperature since 1671 based on high-latitude tree-ring data from North America. Clim Chang 14:39–59. https://doi.org/10.1007/BF00140174

Kitzberger T, Veblen TT (2003) Influences of climate on fire in northern Patagonia, Argentina. In: Veblen TT, Baker WL, Montenegro G, Swetnam TW (eds) Fire and climatic change in temperate ecosystems of the western Americas. Ecological studies (analysis and synthesis), vol 160. Springer, New York, NY

LaMarche VC, Holmes RL, Dunwiddie PW et al (1979) Tree-ring chronologies of the Southern Hemisphere. University of Arizona

Larsen DR, Hann DW (1987) Height-diameter equations for seventeen tree species in Southwest Oregon. Forest Research Laboratory, College of Forestry, Oregon State University, Corvallis

League K, Veblen T (2006) Climatic variability and episodic *Pinus ponderosa* establishment along the forest-grassland ecotones of Colorado. For Ecol Manag 228:98–107. https://doi.org/10.1016/j.foreco.2006.02.030

Lebourgeois F, Eberle P, Mérian P et al (2014) Social status-mediated tree-ring responses to climate of *Abies alba* and *Fagus sylvatica* shift in importance with increasing stand basal area. For Ecol Manag 328:209–218. https://doi.org/10.1016/j.foreco.2014.05.038

Mast J, Veblen TT, Linhart YB (1998) Disturbance and climatic influences on age structure of Ponderosa Pine at the pine/grassland ecotone, Colorado Front Range. J Biogeogr 25(4):743–755. https://doi.org/10.1046/j.1365-2699.1998.2540743.x

Mehtätalo L, De Miguel S, Gregoire TG (2015) Modeling height-diameter curves for prediction. Can J For Res 45(7):826–837. https://doi.org/10.1139/cjfr-2015-0054

Molowny-Horas R, Suarez ML, Lloret F (2017) Changes in the natural dynamics of *Nothofagus dombeyi* forests: population modeling with increasing drought frequencies. Ecosphere 8:1–17. https://doi.org/10.1002/ecs2.1708

Mundo IA, Roig JFA, Villalba R et al (2012) *Araucaria araucana* tree-ring chronologies in Argentina: spatial variations and climate influences. Trees-Struct Funct 26:443–458. https://doi.org/10.1007/s00468-011-0605-3

Mundo IA, Kitzberger TF, Roig JA et al (2013) Fire history in the *Araucaria araucana* forests of Argentina: human and climate influences. Int J Wildland Fire 22:194–206. https://doi.org/10.1071/WF11164

Munné-Bosch S (2015) Sex ratios in dioecious plants in the framework of global change. Environ Exper Bot 109:99–102. https://doi.org/10.1016/j.envexpbot.2014.08.007

Muñoz AA, Barichivich J, Christie DA et al (2014) Patterns and drivers of *Araucaria araucana* forest growth along a biophysical gradient in the northern Patagonian Andes: linking tree rings with satellite observations of soil moisture. Austral Ecol 39:158–169. https://doi.org/10.1111/aec.12054

North M, Chen J, Oakley B et al (2004) Forest stand structure and pattern of old-growth western hemlock/Douglas-fir and mixed-conifer forests. For Sci 50:299–311

Nuñez C, Nuñez M, Kitzberger T (2008) Sex-related spatial segregation and growth in a dioecious conifer along environmental gradients in Northwestern Patagonia. Ecoscience 15:73–80. https://doi.org/10.2980/1195-6860(2008)15[73:SSSAGI]2.0.CO;2

Oliver CD, Larson BC (1996) Forest stand dynamics: updated edition. Wiley, New York

Ortiz PL, Artista M, Talavera S (2002) Sex ratio and reproductive effort in the dioecious *Juniperus communis* subsp. *alpina* (Suter) Celak. (Cupressaceae) along an altitudinal gradient. Ann Bot 89:205–211. https://doi.org/10.1093/aob/mcf028

Pinheiro J, Bates D, DebRoy S et al (2017) nlme: linear and nonlinear mixed effects models. R package version 3.1–131, https://CRAN.R-project.org/package=nlme.

R Core Team (2016) R: a language and environment for statistical computing. R Foundation for Statistical Computing, Vienna. https://www.R-project.org/

Rebertus AJ, Burns BR, Veblen TT (1991) Stand dynamics of *Pinus flexilis*-dominated subalpine forests in the Colorado Front Range. J Veg Sci 2:445–458. https://doi.org/10.2307/3236026

Rechene C (2000) Los bosques de *Araucaria araucana* en Argentina. Estudios silvícolas. CIEFAP, Argentina & Univ Técnica de Munich

Roig FA (1998) La vegetación de la Patagonia. In: Correa M (ed) Flora Patagónica. Colec Cient INTA. Tomo VIII (1), Buenos Aires, pp 48–166

Roig FA, Villalba R (2008) Understanding climate from Patagonian tree rings. In: Rabassa J (ed) The late Cenozoic of Patagonia and Tierra del Fuego developments in quaternary science, 11, (Series Ed. Van Der Meer Jjm). Elsevier, Amsterdam, pp 411–438

Roig FA, Hadad M, Moreno C et al (2014) Hiatos de regeneración del bosque de *Araucaria araucana* en Patagonia: vinculaciones al uso de tierras y desertificación regional. Zonas áridas 15:326–342

Romme WH, Knight DH (1982) Landscape diversity: the concept applied to Yellowstone Park. Bioscience 32:664–670. https://doi.org/10.2307/1308816

Sanguinetti J, Kitzberger T (2009) *Araucaria araucana* temporal and spatial seedling establishment patterns: masting, seed predation and understory vegetation effects. Rev Chil Hist Nat 82:319–335

Saunders MR, Wagner RG (2008) Height-diameter models with random coefficients and site variables for tree species of Central Maine. Ann For Sci 65:1–10. https://doi.org/10.1051/forest:2007086

Schmidt H (1977) Dinámica de un bosque virgen de *Araucaria*-Lenga (Chile). Bosque 2:3–11

Schulman E (1954) Longevity under adversity in conifers. Science 119:396–399. https://doi.org/10.1126/science.119.3091.396

Schulman E (1956) Dendroclimatic changes in semiarid America. University of Arizona Press, Tucson

Seoane J (2014) ¿Modelos mixtos (lineales)? Una introducción para el usuario temeroso. Etología 24:15–37

Sinclair JP, Emlen J, Freeman DC (2012) Biased sex ratios in plants: theory and trends. Bot Rev 78:63–86. https://doi.org/10.1007/s12229-011-9065-0

Soares P, Tomé M (2002) Height-diameter equation for first rotation eucalypt plantations in Portugal. Forest Ecol Manag 166:99–109. https://doi.org/10.1016/S0378-1127(01)00674-0

Stokes M, Smiley T (1968) An introduction to tree-ring dating. University of Chicago Press, Chicago

Suárez ML, Villalba R, Mundo I et al (2015) Sensitivity of *Nothofagus dombeyi* tree growth to climate changes along a precipitation gradient in northern Patagonia, Argentina. Trees-Struct Funct 29:1053–1067. https://doi.org/10.1007/s00468-015-1184-5

Svensson JS, Jeglum JK (2001) Structure and dynamics of an undisturbed old-growth Norway Spruce forest on the rising Bothnian coastline. For Ecol Manag 151:67–79. https://doi.org/10.1016/S0378-1127(00)00697-6

Temesgen H, Gadow KV (2004) Generalized height-diameter models – an application for major tree species in complex stands of interior British Columbia. Eur J For Res 123:45–51. https://doi.org/10.1007/s10342-004-0020-z

Veblen TT (1982) Regeneration patterns of *Araucaria araucana* forests in Chile. J Biogeogr 9:11–28. https://doi.org/10.2307/2844727

Veblen TT, Kitzberger T, Raffaele E et al (2003) Fire history and vegetation changes in northern Patagonia, Argentina. In: Veblen TT, Baker WL, Montenegro G, Swetnam TW (eds) Fire and climatic changes in temperate ecosystems of the Western Americas ecological studies. Springer-Verlag, New York, pp 265–295

Venables WN, Ripley BD (2002) Random and mixed effects. In: Modern applied statistics with S. Springer, New York, pp 271–300

Villalba R, Veblen TT (1997) Regional patterns of tree population age structures in Northern Patagonia: climatic and disturbance influences. J Ecol 85:113–124. https://doi.org/10.2307/2960643

Villalba R, Lara A, Masiokas MH et al (2012) Unusual Southern Hemisphere tree growth patterns induced by changes in the Southern Annular Mode. Nat Geosci 5:793–798. https://doi.org/10.1038/ngeo1613

Wang T, Liang Y, Ren H et al (2004) Age structure of *Picea schrenkiana* forest along an altitudinal gradient in the central Tianshan Mountains, Northwestern China. For Ecol Manag 196:267–274. https://doi.org/10.1016/j.foreco.2004.02.063

Xing P, Zhang QB, Baker PJ (2012) Age and radial growth pattern of four tree species in a subtropical forest of China. Trees-Struct Funct 26:283–290. https://doi.org/10.1007/s00468-011-0590-6

Zamorano-Elgueta C, Cayuela L, González-Espinosa M et al (2012) Impacts of cattle on the South American temperate forests: challenges for the conservation of the endangered monkey puzzle tree (*Araucaria araucana*) in Chile. Biol Conserv 152:110–118. https://doi.org/10.1016/j.biocon.2012.03.037

Chapter 14
Dendroecology Applied to Silvicultural Management in the Southern Patagonian Forests: A Case Study from an Experimental Forest in Tierra del Fuego, Argentina

Ignacio A. Mundo, Dino A. Palazzini, Antonio J. Barotto, Guillermo J. Martínez-Pastur, and Marcelo D. Barrera

Abstract Most of radial growth studies of *Nothofagus pumilio* (locally known as 'lenga') in Tierra del Fuego have focused on mature forests and climate relations, but little is known about growth dynamics after harvesting and even less after applying intermediate treatments. The objective of this chapter is to analyze the effect of thinning and pruning on radial growth of *N. pumilio* in Tierra del Fuego (Argentina) in the Aguas Blancas experimental long-term forest, which was harvested in 1967 and had different combinations of interventions in 1984–1985 and 1999. Three treatments were analyzed: (i) one thinning (1 T, thinned in 1984–85), (ii) two thinnings (2 T, thinned in 1984 and 1999), and (iii) two thinnings with pruning of 2/3 of the living crown (2 T + P, thinned in 1984 and 1999 with pruning). Using standard dendrochronological methods, ring widths were cross-dated, measured and the mean ring width (RW), basal area increment (BAI) and the corresponding

I. A. Mundo (✉)
Laboratorio de Dendrocronología e Historia Ambiental, Instituto Argentino de Nivología, Glaciología y Ciencias Ambientales (IANIGLA), Consejo Nacional de Investigaciones Científicas y Técnicas (CONICET), Mendoza, Argentina

Facultad de Ciencias Exactas y Naturales, Universidad Nacional de Cuyo, Mendoza, Argentina
e-mail: iamundo@mendoza-conicet.gob.ar

D. A. Palazzini · A. J. Barotto · M. D. Barrera
LISEA, Facultad de Ciencias Agrarias y Forestales, Universidad Nacional de La Plata, La Plata, Argentina
e-mail: dino.palazzini@agro.unlp.edu.ar; jose.barotto@agro.unlp.edu.ar; mbarrera@agro.unlp.edu.ar

G. J. Martínez-Pastur
Centro Austral de Investigaciones Científicas (CADIC), Consejo Nacional de Investigaciones Científicas y Técnicas (CONICET), Ushuaia, Argentina
e-mail: gpastur@conicet.gov.ar

M. Pompa-García, J. J. Camarero (eds.), *Latin American Dendroecology*,
https://doi.org/10.1007/978-3-030-36930-9_14

percentages of change in growth (PGC) in 5-year windows were calculated. The three treatments showed significant increments in PGC in both RW (129%) and BAI (419%) after the 1984–1985 thinning. However, from 1999 there were significant differences in periodic growth between the three treatments with the highest RW and BAI in the thinned with pruning treatment. In these forests, dendrochronology emerges as a complementary method to quantify and to reconstruct the individual response of trees affected by silvicultural treatments.

Keywords Intermediate treatments · Thinning · Pruning · *Nothofagus pumilio* · Forest management

14.1 Introduction

Forest management, whatever its objectives and modalities, substantially modifies the original forest structure. Silvicultural interventions produce changes in microclimatic conditions, such as increased solar radiation input, variations in the redistribution of the intercepted rainfall (Caldentey et al. 1998; Otero et al. 1998; Martínez-Pastur et al. 2011), changes in temperature, nutrient cycling (Thibodeau et al. 2000), moisture availability (Fayle 1983) and air humidity (Barg and Edmonds 1999). These changes, according to their intensity and the spatial distribution of the silvicultural interventions, influence and stimulate the growth of the remaining or residual trees (Thorpe et al. 2007; Vincent et al. 2009; Peri et al. 2013; Venturoli et al. 2015; Pamerleau-Couture et al. 2015). In this way, improved radial growth of residual trees generally leads to higher financial returns at the stand level despite the reduced tree density (Liu et al. 2007).

Dendroecology (Fritts and Swetnam 1989) is a useful tool to evaluate different processes of forest dynamics, e.g. regeneration, establishment, maturity or senescence of trees. In addition, it is used to date and quantify the effect of past disturbances in the tree canopy which can be translated into periods of suppression or releases of tree rings (Schweingruber 1996). Such information is required to build growth and yield models useful when predicting forest responses after the application of silvicultural treatments based on its past responses. In this way, radial growth releases due to thinning interventions and detected by dendrochronological methods were reported for several species in different parts of the world (Misson et al. 2003; Pérez-de-Lis et al. 2011; Ruano et al. 2013). In particular for Tierra del Fuego (Argentina), Franco (2015) analyzed the response to different combination of thinnings from below in an experimental *Nothofagus betuloides* secondary forest at Estancia Moat, located close to Beagle channel. She found that different thinning intensities produced differential effects on radial growth, where heavy thinning treatments resulted in higher individual radial growth rates than the control treatments.

Nothofagus pumilio (Poepp. et Endl.) Krasser (locally known as 'lenga') is the most important native tree forest species in economic terms of the sub-Antarctic forests of Argentina and Chile. On the Isla Grande de Tierra del Fuego (Argentina), the silvicultural systems have varied in the last decades, from high grading (Mutarelli and Orfila 1973), shelterwood cuts (Martínez-Pastur et al. 2000), and recently, variable retention harvesting (Martínez-Pastur et al. 2009). However, there are few examples of intermediate treatments that improve and lead the second growth forests (Martínez-Pastur et al. 2001; Peri et al. 2002, 2013), because forests are generally abandoned after the first cuts. Recently, the Dirección General de Bosques de Tierra del Fuego, within the framework of National Law No. 26,331, has promoted the large-scale implementation of these intermediate treatments in secondary forests of Tierra del Fuego (e.g. Martínez-Pastur et al. 2013). In the Argentinean forests of Tierra del Fuego, *N. pumilio* has been used in dendroclimatic studies because of its radial growth sensitivity to annual climate variability (Boninsegna et al. 1989; Gutiérrez 1992; Mundo et al. 2007; Massaccesi et al. 2008). However, dendrochronological methods have not been used until now for silvicultural management.

Intermediate treatments are performed to ensure the composition, stem quality, spacing and desirable characteristics within a stand (Daniel et al. 1982; Smith et al. 1998). The positive effects of thinning are to increase net productivity and the stand quality as well as to favor the growth of the remaining trees, i.e. those with the best characteristics for the management objective. However, so far, the effects of thinning, and even pruning, on radial growth to sustain their inclusion in silvicultural systems in the *N. pumilio* forests of Tierra del Fuego have not been quantified through tree-ring methods yet. In consequence, the objective of this study was to analyze the effect of thinning and pruning on *N. pumilio* radial growth through dendrochronological techniques in a long-term experimental stand.

14.2 Materials and Methods

14.2.1 Study Site

This study was based on samples from plots located in the Aguas Blancas long-term experimental forest (Fig. 14.1, 54°36′35" S; 67°15′43" W; 215 m a.s.l.) in the department of Ushuaia (Tierra del Fuego province, Argentina), 14 km from the town of Tolhuin. These plots are part of the PEBANPA network (Peri et al. 2016). The forest type corresponds to *N. pumilio* pure-type based on the classification of Uriarte and Grosse (1991) and to site classes II and III according to the classification proposed by Martínez-Pastur et al. (1997), where dominant heights fluctuate between 22 and 26 m.

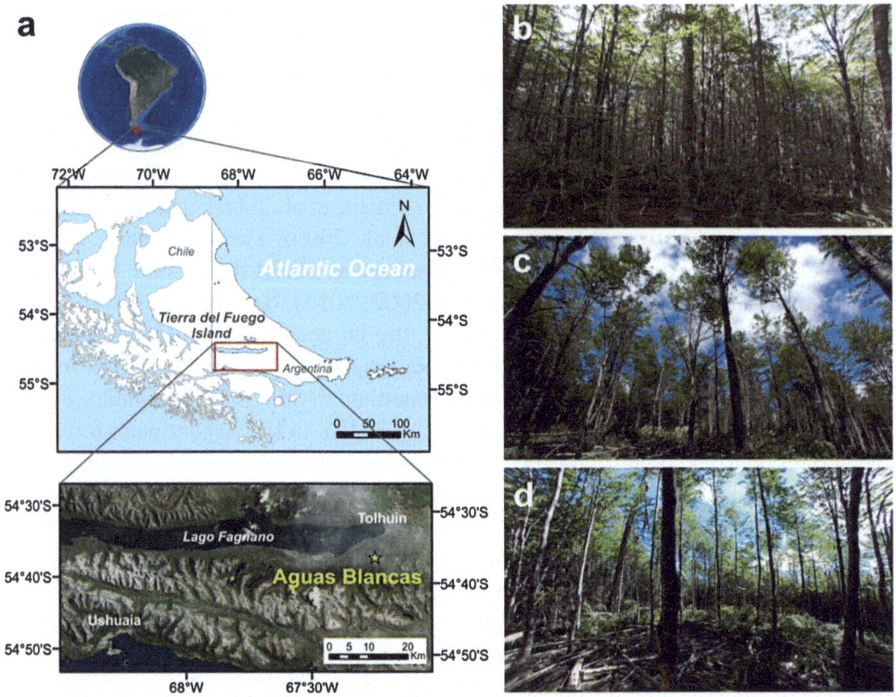

Fig. 14.1 (**a**) Map showing the location of study site in Tierra del Fuego, Argentina. On the right, photos (taken in 2010) of the different treatments analyzed in this study: (**b**) one high-intense thinning (1 T, only thinned in 1984–1985), (**c**) two thinnings (2 T, thinned in 1984–1985 and 1999) and (**d**) two thinnings with pruning (2 T + P, thinned in 1984–1985 and 1999 plus pruning)

Between 1965 and 1967, the National Forestry Administration and the University of Buenos Aires installed a 16-hectare trial in the Aguas Blancas experimental forest where single-tree selective cutting was carried out previously. At that moment, strip clearcuts were performed on alternate strips of 30–40 m wide and 400 m long. Natural regeneration was installed homogeneously and abundantly in all the strips. In the 1984–1985 season, the former National Forest Institute (IFONA) performed different pre-commercial low thinning systems on the regeneration (saplings of approximately 2 m height, 2.5 cm at DBH and 80,000–100,000 ind.ha^{-1}). In 1999, the Centro Austral de Investigaciones Científicas (CONICET) and the Provincial Forest Administration (Dirección de Bosques) of Tierra del Fuego evaluated the first trial on a 2350 m^2 area where 5% of the original basal area (BA) had been left. Subsequently, new treatments were installed on this thinned forest and unmanaged sector within the same strip. A second low-intense thinning was performed leaving 68–75% of the BA left by the thinning carried out in 1984–1985. In addition, the same low-intensity thinning with pruning up to 4–5 m in height (2/3 of the living crown) was applied on another sector of the same strip. In consequence, in this study, the following treatments were analyzed: (i) one high-intense thinning (1 T, only thinned in 1984–1985), (ii) two thinnings (2 T, thinned in 1984–1985 and 1999), and (iii) two thinnings with pruning (2 T + P, thinned in 1984–1985 and 1999 plus pruning). In each of these treatments, ring width patterns of 20 individuals were analyzed by means of transverse sections taken at breast height (1.30 m) during an intervention in 2011 (not analyzed in this study). The mean DBH (SE) of the 20 sampled in each treatment was 9.7 cm (0.5) for 1 T, 10.8 cm (0.3) for 2 T and 11.6 cm (0.5) for 2 T + P.

14.2.2 Dendrochronological Methods

The processing of cross-sections followed standard dendrochronological procedures (Stokes and Smiley 1968). Samples were air-dried and sanded to create polished surfaces to facilitate the identification of the annual rings under a stereomicroscope (10x–50x). Two perpendicular radii were analyzed on each sample and the growth rings were identified and visually cross-dated. According to Schulman's (1956) convention for the Southern Hemisphere, calendar years of annual rings were assigned to the year in which ring formation begun. The ring widths (RW) were measured using a Velmex UniSlide tablet connected to a Metronics Quick-Check QC-1000 digital counter with an accuracy of 0.01 mm. COFECHA software (Holmes 1983) was used to statistically verify the dating quality of the series and to calculate the correlation between series and mean sensitivity. Basal area increment (BAI) was calculated from RW series assuming circular cross sections after averaging the two radii for each tree. BAI is usually preferred over the RW series since this variable provides a more precise quantification of wood production due to the ever increasing diameter of a living tree (Visser 1995; Pedersen 1998). Then, individual records of both variables were combined to obtain a mean chronology per treatment. In addition, the methodology developed by Nowacki and Abrams (1997) was used to detect growth releases. This method calculates the percentage of growth change (PGC) year by year over the series, in this case individual and average per treatment, using the following equation: $[(M2-M1)/M1] \times 100$ where M1 is the average ring width of the previous 5 years including the calculated year, and M2 is the average of the following 5 years following Fang et al. (2018). A PGC that exceeded 50% and was sustained for at least 3 consecutive years was considered a release.

To compare growth before and after interventions, we calculated 5-year means (periodic annual increments, PAI_5) and statistically analyzed the differences between them for the 2 years of interventions. The mean differences in PAI of pre- and post-intervention periods (ΔPAI_5) for each treatment were calculated in the context of pretest-posttest analyses (Dimitrov and Rumrill Jr 2003) and then compared with one-way ANOVAs and HSD Tukey post-hoc tests. Before performing this procedure, the series were checked for normality using the Shapiro-Wilk test. Residuals normality and homoscedasticity were also checked through Kolmogorov-Smirnov and Bartlett tests, respectively. Since the ΔPAI in BAI for the 1984 did not fulfilled this assumption, the non-parametric Kruskal-Wallis test was used to detect differences between means.

To estimate the effect of the second intervention in relation to what was expected with an unaffected control situation, the radial growth pattern of 1 T treatment was used as a reference pattern. Thus, the percentage differences of 2 T and 2 T + P treatments in relation to 1 T were calculated. In addition, paired student's *t* tests were performed between 1 T, 2 T and 2 T + P mean annual increments.

All statistical analyses were conducted on R version 3.5.3(R Development Core Team 2019).

14.3 Results

14.3.1 Radial Growth Patterns

To analyze the effect of thinning and pruning on radial growth in the experimental long-term stand at Aguas Blancas, ring widths were measured for each radii to create individual RW and BAI series and then combined into treatment mean series (Table 14.1 and Fig. 14.2). Firstly, all mean series had almost the same starting year, which corroborated the existence of a post-harvesting cohort in the three treatments.

The first thinning in 1984, which was very intense since it reduced the occupation level to 6% of the initial density, generated an increase in the average annual increment for the three treatments. This was reflected by large peaks in the radial growth patterns associated to this event, both in ring width and basal area increment in all three treatments. The second intervention of 1999, which left 56% of the density of individuals in 2 T and 2 T + P treatments, could only be seen as a peak, smaller than that of 1984, in the RW and BAI series of 2 T + P treatment (Fig. 14.2e, f), but with almost no changes in 2 T (Fig. 14.2c, d).

In terms of common variability between individual series per treatment, the average inter-correlation per individual series before the first intervention was relatively low ($r = 0.274 \pm 0.031$) and it increased to 0.719 ± 0.023 between-interventions period (1985–1999). Then, after the second intervention, the 2 T and 2 T + P treat-

Table 14.1 Structural characteristics (density), radial growth (ring width) and inter-annual variation in growth (mean sensitivity) for the three treatments during the three periods defined by the two interventions (1984 and 1999)

	Pre-first cut (pre-1984)			Between cuts (1985–1999)			Post-second cut (2000–2009)		
Treat.	Density (ind/ha)	Mean RW (mm)	MS	Density (ind/ha)	Mean RW (mm)	MS	Density (ind/ha)	Mean RW (mm)	MS
1 T	80,000	1.25 ± 0.02	0.23	4426	1.79 ± 0.04	0.37	4426	0.89 ± 0.02	0.38
2 T	80,000	1.26 ± 0.02	0.26	4426	1.66 ± 0.04	0.33	2478	1.15 ± 0.03	0.36
2 T + P	80,000	1.22 ± 0.03	0.26	4426	1.91 ± 0.05	0.34	2478	1.39 ± 0.04	0.30

Treat treatment, *RW* ring width, *MS* mean sensitivity

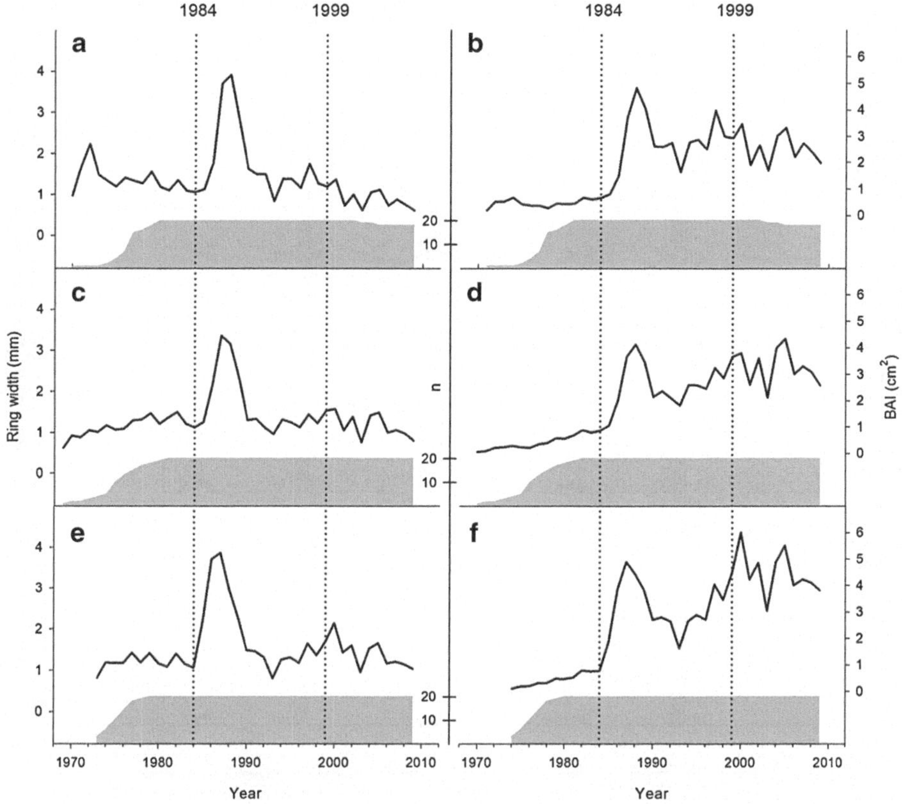

Fig. 14.2 Average ring width (RW, left panels) and basal area increment (BAI, right panels) for one thinning (1 T; **a** and **b**), two thinnings (2 T; **c** and **d**) and two thinnings with pruning (2 T + P; **e** and **f**) treatments. The central axis indicates the number of trees (n). The vertical dotted lines indicate the years of the two interventions (1984 and 1999). The grey shaded areas at the bottom of each panel represent the sample size (average series per tree) per year

ments showed a greater increase, reaching *r* values of 0.733 and 0.749 respectively. The 1 T treatment, which had no intervention in 1999, showed a slight decrease in the mean inter-correlation from 0.764 to 0.722. On the other hand, the mean sensitivity, as a measure of the degree of inter-annual variation, also showed an increase associated with the application of the intermediate treatment.

To evaluate the annual variation in each treatment and to quantify growth releases, the annual percentages of growth change (PGC) in RW and BAI series were calculated (Fig. 14.3). For the 1984 intervention, the average PGC between treatments was 129% (SE = 20%), and it was recorded as individual releases on 57% of the trees in each treatment. However, the PGC for BAI increased to 419% (SE = 39%) at same year. Neither RW nor BAI showed a PGC greater than 50% nor individual release associated with the interventions carried out in 1999 were detected. Only a light growth release was found in 1996 for the three treatments.

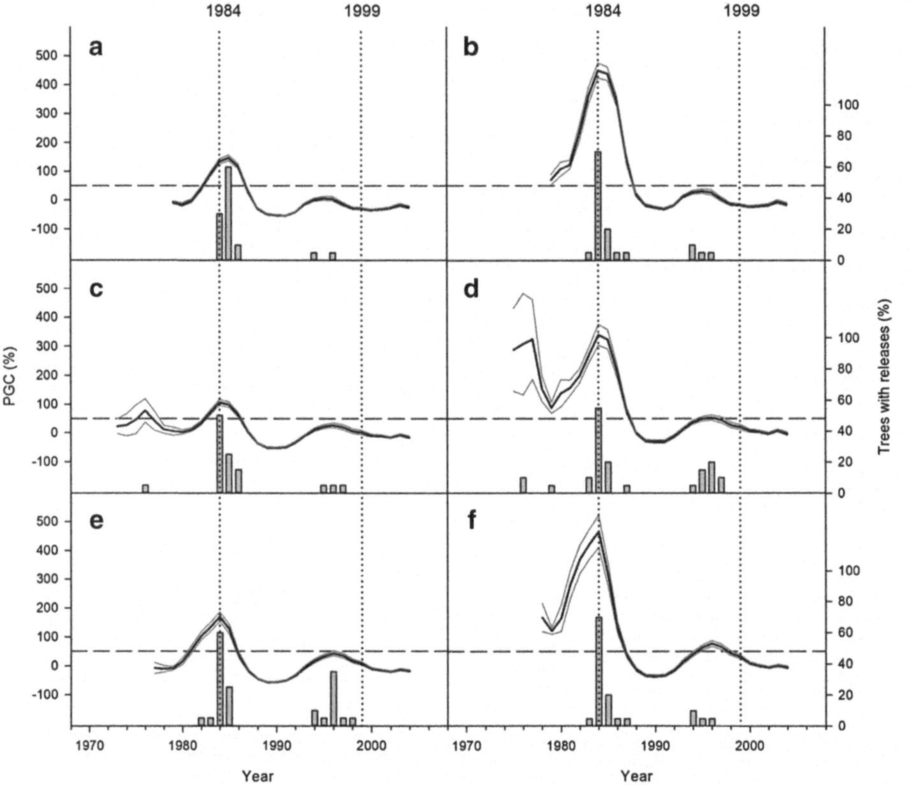

Fig. 14.3 Percentages of growth change (PGC) in mean ring width (RW, left panels: **a**, **c** and **e**) and basal area increment (BAI, right panels: **b**, **d** and **f**) (black lines) with their respective standard errors (grey lines) for the three treatments: 1 T, one thinning (**a**, **b**); 2 T, two thinnings (**c**, **d**) and 2 T + P, two thinnings with pruning (**e**, **f**). Cut lines indicate 50% PGC threshold for release detection. The bars show the individual trees released (PGC >50%) in each treatment. The vertical dotted lines indicate the two interventions (years 1984 and 1999)

14.3.2 Differences in Periodic Growth Due to the Interventions

Growth changes associated with the two interventions were analyzed by calculating 5-year periodic annual differences (ΔPAI_5) and by comparing them between treatments (Fig. 14.4). In the case of the 1984 strong thinning, significant differences were found in RW ($F = 7.994$, $p < 0.001$), but not for BAI ($H = 5.318$, $p = 0.07$). The paired comparison of means indicated that 1 T treatment was not different from 2 T + P and 2 T, respectively, with the highest and lowest growth which were significantly different from each other. In opposition, the 1999 intervention generated significant differences in both RW ($F = 14.008$, $p < 0.001$) and BAI ($F = 16.729$, $p < 0.001$). In this last intervention, 1 T presented negative differences and differed

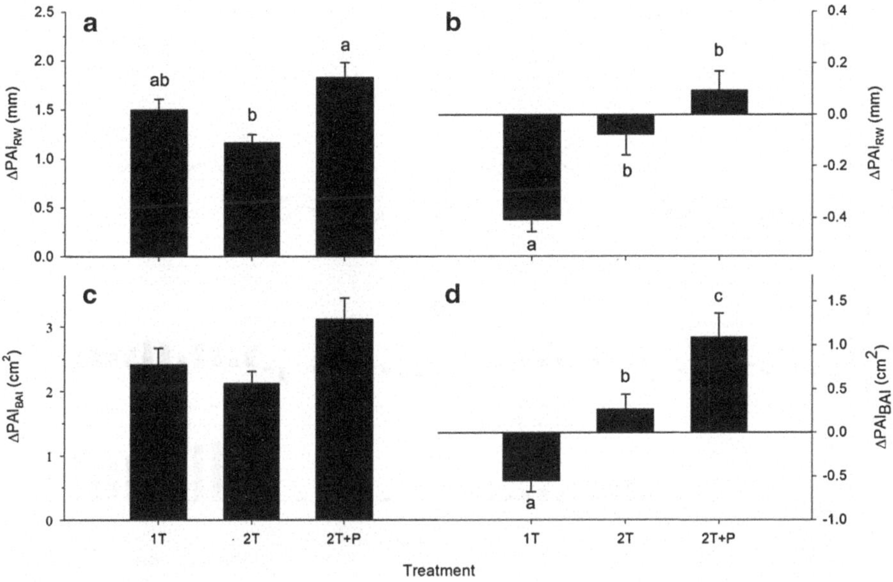

Fig. 14.4 Differences in periodic annual increment (ΔPAI) for 5 years before and after the 1984 (left panels: **a** and **c**) and 1999 (right panels: **b** and **d**) interventions in ring width (RW, upper panels: **a** and **b**) and basal area increment (BAI, lower panels: **c** and **d**) for each treatment (1 T, one thinning; 2 T, two thinnings and 2 T + P, two thinnings with pruning). Bars represent mean difference with standard error bars. Means with the same letter are not significantly different at $p < 0.05$ based on ANOVA and HSD Tukey post-hoc tests. Kruskal-Wallis test was used to detect differences between medians in panel c ($H = 5.318$, $p = 0.07$)

from 2 T and 2 T + P in both RW and BAI. The treatments 2 T and 2 T + P had positive values in ΔPAI_5, but differed significantly from each other with a greater increase in 2 T + P treatment.

In order to evaluate the effect of the second intervention compared to what was expected without any type of treatment, the percentage differences in the annual growth of the 2 T and 2 T + P treatments in relation to the 1 T treatment were analyzed (Fig. 14.5). No percentage changes lower than −25% or greater than 15% were found for 2 T and 2 T + P during the years prior to the 1999 intervention (Fig. 14.5a, b). After the intervention, positive differences were found with peaks in 2001 with greatest percentage difference of 96% and 121% in RW and BAI, respectively. The difference corresponding to 2 T + P was found to be greater than 2 T in all situations. Paired *t* tests evidenced that these differences were of lesser magnitude and limited to 2 years for 2 T treatment (Fig. 14.5c, d) while in 2 T + P the increase in growth lasted until 10 years after the last intervention (Fig. 14.5e, f).

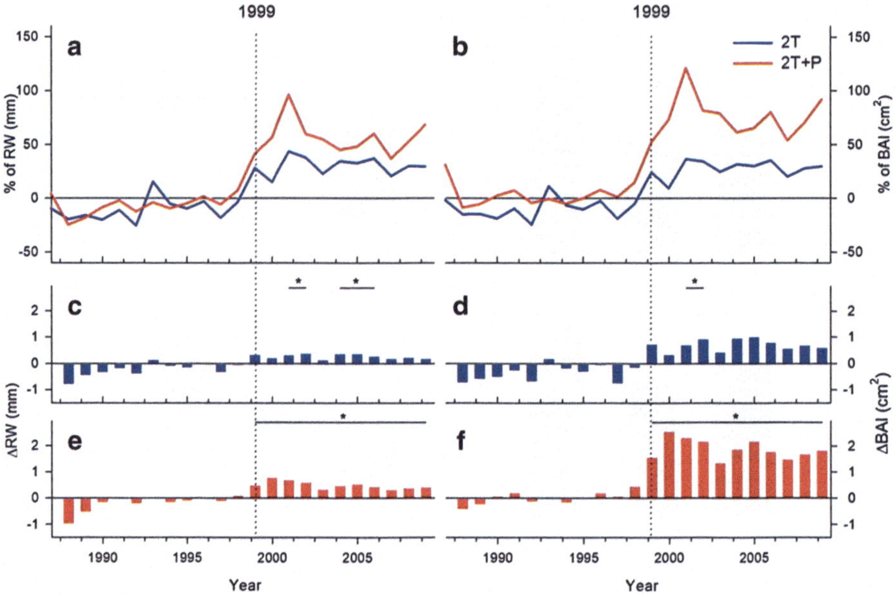

Fig. 14.5 Annual average differences of the two-thinning (2 T, blue lines and blue bars) and two-thinning plus pruning (2 T + P, red lines and red bars) treatments with respect to the one thinning treatment (1 T) for ring width (RW, left panels) and basal area increment (BAI, right panels) analyzed for 10 years before and after the second intervention. Differences are expressed as a percentage of the 1 T treatment (**a**, **b**) or in the same units (**c**–**f**). The vertical dotted lines indicate the date of the second intervention (1999). On the lower plots (c, d, e and f), segments capped with an asterisk correspond to intervals with index differences statistically different ($p < 0.05$) from the long-term mean based on paired t tests

14.4 Discussion

This long-term case study showed that the dendrochronological techniques widely and mainly used to detect growth releases due to disturbances can be applied to quantify and to reconstruct the effect of intermediate treatments in *N. pumilio* forests. In addition, as it was expected, higher growth rates were found as a consequence of highly intensive thinning interventions. On the other hand and using dendrochoronological methods for the first time in this species under silvicultural management, a significant increase in radial growth was detected as a response to the combined effect of thinning and pruning. Moreover, BAI appears to be a more sensible parameter than RW to detect the effect of intermediate treatments.

The effect of the intermediate treatments performed in 1984 and 1999 were verified both visually through peaks in their growth patterns and increases in the PGC for the three treatments. In the three treatments analyzed, in terms of trends, considerable increases in RW and BAI were observed as a consequence of the heavy thinning that these plots had in 1984 (Fig. 14.2) These radial growth releases are due to the fact that thinning, as an intermediate type of silvicultural treatments, produces the release of resources (Smith et al. 1998) and, consequently, has a positive effect on the growth of the remaining trees. In relation to the 1999 intervention, the plot that received thinning and pruning (2 T + P) showed a higher growth, in relation to the 1 T treatment, 2 years after the intervention (2001). The treatment that only received thinning on 1999 (2 T) also showed an increase in the percentage difference with respect to treatment without intervention. However, this increase did not reach the magnitude detected in 2 T + P. This could be due to the fact that the elimination of basal branches coincided with those less efficient in terms of photosynthetic balance and, consequently, leaving remnant branches with greater photosynthetic capacity that use more efficiently the higher amount of light irradiance resulting from the simultaneous thinning of individuals.

In relation to the second intervention that took place in 1999, in the 2 T treatment, which only had a low intense thinning, no changes were observed in RW growth, though there was a slight increase in BAI. On the contrary, treatment with thinning and pruning (2 T + P) showed an increase in both variables. Therefore, the repetition of thinnings 15 years apart did not translate into significant changes in radial growth. However, the addition of pruning produced the increase of radial growth, both measured in RW and BAI, for 10 years after the intervention in comparison to the control situation. This fact would be explained by a greater accumulation of biomass in the stem, possibly due to a greater efficiency in the capture of resources by the remaining individuals who suffered a rejuvenation of the crown. Other authors also detected an increase in radial growth due to the application of pruning in other species from the northern hemisphere (Stein 1955; Staebler 1963). Recently, Giantomasi et al. (2015) also observed that *Prosopis flexuosa* trees increased radial growth after pruning in the Monte region of western-central Argentina.

The growth rate recorded before the first intervention in the three treatments is similar to that observed by Barrera et al. (2000) in self-thinning young forests located in Valle de Andorra, Tierra del Fuego. The percentage of growth change quantified in this study (with maximum values of 155% in RW and 449% in BAI) was parallel to those reported by Franco (2015) (360% in RW and 431% in BAI) for thinned *N. betuloides* trees in Tierra del Fuego, and even surpassed the value (80% in BAI) reported by Pérez-de-Lis et al. (2011) for thinned *Pinus canariensis* trees. However, these values are similar to those reported in other thinning studies on *N. pumilio* conducted in Southern Patagonia (Peri et al. 2002, 2013).

On another subject, after the first intervention, the mean inter-correlation values, as an expression of the degree of agreement and common signal between ring width series, in comparison to those values reported by Mundo et al. (2007) and Massaccesi et al. (2008) were very high. This would indicate that the application of thinning and pruning treatments had a very homogeneous effect on the remaining trees, they were not damaged and they responded very homogeneously. In addition, the increase in average sensitivity after the first intervention in the three treatments would indicate that these trees had a quite constant radial growth before the opening of the canopy and that once competition had been reduced due to the thinning; the remaining individuals began to show radial patterns with greater inter-annual variability more linked to climatic annual variability than to the competition for resources.

Finally, this research could be used as reference for future studies to quantify the intensity of past interventions using these same dendrochronological methods presented in this chapter in complementation with reports of exploitation or measurements of forest structure. In this way, by calibrating different thinning intensities with releases in RW and BAI, it would be possible to estimate past thinning and pruning intensities on stands where this type of information is unknown. In addition, it would be very useful to carry out anatomical studies on the samples analyzed in this study or in other sites with known intermediate treatments to better understand the response of *N. pumilio* in terms of xylem transportation and ecophysiology linked to silvicultural management.

Acknowledgments This chapter was partly funded by SECYT-Universidad Nacional de La Plata (award code A282). The authors are very grateful to the editors of this book for their invitation to write a chapter. We also thank two anonymous reviewers for helping to improve the final version of this manuscript.

References

Barg AK, Edmonds RL (1999) Influence of partial cutting on site microclimate, soil nitrogen dynamics, and microbial biomass in Douglas-fir stands in western Washington. Can J For Res 29:705–713. https://doi.org/10.1139/x99-045

Barrera MD, Frangi JL, Richter LL et al (2000) Structural and functional changes in *Nothofagus pumilio* forests along an altitudinal gradient in Tierra del Fuego, Argentina. J Veg Sci 11:179–188. https://doi.org/10.2307/3236797

Boninsegna JA, Keegan J, Jacoby GC et al (1989) Dendrochronological studies in Tierra del Fuego, Argentina. Quat S Am A 7:305–326

Caldentey J, Schmidt H, Ibarra M et al (1998) Modificaciones microclimáticas causadas por el uso silvícola de bosques de lenga (*Nothofagus pumilio*), en Magallanes, Chile. In: Actas Primer Congreso Latinoamericano de IUFRO, IUFRO-CONAF, Valdivia, p 12

Daniel TW, Helms JA, Baker FS (1982) Principios de silvicultura. McGraw-Hill, México DF

Dimitrov DM, Rumrill PD Jr (2003) Pretest-posttest designs and measurement of change. Work 20:159–165

Fang O, Alfaro RI, Zhang QB (2018) Tree rings reveal a major episode of forest mortality in the late 18th century on the Tibetan Plateau. Glob Planet Change 163:44–50. https://doi.org/10.1016/j.gloplacha.2018.02.004

Fayle DCF (1983) Differences between stem and root thickening at their junction in red pine. Plant Soil 71:161–166. https://doi.org/10.1007/BF02182651

Franco MG (2015) Respuesta del crecimiento radial de *Nothofagus betuloides* "guindo" a diferentes intensidades de raleo en Tierra del Fuego, Argentina. Undergraduate thesis, Universidad Nacional de La Plata, Argentina

Fritts HC, Swetnam TW (1989) Dendroecology: a tool for evaluating variations in past and present forest environments. Adv Ecol Res 19:111–188. https://doi.org/10.1016/S0065-2504(08)60158-0

Giantomasi MA, Alvarez JA, Villagra PE et al (2015) Pruning effects on ring width and wood hydrosystem of *Prosopis flexuosa* DC from arid woodlands. Dendrochronologia 35:71–79. https://doi.org/10.1016/j.dendro.2015.07.002

Gutiérrez E (1992) Dendrochronological analysis of *Nothofagus pumilio* along an altitudinal gradient in a *Nothofagus* forest in Tierra del Fuego. Argentina. Lundqua Rep 34:137–141

Holmes RL (1983) Computer-assisted quality control in tree-ring dating and measurement. Tree-Ring Bull 43:69–78

Liu C, Ruel JC, Zhang SY (2007) Immediate impacts of partial cutting strategies on stand characteristics and value. For Ecol Manag 250:148–155. https://doi.org/10.1016/j.foreco.2007.05.010

Martínez-Pastur G, Peri PL, Vukasovic RF et al (1997) Site index equation for *Nothofagus pumilio* Patagonian forest. Phyton 61:55–60

Martínez-Pastur G, Cellini JM, Peri PL et al (2000) Timber production of *Nothofagus pumilio* forests by a shelterwood system in Tierra del Fuego (Argentina). For Ecol Manag 134:153–162. https://doi.org/10.1016/S0378-1127(99)00253-4

Martínez-Pastur G, Cellini JM, Lencinas MV et al (2001) Modificación del crecimiento y de la calidad de fustes en un raleo fuerte de un rodal en fase de crecimiento óptimo inicial de *Nothofagus pumilio* (Poepp. et Endl.). Ecol Austral 11:95–104

Martínez-Pastur G, Lencinas MV, Cellini JM et al (2009) Timber management with variable retention in *Nothofagus pumilio* forests of Southern Patagonia. For Ecol Manag 258:436–443. https://doi.org/10.1016/j.foreco.2009.01.048

Martínez-Pastur GJ, Cellini JM, Lencinas MV et al (2011) Environmental variables influencing regeneration of *Nothofagus pumilio* in a system with combined aggregated and dispersed retention. For Ecol Manag 261:178–186. https://doi.org/10.1016/j.foreco.2010.10.002

Martínez-Pastur G, Favoretti S, Paredes D et al (2013) Propuestas de cosecha para bosques secundarios y floreados de *Nothofagus pumilio* en Tierra del Fuego. In: Actas de las II Jornadas Forestales de Patagonia Sur y II Congreso Internacional Agro-forestales Patagónico. Calafate, p 14

Massaccesi G, Roig FA, Martínez-Pastur GJ et al (2008) Growth patterns of *Nothofagus pumilio* trees along altitudinal gradients in Tierra del Fuego, Argentina. Trees-Struct Funct 22:245–255

Misson L, Vincke C, Devillez F (2003) Frequency responses of radial growth series after different thinning intensities in Norway spruce (*Picea abies* (L.) Karst.) stands. For Ecol Manag 177:51–63. https://doi.org/10.1016/S0378-1127(02)00324-9

Mundo IA, Barrera MD, Roig FA (2007) Testing the utility of *Nothofagus pumilio* for dating a snow avalanche in Tierra del Fuego, Argentina. Dendrochronologia 25:19–28. https://doi.org/10.1016/j.dendro.2007.01.001

Mutarelli E, Orfila E (1973) Algunos resultados de las investigaciones de manejo silvicultural que se realizan en los bosques Andino-Patagónicos de Argentina. Rev For Argent 17:69–75

Nowacki GJ, Abrams MD (1997) Radial-growth averaging criteria for reconstructing disturbance histories from presettlement-origin oaks. Ecol Monogr 67:225–249. https://doi.org/10.1890/0012-9615(1997)067[0225:RGACFR]2.0.CO;2

Otero L, Contreras A, Barrales L (1998) Cortas de protección en claros de bosque: efectos microclimáticos. In: Actas Primer Congreso Latinoamericano de IUFRO. IUFRO-CONAF, Valdivia, p 8

Pamerleau-Couture É, Krause C, Pothier D et al (2015) Effect of three partial cutting practices on stand structure and growth of residual black spruce trees in North-Eastern Quebec. Forestry 88:471–483. https://doi.org/10.1093/forestry/cpv017

Pedersen BS (1998) The role of stress in the mortality of midwestern oaks as indicated by growth prior to death. Ecology 79:79–93. https://doi.org/10.1890/0012-9658(1998)079[0079:TROSIT]2.0.CO;2

Pérez-de-Lis G, García-González I, Rozas V et al (2011) Effects of thinning intensity on radial growth patterns and temperature sensitivity in *Pinus canariensis* afforestations on Tenerife Island, Spain. Ann For Sci 68:1093–1104. https://doi.org/10.1007/s13595-011-0125-8

Peri PL, Martínez-Pastur G, Vukasovic R et al (2002) Thinning schedules to reduce risk of windthrow in *Nothofagus pumilio* forests of Patagonia, Argentina. Bosque 23:19–28

Peri PL, Martínez-Pastur G, Monelos L (2013) Natural dynamics and thinning response of young lenga (*Nothofagus pumilio*) trees in secondary forests of Southern Patagonia. Bosque 34:273–279

Peri PL, Lencinas MV, Bousson J et al (2016) Biodiversity and ecological long-term plots in Southern Patagonia to support sustainable land management: the case of PEBANPA network. J Nat Conserv 34:51–64. https://doi.org/10.1016/j.jnc.2016.09.003

R Development Core Team (2019) R: A language and environment for statistical computing. R Foundation for Statistical Computing, Vienna. https://www.R-project.org/

Ruano I, Rodríguez-García E, Bravo F (2013) Effects of pre-commercial thinning on growth and reproduction in post-fire regeneration of *Pinus halepensis* mill. Ann For Sci 70:357–366. https://doi.org/10.1007/s13595-013-0271-2

Schulman E (1956) Dendroclimatic changes in semiarid America. University of Arizona Press, Tucson

Schweingruber FH (1996) Tree rings and environment: dendroecology. Haupt, Vienna

Smith DM, Larson BC, Kelty MJ et al (1998) The practice of Silviculture: applied forest ecology, 9th edn. Wiley, New York

Staebler GR (1963) Growth along the stems of full crowned Douglas-fir trees after pruning to specified heights. J For 61:124–127

Stein WI (1955) Pruning to different heights in young Douglas-fir. J For 53:352–355

Stokes M, Smiley T (1968) An introduction to tree-ring dating. University of Chicago Press, Chicago

Thibodeau L, Raymond P, Camiré C et al (2000) Impact of precommercial thinning in balsam fir stands on soil nitrogen dynamics, microbial biomass, decomposition, and foliar nutrition. Can J For Res 30:229–238. https://doi.org/10.1139/x99-202

Thorpe HC, Thomas SC, Caspersen JP (2007) Residual-tree growth responses to partial stand harvest in the black spruce (*Picea mariana*) boreal forest. Can J For Res 37:1563–1571. https://doi.org/10.1139/X07-148

Uriarte A, Grosse H (1991) Los bosques de lenga. Una orientación para su uso y manejo. Recopilación bibliográfica. CORFO-INFOR, Concepción

Venturoli F, Franco AC, Fagg CW (2015) Tree diameter growth following silvicultural treatments in a semi-deciduous secondary forest in Central Brazil. Cerne 21:117–123. https://doi.org/10.1590/01047760201521011204

Vincent M, Krause C, Zhang SY (2009) Radial growth response of black spruce roots and stems to commercial thinning in the boreal forest. Forestry 82:557–571. https://doi.org/10.1093/forestry/cpp025

Visser H (1995) Note on the relation between ring widths and basal area increments. For Sci 41:297–304. https://doi.org/10.1093/forestscience/41.2.297

Chapter 15
Tree-Growth Variations of *Nothofagus antarctica* Related to Climate and Land Use Changes in Southern Patagonia, Argentina

Evangelina Soledad Vettese, Ricardo Villalba, Ivonne Alejandra Orellana Ibáñez, and Pablo Luis Peri

Abstract Isolated forest patches of *Nothofagus antarctica* (ñire) are frequent in the Patagonian forest-steppe ecotone. These remnants, also called *relicts* (R), are separate from the *continuous* forests (C). Over the past century, these ecotonal forests have been impacted by anthropogenic activities, including fires, logging, and cattle ranching. In order to identify in *N. antarctica* ring-width records the variations in tree growth associated with documented changes in land use, five sites were selected in Santa Cruz, Argentina. In each site, increment cores from R and C were collected. We developed individual chronologies, and the relationships between regional climate variations and *N. antarctica* growth were established for each forest type and site. The similarities/differences between site-paired chronologies (R-C) were estimated by calculating moving correlation coefficients lagged by 1 year. *N. antarctica*

Electronic supplementary material The online version of this chapter (https://doi.org/10.1007/978-3-030-36930-9_15) contains supplementary material, which is available to authorized users.

E. S. Vettese (✉)
Centro de Investigación y Transferencia Santa Cruz (CIT – Santa Cruz – CONICET), Río Gallegos, Santa Cruz, Argentina

R. Villalba
Instituto Argentino de Nivología, Glaciología y Ciencias Ambientales (IANIGLA – CONICET), Mendoza, Argentina
e-mail: ricardo@mendoza-conicet.gob.ar

I. A. O. Ibáñez
Facultad de Ciencias Naturales y Ciencias de la Salud– Universidad Nacional de la Patagonia San Juan Bosco (FCNyCS – UNPSJB – Esquel), Esquel, Chubut, Argentina

P. L. Peri
Instituto Nacional de Tecnología Agropecuaria – Estación Experimental Agropecuaria Santa Cruz (INTA EEA – Santa Cruz). Universidad Nacional de la Patagonia Austral (UNPA – CONICET), Río Gallegos, Santa Cruz, Argentina
e-mail: peri.pablo@inta.gob.ar

M. Pompa-García, J. J. Camarero (eds.), *Latin American Dendroecology*,
https://doi.org/10.1007/978-3-030-36930-9_15

regional growth was directly related to precipitation during the current growing season (November–December; $r = 0.34$, $n = 62$, $p < 0.01$), and inversely related to temperature (December–March; $r = -0.58$, $n = 62$, $p < 0.001$). Since the middle of the twentieth century, a progressive decrease has been recorded in regional radial growth, consistent with an increase in summer temperature and a decrease in spring precipitation. In the context of this regional response of *N. antarctica* to climate, differences in growth patterns between R and C were associated with past changes in land use. Overall, the largest differences between R and C chronologies were concurrent with the settlement of cattle ranches and the associated use of forests. Conversely, similarities between R and C records increased after the establishment of protected areas and during the implementation of similar management practices in both forest types. Our research provides the first dendrochronological records from *Nothofagus antarctica* for the Argentinean Patagonia and represents one of the first efforts to identify in tree-rings past changes in livestock practices in southern South America.

Keywords Forest-steppe ecotone · Relict and continuous forest · Dendrochronology · Climate-tree growth relationship · Historical and cultural use of the land · Dendroclimatology

15.1 Introduction

The genus *Nothofagus* is widely distributed in the continental and insular sectors bordering the South Pacific Ocean, particularly in the mid-latitudes of Australasia and South America (Veblen et al. 1996). *Nothofagus antarctica* (G. Forster) Oerster (ñire or ñirre) and *Nothofagus pumilio* (Poepp. and Endl.) Krasser (lenga) have the most extensive geographic distributions of the genus in South America, covering 2200 km along the Andes from 35° 36' S in Central Chile to 55° S in Tierra del Fuego (Donoso 1993; Lara et al. 2001). The altitudinal range of these species extends from sea level to the upper tree line (Veblen et al. 1996; Roig 1999).

It is known that the Andes induce a marked precipitation gradient across Patagonia (Paruelo et al. 1998). The *Nothofagus* forests are located in the western, most humid areas of this gradient (Moore 1983; Donoso 1993). Along the west-east precipitation gradient in southern Patagonia, *N. antarctica* forests develop within a precipitation range that varies between 3000 mm to less than 800 mm over a west-to-east distance of only 50 km (Veblen et al. 1996; Paruelo et al. 1998). In response to its high phenotypic plasticity, *N. antarctica* presents a large ecological amplitude (Ramírez et al. 1985) growing from wet meadows (mallines) to dry areas in the forest-steppe ecotone. Furthermore, *N. antarctica* grows at flat-low valley sites as well as on steep slopes with strong and persistent winds (Veblen et al. 1996; Donoso et al. 2006; Kreps et al. 2012). In addition, *N. antarctica* tolerates winter frosts up to −22 °C (Alberdi 1995). In consequence, *N. antarctica* shows a great adaptive environmental capacity and multiple phenotypic responses. Under suitable site conditions (moderate temperature, adequate humidity, well-drained and fertile soils), *N. antarctica* is an erect tree of up to 20 m in height (Marazana et al. 1984; Ramírez et al. 1985; Veblen et al. 1996). By contrast, under extreme soil moisture conditions or at the upper treeline, it becomes a small tree or shrub of approximately 2–3 m height. Shrubby individuals (krumholz) are common in upper tree lines due to abundant snow deposition in winter (Pérez Moreau 1959; Veblen et al. 1996; Donoso et al. 2006). In the forest-steppe ecotone, it is common to observe trees up to 10 m coexisting with individuals less than 3 m in height (Marazana et al. 1984; Ramírez et al. 1985).

In the forest-steppe ecotone, *N. antarctica* usually forms the eastern boundary of the arboreal vegetation. In this ecotone, trees group into a mosaic of small monospecific patches of forests that alternate with the steppe in elevated areas and with meadows or mallines in depressions (Roig 1999; Peri and Ormaechea 2013). These remaining patches, called *relicts* (R), are marginal forest communities spatially disconnected with the *continuous forests* (C) located towards the west over the Andes mountain range (Fig. 15.1).

The ecotones are considered very dynamic habitats where the initial impacts of environmental changes become more evident (di Castri et al. 1988; Neilson 1991). In this sense, several authors (Iverson et al. 2004; Hampe and Petit 2005) claim that the populations of tree species located on their distribution margins are very sensitive to environmental variations. Due to the high perimeter/surface ratio, trees at the

Fig. 15.1 Photograph showing sectors of the relict (*black arrow*) and continuous (*grey arrow*) forests of *N. antarctica* in Estancia Cancha Carrera (site 3). The relict forests are represented by isolated patches bordered by steppe vegetation, while the most developed continuous forest extends all along the foothills of the Andes

ecotonal patches are exposed to more severe stress and higher rates of change than trees in continuous forests (Murcia 1995).

Ecotones have often been heavily impacted by human activities, particularly at the low elevation forest-grassland transitions. Changes in cultural land use have greatly influenced their dynamics (Kitzberger 2012). In southern Patagonia, the *N. antarctica* relicts have been subjected to different cultural uses, and only 14% of these are included in protected areas. The European immigration in the late-nineteenth and early-twentieth centuries caused the demise of the native Aonikenk hunter-gatherer population and the rapid establishment of livestock farming (Bandieri 2005). The new socio-economic system gradually modified the forests in the region (Barbería 1994, 1995; Schweitzer 2011). The *N. antarctica* forests were intensely used during the European colonization inducing a substantial reduction in the total area of the ecotonal forests. Among the various anthropogenic causes related to the decrease in *N. antarctica* forest extension were the opening of new roads, unplanned settlements, intensive grazing, forest fires, and logging (Montaña 1982; Manacorda and Bonvissuto 2001). Livestock causes the browsing of arboreal species and limits the natural regeneration of the forest. Livestock also facilitates the introduction of exotic plants that compete with native species (Christensen and Emborg 1996; SAyDS 2005). There have also been registered increments of certain understory taxa resistant to browsing, such as the *Berberis* sp. (Rajchenberg et al. 2014). As a consequence, the overconsumption and unsustainable use of natural resources led to the accelerated transformation of the environment (Schweitzer 2011). Concurrent with these vegetation changes, a general warming trend at the extreme south of South America (45–55° S) has become evident during the last century (Rosenblüth et al. 1997; Villalba et al. 2003; SAyDS 2005) concurrent with a decrease in precipitation (Aravena and Luckman 2009). These climatic changes, in combination with other local anthropic factors, may have also contribute to the fragmentation of native forests along the region.

It is well known that tree rings can record information about changes in the environmental factors affecting tree growth. However, few studies have been conducted in South America to determine variations in tree rings associated with past changes in land use related to livestock (but see Morales et al. 2005; Morales and Villalba 2006; Mundo et al. 2017). The objectives of this study were to use ring-width chronologies from *N. antarctica* to identify the species response to climatic variations and determine the influence of different land use practices between relict and continuous forests in Santa Cruz, Argentina. To reach these goals, we first established the relationships between *N. antarctica* tree growth and the regional climate. Based on these relationships, we postulated that historical changes in stocking density and harvesting intensity have led to important differences in the radial growth of *N. antarctica* at isolated and continuous forests in the eastern limit of its natural distribution. Changes in cultural land use may provide clues to understand the differences observed in tree growth between nearby stands. Although more than 300 tree-ring chronologies have been developed from different species in Argentina (Villalba et al. 1997; Villalba et al. 2003; Roig and Villalba 2008; Boninsegna et al. 2009), to the best of our knowledge, there are no dendrochronological records from *N. antarctica* trees. In consequence, the present study provides the first chronologies for *N. antarctica* in the Argentinean Patagonia.

15.2 Methods

15.2.1 Study Area

Located in southern South America, Patagonia presents a cold climate. In Santa Cruz, there are two main geo-climatic units: the Andes dominated by ice and forests with relatively abundant rainfalls to the west, and the Patagonian steppe dominated by grasslands with dry climate to the east (Díaz 2005). The difficulty of analyzing climatic variability in the region lies in the lack of weather stations, or the short extension of available records (Paruelo et al. 1998; Díaz 2005; Garreaud et al. 2013; Soto-Rogel and Aravena 2017). In the study area, the mean annual temperature is 6 °C, with a mean maximum of 17 °C in January and a mean minimum of −5 °C in July (Kreps et al. 2012). The reported absolute minimum temperature for the region is −20 °C (Paruelo et al. 1998). Total annual rainfalls at the study sites is estimated around 400 mm ± 150 mm. In this region, strong westerly winds prevail throughout the year but intensify during the spring and summer (Coronato and Del Valle 1993; González and Vera 2010).

We selected pairs of *relict* (R) and *continuous forests* (C) of *N. antarctica* in five sites in southwestern Santa Cruz province, Argentina. These forests are distributed in a latitudinal range of 467 km, between 47° 04′ W and 51° 05′ S in southern Patagonia (Fig. 15.2). Table 15.1 summarizes the most relevant climate data and the location of each study site.

15.2.2 Tree-Ring Data and Chronology Development

Dendrochronological samplings were conducted during the interval 2012–2016 in *relict* (R) and *continuous* (C) forests at five study areas (Fig. 15.1). In each site, 20 dominant individuals were selected, and 2 cores were obtained by tree using increment borers. Even though individuals with the best health status were chosen, in many cases only one sample could be recovered per tree. *N. antarctica* commonly presents tortuous stems of poor sanitary status with rotten centers. For the most part, samples were taken close to the root collar of the trees. However, when sampling at the collar was impeded by logistical problems, cores were extracted at the lowest possible height on the stem.

The samples were processed according to conventional techniques used in dendrochronology (Stokes and Smiley 1968). Tree rings were dated under a binocular magnifier, and the ring widths were measured with an accuracy of 0.001 mm using

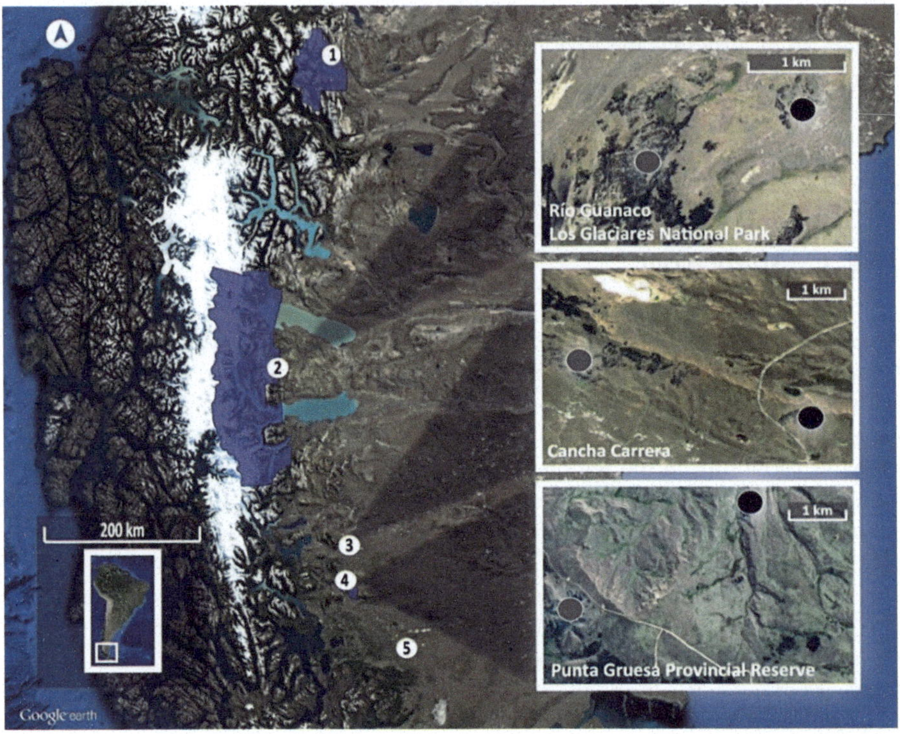

Fig. 15.2 Geographical distribution of the study sites are indicated by *white dots*: (**1**) Perito Moreno National Park, (**2**) Río Guanaco – Los Glaciares National Park, (**3**) Estancia Cancha Carrera, (**4**) Punta Gruesa Natural Reserve and (**5**) Estancia Rincón de los Morros. Protected areas (shown in *shaded blue*). Satellite images (framed *in white*) for the three study sites in which good-quality chronologies were developed are also included. Relict and continuous forests in these three images are indicated by *black* and *grey dots*, respectively

a Velmex machine (Fig. 15.3). After that, the quality of the measurements was examined using the COFECHA software (Holmes 1983). Finally, the ARSTAN program (Cook 1985) was used to develop raw, standard and residual chronologies from each pair of *N. antarctica* stands (R and C) at the five study sites. In addition, all series were combined to develop three additional chronologies: two independent chronologies for relict (R) and continuous (C) forests, and a regional chronology including all series from the five study areas. This set of chronologies was developed to validate different hypotheses. Thus, to identify differences in tree growth between relicts and continuous forests due to changes in land use, we developed chronologies in both forest types at each study area. Regional scale chronologies were elaborated to determine the existence of a common growth signal among all sampled trees and to establish differences between relicts and continuous forests.

Table 15.1 Site characteristics of tree-ring collections

Study sites	Site code	Forest type	Mean annual temperature (°C) [a]	Total annual precipitation (mm)[a]	Longitude	Latitude	Altitude (m.a.s.l.)
Perito Moreno national park	**1**	**R**	4.9	252	72° 8′47.82”W	47°43′40.91”S	971
	1	**C**	5.0	252	72° 9′16.07”W	47°43′39.86”S	886
Río Guanaco, Los Glaciares National Park	**2**	**R**	5.5	516	72°49′10.42”W	49°55′22.79”S	708
	2	**C**	5.5	528	72°50′32.42”W	49°55′37.01”S	692
Estancia Cancha Carrera	**3**	**R**	5.3	228	72° 6′56.25”W	51°12′38.15”S	431
	3	**C**	6.0	223	72°13′41.88”W	51°11′49.47”S	312
Punta Gruesa natural reserve	**4**	**R**	4.5	304	72° 8′49.84”W	51°29′28.07”S	534
	4	**C**	4.1	324	72° 9′10.55”W	51°31′6.65”S	577
Estancia Rincón de los Morros	**5**	**R**	6.3	220	71°30′27.68”W	51°55′10.15”S	139
	5	**C**	9.4	300	71°32′57.92”W	51°54′40.11”S	119

The name and codes of the study sites together with the forest types, the mean annual temperature, the total annual precipitation and the geographical locations are indicated

R relict forest, *C* continuous forest

[a]www.worldclim.org; Hijmans et al. (2005)

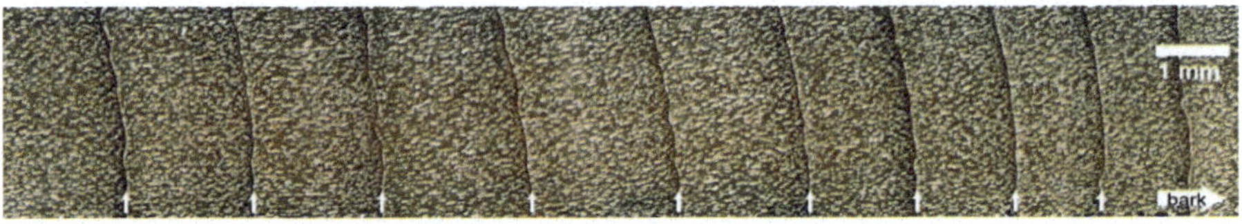

Fig. 15.3 An incremental core of *N. antarctica* showing the annual bands (*white arrows*), clearly discernible in the wood of this species. Although this is a diffuse porous species, a common feature of most *Nothofagus* species, the vessels at the beginning of the growing season are larger and their cell walls thinner than those formed later in the growing season. Fibers are abundant at the end of the growing season and show thick cell walls

15.2.3 Climate-Tree Growth Relationship

To determine the climate influences on the *N. antarctica* radial growth, the residual ring-width chronologies from the three well-replicated sites (sites 2, 3 and 4; Fig. 15.2) were compared with the instrumental temperature and precipitation records available for the region. The correlation function is a simple method for comparing the interannual variations in tree-ring chronologies with monthly interannual fluctuations in climate (Blasing et al. 1984). The statistical relationship between ring-width and climate variable was examined over the common 1952–2014 period (62 years) between the chronologies and the instrumental data. In addition, since the growth in a given year may be influenced by the climatic conditions during the previous years (Fritts 1976), the comparison period analyzed comprised 20 months, starting in September of the previous year until April of the year in which the rings were formed. Based on the results of the correlation function comparing variations between the regional chronology and the regional climate, we proceeded to group the monthly temperature and precipitation records into seasonal means including the months most strongly related to growth. To carry out these calculations, the Microsoft Excel spreadsheet (2013) was used.

15.2.4 Cultural Land Use Changes

Land uses changes in the forest-steppe ecotone in Santa Cruz were compiled from several historical documents, including the Memories from the National Park Administration, scientific publications on regional anthropological and archaeological studies, management plans of protected areas, technical reports and documents from the Santa Cruz Cadastral Department. Bibliography and files deposited at the Provincial Historical Archive in Río Gallegos were also consulted. Open interviews were also conducted among the residents and cattle ranch administrators.

Finally, the standard version of the tree-ring width chronologies from *N. antarctica*, obtained for R and C forests at each site, were compared by calculating the correlation coefficient in segments of 20 years lagged by 1-year between consecutive segments. To carry out these calculations, the Microsoft Excel spreadsheet (2013) was used. Finally, we analyzed the consistency between historical changes in land use and in growth patterns.

15.3 Results

15.3.1 Chronology Development

Eleven chronologies from *N. antarctica* were developed in Santa Cruz, one from each site and forest type (R2, C2, R3, C3, R4, C4, R5, and C5), with the exception of site 1 (Perito Moreno National Park), where samples were not cross-dated. Two other chronologies were developed using all the cross-dated series of each forest type (*relicts* and *continuous* respectively) and, finally, a *regional* chronology with all cross-dated series in southwestern Santa Cruz (Fig. 15.4).

The longest chronology, extended for 206 years, is from the Cancha Carrera relict forest (R3). On the other hand, the shortest chronology is from Rincón de los Morros continuous forest (C5), and it extended for 42 years. The mean length of all chronologies per site was 122 ± 54 years long (Table 15.2).

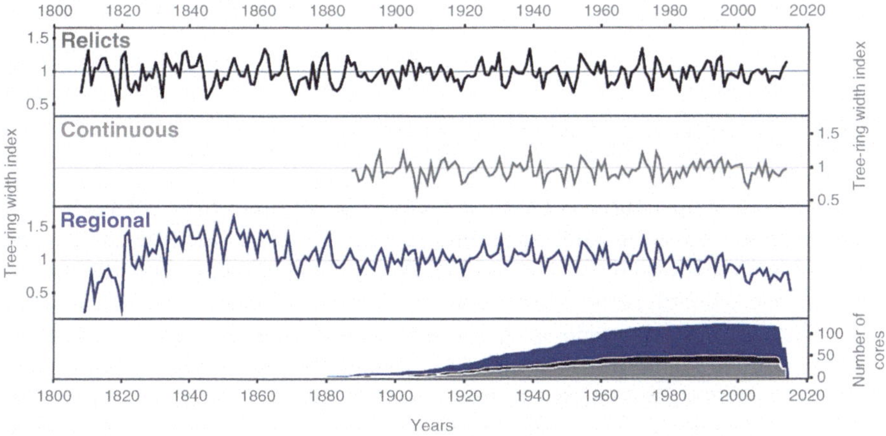

Fig. 15.4 Composite *N. antarctica* width chronologies in their standard versions. The left and right axes represent the tree-ring width index (dimensionless) and the number of series per year (replication), respectively; in *black* is represented the regional relict forest (R); in *light grey* the regional continuous forest (C) and in *blue* the regional chronology. Forest type and site codes are listed in Table 15.1

Table 15.2 Statistics for the 11 *N. antarctica* chronologies developed in Santa Cruz, Argentina

Forest type and site codes	Number of trees bored	Total series number	% Correlated series	Inter-correlation coefficient[a]	Period (years)	EPS[b]	R bar[b]
R2	86	56	36	0.510	1819–2013	0.936	0.370
C2	72	42	44	0.557	1914–2013	0.941	0.389
R3	61	40	48	0.608	1807–2013	0.945	0.439
C3	77	46	32	0.567	1888–2013	0.934	0.398
R4	30	23	53	0.578	1909–2011	0.897	0.397
C4	25	24	48	0.595	1886–2011	0.899	0.489
R5	32	24	50	0.645	1945–2011	0.935	0.510
C5	21	20	29	0.585	1971–2011	ND	ND
Relicts	**172**	**56**	**33**	**0.509**	**1807–2013**	**0.958**	**0.328**
Continuous	**154**	**37**	**24**	**0.506**	**1886–2013**	**0.920**	**0.256**
Regional	**326**	**137**	**41**	**0.4783**	**1807–2014**	**0.970**	**0.224**

Forest type and site codes are shown in Table 15.1
ND no data
[a]From COFECHA
[b]From ARSTAN

The composite R chronology was considerably longer than the composite C chronology (206 and 125 years, respectively). The regional chronology began in 1807 with 1 sample, reached 10 samples in 1888, a maximum of 133 samples between 1992 and 1995, and ended with 73 samples in 2014 (Fig. 15.5).

15.3.2 Climate-Tree Growth Relationship

The patterns of temperature variability are relatively uniform across our study area. Thus, the most extensive local temperature record in Torre Glacier (from 2002 to the present) is near the town of El Chaltén. This is significantly correlated with gridded data from ERA-Interin across most southern Patagonia (Fig. 15.S1). The extensive field of spatial correlation includes the Argentinean weather stations of Esquel and Río Gallegos and Punta Arenas in Chile (Fig. 15.S1). Located to the north, east and south of the study sites, respectively, these weather stations provide the most extensive and homogeneous records in southern Patagonia. Therefore, the mean from the monthly temperature deviations from these three stations were composed in a single regional record, and it was used to compare with tree-ring variations. All temperature series were normalized over the 1952–2014 common period. The use of normalized deviations to develop the regional temperature record implies that each meteorological station has similar contribution in the regional mean temperature.

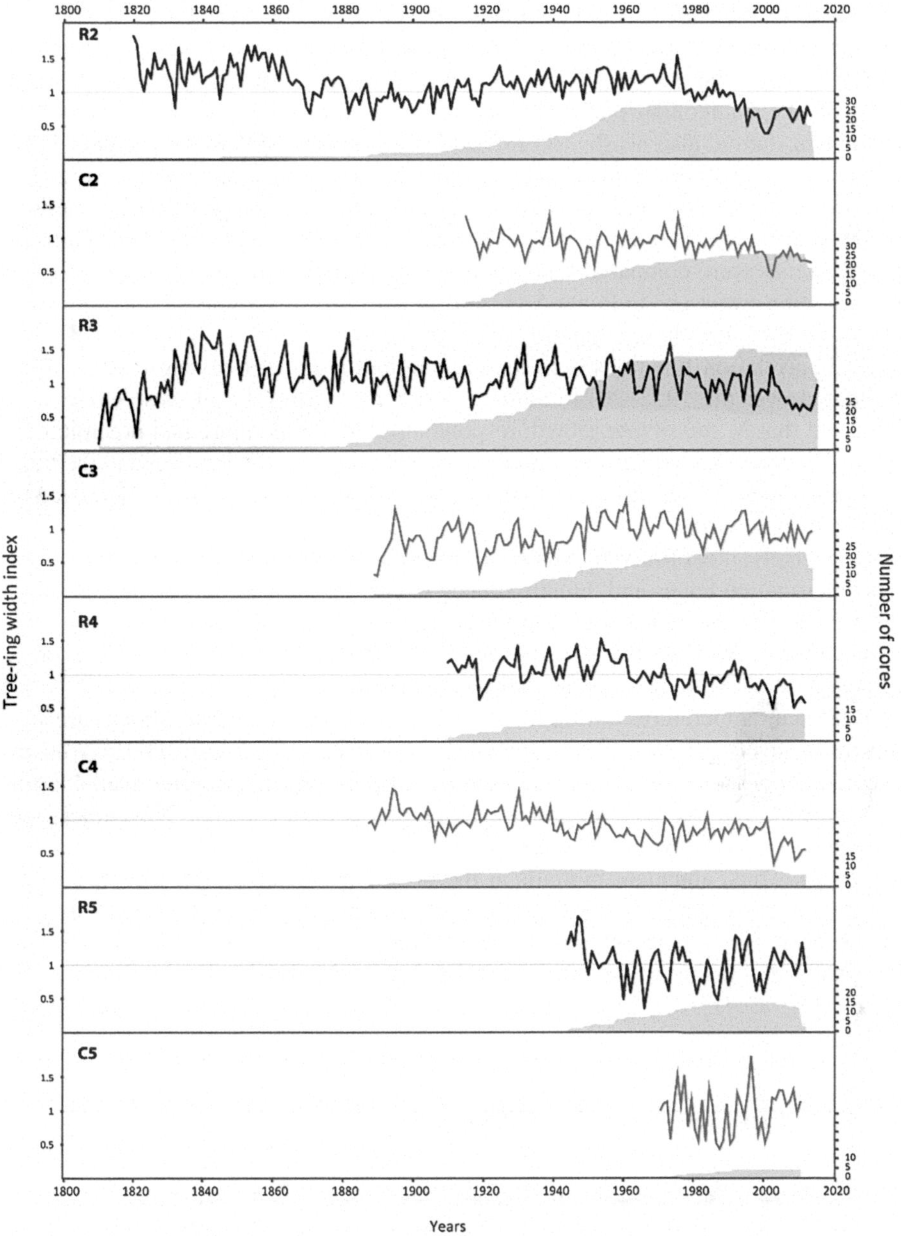

Fig. 15.5 *N. antarctica* tree-ring width chronologies in their standard version. The left axis represents the tree-ring width index (dimensionless); the right axis represents the number of series per year (replication); in *black* are represented the relict forests (R), and in *light grey* are represented the continuous forests (C). Forest type and site codes are listed in Table 15.1

Unlike the extended spatial correlation pattern of temperature for Torre Glacier, the spatial pattern of precipitation is much more limited. The abundant precipitation on the western and eastern slopes of the Andes sharply contrasts with the reduced rains on the Patagonian plateau. Thus, the local precipitations measured at Los Huemules station, also nearby to El Chaltén town (from 2006 to the present), have a much more spatially reduced representability pattern than temperature (Fig. 15. S2). Because there are no long (>20 years) and homogeneous precipitation records in the forest-steppe ecotone in Santa Cruz province, the monthly precipitation from Punta Arenas were compared with the *N. antarctica* chronologies here developed. Precipitation variations in Punta Arenas from December to August are significantly related to ERA-Interin precipitation over the study sites (Fig. 15.S3).

The correlation functions between climate and regional tree-ring records were estimated over the 1952–2014 common period (62 years). In all these cases, it is observed that *N. antarctica* growth responds to both temperature and precipitation levels. The previous growing season temperatures are directly associated to the tree-ring growth, while the current growing season temperatures are inversely related to the tree-ring growth.

Conversely, it is observed that *N. antarctica* growth is inversely related to previous precipitation levels and directly related to current growing season precipitation levels. There is even a negative trend in *N. antarctica* growth due to rain during the winter (July). Besides, the R responses, compared to the C responses, are most closely associated to the regional pattern (Fig. 15.6).

Regional temperatures during the growing season (December to March) are significantly inverse related to tree growth ($r = -0.58$, $n = 62$, $p < 0.001$). Above-average temperatures increase evapotranspiration, reducing the water available for tree growth. In contrast, the accumulated precipitation during November and December was positively correlated with *N. antarctica* growth ($r = 0.34$, $n = 62$, $p < 0.01$). Thus, abundant rainfalls at the beginning of the growing season favor tree-ring growth. Since the 1950s to the present, a negative trend has been recorded in the *N. antarctica* radial growth. These pattern is consistent with a regional increase in the mean annual temperature and, consequently, with a regional water deficit (Fig. 15.7).

15.3.3 Cultural Land Use Changes

The local historical documents for each site showed substantial changes in land use over time (Table 15.3). These land use changes were used to interpret the differences in radial growth variations between the pairs of R and C chronologies at each site (2–5).

At each sampling site, we registered several periods with opposite trends in radial growth between the R and C chronologies (Fig. 15.8). In general, the largest differences between the growth patterns from R and C occurred in periods concurrent with the settlement of the estancias, i.e. large rural establishments for the

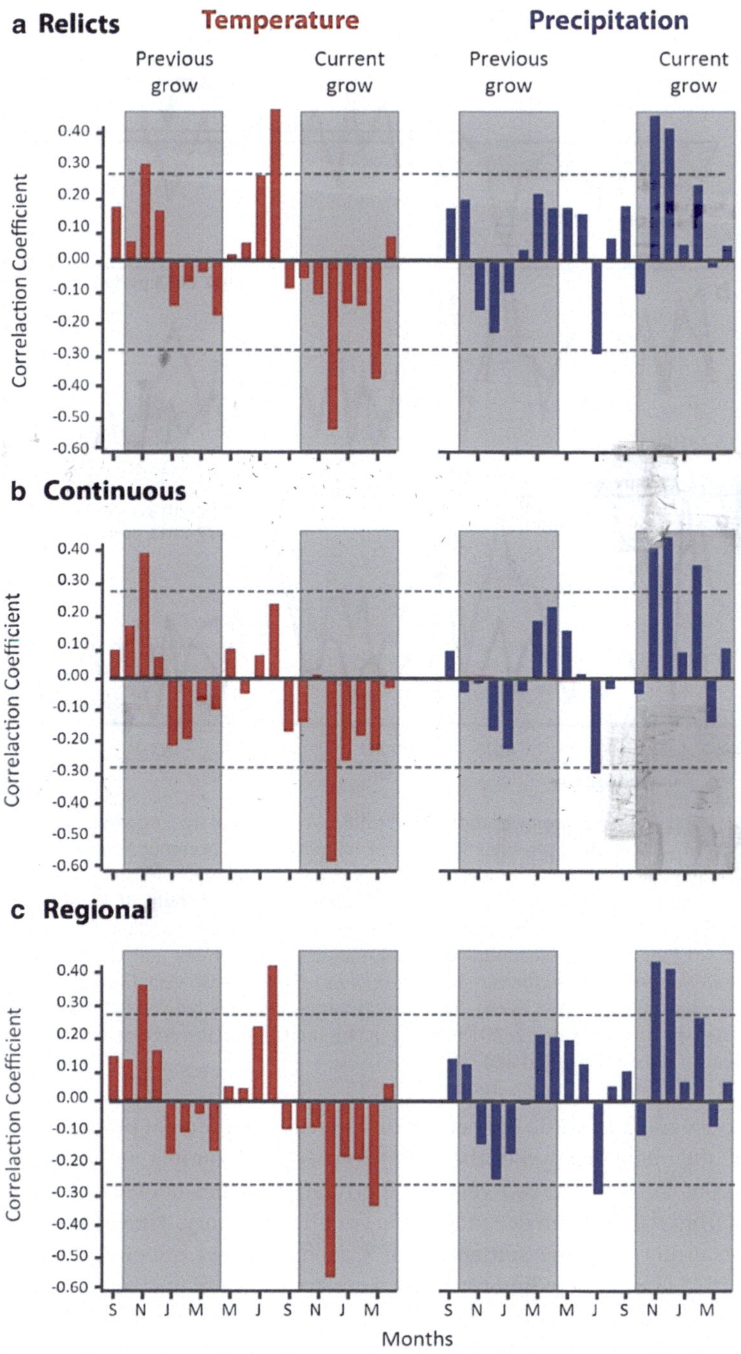

Fig. 15.6 Correlation functions between *N. antarctica* tree growth and regional climate. The red bars represent the relationship with temperature, and the blue bars represent the relationship with precipitation. (**a**) Relict forests. (**b**) Continuous forests. (**c**) Forests in the entire study area (*regional*). Dotted lines indicate 95% confidence levels

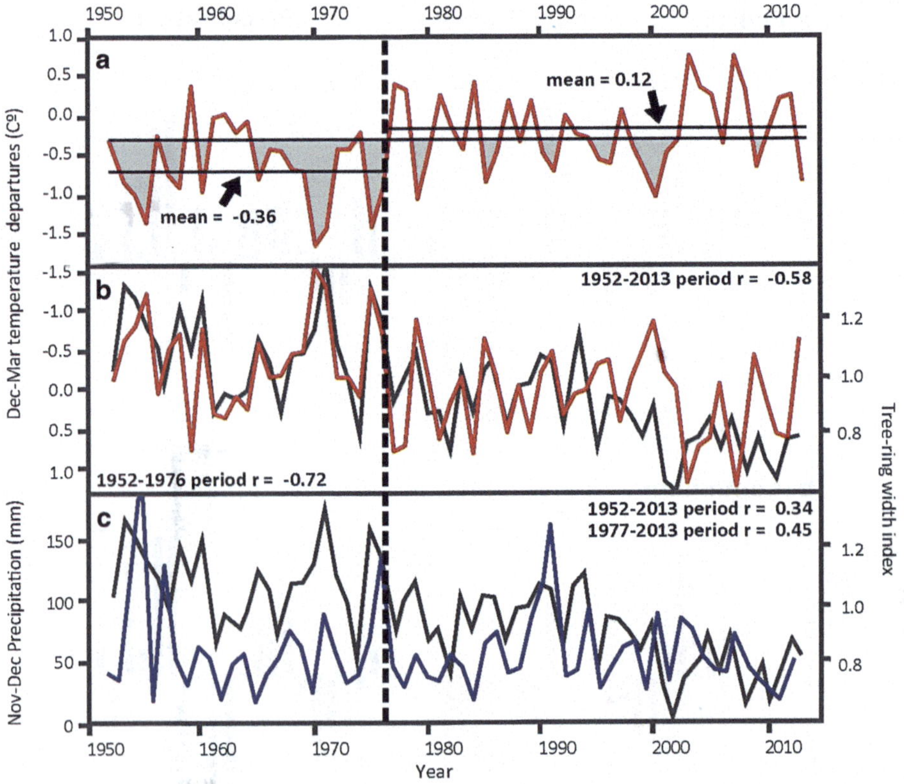

Fig. 15.7 Comparison between variations in *N. antarctica* regional tree growth and climate during the 1952–2013 period. (**a**) Variations in summer temperature (December to March; *red line*) in southern Patagonia showing the abrupt increase close to 0.5 °C around 1976/77. The temperature means for the periods before and after the 1976/77 climatic jump are indicated. (**b**) Regional temperature deviations from December to March (*red line*) and *N. antarctica* tree-ring width variations (*grey line*). (c) Punta Arenas accumulated precipitation for November and December (*blue line*) and *N. antarctica* tree-ring width variations (*grey line*). To facilitate visual comparison, the temperature axis has been inverted in (**b**). The correlation coefficients between climate and tree-ring variations are shown for the 1952–2013 period in (**b**) and (**c**), and the periods before and after the 1976/77 temperature jump in (**b**) and (**c**), respectively

extensive breeding of cattle or sheep. This extensive settlement process was associated with the intensification of the use of forests. For example, in Estancia Cancha Carrera (site 3), we observed very low correlation values between R and C chronologies from the first two decades of the twentieth century. These low correlations were concurrent with the settlement of Cancha Carrera affecting the C forests. Around 1945 similar breeding practices were applied in both stands (R and C), and after that period, correlation values increased sharply (Fig. 15.8b). Conversely, records were similar since the establishment of governmental protected areas, or during periods when the intensity of land use was similar in R and C. For example, towards 1980 in Punta Gruesa Natural Reserve (site 4) the anthropic activity

Table 15.3 Historical changes in cultural land use in each study site

Study site	Code site	Historical events
Perito Moreno national park	1	Between 1910 and 1937, the settlers based their economy on the intensive extraction of firewood for sale and on cattle ranching. R and C were strongly intervened. In 1937, Perito Moreno National Park was created. In 1975, the extraction of firewood was interrupted. In 1982, the cattle were removed.
Río Guanaco, Los Glaciares National Park	2	In 1926, Estancia La Irene was built with sawmills and livestock activity in R and C. In 1937, Los Glaciares National Park was created; the woody industry was forbidden and precarious occupation and pasture permissions for cattle were assigned to the owners. Livestock use intensified between 1965 and 1985. In 2010, the cattle were definitively retired.
Estancia Cancha Carrera	3	In 1895, sawmills were established, and livestock activity began in C. R remained without human intervention. In 1945, in C the stocking density increased, and in R the intervention began. Since 2000, in C there were no changes of use. In R, the use intensified by the installation of a precarious cabin and the National Route 40 was paved, with the consequent increase in cattle ranching and wood extraction.
Punta Gruesa natural reserve	4	Between 1880 and 1890, the Sociedad Explotadora de Tierra del Fuego was established. Its economy was based on sheep farming and forest wood extraction in association with small sawmills. In 1940, the Yacimientos Carboníferos Fiscales (National Coal Company) was established. The use of wood increased due to road building, steam engines use, urban population rise, house heating and house building. In 1980, a cattle ranch was established in the area close to R. In 2002, Santa Cruz province created the Punta Gruesa Natural Reserve. R and C cattle use were reduced. In C, there is now a tourist camping area, but there is no access road to the R.
Estancia Rincón de los Morros	5	Since European immigration, the local population based its economy on extensive livestock (cows and sheep) and the intensive extraction of firewood in order to build and heat their homes.

R relict forest, *C* continuous forest

decreased. Since then, there has been a drastic increase in the correlation between the R and C chronologies, which continues after the creation of the Punta Gruesa Natural Reserve in 2002 (Fig. 15.8c).

15.4 Discussion

In this study, we developed the first dendrochronological records for *N. antarctica* in Argentina. The new chronologies contribute to the expansion of the dendrochronological network and provide new knowledge of ecological processes and climatic

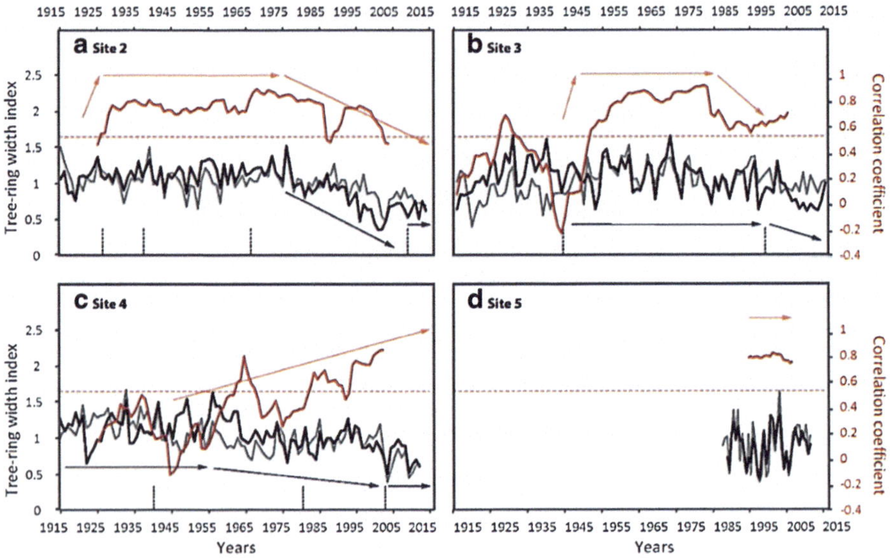

Fig. 15.8 Comparisons between R and C tree-ring width chronologies obtained at each site: (**a**) Río Guanaco – Los Glaciares National Park, (**b**) Estancia Cancha Carrera, (**c**) Punta Gruesa Provincial Natural Reserve and (**d**) Estancia Rincón de los Morros. Growth patterns (relict forests; *black lines*, continuous forests; *grey lines*) and the history of land uses (historical events mentioned in Table 15.3; *grey dotted lines* on the X-axis). Variations in the correlation coefficient between chronologies evaluated in 20 years lagged between them by 1 year (*red lines*). Correlation patterns (*red arrows*). Changes in growth patterns (*grey arrows*). Dotted red lines indicate 99% confidence levels

variability across southern Patagonia (Villalba et al. 2009). The maximum reported age of 259 years for *N. antarctica* in Tierra del Fuego (Ivancich 2013) is a bit longer than those found in our study for Santa Cruz (206 years). This suggests that the maximum ages reached by *N. antarctica* in the southern extreme of South America would be around 200–300 years old. Furthermore, the intercorrelation between the series that integrate the regional chronology is similar to those reported for *N. pumilio* in the same region (Aravena et al. 2002b; Villalba et al. 2003; Suarez 2010).

The collection of samples for providing well-replicated dendrochronological records *of N. antarctica* is problematic. The difficulties are in part due to the stem architecture, which commonly has a sympodial growth, with trunks branched from the base. Minor fires used for opening the forest canopy and browsing at early stages of growth cause morphological changes in *N. antarctica* trees, including branching from the base (Veblen and Lorenz 1987; Aravena et al. 2002a). In addition, browsing affects tree-growth rates by reducing the tree photosynthetic biomass (Veblen et al. 1992; Relva and Veblen 1998; Martínez-Pastur et al. 1999; Pulido et al. 2000; Srur et al. 2013). In addition, these forests usually present deficient phytosanitary conditions (Fig. 15.9). *N. antarctica* trees frequently have hollow trunk centers and abundant rotting wood. There are reports about several species of fungi that cause

Fig. 15.9 The *N. antarctica* relict forest in Punta Gruesa Natural Reserve, Santa Cruz (site 4). The phytosanitary status of the individuals is regular to poor with numerous dead trees and branches

chestnut and white rot in the xylem of several *Nothofagus* species (Rajchenberg et al. 2014). One of the most conspicuous fungus genus is *Cyttaria*, which develops obligatory parasitism with *Nothofagus*. *Cyttaria* hyphae alter the morphology of *N. antarctica*'s vascular system and generate extrusions that totally or partially surround the stem (Ipinza et al. 1989). Another very frequent interaction occurs between *Nothofagus* trees and *Misodendrum* species. Although *Misodendron* presence is not always a direct indication of tree decline (Rodríguez-Catón and Villalba 2018), these hemiparasitic plants may affect the *Nothofagus* radial-growth rates (Velásquez et al. 2012).

The well-defined tree rings in *N. antarctica* make this species suitable for developing good chronological records. However, the sampling effort was high at most sites with only 41% of the collected radii cross-dated and used for the chronology development. In addition to the poor health, the endogenous dynamics of the *N. antarctica* related to site and competition factors rather than climate reduce the interannual common signal in tree growth (Wigley et al. 1984). Some local factors, such as anthropogenic fires, gap canopy dynamics, local defoliation episodes, and differences in the browsing intensity between nearby forest sectors may contribute to the lack of a common signal between trees. In spite of the mentioned limitations, the radial growth of *N. antarctica* is sensitive to regional climatic variability. This characteristic offers the opportunity to use the *N. antarctica* ring-width variations to estimate past changes in environmental fluctuations.

The *N. antarctica* radial growth is positively correlated with spring precipitation. Radial growth is favored by abundant precipitations at the beginning of the growing season (November and December). On the other hand, tree-ring widths are narrower when the temperature during the growing season (December–March) is above average (Fig. 15.6). Our results are consistent with those reported by Lara et al. (2005), Suarez (2010) and Jara (2011) for other *Nothofagus* species in dry Patagonian environments.

The stronger relationship between tree growth and temperature ($r = -0.58$) in comparison with precipitation ($r = 0.34$) is related to the lower spatial representativeness of the Punta Arenas precipitation record in relation to the regional temperature pattern (Figs. 15.S1 and 15.S3). It is known that precipitation is more spatially and temporally variable than temperature (Aravena and Luckman 2009). Therefore, we can assume that precipitation records in the proximity of the sampling sites would provide stronger relationships between rainfall and *N. antarctica* growth.

The regional chronology showed a progressive decrease in the radial growth of *N. antarctica* from the 1950s to present, which is consistent with the increase in summer temperatures since 1977 and the negative trend in precipitation starting in the 1990 across the region (Fig. 15.7). The increase in water deficit, exacerbated by higher summer temperatures and scarce precipitation, reduces tree growth. Similar negative trends in radial growth have been documented in several studies for other *Nothofagus* species in the Patagonian Andes (Villalba et al. 2003; Jara 2011; Villalba et al. 2012; Soto-Rogel and Aravena 2017). Interestingly, the relationships between tree growth and temperature were much stronger prior to the documented climatic jump of 1976/77 (Villalba et al. 2003). As a consequence of the shift from the negative to positive phase of the Pacific Decadal Oscillation in 1976/77, an abrupt increase in temperature was recorded all along the coast of South America, including Patagonia (Villalba et al. 2003; Vuille et al. 2015). Based on the regional temperature variations, this 1976/77 jump was associated with an approximate 0.5 °C increase in summer temperatures (Dec-Mar) throughout our study area (Fig. 15.7). Comparatively to the long-term 1952–2013 relationship between tree growth and climate ($r = -0.58$), it was much stronger during the colder interval from 1952 to 1976 ($r = -0.72$). In contrast, the relationship between tree growth and precipitation becomes stronger during the more recent, relatively warmer interval starting in 1977 ($r = 0.34$ and $r = 0.45$ for the long-term 1952–2013 and 1977–2013 intervals, respectively). These results suggest changes in the intensity of the relationships between climate and *N. antarctica* tree growth modulated by recent climate changes.

The comparative analysis between the R and C chronologies reveals periods of poor relationships between the chronologies, consistent with different forest stand histories (Fig. 15.8; Table 15.3). The most remarkable differences in the *N. antarctica* growth patterns are associated with the local history of cultural land use. For Perito Moreno National Park (site 1), it was not feasible to cross-date the samples. This reflects the impacts of persistent anthropogenic changes since the beginning of the twentieth century when the first cattle ranches were established in the region. At that time, the economy of Estancia El Rincón was based on livestock production and firewood sale. For the intensive extraction of wood from the *N. antarctica* forest, a network of trails was built in the area. During our surveys, abundant tree stumps were observed in the steppe matrix located between R and C, which evidences the recent anthropic fragmentation of these forests. In 1937, the area was declared a Natural Reserve and in 1945 this area became part of the Perito Moreno National Park. However, it was not until 1975 that the cattle were effectively removed from the area, and in 1982 firewood extraction was finally interrupted.

At the Sectional Río Guanaco in Los Glaciares National Park (site 2; Fig. 15.8a), we observed an abrupt increase in the correlation between the R and C chronologies which started in the 1920s. Around 1926 the main facilities for the cattle ranch La Irene were built, and livestock and forest practices were unified in R and C forests. Although Los Glaciares National Park was created in 1937, cattle were not removed from the Rio Guanaco's area. In contrast, in the 1960s the owners increased six times the stocking. This reduced *N. antarctica* growth, affecting R to a greater extent than C forest. In 2010, the cattle were definitively removed, which is consistent with a period of stabilization in tree growth.

In Estancia Cancha Carrera (site 3; Fig. 15.8b), the cattle ranch was established at the end of the nineteenth century. The livestock and forestry activities focused on C, and R remained without human intervention until 1945. Coinciding with this first period, higher growth rates occurred in R than in C. Results showed a significant increase in the correlation after 1945, probably due to the establishment of four paddocks with different use intensities. This subdivision caused an increase in the livestock numbers in C and the first introduction of cattle in R. In 2000, a little ranch was installed in R at the time that the National Route 40 was been paved. All these changes provoked and intensified the use of firewood for heating. The proximity of the forest to the new route facilitates the extraction of firewood from R. These events, coupled with the regional increase in temperature, induced the extremely low growth in R observed during the last decade.

The first records of permanent use in Punta Gruesa Natural Reserve (site 4; Fig. 15.8c) date back to the end of the nineteenth century. At that time, the Sociedad Explotadora de Tierra del Fuego, the most important cattle company in the history of Chilean-Argentinean Patagonia, started working in the area. Historically, C has been more intensively intervened than the R. While R is isolated in a sector without any access trail and with a high slope (>30°), C has been intensely used for the extraction of firewood. Moreover, C was exposed to roads that were previously used to transport livestock to Chile. The relationships between chronologies were poor until the 1940s, when they became more similar to each other. It was precisely in 1940 that the exploitation of the area by Yacimientos Carboníferos Fiscales (National Coal Company) began. As a consequence, logging for trail building started as well as the construction of urban and rural infrastructure. The use of wood for house heating and steam engines also increased. In 1980, a farm breeding livestock was established in the area with a limited number of animals and remains until today. Results showed an increase in the correlation between chronologies since 1945, following the establishment of Yacimientos Carboníferos Fiscales, and the persistent decrease in radial growth, reaching a minimum growth at C in 2003. In 2002, the

Punta Gruesa Natural Reserve was created and logging prohibited in the area. In recent years, the growth rate tends to stabilize.

Finally, the Estancia Rincón de Los Morros (site 5; Fig. 15.8d) represents an area with a permanent history of intense anthropogenic uses and changes. Since the arrival of the Europeans, the economy was based on extensive livestock production and the extraction of wood by selective logging. As a result, the trees present in the area are young (<69 years), so that the chronologies obtained were short and strongly correlated. This led to the assumption that the intensive anthropogenic uses in Estancia Rincón de Los Morros have a lasting impact on the *N. antarctica* longevity and growth pattern.

The presence of livestock in the forest causes persistent browsing, soil erosion in overgrazed sites and damages in the root systems, leading to a decrease in primary forest productivity (Teague and Smit 1992; Quinteros 2018). Browsing also generates changes in the composition, structure and functional diversity of the understory plant community. In particular, it induces the occurrence of a shrub stratum and the increase of browsing resistant invasive species (Relva et al. 2009; Arias-Sepúlveda and Chillo 2017). Other authors have reported dominance of herbs and annual grasses in overgrazed sites (Gowda et al. 2012; Peri et al. 2016). As a consequence of the understory plant community changes, the decomposition rate (Vázquez 2002; Arias-Sepúlveda and Chillo 2017) the nutrient cycling (Relva and Núñez 2014), and the net primary productivity can be altered (Lencinas et al. 2011). In addition, the livestock production includes practices to increase the forage availability by opening the forest canopy with small fires, thinning and dead tree removal (Somlo et al. 1997; Peri et al. 2016; Amoroso et al. 2018). As a result of the increase in livestock density at *N. antarctica* stands, the intensification of human practices in neighboring forests is inevitable. These induces in turn, important changes in the growth patterns not related to climatic variability.

In conclusion, our work reports on the first tree-ring chronologies from *N. antarctica* developed for the Argentinean sector in southern Patagonia. Given the common tortuous form of *N. antarctica* trees in the forest-steppe ecotone and the poor health conditions of these forests, the development of tree-ring chronologies requires a sampling strategy including a large number of trees. Only 41% of the collected samples were cross-dated and included in the chronologies. At the forest-steppe ecotone, the interannual variations in *N. antarctica* radial growth are particularly sensitive to water availability during the growing season. The *N. antarctica* growth increased with the increase in the water supply. However, these relationships between climate and tree growth seem to be unstable and vary according to the prevailing climatic conditions (cold versus warm periods). In the recent warmer decades following the 1976/77 climatic shift in the Pacific Decadal Oscillation (PDO) (Mantua and Hare 2002), the radial growth seems to be more closely related to spring rainfall variations. Finally, the interactions between land use changes and climate variability considered in this study provide a valid approach to interpret past variations in *N. antarctica* tree-ring in the forest-steppe ecotone in the Patagonian Andes foothills.

References

Alberdi M (1995) Ecofisiología de especies leñosas de los bosques hidrófilos templados de chile: resistencia a la sequía y bajas temperaturas. En: Armesto J, Villagrán C, Arroyo M (eds) Ecología de los bosques nativos de Chile. Universidad de Chile, Santiago, p 420

Amoroso M, Chillo V, Alcala V et al (2018) Efecto del manejo silvopastoril sobre la estructura y dinámica poblacional de bosques mixtos de ciprés de la cordillera (*Austrocedrus chilensis*) y coihue (*Nothofagus dombeyi*). Revista Ecosistemas 27(3):33–40. https://doi.org/10.7818/ECOS.1502

Aravena JC, Luckman BH (2009) Spatio-temporal rainfall patterns in southern South America. Int J Climato 129(14):2106–2120. https://doi.org/10.1002/joc.1761

Aravena JC, Carmona M, Pérez C et al (2002a) Cambios en la riqueza de especies arbóreas, estructura de rodales y propiedades del suelo en una cronosecuencia sucesional en el norte de la Isla de Chiloé, Chile. Revista Chilena de Historia Natural 75(2):339–360. https://doi.org/10.4067/s0716-078x2002000200007

Aravena JC, Lara A, Wolodarsky-Franke A et al (2002b) Tree-ring growth patterns and temperature reconstruction from *Nothofagus pumilio* (Fagaceae) forests at the upper tree line of southern Chilean Patagonia. Rev Chil Hist Nat 75(2):361–376. https://doi.org/10.4067/S0716-078X2002000200008

Arias-Sepúlveda JEA, Chillo V (2017) Cambios en la diversidad funcional del sotobosque y la tasa de descomposición frente a diferentes intensidades de uso silvopastoril en el noroeste de la Patagonia, Argentina. Ecol Austral 27(1):029–038

Bandieri S (2005) Asuntos de familia. La construcción del poder en la Patagonia: el caso de Neuquén. Boletín del Instituto de Historia Argentina y Americana Dr. Emilio Ravignani 28:65–94

Barbería EM (1994) El extremo austral sudamericano.: Ocupación y relaciones de los territorios argentinos y chilenos, 1880–1920. Estud Front 33:185–212

Barbería EM (1995) Los dueños de la tierra en la Patagonia Austral: 1880–1920. Universidad federal de la Patagonia Austral, Santa Cruz

Blasing TJ, Solomon AM, Duvick DN (1984) Response functions revisited. Tree-Ring Bull 44:1–15

Boninsegna JA, Argollo J, Aravena JC et al (2009) Dendroclimatological reconstructions in South America: a review. Palaeogr Palaeocl 281(3–4):210–228. https://doi.org/10.1016/j.palaeo.2009.07.020

Christensen M, Emborg J (1996) Biodiversity in natural versus managed forest in Denmark. Forest Ecol Manag 85(1–3):47–51. https://doi.org/10.1016/s0378-1127(96)03749-8

Cook ER (1985) A time series analysis approach to tree-ring standardization. Ph. D. dissertation, University of Arizona, Arizona, p 183

Coronato FR, Del Valle HF (1993) Methodological comparison in the estimate of fluvial erosion in an arid closed basin of northeastern Patagonia. J Arid Environ 24(3):231–239. https://doi.org/10.1006/jare.1993.1021

Di Castri F, Hansen A, Holland M (1988). A new look at ecotones: emerging international projects on landscape boundaries. Biology Internat. Special issue 17, Internet. Union of Biol Sci, pp 1–163

Díaz BG (2005) Uso múltiple de recursos naturales renovables en la Cuenca Río de las Vueltas, provincia de Santa Cruz. (Tesis de Mg. Sci. Manejo de Cuencas Hidrográficas). Universidad Nacional de La Plata. Argentina, p 195

Donoso C (1993) Bosques templados de Chile y Argentina. Variación, estructura y dinámica. Ecología Forestal. Editorial Universitaria, Santiago

Donoso C, Steinke L, Premoli A (2006) *Nothofagus antarctica* (G. Forster) Oerst. ñirre, ñire, ñiré, anís (Tierra Del Fuego) ñirre: de ngërü (Mapudungun): zorro. En: Donoso C (Ed.), Las especies arbóreas de los bosques templados de Chile y Argentina. Autoecología. Ed. Cuneo, M. Santiago, Chile

Fritts HC (1976) Tree rings and climate. Academic Press, London
Garreaud R, Lopez P, Minvielle M et al (2013) Large-scale control on the Patagonian climate. J Clim 26(1):215–230. https://doi.org/10.1175/JCLI-D-12-00001.1
González MH, Vera CS (2010) On the interannual wintertime rainfall variability in the southern Andes. Int J Climatol 30(5):643–657. https://doi.org/10.1002/joc.1910
Gowda JH, Kitzberger T, Premoli AC (2012) Landscape responses to a century of land use along the northern Patagonian forest-steppe transition. Plant Ecol 213(2):259–272. https://doi.org/10.1007/s11258-011-9972-5
Hampe A, Petit RJ (2005) Conserving biodiversity under climate change: the rear edge matters. Ecol Lett 8(5):461–467. https://doi.org/10.1111/j.1461-0248.2005.00739.x
Hijmans RJ, Cameron SE, Parra JL et al (2005) Very high resolution interpolated climate surfaces for global land areas. Int J Climatol 25:1965–1978. https://doi.org/10.1002/joc.1276
Holmes RL (1983) Computer-assisted quality control. Tree-Ring Bull 43:69–78
Ipinza R, Pérez F, Kappes A (1989) *Cyttaria espinosae* Lloyd., un hongo de interés en fitopatología, alimentación y evolución de los bosques de *Nothofagus* en Chile. Bol Soc Micol Madrid 13:31–48
Ivancich HS (2013) Relaciones entre la estructura forestal y el crecimiento del bosque de *Nothofagus antarctica* en gradientes de edad y calidad de sitio. Tesis doctoral, Facultad de Ciencias Agrarias y Forestales, Universidad Nacional de La Plata, Argentina
Iverson LR, Schwartz MW, Prasad AM (2004) How fast and far might tree species migrate in the eastern United States due to climate change? Glob Ecol Biogeogr 13(3):209–219. https://doi.org/10.1111/j.1466-822X.2004.00093.x
Jara JCL (2011) Relaciones entre el crecimiento de *Nothofagus betuloides* y el clima local y de gran escala en bosques subantárticos de la Isla Navarino. (Tesis de grado). Universidad Austral de Chile, Valdivia
Kitzberger T (2012) Ecotones as complex arenas of disturbance, climate, and human impacts: the trans-Andean forest-steppe ecotone of northern Patagonia. In: Ecotones between forest and grassland. Springer, New York, pp 59–88
Kreps G, Martínez PGM, Peri PL (2012) Cambio climático en Patagonia sur. Escenarios futuros en el manejo de los recursos naturales. INTA (ed), Santa Cruz, Argentina
Lara A, Aravena JC, Villalba R et al (2001) Dendroclimatology of high-elevation *Nothofagus pumilio* forests at their northern distribution limit in the Central Andes of Chile. Can J For Res 31(6):925–936. https://doi.org/10.1139/x00-208
Lara A, Villalba R, Wolodarsky-Franke A et al (2005) Spatial and temporal variation in *Nothofagus pumilio* growth at tree line along its latitudinal range (35 40′-55 S) in the Chilean Andes. J Biogeogr 32(5):879–893. https://doi.org/10.1111/j.1365-2699.2005.01191.x
Lencinas MV, Martínez PGM, Gallo E et al (2011) Alternative silvicultural practices with variable retention to improve understory plant diversity conservation in southern Patagonian forests. For Ecol Manag 262(7):1236–1250. https://doi.org/10.1016/j.foreco.2011.06.021
Manacorda M, Bonvissuto G (2001) Uso silvopastoril de los bosques de ñire (*Nothofagus antarctica*) en Río Negro, Patagonia Argentina. Revista Forestal Centroamericana 35:41–44
Mantua NJ, Hare SR (2002) The Pacific decadal oscillation. J Oceanogr 58(1):35–44. https://doi.org/10.1023/A:1015820616384
Marazana O, Verzino G, Ocampo JB et al (1984) Situación actual de los nativos de la Provincia de Santa Cruz: descripción manejo y estado sanitario. Informe Técnico Consejo Agrario Provincial, Córdoba p 33
Martínez-Pastur GM, Peri PL, Fernández C et al (1999) Desarrollo de la regeneración a lo largo del ciclo del manejo forestal de un bosque de *Nothofagus pumilio*: incidencia del ramoneo de *Lama guanicoe*. Bosque 20(2):47–53
Montaña C (1982) Las comunidades de ñire (*Nothofagus antarctica* (Forst.) Oerst.) de la cuenca del río Manso Superior (Río Negro, Argentina). Tesis Doctoral, Universidad Nacional de Córdoba, Córdoba, Argentina
Moore DM (1983) The flora of the Fuego-Patagonian cordilleras: its origins and affinities. Rev Chil Hist Nat 56(123–136):123–136

Morales M, Villalba R (2006) Climatic and anthropogenic influences on the dynamics of *Prosopis ferox* forests in the Quebrada de Humahuaca, Jujuy, Argentina. In: Land use change and mountain biodiversity. CRC Press, Boca Raton, pp 275–282

Morales MS, Villalba R, Boninsegna JA (2005) Climate, land-use and *Prosopis ferox* recruitment in the Quebrada de Humahuaca, Jujuy, Argentina. Dendrochronologia 22(3):169–174. https://doi.org/10.1016/j.dendro.2005.05.004

Mundo IA, Villalba R, Veblen TT et al (2017) Fire history in southern Patagonia: human and climate influences on fire activity in *Nothofagus pumilio* forests. Ecosphere 8(9). https://doi.org/10.1002/ecs2.1932

Murcia C (1995) Edge effects in fragmented forests: implications for conservation. Trends Ecol Evol 10(2):58–62. https://doi.org/10.1016/S0169-5347(00)88977-6

Neilson RP (1991) Climatic constraints and issues of scale controlling regional biomes. In: Ecotones. Springer, Boston, pp 31–51

Paruelo MJ, Beltrán A, Jobbágy E et al (1998) The climate of Patagonia: general patterns and controls on biotic processes. Ecol Austral 8:85–101

Pérez Moreau R (1959) Reseña Botánica sobre el Lago Argentino. Instituto Nacional del Hielo Continental Patagónico, Buenos Aires

Peri PL, Ormaechea S (2013) Relevamiento de los bosques nativos de ñire (*Nothofagus antarctica*) en Santa Cruz: base para su conservación y manejo. Ediciones INTA, Buenos Aires

Peri PL, Hansen NE, Bahamonde HA et al (2016) Silvopastoral systems under native forest in Patagonia Argentina. In: Silvopastoral systems in southern South America. Springer, pp 117–168

Pulido FJ, Díaz B, Martínez-Pastur GM (2000) Incidencia del ramoneo del guanaco (*Lama guanicoe* Müller) sobre la regeneración temprana en bosques de lenga [*Nothofagus pumilio* (Poepp et Endl) Krasser] de Tierra del Fuego, Argentina. Forest Syst 9(2):381–394

Quinteros CP (2018) Propiedades ecosistémicas de bosques de *Nothofagus pumilio* afectadas por diferente intensidad de uso ganadero en Chubut, Patagonia Argentina. Revista Ecosistemas 27(3):24–32. https://doi.org/10.7818/ECOS.1483

Rajchenberg M, Barroetaveña C, Orellana I (2014) Relevamiento de hongos con fines de conocimiento ambiental, turístico y eventualmente productivo. Centro De Investigación Y Extensión Forestal Andino Patagónico (CIEFAP), Esquel, p 96

Ramírez C, Correa M, Figueroa H et al (1985) Variación del hábito y hábitat de *Nothofagus antarctica* en el centro de Chile. Bosque 6(2):55–73

Relva MA, Núñez MA (2014) Factores que facilitan y retrasan la invasión de coníferas exóticas en la Isla Victoria. Ecol Austral 24(2):145–153

Relva MA, Veblen TT (1998) Impacts of introduced large herbivores on *Austrocedrus chilensis* forests in northern Patagonia, Argentina. For Ecol Manag 108(1–2):27–40. https://doi.org/10.1016/s0378-1127(97)00313-7

Relva MA, Westerholm CL, Kitzberger T (2009) Effects of introduced ungulates on forest understory communities in northern Patagonia are modified by timing and severity of stand mortality. Plant Ecol 201(1):11–22. https://doi.org/10.1007/s11258-008-9528-5

Rodríguez-Catón M, Villalba R (2018) Indicadores del decaimiento en bosques de *Nothofagus pumilio* en el norte de la Patagonia, Argentina. Madera Bosques 24(2):e2421588. https://doi.org/10.21829/myb.2018.2421588

Roig FA (1999) La vegetación de la Patagonia. En: Correa M (Dir.) Flora patagónica. Colección Científica del INTA 8(1):48–166

Roig FA, Villalba R (2008) Understanding climate from Patagonian tree rings. Dev Quat Sci 11:411–435. https://doi.org/10.1016/S1571-0866(07)10021-X

Rosenblüth B, Fuenzalida HA, Aceituno P (1997) Recent temperature variations in southern South America. Int J Climatol 17(1):67–85. https://doi.org/10.1002/(SICI)1097-0088(199701)17:1<67:AID-JOC120>3.0.CO;2-G

SAyDS (2005) Primer Inventario Nacional de Bosques Nativos. Ministerio de Salud y Ambiente de la Nación – Secretaría de Ambiente y Desarrollo Sustentable, Buenos Aires

Schweitzer A (2011) Fronteras, recursos naturales y crisis en la Patagonia sur argentina. In: Integración geoestratégica, seguridad, fronteras y migración en América Latina, editado por Sandoval, J., Álvarez, R. Saavedra, L. Editorial INREDH, Ecuador, pp 33–69

Somlo R, Bonvissuto G, Schlichter T et al (1997) Silvopastoral use of Argentine Patagonian forests. In: Temperate agroforestry systems. CAB International, Wallingford; New York, pp 237–250

Soto-Rogel P, Aravena JC (2017) Potencial dendroclimático de *Nothofagus betuloides* en la Cordillera de Darwin, Tierra del Fuego, Chile. Bosque 38(1):155–168. https://doi.org/10.4067/S0717-92002017000100016

Srur AM, Golluscio RA, Villalba R et al (2013) Grazing-induced morphological and growth rate changes in *Anarthrophyllum rigidum*, a Patagonian leguminous shrub. Dendrochronologia 31(3):223–227. https://doi.org/10.1016/j.dendro.2013.02.002

Stokes MA, Smiley TL (1968) An introduction to tree-ring dating. University of Chicago Press, Chicago

Suarez ML (2010) Tree-ring records from *Nothofagus dombeyi*: a preliminary chronology network in Northern Patagonia, Argentina. Dendrochronologia 28(2):65–72. https://doi.org/10.1016/j.dendro.2009.11.001

Teague WR, Smit GN (1992) Relations between woody and herbaceous components and the effects of bush-clearing in southern African savannas. J Grassl Soc South Afr 9(2):60–71. https://doi.org/10.1080/02566702.1992.9648301

Vázquez DP (2002) Multiple effects of introduced mammalian herbivores in a temperate forest. Biol Invasions 4(1–2):175–191. https://doi.org/10.1023/A:1020522923905

Veblen TT, Lorenz DC (1987) Post-fire stand development of *Austrocedrus-Nothofagus* forests in northern Patagonia. Vegetatio 71(2):113–126. https://doi.org/10.1007/BF00044825

Veblen TT, Kitzberger T, Lara A (1992) Disturbance and forest dynamics along a transect from Andean rain forest to Patagonian shrubland. J Veg Sci 3(4):507–520. https://doi.org/10.2307/3235807

Veblen TT, Donoso C, Kitzberger T (1996) Ecology of southern Chilean and Argentinean *Nothofagus* forests. In: Veblen TT, Hill R, Read J (eds) The ecology and biogeography of *Nothofagus* forests. Yale University Press, Connecticut, pp 293–353

Velásquez C, Henríquez JM, Aravena JC (2012) Damage caused by mistletoe *Misodendrum Punctulatum* Banks Ex Dc. on architecture and radial growth of *Nothofagus pumilio* (Poepp. et Endl.) Krasser forests of southern Chile. Austral Ecol 37(7):816–824. https://doi.org/10.1111/j.1442-9993.2011.02342.x

Villalba R, Boninsegna JA, Veblen TT et al (1997) Recent trends in tree-ring records from high elevation sites in the Andes of northern Patagonia. In: Climatic change at high elevation sites. Springer, Dordrecht, pp 193–222

Villalba R, Lara A, Boninsegna JA et al (2003) Large-scale temperature changes across the southern Andes: 20th century variations in the context of the past 400 years. Clim Chang 59:177–232. https://doi.org/10.1007/978-94-015-1252-7_10

Villalba R, Grosjean M, Kiefer T (2009) Long-term multi-proxy climate reconstructions and dynamics in South America (LOTRED-SA): state of the art and perspectives. Palaeogeogr Palaeoclimatol Palaeoecol 281(3–4):175–179. https://doi.org/10.1016/j.palaeo.2009.08.007

Villalba R, Lara A, Masiokas MH et al (2012) Unusual southern hemisphere tree growth patterns induced by changes in the southern annular mode. Nat Geosci 5:793–798. https://doi.org/10.1038/ngeo1613

Vuille M, Franquist E, Garreaud R et al (2015) Impact of the global warming hiatus on Andean temperature. J Geophys Res Atmos 120(9):3745–3757. https://doi.org/10.1002/2015JD023126

Wigley TM, Briffa KR, Jones PD (1984) On the average value of correlated time series, with applications in dendroclimatology and hydrometeorology. J Clim Appl Meteorol 23(2):201–213. https://doi.org/10.1175/1520-0450(1984)023<0201:OTAVOC>2.0.CO;2

Chapter 16
Different Uncoupling of Growth and Water-Use Efficiency in Two Conifers Inhabiting Chilean Temperate Rainforests

J. Julio Camarero

Abstract Temperate rainforests in southern Chile harbor old-growth and threatened conifer species which are affected by climate warming and droughts. Radial growth and intrinsic water-use efficiency (iWUE), reconstructed using carbon isotope discrimination data, are two relevant proxies of those responses. I compare the changes of these two surrogates in two threatened conifers from Southern Chile (*Fitzroya cupressoides* and *Pilgerodendron uviferum*) by sampling old trees (100–500 years). The variability in growth and iWUE of the period 1900–2010, their coupling and how they responded to climate were assessed. Cool and wet conditions during the growing season enhanced growth of *F. cupressoides* but constrained it in *P. uviferum*. In response to warmer and drier conditions in the late twentieth century, *F. cupressoides* growth decreased but increased in the case of *P. uviferum*. However, both species showed rising iWUE confirming that growth and iWUE were decoupled. The recent enhancement of iWUE was less pronounced in *F. cupressoides* than in *P. uviferum*, which could present a more dynamic gas-exchange response to rising atmospheric CO_2 concentrations than the former species. Comparing growth and iWUE changes through time improves the characterization of the long-term responses of South American long-living conifers to global change.

Keywords Carbon isotopes · *Fitzroya cupressoides* · Patagonia · *Pilgerodendron uviferum* · Valdivian temperate rainforest

16.1 Introduction

Temperate rainforests host some of the largest and oldest living trees in the world, thus illustrating several attributes of old-growth forests including gap dynamics and high biomass density and accumulation (Keith et al. 2009; Larjavaara 2014). In Latin America, the largest area of temperate rainforests is found in southwestern

J. J. Camarero (✉)
Instituto Pirenaico de Ecología (IPE-CSIC), Zaragoza, Spain
e-mail: jjcamarero@ipe.csic.es

M. Pompa-García, J. J. Camarero (eds.), *Latin American Dendroecology*,
https://doi.org/10.1007/978-3-030-36930-9_16

South America, i.e. in the Chilean Patagonia (Veblen and Schlegel 1982). These are the Valdivian temperate rainforests, found between 38° and 45° S, which are structurally complex and open formations dominated by many endemic, evergreen tree and shrub species of Gondwana origin (Armesto and Figueroa 1987; Donoso 1993; Tecklin et al. 2011). These forests are also inhabited by some of the oldest living organisms found on Earth, with ages usually surpassing 500 years and sometimes 1000 years (Armesto et al. 2009). These are conifers of the Cupressaceae family including *Fitzroya cupressoides, Pilgerodendron uviferum and Austrocedrus chilensis* (Veblen et al. 1995; Farjon 2005). In this study I will focus on the first two species which coexist in some Valdivian temperate rainforests.

Some of the oldest *F. cupressoides* individuals have been alive for at least 3500 years and have allowed building the longest tree-ring chronologies for southern America (Lara and Villalba 1993). On the other hand, *P. uviferum* is the southernmost conifer in the world reaching a maximum latitude of 54° S (Soto et al. 2007; Holz 2009), but in its northern distribution range it coexists with *F. cupressoides* (Farjon 1998). Both study species are endangered because they show slow growth rates and poor regeneration (Donoso et al. 2006; Lara et al. 2006), they have been severely exploited for timber (*F. cupressoides*) and their habitats (e.g. peatlands where *P. uviferum* occurs) are prone to destruction and loss due to logging, fire, grazing or conversion to plantations (Veblen et al. 1976; Lara et al. 2003; Soto et al. 2007). *Fitzroya* and *Pilgerodendron* are still threatened by ongoing illegal logging in many Chilean forests (Wolodarsky-Franke and Lara 2005; Soto et al. 2007). Here, retrospective, long-term information on growth and water use is provided, which is helpful to improve the understanding of the ecology and conservation of these two temperate, threatened conifer species, given that they will be exposed to forecasted climate warming (Gutiérrez et al. 2014).

Both *F. cupressoides* and *P. uviferum* form annual rings whose width variability is partially driven by climate, and grow slowly in diameter with average tree-ring widths of adults between 0.4 and 1.0 mm (Schulman 1956; Szeicz 1997; Szeicz et al. 2000; Lara et al. 2003). In *F. cupressoides*, dry conditions during the growing season may reduce growth (Urrutia-Jalabert et al. 2015a). In *P. uviferum*, growth is enhanced by wet conditions as climate dries northwards but it may be constrained by wet conditions at higher latitudes, southwards, where climate conditions are wet and cool (Holz et al. 2018). Such low growth rates allow sampling mature individuals which provide retrospective proxies of tree functioning (growth, water use) encompassing at least the twentieth century.

Reconstructing water-use efficiency can be addressed by using carbon isotope composition ($^{13}C/^{12}C$ or $\delta^{13}C$) of tree-rings which depends on the rates of carboxylation during photosynthesis (A) and stomatal conductance (g) (Farquhar et al. 1982). These two rates determining carbon uptake and water loss through leaves are differently responding to the rising atmospheric CO_2 concentrations (Frank et al. 2015; Voelker et al. 2016). C3 plants discriminate against ^{13}C during diffusion of CO_2 through stomata and assimilation by photosynthesis (Farquhar et al. 1989). Tree-ring $\delta^{13}C$ has been used to reconstruct the intrinsic water-use efficiency (iWUE), i.e. the ratio of carbon gain to water loss (A/g), through time (Saurer et al. 2004, 2014).

The iWUE is linked to the changes in the ratio of the CO_2 concentrations in the sub-stomatal internal cavity of leaves (c_i) and in the atmosphere (c_a) (Farquhar et al. 1989).

Nowadays, c_a is 412 ppm, but it could double by the end of this century (IPCC 2013). Tree-ring $\delta^{13}C$ data show increasing iWUE indicating active or dynamic responses to rising c_a, and suggesting a proportional rise of c_i leading to a near-constant c_i/c_a (Saurer et al. 2004; Frank et al. 2015) which could be associated to growth enhancement (CO_2-fertilization effect). However, no study has tested if this applies to *F. cupressoides* and *P. uviferum* by comparing long-term changes in iWUE and proxies of carbon uptake such as radial growth (but see Camarero and Fajardo 2017; Camarero et al. 2018). Some retrospective studies comparing iWUE and tree-ring width data have found that the iWUE improvement triggered by rising c_a over the twentieth century did not correspond to growth enhancement neither in xeric nor in mesic sites (Silva et al. 2010; Nock et al. 2011; Peñuelas et al. 2011; Linares and Camarero 2012; Gómez-Guerrero et al. 2013; Granda et al. 2014; Lévesque et al. 2014). Notably, previous global synthesis (Peñuelas et al. 2011; Silva and Anand 2013) found no CO_2-related growth enhancement in most study sites despite increasing c_a, but only one out of those study cases corresponded to a temperate southern forest and it was a hardwood species (*Nothofagus pumilio*; Srur et al. 2008). Here the (i) patterns and climate drivers of growth of *F. cupressoides* and *P. uviferum* are determined, and it is tested if (ii) their iWUE and growth were coupled throughout the twentieth century and the early twenty-first century (period 1900–2010). I argue that investigating these patterns and trends of iWUE and growth in long-living conifers of temperate, humid forests will contribute to better understanding how these and similar long-living tree species will respond to rising c_a in mesic conditions. Lastly, the study of long-living species helps understanding how carbon is retained as wood in similar old-growth forests (Körner 2017).

16.2 Materials and Methods

16.2.1 Study Areas, Tree Species and Climate Data

The two study areas are located in southern Chile. The *F. cupressoides* trees were sampled near the Comau fjord, in the slope above the Huinay field station, south of Puerto Montt city (42° 23' S, 72° 25' W, 780 m a.s.l.). The *P. uviferum* were sampled near Río El Pobre and Lago Leal sites (48° 04' S, 73° 04' W, 40 m a.s.l.), south of Cochrane city.

Chilean temperate rainforests are characterized by cold to mild and humid climate conditions (Luebert and Pliscoff 2006), with mean annual temperature of 10–11 °C and annual precipitation of 2500–4200 mm, regularly distributed throughout the year. In the *F. cupressoides* site, mean annual temperature is ca. 10 °C, annual precipitation is ca. 2000–2500 mm, with a very wet winter and a less wet summer, and frosts occur from July to August (Camarero and Fajardo 2017). In the *P. uviferum* site, climate is also temperate and oceanic but conditions are drier and more continental with a mean annual temperature of 8.6 °C and annual precipitation of ca. 800 mm (Camarero et al. 2018).

Both study sites share some tree and shrub species characteristic of Valdivian temperate rainforests and North Patagonian forests such as *Nothofagus betuloides* (Mirb.) Oerst, *Podocarpus nubigena* Lindl., and *Desfontainia spinosa* Ruiz & Pav. However, in the *F. cupressoides* site *N. betuloides, Weinmannia trichosperma* Ruiz & Pav., *P. nubigena*, *Nothofagus antarctica* (G.Forst.) Oerst., and *D. spinosa* are dominant, whereas in the *P. uviferum* site *N. betuloides*, *P. nubigena* and *Gaultheria mucronata* (L.f.) Hook. & Arn. are dominant. Soils in both study areas are acid.

Fitzroya cupressoides I.M. Johnst. (alerce) stands are found in the Coastal Range, with some Mediterranean influence and in the nearby Andes, on poor or volcanic soils (Donoso et al. 1993, 2006). In the Andes, *F. cupressoides* forms old growth stands, often developed following large-scale disturbances such as volcanic eruptions and landslides (Veblen et al. 1995). *Pilgerodendron uviferum* (D. Don) Florin (ciprés de las Guaitecas) appears in the coastal ranges and the Andes where is commonly found in lowlands and bogs with poorly-drained, acid, and usually nutrient–deficient soils (Donoso 1981). The main disturbance of these relatively stable formations is fire (Holz 2009).

To analyze climate data for the 1950–2010 period, I obtained monthly mean temperature and precipitation from the Climate Research Unit (CRU) version 3.22 with gridded data at 0.5° resolution (Harris et al. 2014). Data were extracted for the grids with coordinates 42.0°–42.5° S and 72.0°–72.50° W (*F. cupressoides*) and 48.0°–48.5° S and 73.0°–73.5° W (*P. uviferum*) using the Climate Explorer webpage (https://climexp.knmi.nl/).

16.2.2 Field Sampling and Tree-Ring Width Processing

In each site, 14–20 dominant individuals of each species in a 1-ha large area were selected. The diameter at 1.3 m of height and the total height of each sampled tree was measured using diameter tapes and a clinometer, respectively. Two to three to-the-pith stem cores were extracted per tree at 1.3 m and perpendicular to the slope using 5.15 mm Pressler borers (Haglöf, Långsele, Sweden). Dominant trees of similar size were targeted (Table 16.1) to capture similar growth trends and responses to climate throughout the twentieth century. A third core was extracted in all trees and kept for C isotope analyses.

Cores were dried, mounted and glued in grooved wooden sticks, and then carefully sanded until rings were visible under magnification (×10). Following visual cross-dating, tree-ring widths were measured to the nearest 0.001 mm using a microscope mounted on a Lintab-TSAP measuring device (F. Rinn, Heidelberg, Germany). Calendar dates were assigned to rings following the southern hemisphere tree-ring dating convention that assigns an annual ring to the calendar year in which the annual formation begins (Schulman 1956).

Table 16.1 Characteristics of sampled trees and resulting tree-ring width series. Values are means ± standard error (SE) calculated for all trees and over the common period 1900–2010. Different letters indicate significant ($p < 0.05$) differences according to Mann-Whitney tests.

Species	Diameter at 1.3 m (cm)	Height (m)	No. trees (No. cores)	Tree-ring width (mm)	Timespan (best-replicated timespan)	First-order autocorrelation	Correlation with site mean series
Fitzroya cupressoides	47.6 ± 4.9b	30.1 ± 1.8b	17 (31)	0.58 ± 0.05b	1544–2010 (1840–2010)	0.82 ± 0.02b	0.43 ± 0.02
Pilgerodendron uviferum	26.6 ± 1.1a	11.7 ± 0.5a	20 (34)	0.37 ± 0.02a	1546–2016 (1805–2016)	0.68 ± 0.02a	0.42 ± 0.01

Visual cross-dating and ring-width measurements underwent a quality control with the COFECHA program (Holmes 1983). To quantify growth rates I calculated the basal area increment (BAI), which was computed as:

$$BAI = \pi\left(R^2_{\ t} - R^2_{\ t-1}\right) \tag{1}$$

where R_t and R_{t-1} are the stem radii in years t and t–1, which were computed by considering diameter at coring height, and the distance between the theoretical pith and the innermost rings (Fajardo and McIntire 2012). Tree-ring width series were converted into ring-width indices and mean chronologies for each species to eliminate trends related to stem enlargement and preserve high-frequency growth variability (Fritts 2001). First, the individual series were standardized using a cubic-smoothing spline curve of 30 years with a 50% frequency response cut-off. Second, an autoregressive model was applied to remove the first-order temporal autocorrelation and create residual indices. Third, species residual chronologies were generated by averaging individual series using a bi-weight robust mean. Chronologies or mean series of ring-width indices were built using the ARSTAN program (Cook and Krusic 2007). The two chronologies were well replicated for the common 1950–2010 period since they surpassed the expressed population signal (EPS) threshold of 0.85, which is considered as a reliable criterion of chronology replication (Wigley et al. 1984).

To quantify climate-growth associations, I calculated Pearson correlation coefficients between monthly and seasonal (growing season from November to March) climate variables (mean temperature, precipitation) and the two residual chronologies considering the 1950–2010 period. Correlations were calculated from the prior March up to current March based on previous studies (Szeicz 1997; Lara et al. 2003; Holz et al. 2018).

16.2.3 Analyses of Carbon Isotopes

Wood tissue was used for C isotope analyses as studies comparing whole wood and cellulose show similar iWUE trends (Saurer et al. 2004). For each species, I selected the five trees established before the twentieth century (age 150–250 years.) whose cross-dated tree-ring series presented a high correlation with the mean species series. For the five cores selected per species, rings were separated in decadal groups using scalpels. Then, decadal pools were obtained by mixing the samples of the five selected trees. The resulting wood samples were homogenized to a fine powder using a ball mixer mill (Retsch MM301, Haan, Germany). Wood aliquots (0.001 g) were weighed on a microbalance (AX205 Mettler Toledo, OH, USA) into tin foil capsules and combusted to CO_2 using a Flash EA-1112 elemental analyser interfaced with a Finnigan MAT Delta C isotope ratio mass spectrometer (Thermo Fisher Scientific Inc., MA, USA). Isotope analyses were done at the Stable Isotope Facility (University of California at Davis, USA). Stable isotope ratios were expressed as per mil (‰) deviations using the δ notation relative to Vienna Pee Dee Belemnite (VPDB). The standard deviation for repeated analyses was less than 0.1‰.

16.2.4 Calculating Intrinsic Water-Use Efficiency (iWUE) and Relating it to Growth

I calculated iWUE values from tree-ring wood C isotopes. First, C isotope discrimination in wood ($\Delta^{13}C$) was calculated based on changes in $\delta^{13}C$ of atmospheric CO_2 ($\delta^{13}c_a$) and wood $\delta^{13}C$ ($\delta^{13}c_w$) following Farquhar and Richards (1984):

$$\Delta^{13}C = \left(\delta^{13}c_a - \delta^{13}c_w\right) / \left(1 + \delta^{13}c_w / 1000\right) \tag{2}$$

$\delta^{13}c_a$ was obtained from McCarroll and Loader (2004). To relate $\Delta^{13}C$ and the CO_2 concentration in the sub-stomatal internal cavity of leaves (c_i) I followed Francey and Farquhar (1982) and Farquhar (1983):

$$\Delta^{13}C = a + (b - a)(c_i / c_a) \tag{3}$$

where a is the diffusion fractionation across the boundary layer and the stomata (4.4‰) and b is the Rubisco fractionation (27.0‰). Second, iWUE (in μmol mol^{-1}) was calculated using the equation proposed by Farquhar et al. (1982):

$$\text{iWUE} = c_a\left[1 - (c_i / c_a)\right]0.625 = (c_a - c_i) / 1.6 \tag{4}$$

where 0.625 is the relation among the conductance of H_2O compared to the conductance of CO_2. Rearranging equation (4) allows estimating c_i as a function of c_a. This was done using the equation obtained by Voelker et al. (2016) and relating c_a and c_i / c_a for gymnosperms and considering a wide c_a range (200–700 ppm).Finally, I compared the relative changes in growth (BAI) and iWUE by using Pearson correlations. First, BAI values were converted into decadal means. Second, the rates of change between consecutive decadal values of BAI and iWUE were obtained and related for each species.

16.3 Results

16.3.1 Age, Growth Trends and Annual Variability

The oldest dated *F. cupressoides* individual was at least 467 years old and the mean series length was 209 years, whereas the oldest dated *P. uviferum* individual was at least 471 years old and the mean series length was 240 years (Table 16.1). The tree size, mean growth rate and first-order autocorrelation of ring-width data were higher

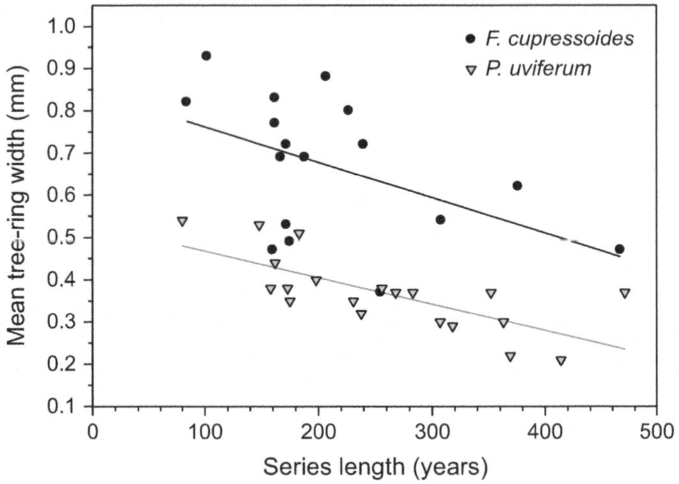

Fig. 16.1 Negative associations observed between the tree-ring series length and the mean ring width in *F. cupressoides* and *P. uviferum* trees

in *F. cupressoides* than in *P. uviferum*, but both species showed similar mean correlations with site mean series (Table 16.1). The series length was negatively associated to the mean ring-width in *F. cupressoides* ($r = -0.49$, $p = 0.046$) and in *P. uviferum* ($r = -0.72$, $p = 0.0003$; see Fig. 16.1). Considering the period 1900–2010, common to both species, the mean (± SE, standard error) basal area increment values were 10.8 ± 0.3 and 1.3 ± 0.1 cm^2 $yr.^{-1}$ in *F. cupressoides* and *P. uviferum*, respectively.

The BAI curves showed a growth divergence starting in the 1990s when *F. cupressoides* BAI started to decrease whereas *P. uviferum* BAI increased and remained stable (Fig. 16.2a). Both species showed also different BAI trends in the early twentieth century, which could be due to some *F. cupressoides* trees showing high BAI values during the 1910s during their juvenile growth phase. The two species chronologies were positively correlated ($r = 0.35$, $p = 0.006$) and their high-frequency growth patterns matched some recent climate extremes affecting southern Chile such as the 2007–2008 drought which corresponded to quite narrow rings in *F. cupressoides* (Fig. 16.2b).

16.3.2 Responses of Growth Variability to Climate

The *F. cupressoides*chronology was significantly ($p < 0.05$) and positively correlated with growing-season precipitation (January, March) and prior June precipitation, and negatively correlated with January temperature and prior March temperature (Fig. 16.3). The *P. uviferum* chronology was positively correlated with growing-season temperatures (December, November to March) and negatively cor-

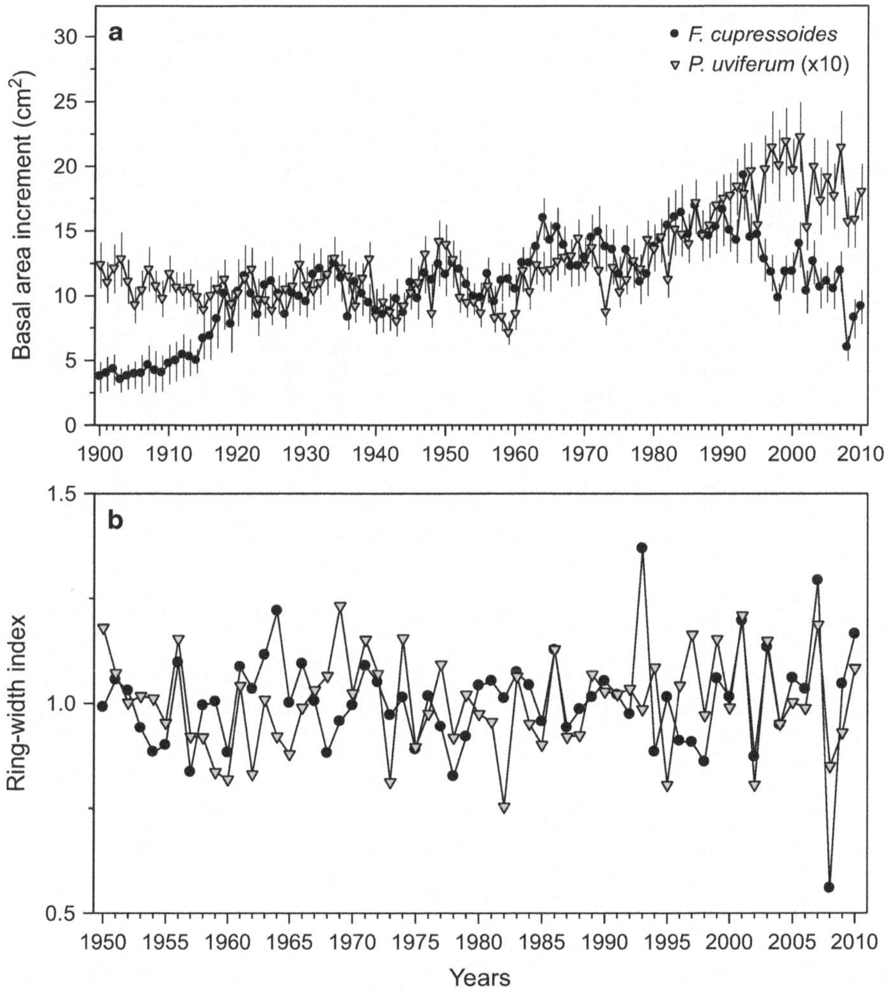

Fig. 16.2 Growth trends (**a**, basal area increment -BAI) and growth variability (**b**, residual ring-width indices) of *F. cupressoides* and *P. uviferum*. In the upper plot, BAI values are means ± SE and *P. uviferum* BAI values are multiplied by 10. The lower plot shows the period 1950–2010 which was used to calculate climate-growth associations

related with precipitation (November, December, November to March). Prior May temperatures were also positively related to this chronology.

16.3.3 Growth and Water-Use Efficiency

The iWUE increased in both species with major relative rises in the 1980s and 1970s in the case of *F. cupressoides* (+4.6%) and *P. uviferum* (+9.3%), respectively (Fig. 16.4). From the 1900s until the 1950s the relative decadal enhancements of

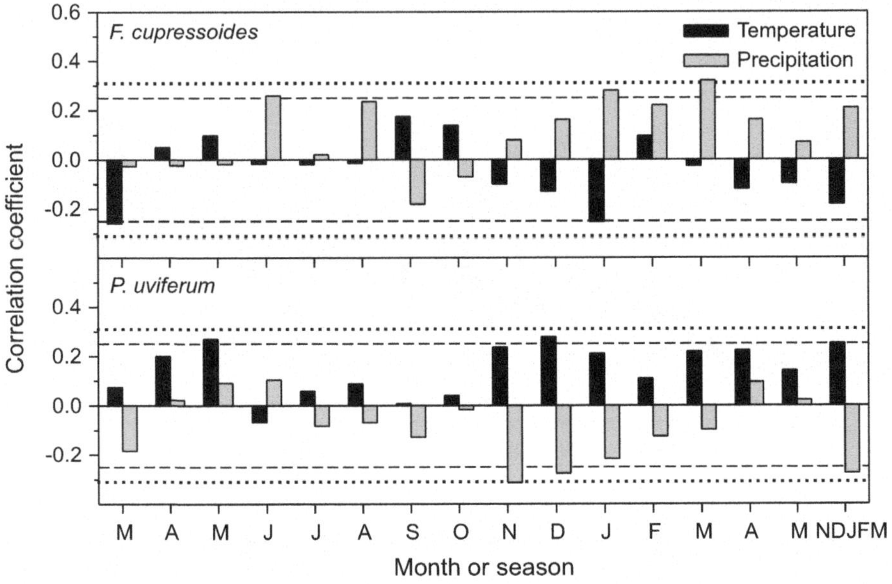

Fig. 16.3 Climate-growth associations calculated for *F. cupressoides* and *P. uviferum*. Bars are Pearson correlation coefficients obtained by relating the mean series of ring-width indices and mean monthly temperatures and total monthly precipitation values. Correlations were calculated from prior March to current May. The last columns show correlations for the assumed growing season (November to March). The dashed and dotted horizontal lines show the 0.05 and 0.01 significance levels, respectively

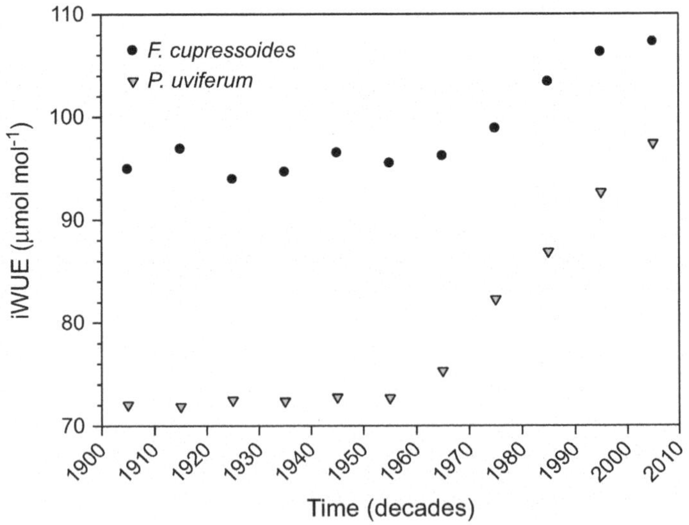

Fig. 16.4 Decadal values of intrinsic water-use efficiency (iWUE) in *F. cupressoides* and *P. uviferum*

iWUE were only +0.1% and + 0.2% in *F. cupressoides* and *P. uviferum*, respectively. From 1960 to 2010, the mean relative decadal enhancements of iWUE were + 2.4% and + 6.1% in *F. cupressoides* and *P. uviferum*, in that order. During that period, when c_a started rising rapidly, the mean iWUE of *F. cupressoides* (mea n ± SE = 102.4 ± 2.1 μmol mol^{-1}) was significantly higher ($t = 3.5$, $p = 0.012$) than that of *P. uviferum* (86.9 ± 3.9 μmol mol^{-1}). In contrast, during the first half of the twentieth century, iWUE was much higher ($t = 47.7$, $p = 0.002$) in *F. cupressoides* (95.4 ± 0.5 μmol mol^{-1}) than in *P. uviferum* (72.4 ± 0.1 μmol mol^{-1}). These iWUE trends suggest a convergence between the two species during the twenty-first century.

The rapid increase of iWUE from the 1960s until the 2010s did not correspond to a rapid increase in growth. For instance, the rapid rise of *P. uviferum* iWUE in the 1970s with respect to the 1960s corresponded to a growth decrease (−0.6%). The relative changes in growth (BAI) and iWUE decadal values were not significantly correlated neither in *F. cupressoides* ($r = -0.30$, $p = 0.40$) nor in *P. uviferum* ($r = 0.48$, $p = 0.17$; Fig. 16.5).

According to the calculated iWUE values, c_i and the c_i/c_a ratio would be higher in *P uviferum* than in *F. cupressoides* (Fig. 16.6). Assuming a linear relationship, the slope of the c_i as a function of rising c_a would be also higher in *F. cupressoides* (0.62 ± 0.03; 95% bootstrapped confidence intervals (0.46, 0.67)) than in *P uviferum* (0.39 ± 0.03; 95% CIs (0.30, 0.45)). Similarly, the slope of the c_i / c_a ratio as a function of rising c_a would be higher in *F. cupressoides* (0.0003 ± 0.0001; 95% CIs

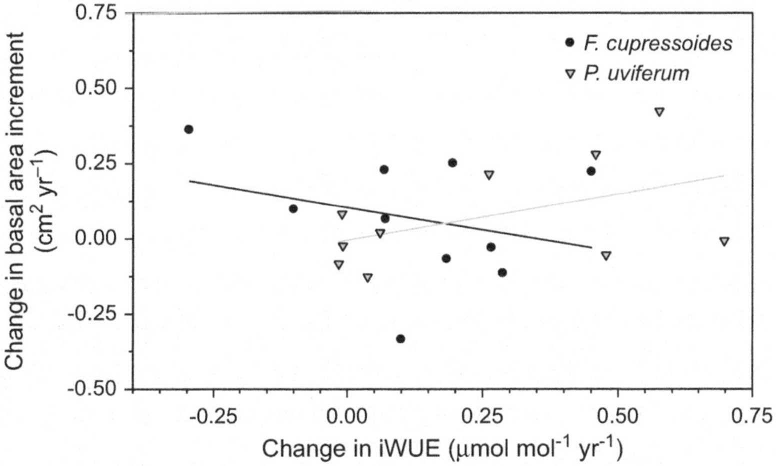

Fig. 16.5 Comparison of relative changes between decades in the intrinsic water-use efficiency (iWUE) vs. growth (basal area increment) of *F. cupressoides* and *P. uviferum*

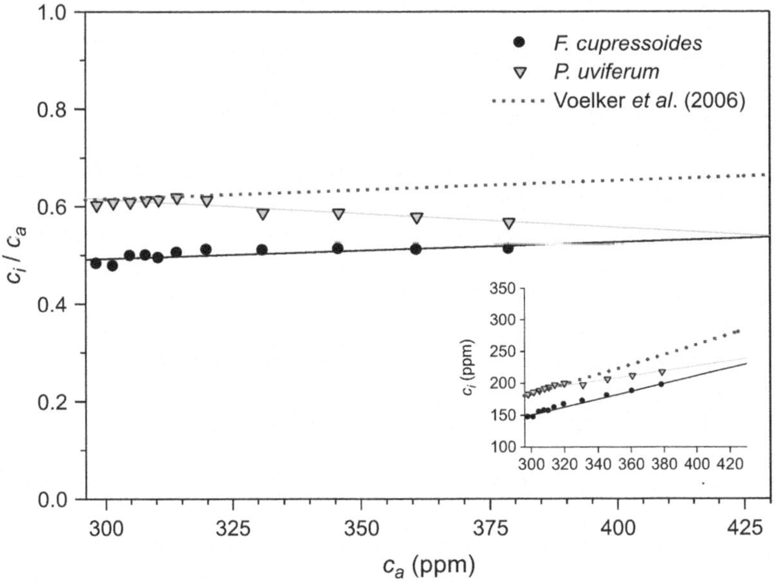

Fig. 16.6 Inferred changes in the ratio between leaf internal CO_2 concentration (c_i) and atmospheric CO_2 concentration (c_a) as a function of changes in c_a. Changes in the c_i / c_a ratio are shown for *F. cupressoides* and *P. uviferum*. The lines show fits corresponding to ordinary least squares regressions of c_i/c_a as a function of c_a. The inset shows the values of c_a and c_i. The dotted lines correspond to the equation (c_i/c_a = 0.00038c_a + 0.502) obtained by Voelker et al. (2016) for gymnosperms

(−0.0001, 0.0005)) than in *P uviferum* (−0.0006 ± 0.0001; 95% CIs (−0.0008, −0.0004)). In fact, the two species c_i/c_a trends are predicted to converge at c_a values of ca. 430 ppm.

16.4 Discussion

The growth (BAI) and water-use (iWUE) patterns of *F. cupressoides* and *P. uviferum* were not coupled throughout the twentieth century, an uncoupling also observed in other gymnosperms from the northern hemisphere (Peñuelas et al. 2011). However, they presented divergent patterns with *P. uviferum* presenting rising growth and iWUE, whereas *F. cupressoides* radial growth tended to decrease or remained stable after the 1990s. This decrease and the higher iWUE values may indicate a more pronounced sensitivity of *F. cupressoides* to recent droughts and the decline in precipitation observed in southern Chile during the last part of the twentieth century (Quintana and Aceituno 2012). This drying trend exacerbated after the 1970s climate shift (Jacques-Coper and Garreaud 2015), and caused a decline in growing-season rainfall reducing growth of some *F. cupressoides* stands which are sensitive to seasonal drought (Urrutia-Jalabert et al. 2015a; Camarero and Fajardo 2017), and even triggered canopy dieback in others (Barichivich 2005). Therefore, those findings and the results presented here indicate that future monitoring efforts should consider drought and warmer conditions as potential stress factors affecting the remnants of old-growth *F. cupressoides* forests in Valdivian temperate rainforests. Tree-ring C and O isotope data obtained from different sites situated along climatic gradients would further contribute to build a long and solid temporal framework of performance for these threatened forests (Camarero and Fajardo 2017). It should be also considered that drought-prone regions may change through time since climate models forecast warmer and drier summers at higher latitudes because rain belts would shift southwards (Fuenzalida et al. 2007).

Age seems to be a better predictor of radial-growth rates in *P. uviferum* than in *F. cupressoides* for the range of series lengths considered here (Fig. 16.1). However, this needs more data from additional trees and sites to be conclusive. Such difference would suggest that competition and stand dynamics are more relevant factors to predict growth in *F. cupressoides* whilst *P. uviferum* stands are mainly disturbed by wildfires. This agrees with previous research on the autoecology of these two species (Donoso et al. 2006; Lara et al. 2006; Holz 2009). This different behavior opens opportunities by managing suppressed *F. cupressoides* trees through selective thinning which could alleviate water shortage (Grant et al. 2013).

The climate-growth associations described for *F. cupressoides* indicate that its growth is sensitive to soil moisture limitations which could be explained by the combined effects of high summer temperature and elevated vapour pressure deficit reducing water availability to grow in sites whose soils show low soil water holding capacity. The radial growth of *F. cupressoides* has been shown to be reduced by dry and sunny conditions during summer which could reduce carbohydrate accumulation (Urrutia-Jalabert et al. 2015b). *F. cupressoides* stem increment was enhanced by wet and cloudy conditions at inter- and intra-annual scales, but warm conditions (high minimum temperatures) during the growing season have also been shown to enhance the growth of this species (Urrutia-Jalabert et al. 2015a, b), which agrees with the fact that its maximum growth rates were observed in early summer (Pérez et al. 2009).

In the case of *P. uviferum*, growth decreased in response to cool-wet conditions during the growing season, which explains its positive growth rates as climate is warming and drying. This climate-growth association was also observed in the core (46–49° S) of its distribution range, in central Patagonia (Roig and Boninsegna 1990; Roig 1991; Szeicz 1997; Szeicz et al. 2000; Holz et al. 2018). However, in the northern Patagonia (39–45° S), where the drought-prone distribution edge of the species is located, growth increased in response to high moisture levels during the current and previous growing seasons (Holz et al. 2018). The positive response to moisture in northern *P. uviferum* forests increased after the 1950s and 1970s shift towards warmer or drier conditions which caused a growth decline of some drought-prone conifers (e.g. *Austrocedrus chilensis*) in southern Chile (Villalba et al. 2003, 2012). Such shift towards drier and warmer conditions could be also responsible of the reported growth decline reported for *F. cupressoides* whilst *P. uviferum* kept more stable growth rates (Camarero and Fajardo 2017, Camarero et al. 2018). Despite both species showed similar year-to-year growth variability and could respond to the same regional drought, it was showed that they presented different long-term growth tendencies and iWUE strategies.

The slopes of c_i and c_i/c_a as a function of c_a were more similar to those derived from Voelker et al. (2016) gymnosperms equation in the case of *F. cupressoides* than in the case of *P uviferum*. Note however that the gymnosperms' equation developed by Voelker et al. (2016) and relating c_i/c_a and c_a was exclusively based on gymnosperms native to the northern Hemisphere. Both study species could show similar c_i/c_a values in the mid-twenty-first century despite their contrasting growth trends, which means that as c_a continues rising they would show converging iWUE values and their growth could become more limited by warmer and drier conditions than in the twenty-first century. This idea should be further explored by using iWUE models and updating tree-ring series.

In mesic conditions, the constant c_i/c_a scenario explains better iWUE increases than other scenarios (Saurer et al. 2004; Seibt et al. 2008). Similar iWUE increases over time (range 8–32%) as those presented here have been observed in other forest types and biomes from the southern (Leavitt 1994; Srur et al. 2008; Urrutia-Jalabert et al. 2015a; Lavergne et al. 2017; Camarero and Fajardo 2017; Camarero et al. 2018; Fajardo et al. 2019) and northern (Gómez-Guerrero et al. 2013; Linares and Camarero 2012; Saurer et al. 2004; Guerrieri et al. 2019) Hemispheres. These iWUE improvements have been mainly attributed to the reduction of g in response to increasing c_a (Saurer et al. 2004; Andreu-Hayles et al. 2011). In one of the few instances corresponding to southern South America, Lavergne et al. (2017) found that the iWUE of *F. cupressoides* increased by ca. 25% since 1952 and they found that its iWUE variability was controlled by both A and g rates. This agrees with what Urrutia-Jalabert et al.(2015a) reported, who found a recent decrease in C isotope discrimination for this species in southern Chile. *F. cupressoides* was responding to the atmosphere CO_2 increase by keeping a constant c_i/c_a but other factors could be also influencing A and/or g leading to a decoupling of iWUE from the c_i/c_a scenario. Using O isotope discrimination ($\delta^{18}O$) Camarero and Fajardo (2017) argued that *F. cupressoides* iWUE changes were less related to changes in g than in A, which could rise in response to warmer conditions and higher c_a whereas coexisting *Nothofagus* species showed more noticeable g changes in response to rising vapor pressure deficit. This could be explained because *Nothofagus* displays better osmotic adjustments than *F. cupressoides* and are able to access deep soil water through the development of extensive root systems (Wang et al. 1998; Piper et al. 2007). Leaf gas-exchange rates (A and g) are more sensitive to changes in vapor pressure deficit in conifer than in broadleaf species, a difference related to narrower hydraulic safety margins (Carnicer et al. 2013).

There could be a change of paradigm from (i) a more static view which associated the increase in iWUE to proportional adjustments of A and g in response to c_a leading to an almost constant c_i/c_a through faster increases of c_i than c_a (Saurer et al. 2004) to (ii) a dynamic leaf gas exchange response to c_a and climate warming

(Voelker et al. 2016). This dynamic response is consistent with higher *A* increases and small changes in *g* modulated by differences in soil moisture among sites (Guerrieri et al. 2019), as it was found for *F. cupressoides*. In other words, c_i would not significantly increase in response to rising c_a, particularly in conifers, and iWUE would not increase as much as expected (Frank et al. 2015). This could correspond to a CO_2-fertilization effect on *A* in mesic sites not subjected to soil moisture limitations (Guerrieri et al. 2019), and also agrees with simulations showing a relatively modest increase of *g* in response to rising c_a and global warming (Cernusak et al. 2019). Such a scenario could explain iWUE trends in *F. cupressoides* (Camarero and Fajardo 2017) but does not support any CO_2-fertilization effect on growth in the two conifer species studied in southern Chile.

16.5 Conclusion

The two study conifers present in Chilean temperate forests did not show couplings between growth rates and iWUE. The *F. cupressoides* forest showed a decline in growth after the 1990s in response to warmer and drier conditions, whereas the *P. uviferum* forest showed stable growth rates. These findings indicate that *F. cupressoides* could follow a dynamic leaf gas-exchange strategy maintaining relatively constant stomatal conductance rates. Forecasted drier and warmer summers in southern Chile might negatively impact *F. cupressoides* growth and water use in the most drought-prone regions. In contrast, *P. uviferum* could benefit if climate warms and becomes drier. According to iWUE data, both species could show similar leaf CO_2 concentrations in the mid-twenty-first century.

Acknowledgments This work was funded by CSIC–Huinay project "Crecimiento y funcionamiento de los bosques valdivianos de *Nothofagus* en respuesta al cambio global" and the Spanish Ministry of Economy "Fundiver" project (CGL2015-69186-C2-1-R) to J.J.C. The author thanks the assistance of Alex Fajardo, Reinhard Fitzek and Juan Pablo Mora and the logistic support of the Huinay Scientific Field Station (Fundación San Ignacio de Huinay-ENDESA). I thank the reviewers (T.E. Kolb, M. Saurer) for improving an earlier version of the manuscript.

References

Andreu-Hayles L, Planells O, Gutiérrez E et al (2011) Long tree-ring chronologies reveal 20th century in-creases in water-use efficiency but no enhancement of tree growth at five Iberian pine forests. Glob Chang Biol 17:2095–2112. https://doi.org/10.1111/j.1365-2486.2010.02373.x

Armesto JJ, Figueroa J (1987) Stand structure and dynamics of rainforests in the Chiloé archipelago. J Biogeogr 14:367–376. https://doi.org/10.2307/2844944

Armesto JJ, Smith-Ramírez C, Carmona MR et al (2009) Old-growth temperate rain forests of South America: conservation, plant-animal interactions, and baseline biogeochemical processes. In: Wirth C, Gleixner G, Heimann M (eds) Old-growth forests: function, fate and value. Springer, Berlin, Heidelberg, New York, pp 367–390

Barichivich J (2005) Muerte apical episódica en bosques de alerce (*Fitzroya cupressoides* [Mol.] Johnston) de la Cordillera de la Costa de Valdivia. Universidad Austral de Chile, Valdivia

Camarero JJ, Fajardo A (2017) Poor acclimation to current drier climate of the long-lived tree species *Fitzroya cupressoides* in the temperate rainforest of southern Chile. Agric For Meteorol 239:141–150. https://doi.org/10.1016/j.agrformet.2017.03.003

Camarero JJ, Gazol A, Sangüesa-Barreda G et al (2018) Coupled climate–forest growth shifts in the Chilean Patagonia are decoupled from trends in water–use efficiency. Agric For Meteorol 259:222–231. https://doi.org/10.1016/j.agrformet.2018.04.024

Carnicer J, Barbeta A, Sperlich D et al (2013) Contrasting trait syndromes in angiosperms and conifers are associated with different responses of tree growth to temperature on a large scale. Front Plant Sci 4:409. https://doi.org/10.3389/fpls.2013.00409

Cernusak LA, Haverd V, Brendel O et al (2019) Robust response of terrestrial plants to rising CO_2. Trends Plant Sci 24:578–586. https://doi.org/10.1016/j.tplants.2019.04.003

Cook ER, Krusic P (2007) ARSTAN 44 A Tree-ring standardization program based on detrending and autoregressive time series modeling with interactive graphics https://wwwldeocolumbiaedu/tree-ring-laboratory/resources/software. Accessed 23 Aug 19

Donoso C (1981) Tipos forestales de los bosques nativos de Chile. Investigación y Desarrollo Forestal (CONAF-PNUD-FAO), Santiago

Donoso C (1993) Bosques templados de Chile y Argentina: variación, estructura y dinámica. Ecología Forestal. Editorial Universitaria, Santiago

Donoso C, Sandoval V, Grez R et al (1993) Dynamics of *Fitzroya cupressoides* forest in southern Chile. J Veg Sci 4:303–312. https://doi.org/10.2307/3235588

Donoso C, Lara A, Escobar B et al (2006) *Fitzroya cupressoides* (Molina) I.M. Johnst. In: Donoso C (ed) Las especies arbóreas de los bosques templados de Chile y Argentina. María Cuneo Ediciones, Valdivia, pp 68–81

Fajardo A, McIntire EJB (2012) Reversal of multicentury tree growth improvements and loss of synchrony at mountain tree lines point to changes in key drivers. J Ecol 100:782–794. https://doi.org/10.1111/j.1365-2745.2012.01955.x

Fajardo A, Gazol A, Mayr C et al (2019) Recent decadal drought reverts warming–triggered growth enhancement in contrasting climates in the southern Andes treeline. J Biogeogr 46:1367–1379. https://doi.org/10.1111/jbi.13580

Farjon A (1998) World checklist and bibliography of conifers. Royal Botanical Gardens, Kew

Farjon A (2005) A monograph of Cupressaceae and Sciadopitys. Royal Botanic Gardens, Kew

Farquhar GD (1983) On the nature of carbon isotope discrimination in C4 species. Aust J Plant Physiol 10:205–226. https://doi.org/10.1071/PP9830205

Farquhar GD, Richards RA (1984) Isotopic composition of plant carbon correlates with water-use efficiency of wheat genotypes. Austral J Plant Physiol 11:539–552. https://doi.org/10.1071/PP9840539

Farquhar GD, O'Leary MH, Berry JA (1982) On the relationship between carbon isotope discrimination and the intercellular carbon dioxide concentration in leaves. Austr J Plant Physiol 9:121–137. https://doi.org/10.1071/PP9820121

Farquhar GD, Ehleringer JR, Hubick KT (1989) Carbon isotope discrimination and photosynthesis. Ann Rev Plant Physiol Mol Biol 40:503–537. https://doi.org/10.1146/annurev.pp.40.060189.002443

Francey RJ, Farquhar GD (1982) An explanation of $^{13}C/^{12}C$ variations in tree rings. Nature 297:28–31

Frank DC, Poulter B, Saurer M et al (2015) Water-use efficiency and transpiration across European forests during the Anthropocene. Nat Clim Chang 5:579–583. https://doi.org/10.1038/NCLIMATE2614

Fritts HC (2001) Tree rings and climate. The Blackburn Press, Caldwell

Fuenzalida H, Aceituno P, Falvey M et al (2007) Estudio de la variabilidad climática en Chile para el siglo XXI. CONAMA, Santiago

Gómez-Guerrero A, Silva LCR, Barrera-Reyes M et al (2013) Growth decline and divergent tree ring isotopic composition ($\delta^{13}C$ and $\delta^{18}O$) contradict predictions of CO_2 stimulation in high altitudinal forests. Glob Chang Biol 19:1748–1758. https://doi.org/10.1111/gcb.12170

Granda E, Rodrigo-Rossatto D, Camarero JJ et al (2014) Growth and carbon isotopes of Mediterranean trees reveal contrasting responses to increased carbon dioxide and drought. Oecologia 174:307–317. https://doi.org/10.1007/s00442-013-2742-4

Grant GE, Tague CL, Allen CD (2013) Watering the forest for the trees: an emerging priority for managing water in forest landscapes. Front Ecol Env 11:314–321. https://doi.org/10.1890/120209

Guerrieri R, Belmecheri S, Ollinger SV et al (2019) Disentangling the role of photosynthesis and stomatal conductance on rising forest water-use efficiency. PNAS 116:16909–16914. https://doi.org/10.1073/pnas.1905912116

Gutiérrez AG, Armesto JJ, Díaz MF et al (2014) Increased drought impacts on temperate rainforests from Southern South America: results of a process-based, dynamic forest model. PLoS One 9:e103226. https://doi.org/10.1371/journal.pone.0103226

Harris I, Jones PD, Osborn TJ et al (2014) Updated high-resolution grids of monthly climatic observations – the CRU TS310 dataset. Int J Climatol 34:623–642. https://doi.org/10.1002/joc.3711

Holmes RL (1983) Computer-assisted quality control in tree ring dating and measurement. Tree Ring Bull 43:69–78

Holz CA (2009) Climatic and human influences on fire regimes and forest dynamics in temperate rainforests in southern Chile. PhD Thesis. University of Colorado, Boulder

Holz A, Hart SJ, Williamson GJ et al (2018) Radial growth response to climate change along the latitudinal range of the world's southernmost conifer in southern South America. J Biogeogr 45:1140–1152. https://doi.org/10.1111/jbi.13199

IPCC- Intergovernmental Panel on Climate Change (2013) Climate change 2013: the physical science basis. Contribution of working group I to the fifth assessment report of the intergovernmental panel on climate change. Cambridge University Press, Cambridge

Jacques-Coper M, Garreaud RD (2015) Characterization of the 1970s climate shift in South America. Int J Climatol 35:2164–2179. https://doi.org/10.1002/joc.4120

Keith H, Mackey BG, Lindenmayer DB (2009) Re-evaluation of forest biomass carbon stocks and lessons from the world's most carbon-dense forests. PNAS USA 106:11635–11640. https://doi.org/10.1073/pnas.0901970106

Körner C (2017) A matter of tree longevity. Science 355:130–131. https://doi.org/10.1126/science.aal2449

Lara A, Villalba R (1993) A 3620-year temperature record from *Fitzroya cupressoides* tree rings in southern South America. Science 260:1104–1106. https://doi.org/10.1126/science.260.5111.1104

Lara A, Gardner MF, Vergara R (2003) The use and conservation of *Fitzroya cupressoides* (Alerce) forests in Chile. Acta Hortic 615:381–385. https://doi.org/10.17660/ActaHortic.2003.615.41

Lara A, Donoso C, Escobar B et al (2006) *Pilgerodendron uviferum* (D. Don) Florin. In: Donoso C (ed) Autoecología de las especies arbóreas de los bosques templados de Chile y Argentina. Marisa Cúneo Ediciones, Valdivia, pp 82–92

Larjavaara M (2014) The world's tallest trees grow in thermally similar climates. New Phytol 202:344–349. https://doi.org/10.1111/nph.12656

Lavergne A, Daux V, Villalba R et al (2017) Improvement of isotope-based climate reconstructions in Patagonia through a better understanding of climate influences on isotopic fractionation in tree rings. Earth Planet Sci Lett 459:372–380. https://doi.org/10.1016/j.epsl.2016.11.045

Leavitt SW (1994) South American tree rings show declining $\delta^{13}C$ trend. Tellus 46B:152–157. https://doi.org/10.3402/tellusb.v46i2.15760

Lévesque M, Siegwolf R, Saurer M et al (2014) Increased water-use efficiency does not lead to enhanced tree growth under xeric and mesic conditions. New Phytol 203:94–109. https://doi.org/10.1111/nph.12772

Linares JC, Camarero JJ (2012) From pattern to process: linking intrinsic water-use efficiency to drought-induced forest decline. Glob Chang Biol 18:1000–1015. https://doi.org/10.1111/j.1365-2486.2011.02566.x

Luebert F, Pliscoff P (2006) Sinopsis bioclimática y vegetacional de Chile. Santiago de Chile Editorial Universitaria, Santiago

McCarroll D, Loader NJ (2004) Stable isotopes in tree rings. Quat Sci Rev 23:771–801. https://doi.org/10.1016/j.quascirev.2003.06.017
Nock CA, Baker PJ, Wanek W et al (2011) Long-term increases in intrinsic water-use efficiency do not lead to increased stem growth in a tropical monsoon forest in western Thailand. Glob Chang Biol 17:1049–1063. https://doi.org/10.1111/j.1365-2486.2010.02222.x
Peñuelas J, Canadell J, Ogaya R (2011) Increased water-use efficiency during the 20th century did not translate into enhanced tree growth. Glob Ecol Biogeogr 20:597–608. https://doi.org/10.1111/j.1466-8238.2010.00608.x
Pérez CA, Carmona MR, Aravena JC et al (2009) Environmental controls and patterns of cumulative radial increment of evergreen tree species in montane, temperate rainforests of Chiloé Island, Southern Chile. Austr Ecol 34:259–271. https://doi.org/10.1111/j.1442-9993.2009.01927.x
Piper F, Corcuera LJ, Alberdi M et al (2007) Differential photosynthetic and survival responses to soil drought in two evergreen *Nothofagus species*. Ann For Sci 64:447–452. https://doi.org/10.1051/forest:2007022
Quintana JM, Aceituno P (2012) Changes in the rainfall regime along the extratropical west coast of South America (Chile): 30-43° S. Atmósfera 25:1–22
Roig FA (1991) Dendrocronología y dendroclimatología del bosque de *Pilgerodendron uviferum* en su área norte de dispersión. Boletín de la Sociedad Argentina de Botánica 27:217–234
Roig FA, Boninsegna JA (1990) Chiloé Island (Chile) summer precipitation reconstructed for 426 years from *Pilgerodendron uviferum* tree-ring chronologies. Lundqua Rep 32:277–280
Saurer M, Siegwolf RTW, Schweingruber FH (2004) Carbon isotope discrimination indicates improving water-use efficiency of trees in northern Eurasia over the last 100 years. Glob Chang Biol 10:2109–2120. https://doi.org/10.1111/j.1365-2486.2004.00869.x
Saurer M, Spahni R, Frank DC et al (2014) Spatial variability and temporal trends in water-use efficiency of European forests. Glob Chang Biol 20:3700–3712. https://doi.org/10.1111/gcb.12717
Schulman E (1956) Dendroclimatic change in semiarid America. University of Arizona Press, Tucson
Seibt U, Rajabi A, Griffiths H et al (2008) Carbon isotopes and water use efficiency: sense and sensitivity. Oecologia 155:441–454. https://doi.org/10.1007/s00442-007-0932-7
Silva LCR, Anand M (2013) Probing for the influence of atmospheric CO_2 and climate change on forest ecosystems across biomes. Glob Ecol Biogeogr 22:83–92. https://doi.org/10.1111/j.1466-8238.2012.00783.x
Silva LCR, Anand M, Leithead MD (2010) Recent widespread tree growth decline despite increasing atmospheric CO_2. PLoS One 5:e115543. https://doi.org/10.1371/journal.pone.0011543
Soto DP, Le Quesne C, Lara A et al (2007) Precarious conservation status of *Pilgerodendron uviferum* forests in their northern distribution in the Chilean Coastal Range. Bosque 28:263–270
Srur AM, Villalba R, Villagra PE et al (2008) Influencias de las variaciones en el clima y en la concentración de CO_2 sobre el crecimiento de *Nothofagus pumilio* en la Patagonia. Rev Chil Hist Nat 81:239–256. https://doi.org/10.4067/S0716-078X2008000200007
Szeicz JM (1997) Growth trends and climatic sensitivity of trees in the North Patagonian rain forest of Chile. Can J For Res 27:1003–1014. https://doi.org/10.1139/x97-048
Szeicz JM, Lara A, Díaz S et al (2000) Dendrochronological studies of *Pilgerodendron uviferum* in southwestern South America. In: Roig FA (ed) Dendrocronología en América Latina. EDIUNC, Mendoza, pp 240–270
Tecklin D, DellaSala DA, Luebert F et al (2011) Valdivian temperate rainforests of Chile and Argentina. In: DellaSala DA (ed) Temperate and boreal rainforests of the world: ecology and conservation. Island Press, Washington, DC, pp 132–153
Urrutia-Jalabert R, Malhi Y, Barichivich J et al (2015a) Increased water use efficiency but contrasting tree growth patterns in *Fitzroya cupressoides* forests of southern Chile during recent decades. J Geophys Res Biogeosci 120:2505–2524. https://doi.org/10.1002/2015JG003098
Urrutia-Jalabert R, Rossi S, Deslauriers A et al (2015b) Environmental correlates of stem radius change in the endangered *Fitzroya cupressoides* forests of southern Chile. Agric For Meteorol 200:209–221. https://doi.org/10.1016/j.agrformet.2014.10.001

Veblen TT, Schlegel FM (1982) Reseña ecológica de los bosques del sur de Chile. Bosque 4:73–115

Veblen TT, Delmastro R, Schlatter J (1976) The conservation of *Fitzroya cupressoides* and its environment in Southern Chile. Env Conserv 3:291–301. https://doi.org/10.1017/S0376892900019081

Veblen TT, Burns BR, Kitzberger T et al (1995) The ecology of the conifers of southern South America. In: Enright NJ, Hills RS (eds) Ecology of the southern conifers. Melbourne University Press, Carlton, pp 142–148

Villalba R, Lara A, Boninsegna JA et al (2003) Large-scale temperature changes across the southern Andes: 20th-century variations in the context of the past 400 years. Clim Chang 59:177–232. https://doi.org/10.1007/978-94-015-1252-7_10

Villalba R, Lara A, Masiokas M et al (2012) Unusual Southern Hemisphere tree growth patterns induced by changes in the Southern Annular Mode. Nat Geosci 5:793–798. https://doi.org/10.1038/NGEO1613

Voelker SL, Brooks JR, Meinzer FC et al (2016) A dynamic leaf gas-exchange strategy is conserved in woody plants under changing ambient CO_2: evidence from carbon isotope discrimination in paleo and CO_2 enrichment studies. Glob Chang Biol 22:889–902. https://doi.org/10.1111/gcb.13102

Wang X-F, Yakir D, Avishai M (1998) Non-climatic variations in the oxygen isotopic compositions of plants. Glob Chang Biol 4:835–849. https://doi.org/10.1046/j.1365-2486.1998.00197.x

Wigley TML, Briffa KR, Jones PD (1984) On the average value of correlated time series with applications in dendroclimatology and hydrometeorology. J Clim Appl Meteorol 23:201–213. https://doi.org/10.1175/1520-0450(1984)023<0201:OTAVOC>2.0.CO;2

Wolodarsky-Franke A, Lara A (2005) The role of "forensic" dendrochronology in the conservation of alerce (*Fitzroya cupressoides* (Molina) Johnston) forests in Chile. Dendrochronologia 22:235–240. https://doi.org/10.1016/j.dendro.2005.05.008

Index

M. Pompa-García, J. J. Camarero (eds.), *Latin American Dendroecology*,
https://doi.org/10.1007/978-3-030-36930-9

D

The manufacturer's authorised representative in the EU is Springer Nature Customer Service Centre GmbH, Europaplatz 3, 69115 Heidelberg, Germany. If you have any concerns regarding our products, please contact ProductSafety@springernature.com

Printed and bound by CPI Group (UK) Ltd, Croydon, CR0 4YY
30/09/2026
02225458-0015